introductory statistics

A Decision Map

introductory statistics

A Decision Map

Second Edition

Thad R. Harshbarger

The City College, City University of New York

Macmillan Publishing Co., Inc.
NEW YORK
Collier Macmillan Publishers
LONDON

Macmillan Publishing Co., Inc.
866 Third Avenue, New York, New York 10022

Collier Macmillan Canada, Ltd.

Library of Congress Cataloging in Publication Data

Harshbarger, Thad R
  Introductory statistics; a decision map.

  Bibliography: p.
  Includes index.
  1. Statistics. I. Title.
HA29.H253 1977   519.5   75-19057
ISBN O-02-350650-4

Printing: 1 2 3 4 5 6 7 8   Year: 7 8 9 0 1 2 3

Preface to the Second Edition

STATISTICS texts fall into the general category of applied mathematics and technology books; formulas and general principles are presented and applied by the reader in certain situations to produce meaningful results.

Most statistics texts are organized around the methods, general principles, or theories they present. Usually rules are also given for the application of these principles to problems that represent real-life situations, but it is left up to the reader to work from the other direction—to decide for a given problem which theory or general principle or formula should be applied.

The first time I taught statistics I discovered that the students had most of their trouble in solving a problem as soon as the problem was presented: they did not know what to do first. By the time we were to t tests I realized that I would have to provide students with some way of choosing statistics to fit problems, because they had such great difficulty doing it for themselves. The figure at the beginning of Chapter 12 is revised from my first classroom display of the relationships among several different statistical formulas. In it I attempted to show when to use each of the formulas to solve problems. The book developed naturally from that first idea.

The central organizing principle behind this book is that of a general schema into which many different kinds of problems might fit. The text itself, through the schema contained in it, attempts to simulate the kind of problem classification that might be used by an expert to determine which method or methods of solution are most appropriate for a given problem. Users of the book, if they know the general nature of the problems they face and the kinds of result that will be useful, can follow the schema to select methods of achieving the desired kinds of result.

I call this kind of schema for classifying and solving problems a "decision map" because of its usefulness as a guide through the maze of statistical procedures to an appropriate method to use for the solution of a given problem. The user approaches the map for a chapter with a specific problem in mind, for which a solution is required. A line or arrow is followed until a decision point is reached. Here a question is posed, and the nature of the problem determines the answer to the question. The answer, in turn, determines which of two pathways is to be followed upon departure from the decision point. The pathway chosen leads to a second decision point, and so on, until the problem has been sufficiently classified by this process so that a method of solution can be given. At this point a box is encountered, which says: "Do this: ...." By doing what the map dictates, the problem can be solved.

Each decision point and direction on the map is keyed to a section of text that explains it in greater detail for those who are unclear about its meaning or implementation. The order in which a section appears in the body of the text need have no relation to the ordering of the decision map, and sections are arranged in whatever order seems best for didactic purposes. The mapping principle of organization, which is mostly confined to illustrations, is super-imposed on a traditional method of presenting the text, so that the book is compatible with either a methods-oriented or a problem-oriented approach to classroom teaching of the material.

The maps can serve several functions:

1. One function is to eliminate from consideration, as quickly as possible, methods inappropriate for the solution of any given problem. By simulating the classification process that an expert might go through in solving a problem, the maps give novices a degree of functional expertise relative to their problems that they could not otherwise obtain so quickly.

I emphasize this aspect in the way that I teach introductory statistics. I stress the entire process of getting information from data: deciding what sta-tistic to use, applying formulas to data, and interpreting results. I give students a large number of problems and insist that they work each one from start to finish. I do not feel obligated to cover every section in the text, because once students know how to solve problems on their own, they can find the material we haven't covered and apply it by themselves.

The text is organized with this point in mind: there are many more tech-niques and methods than can comfortably be covered in a one-semester course. Material can be chosen without concern for gaps in students' background. I customarily assign something from each of the chapters beginning with Chap-ter 10, but I have never assigned one of the later chapters in its entirety.

Also, by referring students to the decision maps, instructors can save the time they might otherwise spend in answering elementary classification ques-tions from former statistics students. I used to get many such questions as, "Should I use correlation or chi square for these data?"; now the few questions of this kind that I get are generally from students whom I have not previously taught.

2. By giving students problems to solve and showing them how to do each one from start to finish, instructors can maintain motivation at a relatively high level. Students like to increase their skills and to understand what they are doing, and they get a feeling of accomplishment from the knowledge that they can take a problem from the beginning and solve it completely. The maps not only help guide students to answers, but also help them quickly eliminate irrelevant solutions.

3. Another function of the maps is to provide a printed representation of the relationships among various methods for solving statistical problems. Visual presentation seems to (a) provide a stable frame of reference into which new approaches can readily be integrated, so students can more easily understand new material in the context of what they already know; (b) emphasize the

similarities and differences among the techniques presented, so students can begin to see statistics as an integrated whole; (c) give students a feeling that they are making systematic and meaningful increments to their understanding as they learn new techniques, rather than just memorizing lists of methods; and (d) encourage students to learn additional methods that will fill gaps in their own decision maps of statistical techniques.

4. Inappropriate methods of solution are eliminated through explicit consideration of the sequence of decisions made by a student in solving a problem. This approach necessitates more emphasis on the assumptions made by each method of solution than is found in the usual text. I found in writing this book that there were many decision questions not explicitly considered by other texts. Therefore, the material in those texts may seem clear and accurate when it is being read, but when users try to apply it, they cannot decide what to do. Also, in other texts, assumptions are often presented at the ends of method sections, so that a person reading things in sequence learns how to apply each technique before finding out when or whether it should be used.

Here, assumptions are used in the decision questions, for classifying problems and selecting methods. This helps make problem solving more efficient. It should also be less likely that a student using the decision maps will violate the assumptions necessary for the correct application of any method or general principle. Alternatively, students will more quickly recognize violations of assumptions when they have been made and should therefore be able to interpret the results in the light of the nature and probable effects of those violations.

The book is intended to be suitable for students in the social sciences and education. Because of the wide range of backgrounds brought to courses in this area, the only background assumed is an elementary course in algebra, and even that is not counted on heavily. On the other hand, students with greater prior knowledge are encouraged to omit those parts of the text with which they are already familiar, in order to waste as little of their time, and motivation, as possible.

Following are some points for instructors to consider in deciding on the appropriateness of the text for a particular statistics course.

1. Level of difficulty: the material is not easy, and I have not tried to make it appear so. I have great faith that any student who can understand high school algebra can understand this book; but I expect that most students will have to read many passages several times before they understand them.

2. Focus: this book is intended for students who will continue to need statistics after they complete the course. It is intended for social science or education majors, and I think would be a waste of time and energy for students in, say, a junior college course required of everyone, or a "statistics for poets" course.

3. Coverage: the entire book can probably be covered well in about six semester hours. I have used it in three- and four-hour introductory courses at The City College, the main issue being which material not to teach.

4. Problems: a small number of homework problems are given, usually at the end of the chapter. Additional problems can be found in the accompanying workbook. Most problems are based on hypothetical data that make the computations easy; that way homework really can be done at home, and, contrary to some criticism, students really are able without trouble to make the transfer to messy data. I have written a short monograph on counstructing problems with easy numbers, and the reference appears in the Bibliography.

5. Mathematical–conceptual orientation: this book is definitely *not* for people who like to teach mathematical statistics. I view a mathematical approach as totally inappropriate for social science and education students. What they need is to be able to get answers, and that's what this book is all about. Statistical theory and conceptual orientations are presented to the extent that they will help students understand what they are doing.

I am indebted to Dave Lynch and Nancy Anderson for reviews of the manuscript for the second edition, and to John Antrobus for many helpful comments that I used in preparing this revision. I also want to repeat my acknowledgment of the valuable help I received from a number of colleagues in the preparation of the first edition: for reviews of early drafts by E. C. Carterette, James Lingoes, and John Milholland; and of later versions by Patrick Ross, David Payne, and Robert Rosenfeld.

A number of authors and publishers granted permission to reprint copyrighted tables and graphs. In each case, acknowledgment is kept with the material reproduced for the convenience of readers who may wish to refer to the original sources.

And to anyone who reviews, uses, or adopts this book: please send me your comments, corrections, reactions, notes, praise, and/or criticism—or just send me your name and address and I will let you know of any minor changes or alterations that may come up after you receive your copy.

T. R. H.

Contents

introductory statistics

A Decision Map

1 Introduction

THE first point that I wish to make is that it is extremely important for you to begin at the beginning of this book. You may be so eager to get to the "important part" that you will be tempted to skip over the first several chapters. This is not advisable, because statistics is organized so that each topic builds on previous topics. If you study the material carefully and in the indicated sequence, you should find it understandable, but if you miss some of the early discussion, you will almost certainly be confused by later topics.

On the other hand, this book has been organized so that you can proceed as quickly as possible to whatever material you consider important. If you use Chapter 2 properly, you will skip over some of it, and the more you already know, the more you will skip. Chapter 3 introduces the formal method of organization of the rest of the book. This organizational form, here called a "decision map," allows you to move quickly to the statistical techniques that interest you, learning a minimum of additional concepts and techniques on the way. If, at a later time, other material becomes important to you, you can again proceed rapidly to the relevant statistical techniques with a minimum of irrelevant learning.

If you use this book because your interest is in the field of educational and psychological statistics as a whole, rather than in solving a particular problem, you can study the text of each chapter without using the decision map at all. However, if you do so, you will have a harder time appreciating the relationships among the methods presented. A better approach to each chapter would be, first, to read the chapter through, and then to work the problems at the end of the chapter, using the decision map in each chapter as a basic resource and rereading the text as much as necessary.

You should not be greatly concerned if you have had difficulty with arithmetic and algebra in the past, for three reasons. First, Chapter 2 contains a test to determine what your strengths and weaknesses are, and instructions for solving the problems that give you difficulty. If the instructions in the chapter are insufficient for you, references are provided for your use. Second, algebraic and arithmetic manipulations are not emphasized in this text. Where derivations appear, they are presented in some detail, so that you should be able to follow them if you take your time and don't expect them to leap into your mind on their own. Third, homework problems have been constructed to be primarily conceptual; in fact, for most problems, you should need more time to decide what to do than to carry our the actual computations. Answers are provided so that you can verify your work immediately. A separate book containing additional problems and solutions is available from the publisher.

You should keep in mind that the concepts covered in this book are difficult, and that there may be considerably less redundancy than you are used to. This text cannot be read like a novel; in many places you may have to reread passages three, four, five, or more times before the material is clear to you. But although the material is difficult, it is by no means impossible; you simply must allow ample time to study it.

2 Arithmetic and Algebra

THIS chapter was written to test your background in arithmetic and algebra, and to tell you whether you are prepared for the materials in the rest of the text. The chapter is divided into five sections, as shown in the table.

Section	Pages	Contents
2-1	3–6	The test
2-2	6–9	Test answers
2-3	9–16	Methods for solving problems that you missed
2-4	16	Additional forms of notation
2-5	16	References for further reading on arithmetic and algebra as applied to statistics

It is important for you to be able to use arithmetic and algebra in following the explanations and solving the problems in this book; hence this chapter. On the other hand, reading all of Section 2-3 could be very boring if you already know a lot of the material. By taking the test and scoring your answers, you can find out how much of the material you need to study. The directions that follow each answer set tell you which of the parts of Section 2-3 you need to study, so you can focus your work on the material that gives you the most trouble. Should you want additional problems to test yourself, you can find them in the workbook. For explanations and practice beyond that, consult the references in Section 2-5.

2-1 The test

I

Add each of the following.

1. 14
     0
    ———

2. $^-9$
     3
    ———

3.   23
     12
    $^-14$
    $^-2$
    ———

4. $^-7$
    $^-42$
     6
    18
    ———

3

5. 7 6 ⁻2 ⁻4	**6.** ⁻12 ⁻6 4 1	**7.** ⁻9 ⁻4	**8.** 6 ⁻2
9. 0 ⁻23	**10.** 1253 16	**11.** ⁻6 ⁻5	**12.** .826 1.02
13. ⁻8 14	**14.** .0040 7.8	**15.** 18 ⁻26	**16.** 4 8

II

In each of the following, subtract the second from the first.

1. 0 14	**2.** ⁻6 ⁻1	**3.** ⁻8 2	**4.** 14 ⁻26
5. 41 ⁻8	**6.** ⁻14 60	**7.** ⁻26 0	

III

Multiply each of the following.

1. 43 0	**2.** ⁻9 ⁻31	**3.** 114 32	**4.** ⁻6 14
5. .086 ⁻12	**6.** 0 16	**7.** ⁻12 ⁻4	**8.** 207 4
9. 18 ⁻2	**10.** .931 .6		

11. $(17)(0) =$

12. $(\frac{7}{12})(\frac{4}{5}) =$

13. $(^-2)(^-21) =$

14. $(\frac{2}{3})(\frac{1}{8}) =$

15. $(^-8)(4) =$

16. $(\frac{8}{15})(7) =$

17. $(.007)(.11) =$

18. $(16)(3) =$

19. $(\frac{3}{4})(3) =$

IV

Divide each of the following.

1. $^-8/4 =$ **2.** $\frac{4}{1}/\frac{1}{2} =$ **3.** $16/0 =$ **4.** $23/^-4$

5. $^-26/^-5 =$ **6.** $0/6 =$ **7.** $.006/.02 =$ **8.** $^-4/0 =$

9. $0/283 =$ **10.** $4.002/.02 =$ **11.** $\frac{1}{5}/\frac{3}{4} =$ **12.** $^-9/^-3 =$

13. $^-84/7 =$ **14.** $6/^-2 =$

V

Carry out each of the following computations. Use the table below when necessary.

1. $2^3 =$ **2.** $\sqrt{16} =$ **3.** $\sqrt{2500} =$ **4.** $3^2 =$

5. $(3^2)^2 =$ **6.** $\sqrt{9/16} =$ **7.** $\sqrt{89/81} =$ **8.** $(4^3)^2 =$

9. $\sqrt{7744} =$ **10.** $\sqrt{88} =$ **11.** $\sqrt{12/27} =$

12. If $\sqrt{21} = 4.5826$, what is $(458.26)^2$?

13. If $(34)^2 = 1156$, what is $\sqrt{11.56}$?

n	n^2	$\sqrt{n}$	$1/n$	$1/\sqrt{n}$
86	7396	9.2736	0.011628	0.1078
87	7569	9.3274	0.011494	0.1072
88	7744	9.3808	0.011364	0.1066
89	7921	9.4340	0.011236	0.1060
90	8100	9.4868	0.011111	0.1054

VI

Evaluate each of the following.

1. $(5 + 2)(3) - 4 =$ **2.** $[(6)(7) - 2]/5 =$

3. $[(27 + 3)/6] - 2 =$ **4.** $[(27/3) - 5](4) =$

VII

Simplify each of the following.

1. $X = \dfrac{5T + 10}{25}$ **2.** $X = \dfrac{(A + 4)(A - 2)(F + 7)}{(F - 7)(A + 4)}$

Remove fractions from numerator and denominator.

3. $X = \dfrac{M/LQ}{T + 47}$ **4.** $X = \dfrac{3F + R/4}{S - L/4}$

Find equivalent expressions that have no squared terms and no fractions in their numerators or denominators.

5. $X = \dfrac{A^2 + Y^2}{A + Y}$ **6.** $X = \dfrac{A^2 + Y^2}{A - Y}$

VIII

Solve each of the following for X.

1. If $\dfrac{X}{B} = C$, $X =$ **2.** If $14X = 42$, $X =$

3. If $X - 27 = 4$, $X =$ **4.** If $X + 483 = 5$, $X =$

5. If $(X + 5) + T - 5 = 4$, $X =$

6. If $X^2 = 16A^2$, $X =$

7. If $\dfrac{X}{BY + 2} = 4$, $X =$

8. If $ABCX = M$, $X =$

9. If $X - ADL = T$, $X =$

10. If $X + RA = Y$, $X =$

11. If $(X + Y) - (Y - 3) = A$, $X =$

12. If $\sqrt{AX} = 8$, $X =$

IX
Solve each of the following.

1. If $\dfrac{A}{B} = \dfrac{6A}{X}$, $X =$

2. If $\dfrac{A(B + 2C)}{3} = \dfrac{X}{9}$, $X =$

3. If $X = \dfrac{SQ/M}{B + FR}$, $B = SL/2$, and $F = TSQ$, eliminate B and F from the equation for X and simplify.

4. If $X = A + RT$ and $R = T^2 - X$, express X in terms only of A and T.

2-2 Answers to test problems and directions for Section 2-3

I. Answers

1. 14	**2.** $^-6$	**3.** 19	**4.** $^-25$
5. 7	**6.** $^-13$	**7.** $^-13$	**8.** 4
9. $^-23$	**10.** 1269	**11.** $^-11$	**12.** 1.846
13. 6	**14.** 7.8040	**15.** $^-8$	**16.** 12

Directions: If you missed any of these, read explanation 1 in Section 2-3.

If you missed:	10 or 16	7 or 11	2, 8, 13, or 15	3, 4, 5, or 6	1 or 9	12 or 14
Read explanation:	2a	2b	2c	2d	2e	8

II. Answers

1. $^-14$	**2.** $^-5$	**3.** $^-10$	**4.** 40
5. 49	**6.** $^-74$	**7.** $^-26$	

Directions: If you missed any of these, read explanation 3 in Section 2-3.

If you missed:	1 or 7	2	3 or 6	4 or 5
Read explanation:	2e	2c	2b	2a

III. Answers

1. 0   **2.** 279   **3.** 3648   **4.** $^-84$

5. $^-1.032$ **6.** 0   **7.** 48   **8.** 828

9. $^-36$ **10.** .5586   **11.** 0   **12.** $\frac{28}{60}\left(=\frac{7}{15}\right)$

13. 42   **14.** $\frac{2}{24}\left(=\frac{1}{12}\right)$ **15.** $^-32$ **16.** $\frac{56}{15}$

17. .00077   **18.** 48   **19.** $\frac{9}{4}$

Directions

If you missed:	1, 6, or 11	2, 7, or 13	3, 8, or 18
Read explanation:	6	4a	4a

If you missed:	4, 9, or 15	5, 10, or 17	12 or 14	16 or 19
Read explanation:	4b	9	12	11

IV. Answers

1. $^-2$ **2.** 8   **3.** Undefined   **4.** $^-5.75$

5. 5.20   **6.** 0   **7.** .3   **8.** Undefined

9. 0   **10.** 200.1   **11.** $\frac{4}{15}$ **12.** 3

13. $^-12$ **14.** $^-3$

Directions

If you missed:	3, 6, 8, or 9	7 or 10	1, 4, 13, or 14	5 or 12	2 or 11
Read explanation:	7	10	5b	5a	13

V. Answers

1. 8   **2.** 4   **3.** 50   **4.** 9

5. 81   **6.** $\frac{3}{4}$ **7.** 9.4340/9 = 1.0482 (approx.)

8. $(64)^2 = 4096$ **9.** 88   **10.** 9.3808

11. $\sqrt{\frac{4 \cdot 3}{9 \cdot 3}} = \sqrt{\frac{4}{9}} = \frac{2}{3}$ **12.** $(458.26)^2 = 21(100)^2 = 210,000$

13. $\dfrac{1156}{100} = \dfrac{34^2}{100} = 3.4^2$

$\sqrt{11.56} = 3.4$

7

Directions

If you missed:	1, 4, 5, or 8	2 or 10	9 or 12	3 or 13	6, 7, or 11
Read explanation:	14	15a	15b	15d	15c and 16

VI. Answers
1. 17　　　　　**2.** 8　　　　　**3.** 3　　　　　**4.** 16

Directions: If you missed any of these, read explanation 17 in Section 2-3.

VII. Answers

1. $X = \dfrac{T + 2}{5}$

2. $X = \dfrac{(A - 2)(F + 7)}{F - 7}$

$$= \dfrac{AF + 7A - 2F - 14}{F - 7}$$

3. $X = \dfrac{M}{LQ(T + 47)}$

$$= \dfrac{M}{LQT + 47LQ}$$

4. $X = \dfrac{4(3F + R/4)}{4(S - L/4)}$

$$= \dfrac{12F + R}{4S - L}$$

5. $X = \dfrac{A^2 + 2AY + Y^2 - 2AY}{A + Y}$

$$= \dfrac{(A + Y)^2 - 2AY}{A + Y}$$

$$= A + Y - \dfrac{2AY}{A + Y}$$

6. $X = \dfrac{A^2 - 2AY + Y^2 + 2AY}{A - Y}$

$$= \dfrac{(A - Y)^2 + 2AY}{A - Y}$$

$$= A - Y + \dfrac{2AY}{A - Y}$$

Directions

If you missed:	1 or 2	3 or 4	5 or 6
Read explanation:	18a	18b	18c and 19e

VIII. Answers

1. BC　　　　　**2.** $\frac{42}{14} = 3$　　　　　**3.** $4 + 27 = 31$

4. $5 - 483 = (-478)$　　　**5.** $4 - T$　　　　　**6.** $\sqrt{16A^2} = \pm 4A$

7. $4(BY + 2) = 4BY + 8$　　**8.** M/ABC　　　**9.** $T + ADL$

10. $Y - RA$　　　　**11.** $A - 3$　　　　**12.** $64/A$

Directions

If you missed:	1 or 7	2 or 8	3 or 9	4 or 10	5 or 11	6 or 12
Read explanation:	19a	19b	19c	19d	19e	19f

IX. Answers

1. $X = 6B$

2. $X = 3A(B + 2C)$

3. $X = \dfrac{2Q}{M(L + 2TQR)}$

4. $X = \dfrac{A + T^3}{1 + T}$

Directions: If you missed any of these, see solutions in explanation 20, Section 2-3.

X

If you miss any problems and find the explanations of Section 2-3 to be insufficient, see Section 2-5 for additional references.

2-3   Explanations for solving arithmetic and algebra problems

1. Numbers that are preceded by either a plus $(+)$ or a minus $(-)$ sign are called *signed numbers*. A number that is preceded by a plus sign is positive; such numbers should be treated just as any unsigned number is treated. A number that is preceded by a minus sign is negative; negative numbers require special treatment. The four paragraphs that follow give instructions for performing arithmetic operations with signed numbers.
   EXAMPLES   4, $^{+}5$, 3564, and $^{+}2953$ are positive; $^{-}926$, $^{-}3705$, and $^{-}904$ are negative.

2. The following instructions are for the *addition* of signed numbers.
   (a) To add two or more positive numbers, add the numbers without regard to sign. The result will be positive; it may be preceded by a plus sign or, equivalently, may be left without any sign.
   EXAMPLES $^{+}4 + {}^{+}5 = {}^{+}9$; $6 + 2 = 8$

   (b) To add two or more negative numbers, add the numbers without regard to sign and place a minus sign in front of the sum.
   EXAMPLE $^{-}4 + {}^{-}5 = {}^{-}9$

   (c) To add two numbers with different signs, first subtract the smaller from the larger, irrespective of sign. If the larger number is positive, the result is also positive; thus the result may be left without sign,

9

or it may be given a plus sign. If the larger number is negative, the result is negative, and it must be given a minus sign.

EXAMPLES $^+5 + {}^-3 = {}^+2$; $^-5 + {}^+3 = {}^-2$; $6 + {}^-4 = 2$

(d) To add more than two numbers of mixed signs, first add all the positive numbers, according to (a). Then add all the negative numbers, according to (b). Finally, add the two sums, according to (c).

EXAMPLE $^+5 + {}^-4 + {}^+6 + {}^-3 + {}^-2 = ({}^+11) + {}^-4 + {}^-3 + {}^-2$

$$= ({}^+11) + ({}^-9)$$

$$= {}^+2$$

(e) Adding zero to a number leaves the number unchanged; subtracting zero from a number also leaves it unchanged.

EXAMPLES $^+5 - 0 = {}^+5$; $0 + {}^-364{,}890 = {}^-364{,}890$

3. To subtract signed numbers, change the sign of the number to be subtracted and then add, following the directions under explanation 2.

EXAMPLES $^-14 - {}^-3 = {}^-14 + {}^+3 = {}^-11$;
$^+5 - {}^-4 = {}^+5 + {}^+4 = {}^+9$

4. To multiply signed numbers, first multiply without regard to sign. Then look at the signs of the numbers multiplied.
   (a) If both numbers were positive or both were negative, the result (called the "product") is positive.
   (b) If one was positive and one negative, the product is negative.

   EXAMPLES $(4)(5) = ({}^-4)({}^-5) = 20$; $(4)({}^-5) = ({}^-4)(5) = {}^-20$

5. To divide one signed number by another, first divide without regard to signs. Then,
   (a) If both were positive or both negative, the result (the "quotient") is positive.
   (b) If one was positive and one negative, the quotient is negative.

   EXAMPLES $20/4 = {}^-20/{}^-4 = 5$; $20/{}^-4 = {}^-20/4 = {}^-5$

6. If you multiply anything by zero, the product is always zero.

7. If you divide zero by any nonzero number, the quotient is always zero. Mathematicians call division by zero an undefined, or impossible, or meaningless operation.

   EXAMPLES $0/4 = 0$; $0/{}^-5 = 0$; $14/0$ and $^-23/0$ are undefined

8. To add decimals, place them one above the other so that the decimal points are in the same column. Then proceed with addition as usual. When you have finished, place a decimal point in the sum so that it falls

in a column with the decimal points of the numbers added or subtracted.
EXAMPLE   3.14 + 580.1 + .007 can be arranged: 3.14
$$\begin{array}{r} 3.14 \\ 580.1 \\ .007 \\ \hline 583.247 \end{array}$$
and added: 583.247

9. When multiplying decimals, first multiply without regard to decimal
points. Then count the total number of digits to the right of the decimal
points in the terms multiplied; this is the number of decimal places that
you should mark off in the product.
EXAMPLE   (.05)(.0030) = .000150; here the product of the numbers without regard to decimals is 5(30) = 150. There are two decimal places in
the first term of this product (.05) and four in the second (.0030). Therefore, the result has 2 + 4 = 6 decimal places (.000150).

10. When dividing decimals, place the numbers in this form:

$$\text{divisor}\,\overline{)\,\text{dividend}\,}^{\text{quotient}}$$

Move the decimal point to the right in the divisor until you are dividing
by a whole number; then move the decimal point in the dividend the same
number of places to the right. Place the decimal point in the quotient
directly above the new one in the dividend. Then divide as usual.

EXAMPLE $1.5\overline{)\,.045\,}^{.03}$ becomes $15\overline{)\,.45\,}$

11. To multiply a fraction by a whole number, multiply the numerator of the
fraction by that number. To divide a fraction by a whole number, multiply
the denominator of the fraction by that number.
EXAMPLES $\frac{4}{5}(7) = \frac{28}{5}$; $\frac{4}{5} \div 7 = \frac{4}{35}$

12. To multiply two fractions, first multiply the two numerators. This is the
numerator of the product; then multiply the two denominators: this is
the denominator of the product. Express the result as a fraction.
EXAMPLE $\frac{4}{5}(\frac{2}{3}) = (4 \times 2)/(5 \times 3) = \frac{8}{15}$

13. To divide one fraction by another, invert the divisor and multiply following explanation 12.
EXAMPLE $\frac{4}{5} \div \frac{2}{3} = \frac{4}{5} \times \frac{3}{2} = \frac{12}{10} = \frac{6}{5}$

14. Consider an example: 2^3. This is a mathematical way of saying "Raise
the number two to the third power." The number two is called the "base"
and the number three the "exponent." In order to evaluate this expression,
list the number 2 three times, then multiply the three numbers. The result

is $2 \times 2 \times 2 = 4 \times 2 = 8$. If the expression had been 4^2, you would list 4 twice and multiply the two numbers. The answer would be $4 \times 4 = 16$.

Some common expressions and problems involving exponents include the following.

(a) A negative exponent indicates a power in the denominator of a fraction. For example,

$$6^{-3} = \frac{1}{6^3} = \frac{1}{216}; \quad 2^{-4} = \frac{1}{2^4} = \frac{1}{16}$$

(b) To add expressions having exponents, first evaluate the expressions, then add. Thus, $2^2 + 2^3 = 4 + 8 = 12$; $5^2 + 2^4 = 25 + 16 = 41$.

(c) To multiply or divide expressions having different bases, evaluate each expression first, then multiply or divide. Thus, to multiply $(2^3)(3^3)$, first find $2^3 = 8$ and $3^3 = 27$. Then multiply $8(27) = 216$. Similarly, $4^2/5^2 = \frac{16}{25} = .64$.

(d) The product of two or more expressions having the same base is equal to the base raised to the sum of the powers of the expressions multiplied. Thus, $(3^2)(3^2) = 3^{2+2} = 3^4 = 81$; $(2^2)(2^3) = 2^{2+3} = 2^5 = 32$.

(e) To divide expressions having the same base, subtract the exponent of the denominator from the exponent of the numerator. The result is the base raised to this difference in powers. For example, $4^3/4^1 = 4^{3-1} = 4^2 = 16$; $5^2/5^3 = 5^{2-3} = 5^{-1} = \frac{1}{5}$.

(f) To find the power of a power, multiply the exponents. Thus, $(2^2)^2 = 2^{2 \times 2} = 2^4 = 16$; $(5^2)^3 = 5^6 = 15625$.

15. To find the square root of a number, find the number which, when multiplied by itself, gives that number. A square root is indicated by a symbol that looks like a check mark with a horizontal line attached to it: $\sqrt{}$. The number for which you are to find the square root appears under this horizontal line. A computational method is available in arithmetic books and some introductory statistics texts, but it is long, tedious, and easily forgotten. Easier methods include the following.

(a) Look up the square root in Table A-2 of the Appendix: If the number for which the square root is wanted is a whole number less than 1000, look up that number in the column labeled n, then find its square root on the same row, in the column labeled $\sqrt{n}$.

(b) If the number is greater than 1000 but less than 1,000,000, look up the number in the column labeled n^2 of Table A-2, then find its square root in the column labeled n.

(c) Finding the square root of a number can sometimes be simplified if you realize that it is the product or quotient of numbers some of which are perfect squares. Then you can take the square root of the perfect squares, leaving the remaining computations somewhat easier to carry out. The general rule is $\sqrt{A \cdot B} = \sqrt{A}\sqrt{B}$. It is generally easier to take the square root of a small number than of a large one;

so if either A or B is a perfect square, extracting the square root of the other is easier than obtaining the square root of the product.

EXAMPLES $\sqrt{300} = \sqrt{3(100)} = \sqrt{3}\sqrt{100} = 10\sqrt{3}$;

$$\sqrt{32} = \sqrt{16(2)} = \sqrt{16}\sqrt{2} = 4\sqrt{2}$$

(d) A special case of method 15c occurs when you want the square root of a decimal fraction. The fraction can be considered to be a product of a whole number and an even power of $\frac{1}{10}$. You can look up the square root of A, the whole number, in Table A-2; and the square root of B is just $\frac{1}{10}$ to half its original power.

EXAMPLES

$$\sqrt{2.65} = \sqrt{(265)(\tfrac{1}{100})} = \sqrt{265}\sqrt{\tfrac{1}{100}} = (16.2788)(\tfrac{1}{10}) = 1.62788;$$

$$\sqrt{5.336} = \sqrt{(53,360)(1/10,000)} = \sqrt{53,360}\sqrt{1/10,000}$$

$$= (\text{about } 231)(1/100) = 2.31$$

(e) If you need greater accuracy than Table A-2 permits, one way to achieve it is by following these steps.

    (1)   Get the most accurate estimate you can from the table.

    (2)   Divide this square-root estimate into the number and find the quotient.

    (3)   Find the average of the divisor and the quotient (add them and divide by 2). This is a more accurate estimate of the number than either the first estimate or the quotient.

    (4)   Continue this process until your result is as accurate as you wish it.

EXAMPLE Find $\sqrt{45.53}$. Since $\sqrt{45} = 6.71$, begin with this:

$$\begin{array}{r} 6.77 \\ 6.71)\overline{45.43} \end{array}$$

Then average 6.71 and 6.77 to get 6.74. This process could be repeated by dividing 6.74 into 45.43, and so on, to get an even more accurate estimate of the square root of 45.53. The present degree of accuracy could have been achieved from the table alone, by a process similar to (c), since the squares of all whole numbers between 670 and 680 are tabled.

16. To square a fraction, square the numerator and let this be the numerator of the result; square the denominator of the fraction and let this be the denominator of the result.

EXAMPLE $(\tfrac{4}{5})^2 = 4^2/5^2 = \tfrac{16}{25}$

    To find the square root of a fraction, find the square root of the numerator and let this be the numerator of the result; find the square root of the denominator and let this be the denominator of the result.

EXAMPLE $\sqrt{\frac{4}{9}} = \sqrt{4}/\sqrt{9} = \frac{2}{3}$;

$$\sqrt{\frac{75}{64}} = \sqrt{75}/\sqrt{64} = \sqrt{25(3)}/8 = 5\sqrt{3}/8 = \frac{5}{8}\sqrt{3}$$

17. To decide in which order to perform a set of operations, first perform the operations within the innermost set of parentheses; then work outward until all parentheses and brackets have been cleared.

EXAMPLE
$$\begin{aligned}
[(3 + 4)(2) - 8]/2 &= [7(2) - 8]/2 \\
&= (14 - 8)/2 \\
&= \tfrac{6}{2} \\
&= 3
\end{aligned}$$

18. To aid in evaluating a fraction, you may do any of the following without changing the value of the fraction.

    (a) Multiply both numerator and denominator by the same value.

$$\frac{\frac{1}{4}}{\frac{1}{2}} = \frac{2(\frac{1}{4})}{2(\frac{1}{2})} = \frac{\frac{2}{4}}{\frac{2}{2}} = \frac{\frac{1}{2}}{1} = \frac{1}{2}$$

$$\frac{T/R}{MPQ} = \frac{TR/R}{MPQR} = \frac{T}{MPQR}$$

    (b) Divide both numerator and denominator by the same value.

$$\tfrac{6}{8} = \tfrac{6/2}{8/2} = \tfrac{3}{4}$$

$$\frac{6A + 36RA^2}{18AT} = \frac{(6A + 36RA^2)/6A}{18AT/6A} = \frac{1 + 6RA}{3T}$$

$$(R - 1)(T + 2)/(R + 1)(T + 2) = (R - 1)/(R + 1)$$

    (c) Add and subtract the same value to either numerator or denominator.

$$\tfrac{7}{16} = [(7 + 1) - 1]/16 = (8 - 1)/16$$

19. You can perform the following operations on both sides of an equation without destroying the equality:

    (a) Multiply both sides by the same value.

    If $X/TQ = 4A$, then $X = 4ATQ$.

    If $X/(5Y + 3) = 2$, then $X = 2(5Y + 3) = 10Y + 6$.

    (b) Divide both sides by the same value.

    If $8Y = 24$, $Y = \frac{24}{8} = 3$.

$$\text{If } X(3 + 5T) = 12, \text{ then } X = 12/(3 + 5T).$$

(c) Add the same value to both sides.

$$\text{If } X - 14 = 22, \text{ then } X = 22 + 14 = 36.$$
$$\text{If } X - (T + 3) = 0, \text{ then } X = T + 3.$$

(d) Subtract the same value from both sides.

$$\text{If } Y + 22 = 43, \text{ then } Y = 43 - 22 = 21.$$
$$\text{If } X + Y + 14 = Z, \text{ then } X + Y = Z - 14.$$

(e) Add and subtract the same value to either side of the equality.

$$\text{If } (X + 22) + (R - 22) = 2, \text{ then } X + R = 2.$$
$$\text{If } X^2 + Y^2 = 2Z, \text{ then } (X^2 + 2XY + Y^2) - 2XY = 2Z,$$
$$\text{and } (X + Y)^2 - 2XY = 2Z.$$

(f) Take the same power or root of both sides.

$$\text{If } Y^2 = 25(T + Q)^2, \text{ then } Y = \text{either } 5(T + Q) \text{ or } (-5)(T + Q).$$
$$\text{If } X^2 = 15A^2, \text{ then } X = A\sqrt{15}.$$
$$\text{If } \sqrt{2AX} = 3, \text{ then } 2AX = 9, \text{ and } X = 9/2A.$$

20. (a)

$\dfrac{A}{B} = \dfrac{6A}{X}$	(given)
$\dfrac{AX}{B} = \dfrac{6AX}{X}$	(19a)
$\dfrac{ABX}{B} = \dfrac{6ABX}{X}$	(19a)
$\dfrac{ABX}{AB} = \dfrac{6ABX}{AX}$	(19b)
$X = 6B$	(18b)

(b)

$\dfrac{A(B + 2C)}{3} = \dfrac{X}{9}$	(given)
$\dfrac{9A(B + 2C)}{3} = \dfrac{9X}{9}$	(19a)
$3A(B + 2C) = X$	(18b)

(c) $X = \dfrac{SQ/M}{B + FR}, \quad B = \dfrac{SL}{2}, \quad F = TSQ$ \qquad (given)

$$X = \dfrac{SQ/M}{SL/2 + TSQR} \qquad \text{(substitution)}$$

$$X = \frac{Q/M}{L/2 + TQR} \qquad (18b)$$

$$X = \frac{2Q/M}{L + 2TQR} \qquad (18a)$$

$$X = \frac{2Q}{M(L + 2TQR)} \qquad (18a)$$

(d) $X = A + RT,\ R = T^2 - X$ (given)

$\qquad X = A + (T^2 - X)T$ (substitution)

$\qquad X = A + T^3 - XT$ (multiplication)

$\qquad X + TX = A + T^3$ (19b)

$\qquad X(1 + T) = A + T^3$ (removing common term)

$$X = \frac{A + T^3}{1 + T} \qquad \text{(19b and 18b)}$$

2-4 Some additional notation

The following expressions are listed here so that you will be able to find them easily if you need them later on in this book. It would be best if you did not try to learn them until you actually need them. In addition to these expressions, a listing of the Greek alphabet appears as Table A-1 of the Appendix.

1. $A > B$ A is greater than B.
2. $A < B$ A is less than B.
3. $A \geq B$ A is greater than or equal to B.
4. $A \doteq B$ A is approximately equal to B.
5. $|A|$ This is read "the absolute value of A." The absolute value of a number always has the same *magnitude* as that number, but its *sign* is always positive. For example, $|5| = |{}^-5| = 5$.

2-5 References

Additional problems can be found in the workbook that accompanies this text. Useful books on arithmetic and algebra include Baggaley (1969); Games and Klare (1967), Appendix A; Hart (1966), Chapters 1–4; Ray (1962), Chapter 2; and Walker (1951), Chapters 1–13. Complete citations can be found in the Bibliography at the end of the book.

3 Flow Charts and Decision Maps

CHAPTER 2 was intended to inform you of the kinds of arithmetic and algebra that will be needed in the rest of this course, and to give you an idea of what you need to review before going on. Unless you had forgotten a great deal about elementary arithmetic and algebra, you probably did not have to read very much of Section 2-3; Chapter 2 was designed to save you the trouble of reading a lot of material that you already knew. You took the test, scored it, and asked yourself a number of questions about any of the items you missed. Only when you had missed one or more of a particular type of problem was it advisable for you to read an explanation of that type of problem.

The structure of Chapter 2 can be diagrammed in a convenient form known to computer programmers as a *flow chart*. Such charts are used to outline the logic of complicated computer programs so that the programs may be more easily written and read. I call such a diagram a *decision map* when it conveys information directly to human beings for use in their own decision making, and when the boxes in the diagram are explicitly keyed to the accompanying text. To call them decision maps emphasizes the idea that there diagrams are intended to organize and direct your thoughts toward the goal of efficient learning and problem solving. Part of Chapter 2 has been represented in the decision map shown as Figure 3-1.

Notice that two kinds of boxes are used in this decision map: rectangles and diamonds. A rectangle will always be used to present information or to give an instruction. For example, the first rectangle at the top of the illustration says "Take test." This is an abbreviated way of saying that the first thing you must do in Chapter 2 is to take the arithmetic and algebra test. An arrow directs you to the next instruction, which is "Score test." Arrows are used in the map to show the direction of movement. They go in one direction only, and you must follow the directions indicated by the arrows, which usually begin at the top left-hand side of the page.

After scoring your test, you move on to a diamond-shaped box containing the question, "Any errors in part I?" Diamond-shaped boxes will always be used to indicate *decision points,* places where you must make a choice. Decision points always ask you questions to which you must give yes-or-no answers on the basis of information you already have. The question in this particular diamond asks you, in an abbreviated way, whether you made any errors in part I of the arithmetic and algebra test (i.e., in the part on addition).

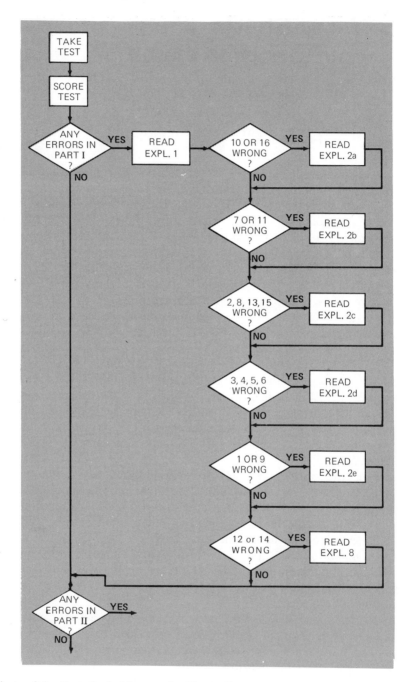

Figure 3-1 *Part of a decision map for Chapter 2.*

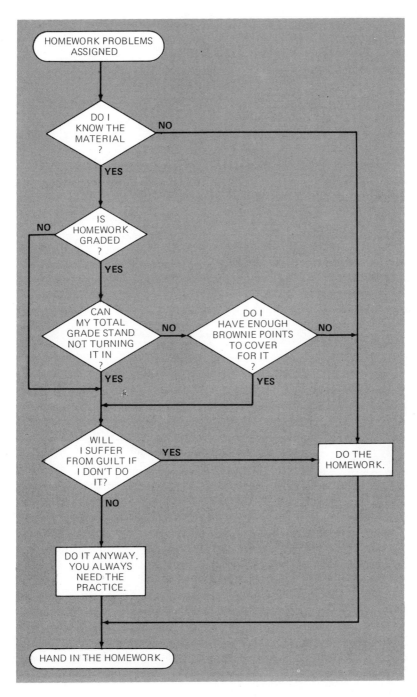

Figure 3-2 *Flow chart for deciding what to do about statistics homework assignments.*

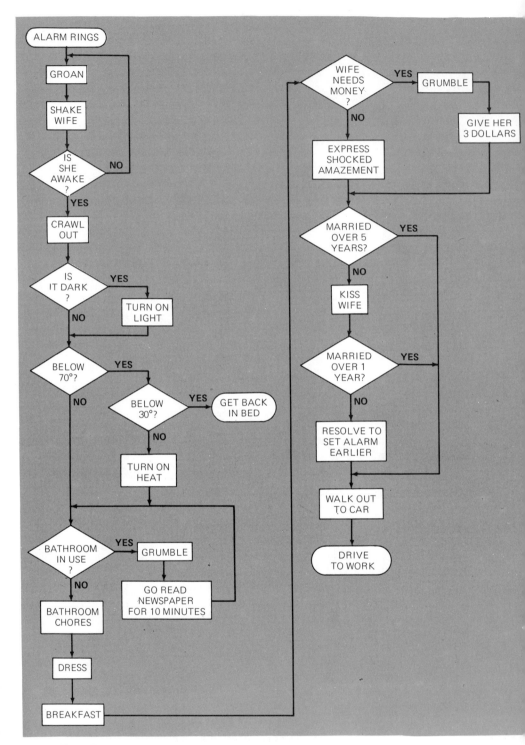

Figure 3-3 *How to get to work in the morning.*

If you did not make any errors in this part of the test, follow the arrow that leads from the bottom of the diamond to another decision point having to do with part II of the test. If you did make errors in part I, follow the arrow that leads from the right-hand side of the diamond to the instruction to "Read explanation 1." When you have read explanation 1 in Section 2-3, proceed to the next decision point, which asks whether you missed either of problems 10 or 16. If you did not answer either problem incorrectly, go on to the next decision point. If you did miss one of these problems, you must read number 2a before continuing. The decision map proceeds in this manner. For each type of problem you missed in part I, you must read an appropriate instructional section.

To diagram Chapter 2 fully we would have to complete this map by showing all the remaining parts of the chapter. I have not done so because the diagrams for the remaining parts are similar to the diagram presented here.

The next example of a flow chart should help you to decide whether or not you should do your statistics homework. Figure 3-2 illustrates three additional characteristics of decision maps. First notice that, although in the previous map answering "Yes" always led you to the right, and answering "No" to the bottom of the page, here the same arrangement is not used consistently. To prevent arrows from crossing one another, "Yes" and "No" responses are permitted to leave a decision point from either side or from the bottom of the diamond. Second, notice that only one arrow leads to any instruction or decision point; the lines of flow diverge from a decision point, depending on your answer, but only one arrow leads out of an instruction box. Finally, notice that each of the charts begins with a special box that has rounded ends. This helps you keep track of the terminal points on the map.

The next flow chart, Figure 3-3, was found on a bulletin board at the University of Michigan computing center. It is directed to the husband in a traditional nuclear family and tells him how to get to work in the morning. It is slightly more complex than the previous diagrams, and it illustrates another point. It contains two *loops*—sections in which you may, if necessary, retrace part of a path that you have gone through once before. The first of these loops is begun by the decision, "Is she awake?" This loop involves, first, groaning and, second, shaking your wife until she awakens. The second loop is initiated by a test of whether the bathroom is in use.

Each of the remaining chapters (except for Chapter 7) begins with a decision map keyed to the body of that chapter. The purpose of the map is to help you find your way through the chapter quickly when you are solving a problem.

The boxes on each map contain the important information for solving problems using that chapter. However, the boxes are often so small that they can contain the questions and instructions only in highly abbreviated form; and you may frequently need further information before you answer the question or carry out the instruction in a given portion of a decision map. To provide you with this information, many of the boxes in the maps that follow will be keyed to sections in the body of the text. When you do not understand or cannot remember the full details of a given instruction or question, look for a small number slightly above and to the left of the box. The number shown is the number of that section in the text which explains the question or instruction more fully. As you become more familiar with the material, you will be able to forego reading more and more sections of the text, and you will save more and more time in solving problems.

You can also consider the map for each chapter as a kind of outline of the chapter, and use it either to organize your study or for review of what you know. Because of the structural organization of the maps, they can not only remind you of the content of the chapters, but also remind you of some of the similarities and differences among the methods presented.

Homework

For each of the following problems, make a reasonable assumption about the amount of detail required to solve the problem. Show that you understand the problem and the method of solution, but keep unnecessary details to a minimum.

1. The first decision map in this chapter showed how to take part I of the arithmetic and algebra test. Make up a continuation of this map for part II of the test.
2. Johnny Goodfellow often drives from his home at 616 4th Street to the home of his friend, Rex Carrs, at 408 6th Street. He invariable takes one of two routes: the shortest way, or the way that requires the least number of turns at corners. Make a flow chart that describes what he must do to get there (but not to return). Notice that Center Street is a through street. Stop signs and traffic signals are shown on the street map.

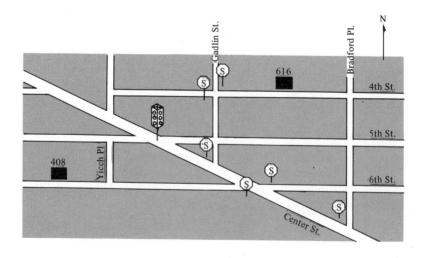

3. We wish to build a machine that will translate Morse code into English. We have two basic components, one of which will answer the question "Is this a dot?" and the other the question "Is this a dash?" Make a flow chart that will allow a machine constructed of such components to distinguish among the following codes:

English	Morse
A	• —
B	— • • •
C	— • — •
D	— • •
E	•
F	• • — •

4. In a study of gambling behavior, subjects roll two dice one at a time; they win if the sum appearing on the two dice is exactly 9. They need not roll both dice, and they map stop the experiment at any time. The experimenter's assistant made the flow chart below of an optimal dice-rolling decision sequence.

(a) Criticize this flow chart.

(b) Produce a better one.

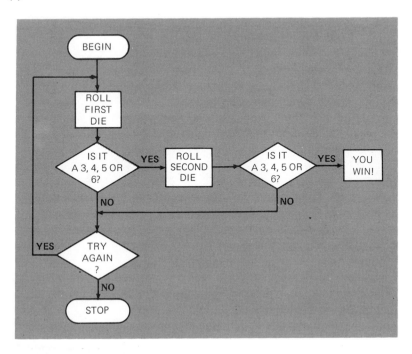

5. Make up your own decision map for some problem or activity. Include at least one or two decision points and a few instruction boxes, but don't make it too complicated. If it is not self-explanatory, include a brief note to explain to the reader what is happening.

Additional problems can be found in the workbook that accompanies this text.

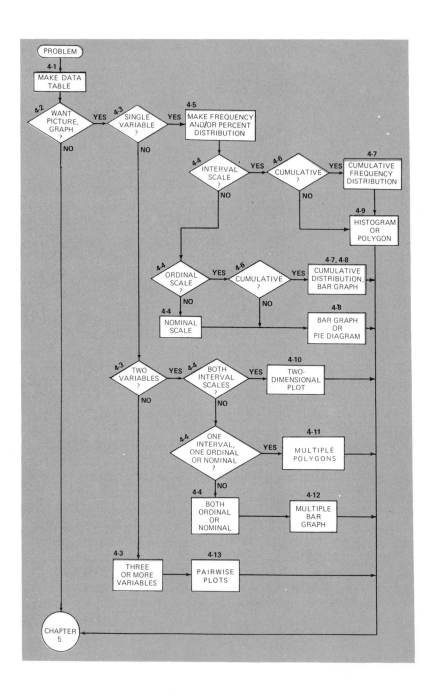

4 Graphing

THE previous chapters have attempted to prepare you for statistics by examining your background in arithmetic and algebra and by explaining the manner in which this text is organized. Understanding the background material is extremely important for understanding the rest of this text, so if you haven't read the previous chapters, you should do so before continuing.

With this chapter we actually begin the study of statistics, by looking at ways of organizing and representing research data we have collected. Later chapters will indicate other ways in which research results can be arranged and manipulated to allow us to draw meaningful conclusions about the research.

The decision map for Chapter 4 appears at full size on the opposite page. A small-scale schematic version of this map appears below; we will call such schematic diagrams *regional maps*. We will use similar regional maps to analyze the structure of later chapters. This regional map shows us that the decision map for Chapter 4 is divided in three main re-regions, and that the regions are distinguished according to the number of variables in the problems they tell you how to solve.

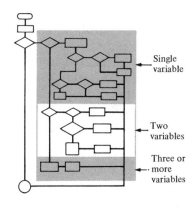

This and later chapters can be studied in any of at least three ways: (1) you can ignore the decision map and read the text straight through; (2) you can come to the chapter with a problem of your own, which will determine your course of study, in conjunction with the maps provided for each chapter; or (3) you can begin by choosing a problem from the homework assignment at the end of each chapter and following it through, using the map at the beginning of the chapter. In either of the last two cases, you will read only those sections of this chapter which lead to a solution of your problem and which you do not already understand. For example, if you wished to construct a graph of scores on a nationally administered intelligence test, you would probably use the method of Section 4-9. But first you would have to read and apply the information in Sections 4-1 through 4-6, and possibly 4-7. You would, however, be spared reading any of the other sections in the chapter in the course of obtaining this particular graph.

4-1 Data table

A. Data

In order to begin any statistical analysis, you will need to collect some data of interest to you and to organize them in some systematic fashion. Data constitute information—any information at all that you both want and are able to collect. A data table is generally the most useful first form of organization for your data, although in some instances it may be possible to skip the data table and begin with the forms of presentation discussed later in this chapter.

As a start, let us assume that you already have a research problem and know the kind of results you want before you begin following the decision map for this chapter. The assumption that the data have been collected first is only for didactic purposes. As you become more proficient in statistics, this assumption will hold for you less and less, because you will grow to rely on your statistical knowledge in developing your hypotheses and planning your research designs. Actual collection of your data will tend to occur only after you have considered most of the major steps to be followed in the analysis. You will tend to use the decision map and your own knowledge in designing your research, and you will use the methods referred to in the map in order to carry out the mechanics of the analysis.

The data of social science consist of measurements on subjects or groups of subjects with respect to some attribute or attributes that they are thought to exhibit in varying degrees. The attributes with respect to which individuals differ are called *variables*. If there are attributes with respect to which the individuals in a group do not differ, those attributes are called *constants* for the individuals in the group.

Scores for intelligence, height, weight, number of brothers and sisters, agreeableness, and neuroticism could be expected to be variables for an unselected group of American college students seated in a single classroom. Country of residence, present room temperature, and the number 4 would be constants for these same people.

Variables are usually signified by the last letters of the alphabet (e.g., W, X, Y, and Z). Constants are usually indicated by the first letters of the alphabet (e.g., A, B, C, and D). This convention is not always followed, however, and one convenient mnemonic device is to let the first letter of the name of a variable represent the variable (e.g., "Let H stand for Halitosis Rating," or "Let A be age in years").

B. Form of the Data Table

A data table is a listing of each subject's score on each variable that is of interest and for which scores have been obtained. An example is presented in Table 4-1. There is a standard form in which such tables are usually laid out, for reasons of convenience and readability.

Table 4-1 *Hypothetical intake data for 25 children examined at a child guidance clinic*

Sub-ject	Sex	Number of siblings	Religion	Age	Principal wage earner	Major symptom	Prognosis in therapy
1	F	1	Prot.	6	Operative	Enuresis	Moderate
2	F	2	Prot.	11	Professional	Enuresis	Favorable
3	F	3	Other	12	Service	Enuresis	Poor
4	M	0	Prot.	10	Operative	Lying	Poor
5	M	4	Other	6	Professional	Selfish	Poor
6	F	2	Other	5	Laborer	Truant	Moderate
7	M	3	Jewish	11	Operative	Enuresis	Questionable
8	M	1	Catholic	14	Operative	Learning	Questionable
9	M	2	Jewish	13	Craftsman	Lying	Poor
10	M	1	Prot.	7	Craftsman	Depressed	Moderate
11	M	1	Prot.	7	Professional	Selfish	Poor
12	F	2	Other	9	Semiprofessional	Stealing	Favorable
13	F	2	Catholic	14	Semiprofessional	Selfish	Moderate
14	M	0	Prot.	5	Laborer	Fighting	Moderate
15	M	3	Other	9	Clerical	Truant	Favorable
16	M	0	Catholic	5	Operative	Enuresis	Poor
17	M	0	Other	11	Operative	Depressed	Questionable
18	F	4	Prot.	5	Clerical	Enuresis	Poor
19	M	0	Catholic	14	Operative	Selfish	Favorable
20	F	2	Other	8	Clerical	Truant	Poor
21	M	0	Other	6	Operative	Selfish	Moderate
22	M	3	Prot.	11	Craftsman	Enuresis	Favorable
23	F	0	Catholic	8	Clerical	Truant	Favorable
24	M	3	Catholic	14	Craftsman	Fighting	Moderate
25	F	0	Catholic	12	Laborer	Learning	Moderate

1. List subjects down the left-hand column, either by name or some other identifying label (in this case, subject number). Label this column "sub-jects." Although names are sometimes used in a data table, as in a teacher's grade book or a summary file in a psychological clinic, the decision to use actual names depends on who will see the data table. Most data tables used in research use codes or numbers instead of names, to protect subjects' rights to privacy (see *Ethical principles in the conduct of research with human participants.* American Psychological Association, 1973).
2. List variables across the top row, either by name, as in this case, or by some other method of identification. Label this row "variables," if there are more than one. Otherwise, assume that the variable name will be sufficient.

3. Fill in each cell of the table with the score identified by its row and column labels.

C. Rounding

Often data will be collected which are more precise than is needed for purposes of analysis; for example, when heights can be measured to the nearest sixteenth of an inch, but only whole-inch values must be reported. When this occurs, you may wish to round the scores as you enter them in a data table (although generally it is a good idea to record data as accurately as possible; thus, you'll have the information in case you need it). It is best to select an explicit rule for rounding. Three different rules are possible: you can round either to the nearest reportable score (in the example above, a whole-inch value), or you can round up or down. Whatever decision you make about rounding will affect the data you carry on for further analysis; and it is possible to distort the data and influence the outcomes of research by your procedure.

When dealing with ages, the custom usually followed is to round down. For example, a person is given his age at his last birthday, and the fraction of a year since that date is simply ignored. By contrast, children's waist measurements are usually rounded up by tailors to leave room in the clothing for growth.

If you consistently round ages up to the next year, and everyone else rounds their data about ages down, then the average age for your subjects may appear to be greater than the averages for other researchers, even when there is no difference. For this reason, it is wise to know what the usual procedures are, and to follow them unless you have specific reasons not to; and if you use an unusual rounding rule, to report it so others can be aware of it in interpreting your findings.

Most measurements are rounded off to the *nearest* scale value. Thus, if we were to deal in common fractions, and were rounding off to the nearest whole number, the values between 9 and $9\frac{1}{2}$ would be rounded down to 9, and the values between $9\frac{1}{2}$ and 10 would be rounded up to 10. If we were dealing with decimal fractions and rounding to the nearest tenth, the values between 3.20 and 3.25 would be rounded to 3.2, and the values between 3.25 and 3.30 would be rounded to 3.3.

This leaves only one more problem: what about a score that is exactly midway between two scale values (as far as we can tell)? What about the $9\frac{1}{2}$; and what about the 3.25: should we round them up or down?

Consistently rounding either up or down could produce rounded data that systematically distort the picture that we would get from the unrounded data. Consistently rounding down, for example, could yield rounded data that are in general lower than the original scores; and the effect can be important if there are a lot of borderline raw scores. Therefore, I like the rule of rounding to the nearest even score. If we were to follow this rule above, we would round up the $9\frac{1}{2}$ to the nearest even number, 10; and we would round down the 3.25 to 3.2 rather than up to 3.3.

When we have rounded data, it is sometimes necessary to know the *real limits* of each score or category. These limits are the maximum and minimum values of the data collected that will be recorded as being at that score. Thus, the score "9 years old" has real limits of "9 years, 0 days" and "9 years, 364 days," inclusive; the height 5 feet 10 inches has real limits of 5 feet $9\frac{1}{2}$ inches and 5 feet $10\frac{1}{2}$ inches (again inclusive, because we will round both of these values to the nearest even inch).

If you were looking for hard-and-fast rules for recording data in this section, you probably were disappointed. Instead, you found guidelines, and you will have to use your judgment in applying them. This is true of the course as a whole: although it will often be possible to say that a particular way of recording or analyzing data is clearly misleading, you may be left with two or more possible ways of doing an analysis or presenting the results. The alternatives might seem equally reasonable, and you will have to choose the one that makes the most sense to you.

One way to dodge the issue of rounding is to carry the data through your analyses to so many decimal places that the results will be considerably more accurate than you can ever need. Then it does not matter how you round. The difficulty with this approach is that you end up doing a lot more work, as your computations are messier with more digits to work with. But it is a reasonable option, especially if you have access to a good calculator or computer.

4-2 Do you want a picture or graph?

Generally, when we collect data for some group of individuals on some variable, the scores appear in no particular order, and the result can be difficult to interpret. A first step in summarizing and displaying data is often the preparation of a frequency distribution and/or a graph. These place the scores in a pictorial mode for clarity and ease of conveying information. It may be possible to see a pattern and shape in a frequency distribution or graph that is not evident in a perusal of the raw scores. Often new ideas and insights can be gained by looking at a graph of a distribution, even when one is familiar with the data and has carried out previous statistical analyses on them.

If you want a general decision rule, try "You should almost always make a graph." You will have to decide in each case whether you can afford not to make one. Sometimes it is a good idea to graph your data two or three different ways, particularly if you are having difficulty interpreting or understanding them.

4-3 Number of variables

In Section 4-1 it was indicated that a variable is a characteristic of subjects, an attribute or dimension on which they differ, or vary. When you collect data and construct a data table, you have already made a decision about the number and nature of the variables that interest you. At each stage in the analysis of your data you must reconsider this decision with respect to the kind of analysis that you now contemplate; you must match analyses to the kind of variable or variables you are working with and the kind of information you wish to obtain from them.

For example, Table 4-1 contains data on seven different variables; it is not necessarily the case, however, that you will need to graph all seven simultaneously. Nor is it necessarily true that you will want only one graph of the relationships among these variables, or even that you will want to graph all of them. The number and nature of the graphs you construct should be determined by the nature of your variables and the kind of results you are contemplating.

If you are only interested in the distribution of scores on some single variable, you can graph that variable by itself, and you can graph several variables singly if you are interested only in the distributions of scores on each by itself. But if you are interested in the relationship between two variables, you will probably want to graph the bivariate (two-variable) distribution; more complex interrelationships among variables can be represented using three variable graphs or sets of two-variable graphs.

Different kinds of graphs give different kinds of information about your variables, and you may want to represent scores on your variables in more than one way in order to obtain more information about them. For example, you might want to indicate the relationship between two variables using a bivariate plot and also represent the distribution of each singly. The bivariate plot might give you an idea of the degree of relationship between the two variables. The univariate plots might indicate which statistic to use to calculate the degree of relationship between the two variables, by indicating what assumptions can be made about them.

You may wish to determine what effect changes in one variable have on a second variable. Questions of this kind are often approached by systematically changing the first variable and observing what happens to the second. In this case, the variable manipulated by the experimenter is called the *independent variable* and the one observed for possible changes is called the *dependent variable*.

If you wished to know the effects of three different drugs on patients' anxiety levels, you might administer the drugs, then observe anxiety levels. The independent variable would be "drug," and it would have three levels. The dependent variable would be "anxiety level"; it might or might not show systematic differences from one level to another of the independent variable.

4-4 Kinds of scales

Statistics may be viewed as a collection of procedures for dealing with certain kinds of numbers; procedures for organizing, simplifying, summarizing, and making inferences. When we use statistical procedures to manipulate the numbers associated with properties of real objects and events, we want to ensure, as much as possible, that any conclusions we may draw from our use of statistics will be applicable to those real-world properties. One way to ensure the applicability of statistical conclusions is to be sure that the operations performed on the numbers as part of the statistical analysis could also be performed with the measured attributes.

For example, numbers representing heights of objects can be added legitimately, because the sum of two heights can be obtained physically by placing two objects end to end and measuring their combined height. On the other hand, if we represent blue eye color by the number 1, green eye color by 2, and brown eye color by 3, we are not justified in using the arithmetic statement $1 + 2 = 3$, since this statement is meaningless when applied to the attributes themselves. Blue eye color + green eye color does not equal brown eye color. A conclusion that depended on the additivity of numbers might well be meaningless if applied to eye colors in this manner.

A. Level of Measurement

The numbering system used to represent the properties of objects and events is called a scale. Social scientists distinguish between four kinds of scales having different formal properties: nominal, ordinal, interval, and ratio scales.

The most primitive kind of scale is a *nominal scale.* Different values on a nominal scale merely amount to different names for distinguishable groups; a subject's scale value on a nominal variable simply indicates of which group he is a member. There is no relationship among groups of subjects having different scale values solely on the basis of differences along a nominal variable. An obvious nominally scaled variable is given name: we could set up a data table for any group of subjects and record each person's first name, then graph and analyze the results. Athletes in competitive sports usually have numerals on their uniforms, but these numerals are for identification purposes only. Other examples of nominal variables are sex, race, eye color, city of residence, and animal favored as a house pet. We might assign different numbers to different scale values of a nominal variable, but doing so would not mean that we could legitimately treat those numbers using the ordinary operations of arithmetic. We might be able to manipulate the numbers by, say, adding or multiplying them, but the results would probably be meaningless.

An *ordinal scale* divides subjects into distinguishable groups also, but in addition implies an order of magnitude among the groups on the characteristic measured. Addition and subtraction are still not justifiable with respect to the numbers representing different scores, but the relationships of "greater than" and "less than" are meaningful. Examples of ordinal data are a subject's preference order among potatoes, carrots, and rhubarb; win, place, and show in a horse race; pecking orders in animals; and hardness orders among metals, where the harder of any two metals is that which scratches the other.

An *interval scale,* in addition to the properties of an ordinal scale, has units that are the same size throughout the scale. Because of this property, the arithmetic operations of addition and subtraction can legitimately be performed on this kind of data. Examples of interval scales include the Fahrenheit and centigrade temperature scales: a change of 5° from 10° F to 15° F has the same physical meaning as a change of 5° from 85° F to 90° F; however, the zero points on both scales are arbitrary. For this reason, the arithmetic operations of multiplication and division cannot be performed on the raw scores, though they can be performed on differences between pairs of scores. Thus, it makes no sense to say that 100° F is twice as warm at 50° F, because the zero on the scale is arbitrary.

It is often both important and difficult to decide whether a given set of data is ordinal or interval. Thus, for example, it may be reasonable to consider intelligence (IQ) an interval scale, but many people feel it does not have equal intervals throughout its range: the issue is whether the difference between IQ's of 95 and 100, for example, has the same meaning as the difference between IQ's of 130 and 135.

To resolve such an issue for yourself, you should decide whether you think equal differences among people in the variable that you are trying to measure (in this case, intelligence) are reflected in equal differences in test scores. If you think they are, then you think you have an interval-level variable.

A *ratio scale,* in addition to having equal intervals, also has a meaningful zero point. Because of this, we can say that some raw score on a ratio scale is a multiple or fraction of another. The usual examples of ratio scales come from physical measurements: height, weight, and age all have meaningful zero points. We can say that one object is twice as tall as another if two of the second, placed one on top of the other, would equal the first in height. It is not clear that in any sense an IQ of 150 is equal to two IQ's of 75; one has even greater difficulty fitting other psychological variables to the assumptions of ratio scales.

In this text no methods for organizing or analyzing statistical data will be presented which are appropriate only for ratio scales. For our purposes, no distinction will be made between ratio and interval scales. Therefore, when you find the question "Interval scale?" you should interpret this to mean "Do you have an interval or a ratio scale?"

Perhaps the greatest power and flexibility can be achieved from the use of this text if you read the question "Interval scale?" even more broadly as "Do you want to treat your data as though they are at the interval level of measurement?" or "Do you want to make the assumption that you have an interval scale?" The questions "Ordinal scale?" and "Nominal scale?" should probably best be read in a similar manner. Because interval-level data have all the characteristics of nominal- and ordinal-level data, they may be treated as such, if it appears useful to do so. Similarly, ordinal data may be treated as nominal if the kind of results obtained by doing so may be useful to you. However, treating data as being at a lower level of measurement may be wasteful of information contained in the data and may give misleading results. In some cases it may even make sense to treat data as being at a higher level than you think it actually to be (i.e., to treat ordinal data as interval or nominal as ordinal), but you should realize that such treatment generally demands a good deal of justification to be acceptable to others.

As an example, suppose that the following data are obtained on subjects: agreeableness, as rated by peers; anxiety level, as judged by a clinical psychologist; and last year's income, as reported to the federal government. The first two are probably best considered ordinal data and the third ratio, but if you wish to carry out comparable analyses on all three variables, you might decide to treat reported income as an ordinal variable.

By contrast, it is quite common to treat intelligence test data as interval level for purposes of analysis, although the actual level of measurement of the test scores is open to debate. The practice is justifiable if you assume (1) that the intervals may not be exactly equal, but the differences are small and do not systematically bias any conclusions you might draw from treating the scale as interval; (2) that the results of treating the scores as interval will therefore be meaningful (a conclusion based on familiarity with treating them that way); and (3) that most psychologists are more comfortable with interval-level statistics than the ones you would have to use if you treated the scale as ordinal.

B. Discrete and Continuous Scales

Any set of ordinal, interval, or ratio data may be classified as either *continuous* or *discrete*. A given portion of a scale is continuous if it is possible for an observation to fall at any point within that portion of the scale. It is discrese if there are breaks or gaps in which no observation can occur. Alternatively, we may say that a portion of a scale is continuous if it is possible for a third observation to appear between any two observations occurring within that portion of the scale: the scale is infinitely divisible. The scale is discrete if this condition does not hold. Thus height, weight, sociability rating, sexiness, and degree of neuroticism may be considered to be continuous variables; number of siblings is a discrete variable because it can take on only whole-number values: you can't, for example, have $2\frac{1}{2}$ siblings.

In social science, the most common sources of discrete data are variables whose magnitude is obtained by counting (e.g., number of siblings, number of previous arrests, and number of cigarettes smoked in a given period of time). When the observed scores are discrete, it may or may not mean that the underlying differences among people are also discrete. Scores on a test of mathematical ability may be discrete if they are obtained by counting the number of correct responses to test items and thus are limited to whole-number values. Nevertheless, we will probably be willing to assume that the underlying scale of real differences among subjects is continuous; accordingly, we will treat the observed (raw) scores as though they were also continuous.

C. Dichotomous Variables

There is a special class of variables that have only two levels or scale values; they are called *dichotomous* or *binary*. Examples include sex (male or female), handedness (right-handed or left-handed), and answer to the question, "Did you receive a high school diploma?" (yes or no).

Other variables that are many-valued or continuous may be *dichotomized,* or divided so that only two values are considered. Examples include number of brothers and sisters (none vs. one or more, or none or one vs. two or more, etc.), intelligence (bright = 101 or more, dull = 100 or below), or country of birth (United States vs. other).

The special thing about dichotomous variables is that they may be treated using methods designed for nominal-, ordinal-, or interval-level variables. Thus, sex may be treated at nominal (which it is), ordinal (if you ignore the order when interpreting the results), or interval (after all, there's only one interval, so the intervals can't be unequal). So if sex or any other dichotomous variable is to be analyzed, you can ignore scale questions about that variable in choosing your method of analysis.

Because of the greater flexibility in dealing with dichotomous variables, you may be tempted to dichotomize in many cases. You can do so; but you should be aware of two difficulties with doing it: (1) You lose information when you combine levels on a variable, and you end up treating things that are really different as the same. Thus, although it may make sense in some studies to give the same score to everyone who is above average in intelligence, for many purposes it will be important to make much finer distinctions in level of intelligence. (2) The way in which you dichotomize a variable is generally arbitrary and may affect your results. Suppose that we were to do a study on the relationship between intelligence and ability to learn calculus, using high school seniors; and suppose we ran the study first dichotomizing at an IQ of 85, and then again, dichotomizing at an IQ of 130. Our results for the two runs would be quite different, just because of the point we chose to divide the intelligence scale.

4-5 Frequency and/or percentage distribution

A. Frequency Distribution

A frequency distribution is a first step in summarizing data from a single variable. For convenience, its organization is quite different from that of a data table. Data are listed not by subject but by scale value. Cell entries are not scale values of the variable for each subject but rather numbers of subjects that have a given scale value. For example, the frequency distribution of ages in Table 4-2 is based on the data in column 5 of the data table, Table 4-1.

Here the variable of concern is age of patients. The scale values of that variable are ages in years, listed down the left-hand margin of the table, and the cell entries are numbers of patients at each scale value or age.

This frequency distribution shows the ages of patients in the group more clearly than the previous data table, provided you are not interested in the

Table 4-2   *Frequency distribution of ages of patients (data from Table 4–1)*

Age	Tallies	Frequency
5	||||	4
6	|||	3
7	||	2
8	||	2
9	||	2
10	|	1
11	||||	4
12	||	2
13	|	1
14	||||	4
Total	25	25

ages of particular patients. By comparison with the data table, it represents both a gain (in legibility and comprehensibility) and a loss (in information that relates scale values of the variable to particular subjects). In further analyses, you can decide which method of representing the data to use on the basis of the information and convenience it provides.

B. Grouped Frequency Distribution

When you make a frequency distribution it is sometimes useful to group the data into fewer categories than you originally began with. For example, if you were to record the ages of 20 consecutive persons leaving a large supermarket, you might find they ranged from 2 to 70. Here you would have more than three times as many scale values as subjects—to list the data in the frequency distribution described above would be less parsimonious than to use the original data table, and probably more confusing. If you wanted only a general

Table 4-3   *Grouped frequency distribution of hypothetical ages of subjects leaving a large supermarket*

Age	Tallies	Frequency
0–9	||	2
10–19	|||	3
20–29	⃫⃫⃫	5
30–39	||	2
40–49	||	2
50–59	|	1
60–69	||||	4
70–79	|	1
Total	20	20

representation of the age distribution, you might group scale values into groups of 5 or even 10 years each. This has been done in Table 4-3.

Here again an exchange has been made between the amount of information available and the clarity and simplicity of presentation. It is easier to read a frequency distribution that has 8 lines than one that has 69 lines, but this presentation treats as identical persons who may differ in ages by as much as 10 years. When scale values are grouped in this manner, each grouping is called an *interval,* and the number of scale values in each interval is called the *interval size* or *interval width.*

In Table 4-3 the age variable has been abbreviated in a meaningful and natural way: ages have been rounded (see Section 4-1C) in the way that ages are usually rounded. For example, the interval 10 to 19 includes anyone on or past his tenth birthday but who has not yet had his twentieth birthday. The values 10 and 19 in years are called the *score limits* or *apparent limits* of that interval; the values 10 years, 0 days, and 19 years, 364 days (ignoring the complication of leap years), are called the *real limits* of the same interval. The interval size can be found by subtracting the lower real limits of successive intervals. The size of the present interval is found by subtracting 10 years, 0 days (its lower real limit), from 20 years, 0 days (the lower real limit of the next interval), to get exactly 10 years. The relationship between the real and apparent limits will be different for different sets of scores, depending on the way in which the data have been rounded.

In Table 4-3 the interval size was 10, but in this case the interval size could have been anything from one to, say, twenty. The choice of interval size (and at the same time, the number of intervals) is partly a matter of how much detail of information is needed and of how many intervals it is convenient to look at, and partly an arbitrary matter. Some authorities recommend that you select an interval size so that between 10 and 20 intervals are needed to describe all the data, and some recommend that you use an odd number of scale values per interval for later convenience of graphing. In general, it is best to use as few intervals as provide enough information for your puposes. That is, your choice of interval size and number of intervals should be determined by what use you expect to make of the information contained in the frequency distribution, and you should be able to justify your choice, at least informally. The choice of interval size for Table 4-3 was based on the small sample size, the wide range of ages found, and the fact that it is easy to read a table having intervals of size 10.

The relationship between the interval size and the number of intervals is straightforward and directly related to the *range.* The range itself is defined as the number of scale values needed to describe the data. The range for Table 4-2 is 10: the number of different ages from the youngest to the oldest, inclusive. You find the range by subtracting the smallest scale value from the greatest, and then adding 1. In the case of Table 4-2, range $= 14 - 5 + 1 = 10$; it can also be found by counting. The number 1 is added because the range includes both end values.*

Once the range has been determined, either the interval size or the number of intervals can be used to find the other, using the formula

$$\text{range} = (\text{interval size})(\text{number of intervals})$$

Let us take the hypothetical data of Table 4-3 for example. The ages ranged from 2 to 70; therefore,

$$\text{range} = 70 - 2 + 1 = 69$$

Intervals of size 10 were selected for convenience and because a great deal of precision was not needed. The number of intervals was then found:

$$\text{number of intervals} = \frac{\text{range}}{\text{interval size}} = \frac{69}{10} = 6.9$$

Now, 6.9 is not a convenient number to work with, so we use 7 instead. The range was thus expanded to $10 \times 7 = 70$. Next a starting point was chosen; because of our familiarity with the decimal system, it seemed reasonable to begin intervals at 0, 10, 20, and so on. Finally, because seven intervals beginning at zero would only include through age 69, an eighth interval was added, in order to include the 70-year-old. If we had used intervals of 1 to 10, 11 to 20, 21 to 30, etc., we would not have needed the eighth interval, but we would not have been able to record infants under 1 year old.

Usually the lowest value in an interval is a multiple of the interval size. Thus, if the interval size is six, the intervals would be 0 to 5, 6 to 11, 12 to 17, 18 to 23, etc., having lower limits 0, 6, 12, 18, etc. Although not mandatory, this practice does yield intervals that are easy to remember.

C. Grouped Distributions: Nominal Scales

When data are in nominal-scale form, the notions of interval size and range of scores do not apply. If it is necessary to group data that are in a nominal scale, therefore, the grouping must be on some other rational basis. For example, subjects in a national survey might be scored on state of residency. There are, however, 50 different states, and the comparing of 50 different frequencies might be too complicated and difficult to be very useful. In such a case, you might consider grouping the states according to region of the country, size, population, or some other means, depending on your purposes. Here, again, the method of grouping would have to be justifiable.

---

*An alternative procedure is to define the range as the difference between the upper real limit of the greatest category containing scores and the lower real limit of the smallest category containing scores. See Section 5-10B for further discussion.

D. Proportion or Percent Distribution

Proportion or percent distributions are constructed in the same manner as frequency distributions, except that instead of tabulating frequencies, or numbers of cases falling at each scale value, you list the proportion or percent of cases having each score.

The proportion of cases at a scale value is found by dividing the number or frequency of cases at that value by the total number of cases. Percentages are found by multiplying corresponding proportions by 100 percent. Table 4-4 gives a frequency, proportion, and percentage distribution which corresponds to the third column of Table 4-1.

Table 4-4 *Frequency, proportion, and percentage distribution for number of siblings (data from Table 4-1)*

Number of siblings	Frequency	Proportion of patients	Percentage of patients
4	‖ (2)	$\frac{2}{25} = .08$	$100(.08) = 8$
3	⁙ (5)	$\frac{5}{25} = .20$	$100(.20) = 20$
2	⁙	(6)	$\frac{6}{25} = .24$	$100(.24) = 24$
1	‖‖ (4)	$\frac{4}{25} = .16$	$100(.16) = 16$
0	⁙ ‖‖ (8)	$\frac{8}{25} = .32$	$100(.32) = 32$
Total	25	1.00	100%

Thus, there were eight patients who had no brothers or sisters; patients with no siblings constituted .32 or 32 percent of the sample.

Table 4-4, although useful for didactic purposes, actually provides too much information for most other uses. Usually, you will want to report frequencies *or* percentages *or* proportions, and there is seldom an advantage to indicating the calculations, as done in this table.

Note that all frequencies are divided by the sum of frequencies to obtain proportions. A trick can be very useful here: we can multiply each frequency by the reciprocal of the total instead of dividing by the total. Thus, in Table 4-4, we multiply by $\frac{1}{25}$ instead of dividing by 25. For example, in the first row, $\frac{2}{25} = 2(\frac{1}{25}) = 2(.04) = .08$. This can easily be done, because it is usually easier to multiply than to divide, and because Table A-2 of the Appendix contains the reciprocals of numbers from 1 to 1000, already calculated.

4-6 Do you want a cumulative distribution and graph?

A cumulative distribution is constructed whenever it is desired to represent at each scale value the total frequency or percentage of scores occurring at that scale value and below.

One common use of a cumulative distribution is for reporting the results of an achievement test (such as the first exam in a statistics course). Looking at the cumulative frequency for your score on the test would tell you how many students got the same or a lower score; and subtracting that number from the total number of students in the class would tell you how many people got higher marks on the test.

Another use of a cumulative distribution occurs in connection with the recording of repetitive behavior in the laboratory. A strip of graph paper is moved at a constant speed under a pen. The pen is in turn connected to a recording device in such a manner that each repetition of the behavior of interest moves the pen a little bit farther across the graph paper. The result is a plot of responses as a function of time, with the time given by the speed at which the paper is moving. The graph in Figure 4-1 is an example of this kind of information. The graph shows "superstitious" behavior in a pigeon. Reinforcement was presented periodically without reference to the bird's behavior, yet a particular behavior sequence (the one graphed) was learned, although it was irrelevant. In this example, the slope of the cumulative graph is an indicator of the rate of response. As time progresses, the slope gets steeper, which indicates that the animal is responding more rapidly.

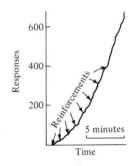

Figure 4-1   *Cumulative record of hopping responses in a pigeon, graphed against time* [Skinner, 1948, 170].

Cumulative distributions are also used in conjunction with percentile ranks. The percentile rank of a person in a group is the percentage of persons in the group who received lower scores than he did.

4-7   Cumulative frequency or percent distribution and percentile ranks

A cumulative distribution of frequencies, percentages, or proportions can be constructed for any ordinal, interval, or ratio scale. Interval or ratio data can then be graphed with the use of a histogram or polygon; cumulative ordinal data can be graphed by means of a bar graph.

A. Constructing a Cumulative Distribution

To construct a cumulative frequency distribution, begin with the frequency distribution constructed in accordance with Section 4-5. Make a second distribution from this in the following manner: the frequency for each scale value of the cumulative distribution is equal to the sum of the frequencies in all the scale values of the noncumulative distribution up to and including the scale value that is being considered.

EXAMPLE   The frequency distribution and cumulative frequency distribution in Table 4-5 are taken from the third column of Table 4-1. In this table, the computations have been included, although you ordinarily would omit them. An alternative computing method is to find each cumulative frequency by adding the frequency at the score to the cumulative frequency for the next lower score. Thus, the cumulative frequency for 4 siblings in Table 4-5 could be found by adding 2 (the frequency at 4) + 23 (the cumulative frequency for 3) = 25. To find the number of people with 3 or fewer siblings, you could add the number having exactly 3 to the number having 2 or fewer to get 5 + 18 = 23.

Table 4-5   *Distribution of subjects by number of siblings*

Number of siblings	Frequency	Cumulative frequency
4	2	8 + 4 + 6 + 5 + 2 = 25
3	5	8 + 4 + 6 + 5 = 23
2	6	8 + 4 + 6 = 18
1	4	8 + 4 = 12
0	8	8
Total	25	

You can calculate cumulative percentages in two different ways: if the percentages are already available for each interval, you can simply sum them as the frequencies are summed in Table 4-5; or you can find the cumulative frequencies and then convert these to cumulative percentages by dividing each by the total number of cases.

Of the two methods, the second is preferable, since in it division is done only in the last step. Each division may involve rounding error, and when the results of several divisions are added, as in adding percentages, there is the possibility that errors will cumulate as well.

Distributions of cumulative proportions and cumulative percentages can be constructed from Table 4-1 according to the method of Section 4-5. Following this section, the data of Table 4-5 have been converted to cumulative proportions and percentages and are presented in Table 4-6. For another example, see Table 4-9 and Figure 4-9.

Table 4-6 *Distribution of hypothetical subjects according to number of siblings (data from Table 4-5)*

Number of siblings	Cumulative frequency	Cumulative proportion	Cumulative percentage
4	25	$\frac{25}{25} = 1.00$	$100(1.00) = 100$
3	23	$\frac{23}{25} = .92$	$100(.92) = 92$
2	18	$\frac{18}{25} = .72$	$100(.72) = 72$
1	12	$\frac{12}{25} = .48$	$100(.48) = 48$
0	8	$\frac{8}{25} = .32$	$100(.32) = 32$

B. Percentile Ranks

Percentile ranks are percentages that indicate a person's status with respect to some group of people. A percentile rank indicates what percent of the group got scores below the given score. Thus, to say that "518 has a percentile rank of 80" is to say that 80 percent of the subjects in the group under consideration got scores below 518.

Percentile ranks are not to be confused with cumulative percentages. This is indicated by example in Table 4-7, where the percentile rank corresponding to a given score (number of siblings) is different from the cumulative percentage. The cumulative percentage is taken from Table 4-7; the percent at each

Table 4-7 *Comparison of cumulative percentages and percentile ranks (data from Table 4-6)*

Score: number of siblings	Cumulative percentage	Percent at this score	Percentile rank
4	100	8	96
3	92	20	82
2	72	24	60
1	48	16	40
0	32	32	16

score is found by subtraction of cumulative percentages. The percentile rank corresponding to a give score is then found by adding the percent below that score (the cumulative percent for the next lower score) to one half the percent at the score.

$$\binom{\text{percentile rank}}{\text{for score } X} = \binom{\text{percent}}{\text{below } X} + \tfrac{1}{2}\binom{\text{percent}}{\text{at } X}$$

For example, the percent of subjects with a score of 3 is 92 (the percent with scores of 3 or less) minus 72 (the percent with scores of 2 or less), giving 20 percent. Half of this percentage is added to the percent having scores of 2 or less, which gives a percentile rank of $72 + \frac{1}{2}(20) = 82$, for a score of 3.

4-8   Bar graphs and pie diagrams

One-variable graphs represent pictorially the data in frequency and percentage distributions. Bar graphs and pie diagrams are especially useful when the data are at an ordinal or nominal level of measurement.

Figure 4-2 is a bar graph of the attitudes of 480 students toward a required faculty loyalty oath, taken in March and April, 1950. The variable "response" forms a nominal scale and is listed on the horizontal axis of the graph. Frequencies corresponding to each scale value are indicated by a vertical bar placed above the score, and percentages are shown by numbers at the tops of the bars.

Bar graphs generally have most or all of the following characteristics, in order to make them clear and unambiguous:

1. Both the scale values and the quantity labels are clearly marked. In Figure 4-2 the scale values are the responses ("approve," "disapprove," "don't know," and "no response"), and the quantity labels are frequency of response, listed along the vertical axis, and percentages, given at the tops of the bars.
2. The length of each bar indicates the quantity for the scale value represented by the bar.
3. Bars are separated by a space to indicate an ordinal or nominal scale.
4. All bars are of the same width.

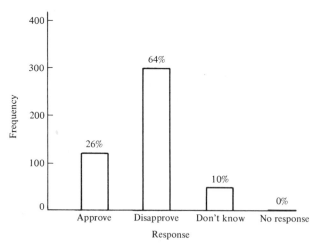

Figure 4-2   *Attitudes of students toward a faculty loyalty oath* [Lipset, 1953, 21].

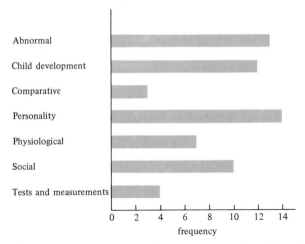

Figure 4-3 *Students in a statistics course who are concurrently taking other psychology courses (hypothetical data). Note that students taking more than one additional psychology course have been entered for all courses. N = 28.*

Figure 4-3 is a second example of a bar graph, showing the number of students in a statistics course who are taking each of several other psychology courses. The scale is "other courses"; it is nominal and appears along the left margin, vertically. To find the number of students taking each course, you read down from the end of the bar to the scale at the bottom of the graph. Notice that all four characteristics of bar graphs are met by this example.

Figure 4-3 differs from Figure 4-2 in that the variable is drawn vertically instead of horizontally, and the frequency scale is along the bottom instead of up the side. However, the labeling makes it clear, so there is no confusion.

A real point of confusion could exist with this graph, though, because the same students may appear on it more than once. (They may take more than one psychology course, in addition to statistics, at the same time.) This is unusual, because in statistics we usually try to place each subject in a single category on any variable used, for purposes of ease of analysis. To avoid confusion, the label for the graph states explicitly that some students have been counted more than once, as would any accompanying text.

An alternative representation would be to have a bar for every combination of courses that appeared in the sample. For example, "Abnormal only"; "Abnormal and child"; "Abnormal and social"; etc. This would lead to other problems, though: more levels of the variable, smaller frequencies for each bar, and a need for the reader to add bars to find the total number of students taking a particular course.

Many of the same data that can be represented by a bar graph can also be presented by means of a pie diagram. Such graphs are seldom found in technical writing but often appear in the popular press, especially when the discussion is about money matters. In a pie diagram, the whole of what is being discussed is represented as a circle, and the proportion of cases that fall in a given category appears as a pie-shaped wedge of the circle. The proportion of the circle given to each category indicates the proportion of cases that fall within that category on the variable graphed. Thus,

$$\frac{\text{number of cases in category A}}{\text{total number of cases}} = \frac{\text{area of category A}}{\text{total area of circle}}$$

$$= \frac{\text{central angle of category A}}{\text{total angle around circle}}$$

The total angle around the center of any circle is 360°, which is used in the denominator of the above formula.

In a questionnaire on religious attitudes and experiences, *Psychology Today* (November, 1974, p. 133) asked the question, "Have you ever felt in close contact with something holy or sacred?" Responses of readers to the four options provided were reported to be as follows:

Option	Percent
No, and I don't care if I ever do	18
No, but I would like to	19
Yes, but it has not had a lasting influence	31
Yes, and it has had a lasting influence	32

These percentages, represented as decimal fractions, are equivalent to the left-hand fraction of the above formula. Thus, for the first response, .18 = $\alpha/360°$, where α is the central angle for the wedge representing the percent of

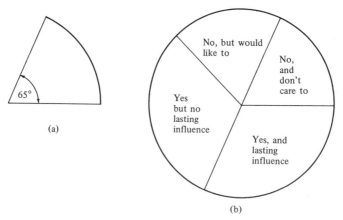

(a)

(b)

Figure 4-4 **(a)** *Central angle of 65°.* **(b)** *Responses of readers to question of whether they had ever felt in contact with something holy or sacred.*

readers who reported never having felt in contact with something holy or sacred and not caring if they ever do. Solving for α, we get $\alpha = (.18)(360°) = 64.8°$. The proportion of *Psychology Today* readers giving this response, then, is represented by a wedge with a central angle of about 65°, as shown in Figure 4-4 (a). Figure 4-4(b) also indicates the other three categories.

The data in Figure 4-4 can be presented as a standard bar graph also; or as a single bar, divided into sections proportional in length to the percentages. This has been done in Figure 4-5.

No, and don't care	No, but would like to	Yes but no lasting influence	Yes, and lasting influence

Figure 4-5   *Responses of readers to question of whether they had ever felt in contact with something sacred or holy.*

4-9   Histogram or polygon

A. Noncumulative Histogram or Polygon

A histogram is used in place of a bar graph when the data are of interval level. It differs from the bar graphs of Figures 4-2 and 4-3 only in that there are no spaces between adjacent bars, to emphasize the continuity and equality of intervals along the scale of the variable.

A histogram is formed directly from a frequency distribution by marking off the horzontal and vertical axes of a graph in accordance with the scale values and frequencies, respectively, of the frequency distribution. Next, bars are drawn above each scale value on the horizontal axis; the heights of the bars indicate the frequency of scores in the intervals. For example, consider the frequency distribution and histogram for the ages of students in an introductory statistics course, given in Table 4-8 and Figure 4-6.

Table 4-8   *Distribution of student ages*

Age	Frequency
27	0
26	1
25	0
24	1
23	1
22	3
21	4
20	7
19	5
18	1
17	0

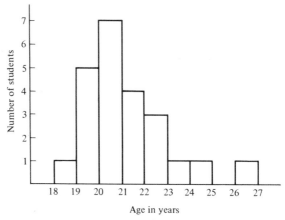

Figure 4-6 *Histogram of student ages.*

Notice in Figure 4-6 that the lower bound of each bar has the age of the students represented by that bar. This is done to emphasize that a person has the same age from birthday to birthday.

For further emphasis on the continuous nature of the underlying scale, we might graph the same data as a frequency or percentage polygon. Here the same axes are maintained, and the data are again taken directly from the frequency or percentage distribution. However, instead of representing the frequency or percentage with a bar, we convey the same information by placing a point above each interval marker on the horizontal scale. The height of this point serves to indicate the percentage or frequency of scores in that category. Once all the points have been plotted and we are sure that there is a point for each interval, we connect the points by straight-line segments to give a picture of the distribution. The data plotted in the histogram of Figure 4-6 have been replotted in this manner in Figure 4-7.

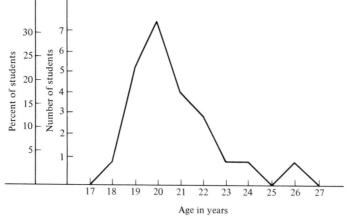

Figure 4-7 *Frequency and percent polygon of student ages.*

In Figure 4-7 the data have been plotted as both a frequency and a percentage polygon, simply by placing two scales in the left-hand margin with the appropriate relationship to one another: percent = 100(frequency)/(total number of cases). Since there were 23 subjects in the sample, a frequency of 1 corresponds to a percentage of 100(1)/23 = 4.3 percent, a frequency of 2 corresponds to a percentage of 100(2)/23 = 8.7%, and so on.

B. Cumulative Histogram or Polygon

A cumulative histogram or polygon is constructed in the same manner as a noncumulative histogram or polygon (see Section 4-9A), except that the data are to be taken from a cumulative frequency distribution (see Section 4-7). The data of Table 4-8 are shown in Table 4-9 as cumulative frequency and proportion distributions and graphed as a cumulative frequency or percent histogram in Figure 4-8.

Table 4-9 *Frequencies, cumulative frequencies, and cumulative proportions of ages of students in a statistics course*

Age	Frequency	Cumulative frequency	Cumulative proportion
27	0	23	1.00
26	1	23	1.00
25	0	22	.96
24	1	22	.96
23	1	21	.91
22	3	20	.87
21	4	17	.74
20	7	13	.56
19	5	6	.26
18	1	1	.04
17	0	0	.00

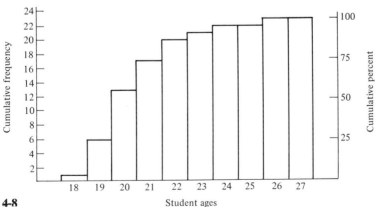

Figure 4-8

Student ages

C. Smoothed and Theoretically Smooth Polygons

Often the particular set of data collected has an unnecessary roughness when presented in graphic form that may be confusing or misleading. It may lead you to think that the origins of the data are very complicated when what you are actually seeing are the results of random error in the data. This unnecessary and confusing randomness can be reduced through smoothing to produce a distribution more regular in appearance.

Once the points of a frequency or percentage polygon have been plotted, you may do the smoothing visually, drawing a freehand curve that fits the points approximately, (i.e., eliminates the rough edges while remaining as faithful to the original points as possible). Figure 4-9 shows how Figure 4-7 might be smoothed in this manner.

Smooth polygons are also used to graph theoretical distributions; such graphs will be used often in Chapters 7 through 15. Some examples are Figures 7-2, 7-6, 7-12, 7-20, and 7-22. A comparison of observed and theoretical distributions is presented in Figure 7-11.

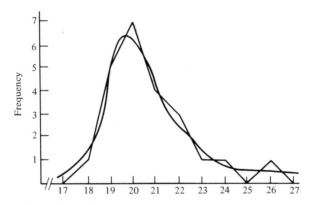

Figure 4-9 *Visual smoothing of Figure 4.7.*

4-10   Two-dimensional plot

A. Plotting Two Variables

In order to graph two interval-level variables simultaneously, begin with a data table (Section 4-1) that has a line for each subject and a column for each variable. Such a data table is given in Table 4-10.

In a two-dimensional plot, each subject is represented as a point on the graph. Arbitrarily select one of the variables to be represented along the horizontal axis and lay out scale values for it. Have enough values so that all scores can be plotted. Lay out scale values for the other variable along the vertical axis, with sufficient range to include all observed values of that variable. Then plot each individual's point on the graph so that its horizontal position indicates its value on the first variable and its vertical position indicates

Table 4-10 *Marks on a test of general science knowledge and grades in an intermediate school course in general science (sixth grade)*

Student	Test score	Course grade	Student	Test score	Course grade
1	17	80	21	15	85
2	15	95	22	15	95
3	14	75	23	14	80
4	14	85	24	13	70
5	13	85	25	13	85
6	12	60	26	12	70
7	12	70	27	11	60
8	11	70	28	11	80
9	11	90	29	10	70
10	10	70	30	10	75
11	10	80	31	9	55
12	9	60	32	9	65
13	9	70	33	9	70
14	9	70	34	9	80
15	8	70	35	8	75
16	7	60	36	7	70
17	7	75	37	7	80
18	6	55	38	5	65
19	5	65	39	5	80
20	4	55	40	3	55

its value on the second variable. This has been done in Figure 4-10 for the data in Table 4-10.

Notice that the vertical axis has been broken to indicate that the intersection of the two axes is not the zero point for grade. When we eliminate the part of

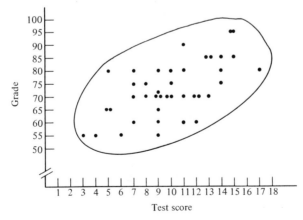

Figure 4-10 *Bivariate plot of scores on a test of general science knowledge and grades in a course in general science* [data from Table 4-10].

the graph between zero and a course grade of 50, we save space without distorting the relationship between the two variables or the values of any of the points along either of the variables.

A line has been drawn around all the points to emphasize the shape of the plot and therefore the relationship between the two variables. This line is not part of the graph, and in most graphs it will not appear at all. However, this shape is one of the things you look for when you interpret the graph, and it may be useful for you to draw it in sometimes for your own clarification.

Note in Figure 4-10 that there are pairs of dots at (15, 95) (where the first number is test score and the second is grade), (13, 85), (10, 70), (12, 70), and (5, 65), and there is a triplet at (9, 70). These represent ties: more than one subject has the same pair of scores. Another way of representing ties, often used in computer-printed graphs, is to replace each pair of dots by a numeral 2, each triplet by a 3, and so on. This could have been done in Figure 4-10 also, but it would be important to include a note with the graph explaining the meaning of the numerals.

Another use of numerals on a graph is to identify some or all of the subjects whose points appear there. This can be done either by writing in the subject's number instead of a point on the graph, or if only a few of the subjects are to be identified, by drawing a line from the subject's point to a space where the subject's identification is written in. Subjects are usually not identified on bivariate graphs, because the information isn't useful and/or because the additional labels take up space and make the graph appear messier.

B. Functional Relationship

When we say Y varies as a function of X, we mean that a change in X is accompanied by a change in Y. From this we may infer that a change in X causes or produces a change in Y; but such an inference is not always justified, and we must be careful when we evaluate the evidence for causality.

If we wish to think of Y as a function of X (whether or not changes in X cause changes in Y), we can write the relationship in either of two ways:

$$Y = f(X) \quad \text{or} \quad Y = fn(X)$$

Read "Y is a function of X" or "Y equals f of X."

But it is very vague to say simply that $Y = f(X)$; we would like to be able to say what the functional relationship is. Does $Y = 3X - 2$? $4X + 3$? $X^2 + \frac{1}{2}$? To find out, we must examine some data.

It is reasonable to expect weight to be a function of height (the taller you are, the heavier you are). Let us examine some data to see if this is true. The data of Table 4-11 were collected on students in a statistics class; they are graphed in Figure 4-11.

A straight line has been drawn through the points in Figure 4-11. This line is an estimate of the functional relationship weight $= f$(height). Its mathematical formula is $W = 4.30H - 144.1$, where W is weight in pounds and H is

Table 4-11 *Heights and weights of 23 statistics students*

Student	Height (inches)	Weight (pounds)	Sex
1	70	150	M
2	67½	120	F
3	66	135	M
4	70	172	M
5	69	145	M
6	69	170	M
7	75	170	M
8	70	170	M
9	65½	125	F
10	64	140	M
11	69	145	F
12	74	165	M
13	71	180	M
14	65	135	M
15	64	125	F
16	73	148	M
17	71	173	M
18	66	140	M
19	70½	200	M
20	73	165	M
21	65½	122	F
22	69	139	F
23	63	140	F

height in inches. The line and its formula indicate a best prediction to make of the weights of students in this sample if you know only their heights. If you were asked to guess the weight of any student in this or a similar sample on the basis of his height only you could read up vertically from the person's height on the

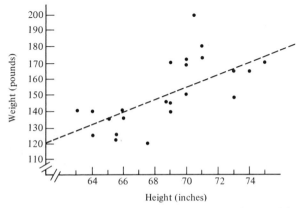

Figure 4-11 *Heights and weights of 23 statistics students* [data from Table 4-11].

horizontal axis to the prediction line (called a *regression line*), then across horonzontally to the scale of weights. The weight found in this manner is the best predicted value for that subject.

For a person 5 feet 5 inches tall, your best guess of his weight on the basis of his height is given by the formula to be

$$W = (4.30)(65) - 144.1$$
$$= 279.3 - 144.1$$
$$= 135.2 \text{ pounds}$$

The regression line gives the same result, by passing through the point at which H = 65 inches and W = 135.2 pounds. Graphic prediction is illustrated by the dashed line in Figure 4-11 running from the height of 65 inches to the regression line, then over to the predicted weight of 135.2 pounds.

This treatment of functional relationships is necessarily casual (in fact, it is not even possible yet to indicate in what sense the regression line gives "best" predictions), because the statistical tools necessary for a more detailed presentation have not been developed. The method used here to calculate the formula for the regression line is treated in detail in Section 15-5. Even when functional relationships can be specified mathematically, however, it is not always legitimate to infer that changes in one of the variables cause changes in the other. For further discussion of this point, you should refer to advanced texts in statistics and experimental design.

4-11   Multiple polygons

> *Relevant earlier sections:*
> **4-5**   Frequency of percent distribution
> **4-9**   Histogram of polygon

A reasonable method for graphing this kind of data is to use several frequency or percent polygons, where a separate polygon is used to represent each value of the nominal- or ordinal-scale variable. Frequency or percent distributions are made for each level of the nominal variable, and the distributions are graphed according to the methods of Section 4-9.

For example, suppose that we want to compare the weights of the male and female students as recorded in Table 4-11. To begin with, we could construct a two-variable frequency distribution, as shown in Table 4-12.

These frequencies can then be graphed using superimposed polygons having the same axes and numbering system. Such a graph is shown in Figure 4-12, where the male and female polygons are distinguished by different kinds of lines.

Table 4-12 *Weights of men and women in a statistics class (data from Table 4–11)*

Weight	Frequency Men	Frequency Women
210–219	0	0
200–209	1	0
190–199	0	0
180–189	1	0
170–179	5	0
160–169	2	0
150–159	1	0
140–149	4	2
130–139	1	2
120–129	0	4
110–119	0	0

Figure 4-12 *Weights of 23 students in a statistics course.*

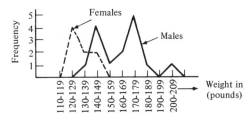

At least four other ways of graphing the data are possible.

1. You could extend the vertical axis below the horizontal one, graphing the frequencies for the women above the horizontal axis and the frequencies for the men below the same line (or the reverse). The problem would be finding a convenient place to write in the weights. In Figure 4-12 they appear below the horizontal axis; if one of the polygons were down there, it might be hard to fit in the numbers.
2. You could make two parallel weight axes, one below the other by an inch or so, and use one for the men and one for the women.
3. You could treat sex as an interval-level variable and graph the results as a two-dimensional plot (Section 4-10). The problem with this idea is that you would sometimes have four and five dots in the same place on the graph, which would be hard to read.
4. You could also make a multiple histogram, instead of multiple polygons. If you decided to do this, it might be useful to have a look at Section 4-13.

A more interesting and only slightly more complicated example is given in Figure 4-13. Here three frequency polygons have been superimposed. Each

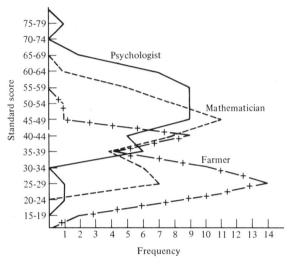

Figure 4-13 *Vocational interests of 50 past presidents of the American Psychological Association* [data from Campbell, 1965, 636].

polygon represents the distribution of standard scores for 50 past presidents of the American Psychological Association (APA) on a different scale of the Strong Vocational Interest Blank, where the higher the standard score, the more a subject is like persons in the given occupation.

Note that it is difficult to see that this graph shows the relationship between two distinct variables since it superimposes three distinct distributions: interest in psychology, interest in mathematics, and interest in farming. To call this a two-variable graph, you would have to call the variables area of interest and standard score in the area of interest (which at best is a little awkward). Nevertheless, the graph is informative. It shows that past presidents of the APA respond more like psychologists than like mathematicians, and more like mathematicians than like farmers.

When the second variable is based on an ordinal scale with several scale values, a set of superimposed polygons may become confusing. It may be reasonable in some cases to use a two-dimensional plot (Section 4-11) for such data, provided that the graph clearly indicates that one variable is an ordinal-and not an interval-level measure.

4-12 Multiple-bar graph

Relevant earlier sections:
4-5 Frequency distribution
4-8 Bar graph

To graph two nominal or ordinal scales simultaneously, begin by constructing a two-way frequency or percent distribution. Such a distribution has been constructed in Table 4-13 for the percent of white adults favoring school integration (data adapted from *Scientific American,* July, 1964, p. 18). Here the nominal variable "area of United States" is listed across the upper margin and the second variable "year the survey was made" is listed down the left-hand margin. Ordinarily, "year" might be considered an interval variable, but here it is treated as ordinal because the difference between 1942 and 1956 is not the same as the difference between 1956 and 1963, and because we do not know whether the change was gradual and continuous from year to year.

Table 4-13 *Percentages of white adults favoring school integration*

Year of survey	Area of United States	
	North (%)	South (%)
1942	40	2
1956	61	14
1963	73	34

These percentages can be graphed in at least three different ways. First, a three-dimensional graph can be drawn, as in Figure 4-14. Here values of the two variables are given on what appears to be a horizontal plane going away from the viewer and to the right; the percentages are represented by vertical bars that rise from the plane of the two variables.

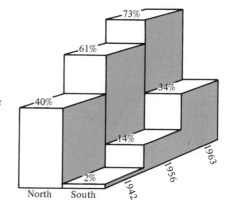

Figure 4-14 *Percentages of white adults favoring school integration.*

A second method for graphing these same data is illustrated by Figure 4-15, where bars are paired in order to allow a two-dimensional representation with minimum distortion.

The pairing of bars in Figure 4-15 emphasizes the comparison of North and South; an alternative representation of the same data appears in Figure 4-16. Here, the grouping of bars emphasizes the increase in percentage favorable to integration from 1942 to 1956 to 1963, in both the North and the South.

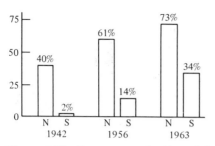

Figure 4-15 *Percentages of white adults favoring school integration.*

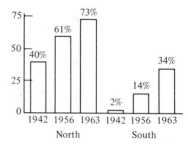

Figure 4-16 *Percentages of white adults favoring school integration.*

4-13 Graphing three or more variables

Relevant earlier sections:
4-10 Two-dimensional plot
4-11 Superimposed polygons
4-12 Multiple bar graph

Research commonly involves collecting data on several variables, not just one or two. Often the hypotheses you want to test are complicated, and you need to collect several scores from each subject; and often additional data are collected merely because it is relatively cheap to get them and because something useful might turn up in them.

Graphing three or more variables can get very complicated and usually can't be done with a single graph. By extension from Section 4-10, you can imagine drawing a three-dimensional plot having all three axes at right angles to one another, and plotting points in the space. But this space is a volume, not a flat surface, and it then becomes difficult to compress the results for display on a printed page.

What is generally done is to graph the variables two at a time and to try to infer the overall set of relationships from the pairwise graphs. The pairwise graphs are constructed according to Sections 4-10, 4-11, and 4-12, and as much as possible you should try to use the same type of graph throughout. A difficulty arises when you represent relationships among a large number of variables. The number of plots increases faster than the number of variables, which makes graphing a big job and interpretation difficult. The number of plots needed is given by the expression

$$\frac{n(n-1)}{2}$$

where n is the number of variables. Thus, when there are three variables, $(3 \times 2)/2 = 3$ plots are needed; when there are 10 variables, $(10 \times 9)/2 = 45$ are needed; and when there are 100 variables, $(100 \times 99)/2 = 4950$ plots are needed. When there are several variables, you will be wisest to plot relationships only among those of greatest interest.

Homework

1. Consider the following set of scores on an internal-level achievement test:

$$
\begin{array}{cccccccccc}
3, & 9, & 18, & 15, & 17, & 18, & 20, & 21, & 23, & 24, \\
10, & 13, & 15, & 17, & 23, & 6, & 10, & 15, & 17, & 4, \\
6, & 4, & 9, & 5, & 5, & 3, & 6, & 7, & 18, & 14, \\
4, & 13, & 18, & 13, & 6, & 8, & 16, & 9, & 10, & 14, \\
10, & 21, & 15, & 12, & 10, & 4, & 5, & 9, & 8, & 8.
\end{array}
$$

Do parts (a)–(e) together, in a single table:
   (a) Make a table that shows the number of subjects getting each score.
   (b) Also show, for each score, the number of people getting that score or less.
   (c) Show the percent of people at each score.
   (d) Show, for each score, the percent of people getting that score or a lower one.
   (e) Present the distribution of percentile ranks.
2. You look at the table you have constructed for problem 1 and decide that there is too much detail for your intended audience. You decide to combine score lines and produce a simpler table. Do this now, repeating steps (a)–(e) of problem 1 for this simpler distribution.
3. For each of the following, indicate which level of measurement it is most reasonable to assume for the data.
   (a) Shoe size
   (b) Distance from home to campus
   (c) Adult intelligence
   (d) Place of birth
   (e) Score on a test of psychological agreeableness
   (f) Grades on a statistics exam
   (g) Number of children in the family
   (h) Age upon completion of high school
   (i) Popularity as judged by peers
   (j) Percent accuracy in a ring-toss game
4. Each of the following scores has one digit too many. Round them appropriately.
   (a) Shoe sizes: 8.8; 9.5; 11.3
   (b) Intelligence: 85.3; 96.8; 101.5; 134.5
   (c) Ages: 6 months 3 days; 9 months 18 days
   (d) Agreeableness rating: 14.26; 26.55; 18.91
   (e) Height: 5 ft $6\frac{1}{2}$ in; 6 ft $1\frac{1}{4}$ in; 5 ft $8\frac{3}{4}$ in.

For each of problems 5 through 9, indicate which kind of graph would seem to be most informative.

5. You teach two sections of introductory statistics and give the same 30-item, multiple-choice midterm exam to both classes. You wish to compare the performance of one class to the other.
6. You have just received a list of all patients at State Mental Hospital and their psychiatric diagnoses. Indicate the relative frequencies of patients in the several categories.
7. Now you wish to compare the graph for problem 4 with a corresponding graph for patients at Private Detention, located right across the street from State Mental.
8. A new tranquilizer is being used at Private Detention which relieves symptoms of paranoid patients pending psychotherapy. Dosage is determined by severity of symptoms; however, the medication contains a poison that builds up in the body over time, becoming fatal after 500 cubic centimeters have been administered. A graph of dosage in each patient's records will make it clear when to stop administering it.
9. An introductory psychology course is divided into four consecutive parts, each taught by a different instructor and having a separate 50-item, multiple-choice exam. Students are also given a multiple-choice test of authoritarianism. You wish to determine whether authoritarianism is more closely related to grades of some instructors than to grades of others.
10. The frequency distribution in the table depicts a continuous, interval-level variable. Provide a single graph on which it will be possible to read both the number of subjects having a particular score and the percentage of subjects having that score or less.

Score	Frequency
9	1
8	3
7	7
6	11
5	14
4	9
3	5

11. Plot the following sociological data on suburbanites:

Primary mode of transportation	Primary source of news	Number of people
Car	TV	10
Train	Newspaper	4
Car	Radio	2
Train	Radio	3
Bus	Newspaper	2
Bus	TV	5
Train	TV	7
Car	Newspaper	6
Bus	Radio	1

12. Show the relationships among the following variables. Assume that all are interval scales.

Person	Ascendency	Masochism	Age
1	4	15	16
2	2	4	18
3	6	17	24
4	8	2	12
5	9	6	19
6	3	8	26
7	1	9	15
8	7	4	21
9	2	5	18

13. In a pilot study of sexual attractiveness, one Norwegian white rat, No. 216T (nicknamed Marge) is kept in isolation for all but 15 minutes a day. During those 15 minutes (on non-estrus days) she is placed in a large cage, with a male in an adjacect cage. The following data are recorded: (1) the male's length, in centimeters; (2) the male's activity level (active/quiet);

(3) the male's shape (fat/average/thin); and (4) whether Marge tries to nuzzle with him (yes/no).

Male	Size	Activity	Shape	Nuzzling
1	11	Active	Thin	No
2	16	Active	Fat	Yes
3	13	Active	Thin	No
4	15	Active	Fat	Yes
5	14	Active	Ave	Yes
6	18	Active	Ave	Yes
7	14	Active	Thin	No
8	15	Active	Ave	Yes
9	11	Quiet	Thin	No
10	14	Active	Fat	Yes
11	14	Quiet	Ave	No
12	12	Quiet	Ave	No
13	18	Active	Fat	Yes
14	14	Quiet	Fat	Yes
15	20	Quiet	Thin	No
16	15	Quiet	Thin	No
17	15	Active	Fat	Yes
18	14	Active	Ave	Yes
19	17	Quiet	Ave	No
20	13	Active	Fat	Yes

From these data, make the following graphs:

(a) A graph of the lengths of the males.
(b) The relationship between a male's size and whether Marge tries to nuzzle with him.
(c) The relationship between a male's activity level and whether Marge tries to nuzzle with him.
(d) The relationship between a male's shape and whether Marge tries to nuzzle with him.

Additional problems appear in the workbook accompanying this text.

Homework Answers

1.

X	f	Cumulative frequency	Percent	Cumulative percentage	Percentile rank
24	1	50	2	100	99
23	2	49	4	98	96
22	0	47	0	94	94
21	2	47	4	94	92
20	1	45	2	90	89
19	0	44	0	88	88
18	4	44	8	88	84
17	3	40	6	80	77
16	1	37	2	74	73
15	4	36	8	72	68
14	2	32	4	64	62
13	3	30	6	60	57
12	1	27	2	54	53
11	0	26	0	52	52
10	5	26	10	52	47
9	4	21	8	42	38
8	3	17	6	34	31
7	1	14	2	28	27
6	4	13	8	26	22
5	3	9	6	18	15
4	4	6	8	12	8
3	2	2	4	4	2

2.

X	f	Cumulative frequency	Percent	Cumulative percentage	Percentile rank
23–24	3	50	6	100	97
21–22	2	47	4	94	92
19–20	1	45	2	90	89
17–18	7	44	14	88	81
15–16	5	37	10	74	69
13–14	5	32	10	64	59
11–12	1	27	2	54	53
9–10	9	26	18	52	43
7– 8	4	17	8	34	30
5– 6	7	13	14	26	19
3– 4	6	6	12	12	6

3. **(a)** Interval   **(b)** Ratio
   (c) Interval or ordinal   **(d)** Nominal
   (e) Ordinal or interval   **(f)** Ordinal, interval, or ratio
   (g) Ratio (discrete)   **(h)** Ratio
   (i) Ordinal or interval   **(j)** Ratio

4. **(a)** 9, 10, 12   **(b)** 85, 97, 102, 134;
   (c) 6 months, 9 months;   **(d)** 14.2, 26.6, 18.9;
   (e) 5 ft 6 in; 6 ft 1 in; 5 ft 8 in. Shoe sizes rounded up to allow room for movement; intelligence and personality scores customarily rounded to nearest value; age and height usually rounded down.

5. Superimposed frequency polygons or multiple bar graph

6. Bar graph   7. Multiple bar graph

8. Cumulative polygon

9. Bivariate graphs of authoritarianism scores with each of the four sets of variable scores

10.

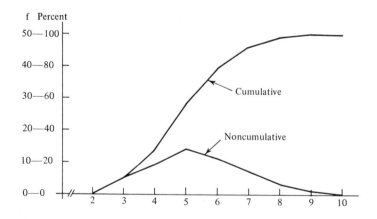

11. (Note that, according to Section 4-12, two other kinds of multiple bar graphs are also possible.)

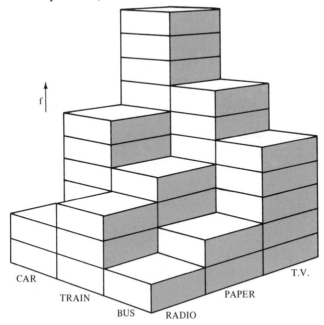

12.

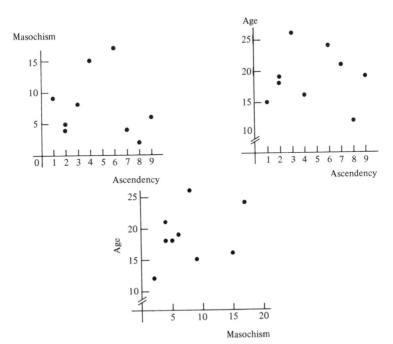

13. (a)

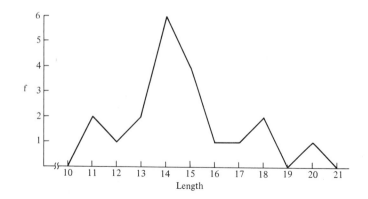

(b)

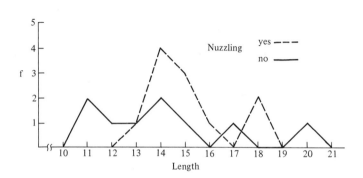

(c)

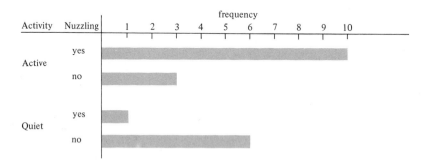

(d)

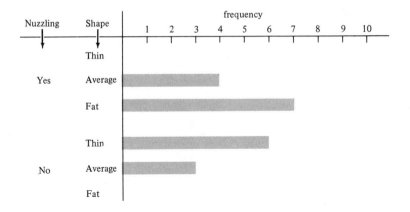

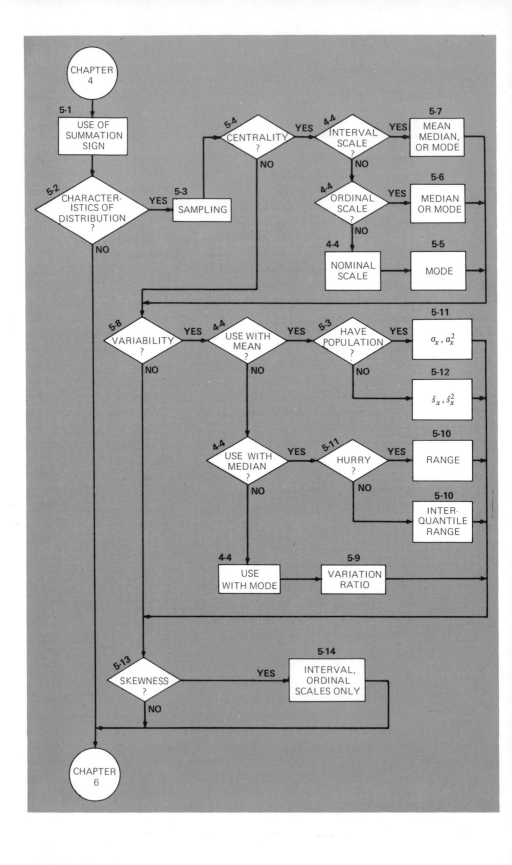

5 Characteristics of a Distribution

THE map for this chapter (opposite page) can be subdivided into conceptual regions in at least two ways, as shown by the two regional maps on this page. The left-hand regional map is divided according to the kind of characteristic you might wish to measure for a distribution. The map on the right is divided into techniques for dealing with different kinds of scales. Sections 5-1 (use of the summation sign) and 5-3 (sampling) appear separately in both figures because they apply regardless of type of scale used or characteristic measured.

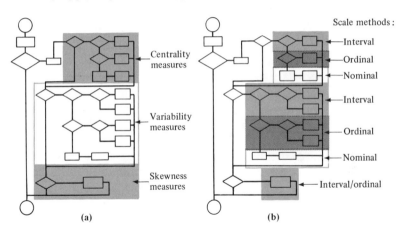

(a)   (b)

As in Chapter 4, you have a choice of ways for studying the material of this chapter: you can read the text straight through, or follow the decision map either to solve your own problem or to solve the problems given at the end of the chapter.

5-1 Use of the summation sign

A. Basic Summation Notation

The summation sign is the capital Greek sigma ($\sum$): it tells you to add together a group of numbers. Use of the summation sign often allows you to represent a considerable amount of addition in abbreviated form. In this section we will discuss the use of the summation sign. The rules for its use will be shown to follow directly from what you already know about algebraic addition of variables and numbers.

Consider this table:

i	X
1	X_1
2	X_2
$\vdots$	$\vdots$
N	X_N

In the table, the i column indicates subject numbers, ranging from 1 for the first subject through N for the last. X is some variable, and X_i is the score of subject i on variable X. Thus, X_1 is the score of the first numbered subject on X, X_{14} is the score of the fourteenth subject on variable X, and so on. The total of the numbers represented by the X's in the table can be stated as "the sum of the X_i's" or "add up the X_i's." This means that you begin with the value of X for the first person in the table and add to it the values of the variable X for all the other persons in the table. In algebraic notation, the words "add up the X_i's" can be replaced by the symbols $X_1 + X_2 + X_3 + \cdots + X_N$, where it has been necessary to use dots because, until N is specified, we do not know how many terms there are in the summation. In statistical notation, the words "the sum of the" and "add up the" can be replaced by the symbol $\sum$. The phrase "the sum of the X_i's" then can be written in shorthand notation as $\sum X_i$, or in many cases simply as $\sum X$.

The subscript is often useful in identifying which numbers are to be summed, particularly if you are not summing over all the numbers in the table. In this case, the *limits of summation* are indicated by symbols placed above and below the summation sign. Let us begin by writing the summation in full: $\sum_{i=1}^{N} X_i$ says "add all the X_i's, beginning with the first one and concluding with the Nth one." The two previous statements, $\sum X$ and $\sum X_i$, are really abbreviations of this more complete statement. Further, by stating the limits of summation, you are not restricted to writing statements about all the numbers in the table. Suppose, for example, that you wish to add only the first 10 numbers: this sum can be written $\sum_{i=1}^{10} X_i$. Similarly, adding all but the first and last the variable X can be written $\sum_{i=2}^{N-1} X_i$.

Working in the other direction, notice that $X_1 + X_2 + X_3 + X_4$ can be written $\sum_{i=1}^{4} X_i$; and $Y_{14} + Y_{15} + Y_{16}$ can be written $\sum_{j=14}^{16} Y_j$ (the subscript doesn't have to be an *i*, but it does have to match the letter under the summation sign).

PROBLEMS   For the data table shown, compute the expressions at the right.

i	*X*	*Y*
1	4	5
2	3	9
3	6	4
4	2	5
5	2	3

A. $\displaystyle\sum_{i=1}^{5} X_i$

B. $\displaystyle\sum Y$

C. $\displaystyle\sum_{i=1}^{3} X_i + \sum_{i=2}^{4} Y_i$

ANSWERS   A. 17;   B. 26;   C. 13 + 18 = 31.

B.   Three Rules

Three rules for the use of the summation sign will be particularly useful in the rest of this book. These rules will now be stated in turn, and an attempt will be made to give an algebraic justification for each.

Rule I
$$\sum_{i=a}^{b} (X_i + Y_i) = \sum_{i=a}^{b} X_i + \sum_{i=a}^{b} Y_i \qquad (5\text{-}1)$$

Suppose that you have obtained scores for subjects on two variables, *X* and *Y*. The rule states that there are two equivalent ways for getting a grand total of all scores. Either you can sum the *X* and *Y* scores for each subject and then sum the totals, or you can sum all the *X* scores, sum all the *Y* scores, and add the two sums.

Let us begin examining this rule with a simple numerical example. Consider the following data table for three subjects and two variables, in which $a = 1$ and $b = 3$:

i	X_i	Y_i	$X_i + Y_i$
1	2	5	7
2	4	4	8
3	1	8	9
$\sum$	7	17	24

The right-hand column is a column of row totals, and the bottom row is a row of column totals. The grand total, 24, can be obtained either by adding the row totals (7 + 8 + 9) or by adding the column totals (7 + 17).

As a second example, let $a = 1$ and $b = 2$; then construct an algebraic table similar to the previous numerical one:

	X_i	Y_i	$X_i + Y_i$
1	X_1	Y_1	$X_1 + Y_1$
2	X_2	Y_2	$X_2 + Y_2$
$\sum$	$X_1 + X_2$	$Y_1 + Y_2$	$X_1 + X_2 + Y_1 + Y_2$

The sum of each column is given in the last row of the table. An examination of the terms shows that the sum of X, given at the bottom of the first column, plus the sum of Y, given at the bottom of the second, is equal to the sum of $(X + Y)$, given at the bottom of the last column, thus confirming the rule.

Finally, let us consider the case when $a = 1$ and $b = N$. A data table for this problem would have the following form:

i	X	Y	$X + Y$
1	X_1	Y_1	$X_1 + Y_1$
2	X_2	Y_2	$X_2 + Y_2$
$\vdots$	$\vdots$	$\vdots$	$\vdots$
N	X_N	Y_N	$X_N + Y_N$

The total for the X column is $\sum X = X_1 + X_2 + \cdots + X_N$; the total for the Y column is $\sum Y = Y_1 + Y_2 + \cdots + Y_N$; and the total for the $(X + Y)$ column is $\sum (X + Y) = (X_1 + Y_1) + (X_2 + Y_2) + \cdots + (X_N + Y_N)$. Then $\sum X + \sum Y = (X_1 + X_2 + \cdots + X_N) + (Y_1 + Y_2 + \cdots + Y_N)$, which, by re-arranging terms and inserting parentheses, can be shown to be identical to $\sum (X + Y)$. The rule is confirmed here also.

Rule I applies not only for addition, but for subtraction as well. If you want to, you can think of the following equation as rule Ia, although I am really only stating it separately for clarity:

$$\sum_{i=a}^{b} (X_i - Y_i) = \sum_{i=a}^{b} X_i - \sum_{i=a}^{b} Y_i$$

No demonstration will be given for the reasonableness of this equation. You can demonstrate it for yourself, following the demonstrations for rule I.

PROBLEMS   Consider the following data table:

i	X	Y	Z
1	4	2	3
2	6	1	5
3	9	0	4
4	5	2	5

A. Find $\sum X$, $\sum Y$, and $\sum Z$.

B. Make an $X + Y$ column and find its sum. Show that $\sum (X + Y) = \sum X + \sum Y$ for these data.

C. Make an $X + Y + Z$ column and find its sum. Show that $\sum (X + Y + Z) = \sum X + \sum Y + \sum Z$.

D. Make an $X - Y$ column and find its sum. Show that $\sum (X - Y) = \sum X - \sum Y$ for these data.

E. Make a $Y + Z$ column and an $X + Z$ column for these data, and find their sums. Show, both numerically and algebraically, that

$$\sum (X + Z) = \sum (X - Y) + \sum (Y + Z)$$

Rule II $\displaystyle\sum_{i=a}^{b} K = (b - a + 1)K$ where K is a constant   (5-2)

This rule states that to find the sum of a constant over any number of subjects, simply multiply the constant by the number of subjects. Here the coefficient $(b - a + 1)$ is the number of subjects added over. This rule states that to find the result of adding a constant any number of times, simply multiply the constant by the number of times it is to be added. Here the coefficient $(b - a + 1)$ is the number of times the constant is to be added, because the summation interval is inclusive (includes both the first and last numbers). Suppose, for example, that we add the scores of subjects 2 through 5; there are four subjects (2, 3, 4, 5), and we obtain the number of subjects by $5 - 2 + 1 = 4$.

Consider this table, where $a = 1$ and $b = N$:

i	K
1	K
2	K
$\vdots$	$\vdots$
N	K

5-1 *Use of the summation sign*

The sum of K from 1 to N is found by adding the same number, K, N times. But from basic arithmetic, it is clear that this is the same as multiplying K by N. Thus, in this example, replace a by 1 and b by N:

$$\sum_{i=a}^{b} K = \sum_{i=1}^{N} K = NK$$

which checks, because $N = (b - a + 1) = (N - 1 + 1)$. As a second example, consider a data table for which $a = 1$, $b = 4$, and $K = 518$:

i	K
1	518
2	518
3	518
4	518
$\sum$	2072

Compare this to $4(518) = 2072$, where we multiply by 4 because $b - a + 1 = 4 - 1 + 1 = 4$.

PROBLEMS For the following data table, solve the problems on the right.

i	X
1	5
2	4
3	9
4	4

A. $\sum X$

B. $\sum 6$ (assume that the limits are 1 and 4)

C. $\sum (X + 6)$

D. $\sum_{i=2}^{4} (X - 5)$

ANSWERS A. 22; B. 24; C. 46; D. 2.

Rule III
$$\sum_{i=a}^{b} KX_i = K \sum_{i=a}^{b} X_i \qquad (5\text{-}3)$$

where again K is a constant

In addition to its other uses, this rule allows you to save some computational time. It states that if you have a summation in which every score shares a common multiplier (K), you can find the sum by factoring out the common term algebraically, adding the quotients, and then multiplying the sum by the common term. Consider the following table:

i	X	KX
a	X_a	KX
$a + 1$	X_{a+1}	KX_{a+1}
$a + 2$	X_{a+2}	KX_{a+2}
$\vdots$	$\vdots$	$\vdots$
b	X_b	KX_b

Written out, the sum of the last column is

$$\sum_{i=a}^{b} KX_i = KX_a + KX_{a+1} + KX_{a+2} + \cdots + KX_b$$

From the right side of this equation, the common term, K, can be factored out algebraically, giving $K(X_a + X_{a+1} + X_{a+2} + \cdots + X_b)$. But the sum in parentheses is none other than $\sum_{i=a}^{b} X_i$. Therefore, the rule has been shown for the general case.

As an example of this rule, consider a data table where you are to find the sum of $7Y$:

k	Y_k	$7Y_k$
1	4	28
2	2	14
3	6	42
4	3	21
$\sum$	15	105

Here the value $7Y = 28 + 14 + 42 + 21 = 105$ is also equal to $7\sum Y = 7(15) = 105$, thus illustrating the rule.

Often you can save work by using the three rules to manipulate the summations before doing any arithmetic. For example, evaluate

$$\sum_{i=1}^{4} \left(\frac{X_i}{2} - 5 \right)$$

where $X_1 = 4$, $X_2 = 2$, $X_3 = 3$, and $X_4 = 3$. This problem can be solved in at least two ways. First, you can make a table:

i	X_i	$X_i/2$	$X_i/2 - 5$
1	4	2	-3
2	2	1	-4
3	3	1.5	-3.5
4	3	1.5	-3.5
Σ			-14

Second, you can do some algebra with the summation sign first:

$$\sum_{i=1}^{4}\left(\frac{X_i}{2} - 5\right) = \sum_{i=1}^{4}\frac{X_i}{2} - \sum_{i=1}^{4} 5$$

$$= \frac{1}{2}\sum_{i=1}^{4} X_i - \sum_{i=1}^{4} 5$$

$$= \frac{1}{2}\sum_{i=1}^{4} X_i - 20$$

$$\sum_{i=1}^{4} X_i = 4 + 2 + 3 + 3 = 12$$

$$\sum_{i=1}^{4}\left(\frac{X_i}{2} - 5\right) = \tfrac{1}{2}(12) - 20 = 6 - 20 = -14$$

Although longer for the present problem, the second method would have saved time if there had been a large number of X scores. It is also a very powerful tool for solving more complicated problems.

PROBLEMS Consider the following data table:

i	X	Y
1	2	5
2	6	3
3	8	4
4	1	3
Σ	17	15

Find:

A. $\sum 3X$

B. $\sum (4X - 2)$

C. $\sum (Y/3 + 1)$

D. $\sum (2X - Y + 2)$

ANSWERS A. 51; B. 60; C. 9; D. 27.

C.   Order of Operations with Summation Signs

You are probably already aware that when parentheses appear in mathematical formulas, you are to carry out all operations within the parentheses before you combine the expression in parentheses with other expressions in the formula. When one set of parentheses appears within another, you evaluate the inner set first. This also applies to summation operations. Thus, when you see the summation

$$\sum_i (X_i + Y_i - 5)$$

all three terms must be summed over i. To carry out the summation, you either evaluate the expression $(X_i + Y_i - 5)$ for each value of i and then add over the values of i, or you perform equivalent operations after simplifying with the use of the three summation rules.

When parentheses are omitted, you may infer that they begin immediately after a summation sign and include all terms in a product. Thus,

$$\sum_i 5X_i = \sum_i (5X_i)$$

$$\sum_i X_i Y_i Z_i = \sum_i (X_i Y_i Z_i)$$

$$\sum XY^2 = \sum (XY^2)$$

where the first of these examples could be simplified still, using rule III.

The inferred parentheses are closed by a plus or minus sign; for example,

$$\sum X + 4 = \sum (X) + 4$$

In this example, only X is summed; after the sum of X is obtained, the number 4 is added to it.

It is important to note that the inferred parentheses do not include the summation sign. For example, the expression

$$\sum_i X_i^2 = \sum_i (X_i^2)$$

says to find X_i^2 for each i, then add over levels of i. This is quite different from $(\sum_i X_i)^2$, which says to add X_i over i, then square the sum. Consider this numerical example:

i	X_i	X_i^2
1	3	9
2	6	36
3	2	4
$\sum_i$	11	49

Here, $\sum_i X_i^2 = \sum_i (X_i^2) = 49$; but $(\sum_i X_i)^2 = (11)^2 = 121$.

PROBLEMS   Consider the following data table:

i	X	Y
1	3	6
2	2	1
3	2	8
4	4	4
5	7	5

Find:

A. $\sum (XY + 3)$

B. $\sum X^2 + 2$

C. $(\sum Y + 1)^2$

D. $\sum (X + 2Y)^2$

ANSWERS   A. 102;   B. 92;   C. 625;   D. 998.

References   The summation notation is discussed in Bashaw (1969), 89–94; Blommers and Lindquist (1960), Chapter 3; Dixon and Massey (1969), Chapter 3; and Hays (1963), 657–666.

5-2   Characteristics of a distribution?

Relevant earlier sections:
4-5   Frequency distribution
4-9   Frequency or percent polygon

Although frequency distributions and graphs are very useful methods for summarizing and representing the scores in a distribution, their usefulness is limited for many purposes. If you wish to describe a distribution quickly and fairly accurately without presenting the entire graph, a verbal characterization will probably be either too long and detailed or too imprecise. Certainly it is difficult to carry out complex mathematical analyses on verbal characterizations of distributions! The problem is further exaggerated when you try to compare two different distributions. In fact, it is even difficult to be precise about the comparison when the graphs are available. And some problems may require that you compare several distributions at the same time.

Certain descriptive characteristics of a set of scores are generally useful for describing it and comparing it to other sets of scores. Among these characteristics are measures of centrality, variability, skewness, and kurtosis. These four kinds of characteristics will now be illustrated for sets of interval-level data. Measures of centrality, variability, and skewness are also defined for ordinal data, and measures of centrality and variability for nominal data.

A measure of centrality is an indicator of the location of a distribution along its score continuum. In Figure 5-1 the centrality of distribution I is greater than that for distribution II, when scores are increasing to the right.

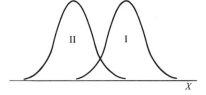

Figure 5-1 *Comparison of two distributions that differ in centrality.*

A measure of variability is an indicator of the extent to which the scores of a distribution are spread out; the greater the differences between high and low scores, the greater the variability. In Figure 5-2 the variability of distribution I is represented as being greater than that of distribution II.

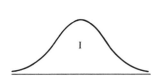
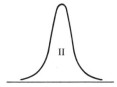

Figure 5-2 *Comparison of two distributions that differ in variability.*

A measure of skewness is an indicator of asymmetry in a distribution. A distribution is said to be positively skewed if the longer tail or small end of it is to the right and negatively skewed if the longer tail is to the left. In Figure 5-3 the skewness of distribution I is positive and that of distribution II is negative.

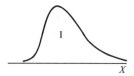
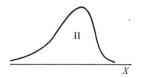

Figure 5-3 *Comparison of two distributions that differ in skewness.*

Kurtosis relates to the degree of peakedness or flatness of a distribution. In Figure 5-4, the kurtosis of distribution I is represented as less than that of distribution II. It is rare to find a problem in behavioral research in which a measure of kurtosis plays a major role; for this reason the concept will not be discussed further. Sections on centrality, variability, and skewness are indicated in the decision map for this chapter.

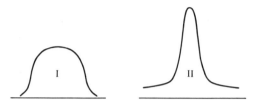

Figure 5-4 *Comparison of two distributions that differ in kurtosis.*

5-3 Populations, samples, and sampling

A. Population and Samples

The distinction between population and sample is a fundamental one in statistics. The relationship is one of whole to part. A population contains all members of a particular category of objects, subjects, or occurrences; a sample contains only some of the members of that category.

Generally, a population is large and a sample is comparatively small; for example, we might obtain a sample of 25 high school seniors in the United States. This sample would be relatively small compared to the population, which would be large and indeterminate but estimable in size. Or we might use a sample of 15 white rats drawn from the population of all white rats. This population contains all white rats that have ever existed or will ever exist. It is definable in that it is possible to specify what characteristics a white rat has that distinguish it from all other kinds of animals.

Populations need not be extremely large or infinite, however. The population of American billionaires is small, as is the population of living astronauts; if necessary, it is possible to get data on the entire population of entering students in American medical schools.

In many cases it may not be possible to say a priori whether a given set of subjects or objects is a population or a sample—to make a judgment, you may need to know for what purpose it is being used. For example, a class of students may be either a population or a sample. It is a sample if the students are considered to be representative of, say, all students in the university; it is a population if no inference is to be made to any larger group of which the class may be considered a part.

In this chapter, you can expect to find a few very small populations used as examples or questions, to make sure you understand the difference in definitions for populations and samples. Throughout the rest of the text, however, it will be assumed that populations are very large relative to samples. This is the way it usually turns out in actual research problems, so the assumption is realistic. The assumption also means that we can use relatively simple formulas: when a sample comprises a noticeable proportion of its population, then the formulas must be modified; and you will have to look in specialized texts for the modified formulas.

B. Sampling

Much of statistics is concerned with the problem of inferring characteristics of populations on the basis of information obtained from samples. This is done for very practical reasons. It is impossible to do an experiment on all white rats, so you do the experiment on a sample of rats and infer from this something about the characteristics of the population. If you want continually to verify the length of life of light bulbs coming off a particular assembly line, you will not test the entire population of bulbs, for then you would have none left to sell. And you simply could not afford to find out how much all high school seniors in the United States know about chemistry. You must be satisfied with the results obtained for a sample drawn from the population of interest.

The *population* characteristic in which you are interested is called a *parameter*. It may be a measure of centrality or variability; a proportion; the degree of relationship between two variables; etc. The *sample* characteristic with which you estimate a parameter is called a *statistic*.

Subjects are included in a sample by the process of *sampling* from a population (i.e., removing subjects from the population until enough have been drawn to complete the sample). The sampling process is carried out in accordance with an implicitly or explicitly stated rule. The nature of the rule used determines the kind of sampling done and, therefore, the kind of sample obtained.

It is sometimes convenient to diagram a sampling process, particularly in connection with complex experimental designs. When this is done, populations will be indicated by large circles and samples by small circles. An arrow will be used to indicate from which population a sample is drawn, as in Figure 5-5.

Figure 5-5 *Example diagram of a sampling process.*

This diagram might equally well represent a national sampling of housewives for an opinion poll, sampling of children from a grade school for a study of toothpastes, or a sampling of white rats for a study of electrical brain stimulation.

It is also legitimate to draw more than one sample from a population. If you wish to compare the effects of two different drugs on the anxiety levels of subjects, you will be most convincing if you compare two samples drawn from the same population. One sample will receive one drug, the other the other drug. The sampling for such an experiment is diagrammed in Figure 5-6. Two things should be noted about this figure. First, both the samples have been specified to be of size N. Second, the sampling process has been specified to be random.

A sampling process is said to be *random* if every possible sample of subjects is just as likely to be selected from the population as every other. A sampling

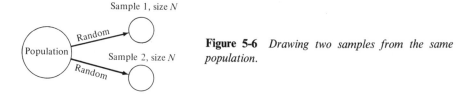

Figure 5-6 *Drawing two samples from the same population.*

method is said to be *nonrandom* or *biased* if some possible samples of subjects are less likely to be chosen than others. Consider the following sampling rule: "Alphabetize the subject names, then select every tenth subject, until twenty subjects have been selected." This method is biased, because it is impossible, using it, to draw a sample consisting of the first 20 subjects in the alphabetized list.

When we are inferring characteristics of populations on the basis of sample data, we cannot avoid making errors. One obvious kind of error is sampling error: no matter how carefully the sample is drawn, it may differ greatly from the population with respect to the characteristic in which we are interested. This cannot be helped, and it cannot even be tested for, unless we examine the entire population (which is usually not possible). However, we can use sample data to make reasonable inferences about populations if we assume that the sampling procedure used is random. For the methods used in this text, it must always be assumed that the sampling is random or unbiased.

C. Methods for Obtaining Random Samples

The basic paradigm for random sampling is that of drawing names from a hat. For each subject in a population, a separate slip of paper is made up with his name on it. The slips of paper are placed in a container and mixed. Slips of paper are then removed blindly from the container, mixing between draws, until the number needed in the sample has been drawn. Subjects whose names have been drawn are then included in the sample.

Alternatively, subjects may be assigned numbers and the numbers selected through the use of a random-number table, such as Table A-3 of the Appendix. The numbers in this table have already been randomly selected by computer; therefore, you can feel confident if you use any of them that there is no systematic relationship among them. On the other hand, if you use the same set of random numbers more than once, you may generate systematic relationships in your data, even though the numbers themselves are random. Therefore, it is a good idea to use some device to ensure that you do not reuse the same numbers. One such device is to pick a page and a direction, close your eyes and put your pencil point on the page. Then read off numbers in the direction you have chosen (perhaps, lower right to upper left, for example) until you have all the numbers you need. If you run off the page start all over again, possibly on a different page and reading in a different direction.

If you need two-digit numbers, simply read them off in pairs; if you need a sign, you might select a digit for each sign, letting odds be positive and evens negative, or let 0 through 4 be positive and 5 through 9 be negative, and so on. Random letters can be selected by making a two-digit equivalence for each letter and selecting two-digit random numbers (perhaps omitting all numbers from 27 through 99 and 00). The sampling method described so far is usually called *simple random sampling*; it is by no means the only sampling method possible.

Sometimes it is useful to divide a population into subpopulations and to sample randomly within each subpopulation, specifying that the proportion of each subsample in the sample will be the same as each subpopulation in the population. Thus, if you know that a population has 55 percent males and 45 percent females, you may randomly sample males and randomly sample females, yet insist that your final sample also have 55 percent males and 45 percent females. This method, known as *stratified random sampling*, provides that, in respect to whatever variables you choose, the sample is constituted similarly to the population.

One of the results of using a method other than simple random sampling is that the formulas for sample estimates of population characteristics presented in this chapter no longer apply. Therefore, in this text, for clarity of presentation, when inferences are to be made about populations on the basis of samples drawn from them, simple random sampling will always be assumed.

References Simple random sampling is discussed in Blommers and Lindquist (1960), 233–246; Dixon and Massey (1969), 37–42; and Games and Klare (1967), 8–14, 211–218. Discussions of other methods of sampling can be found in Guilford (1965), 137–142; Snedecor (1956), Chapter 17; and Walker and Lev (1953), 171–177.

5-4   Centrality?

For most statistical analyses, you will need a measure of the location of one or more distributions of scores along the scale of possible values that your variables could take. A statistic that indicates the typical or average value of a distribution is called a measure of *centrality* or *central tendency*.

A measure of centrality has obvious descriptive uses: it is, in some sense, a value most representative of all the scores in a distribution. It tells you at what point along a scale you can begin most effectively to look for individual scores. And it is very often calculated preliminary to computing other statistics.

Centrality measures are used to compare two or more distributions when it appears that some effect is elevating or depressing scores in one distribution relative to those in another. For example, when we ask whether bright children are taller than children of normal intelligence, we are comparing the two distributions (one for bright children, one for average children) with respect to their centrality measures on the variable "height."

In social science three measures of centrality are in common usage: the mean, the median, and the mode. The mean is defined as the arithmetic average. The median is the point that divides the score distribution into upper and lower halves. The mode is the most frequently occurring score. In this chapter they are classified by type of scale. The mean can only be used with an interval scale; however, the median can be used for either ordinal or interval data, and the mode for all three kinds of scales. When more than one measure of centrality is appropriate, they may give different information. See Section 5-6 for a comparison of median and mode for ordinal data, and Section 5-7 for a comparison of the uses of all three with interval scales.

5-5 Mode

Relevant earlier sections:
4-8 Bar graph
4-9 Histogram or polygon

The mode (*Mo*) is a measure of centrality that can be used with nominal, ordinal, or interval data. It is the only measure of centrality that is defined for a nominal scale.

For a nominal or ordinal scale, or for a discrete or grouped interval scale, the mode is defined as that category or interval having the greatest frequency or percentage of cases. It is the most frequently occurring score in the distribution. For a continuous interval scale, the mode is the point of maximum density (i.e., the point at which the polygon of the distribution reaches its high point or maximum). Figure 5-7 indicates modes for distributions on three different kinds of scales.

It is possible for a distribution to have more than one mode, if the maximum percentage or frequency is attained by two or more intervals. Jung, for example, argued that introversion and extraversion represented different personality types. This has been interpreted by some to indicate that he expected the distribution of scores on a test of introversion-extraversion to be bimodal, as shown in Figure 5-8.

Although the mode is defined as the most frequently occurring value in a distribution, there is nothing in this definition to indicate how much more frequent the mode is than any other value in the distribution. The difference in frequency between the mode and one or more other values may be very slight. Because of this, the mode may be a poor representative of the entire distribution. In fact, it may be an uncommon value, even though it is more common than any other.

Also, if many values of the variable are near the mode in frequency, the mode may vary greatly from sample to sample drawn from the same population. It is not a very stable estimate of centrality.

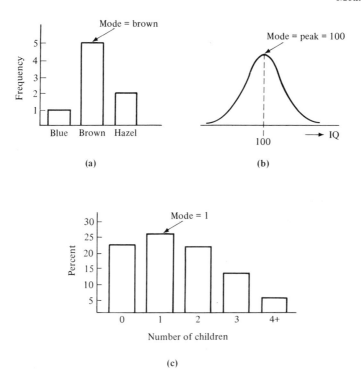

Figure 5-7 *Three different distributions and their modes.* **(a)** *Eye colors (nominal).* **(b)** *IQ (interval, continuous).* **(c)** *Family size (interval, discrete).*

Figure 5-8 *Theoretical bimodal distribution.*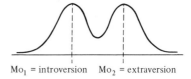

References Brief discussions of the mode can be found in Blommers and Lindquist (1960), 99–101; and Games and Klare (1967), 89–90.

5-6 Median

Relevant earlier sections:
4-5 Frequency distribution
4-9 Histogram

A. Definition and Computation of Median
The median is defined as the point along the scale of a variable which divides the distribution of scores into upper and lower halves. Because there is no order

to nominal-scale values, the median is undefined for nominal data; however, it is used with either ordinal or interval data.

The first step in calculating the median for a set of data is to arrange the data in order of magnitude, from least to greatest. The exact procedure for calculating the median is different for each of three possible distributions, although in each case you are estimating the point that divides the distribution into upper and lower halves.

1. If there are an even number of cases, and exactly half the cases have some value, say X_a or less, and exactly half the cases have some other score, X_b or greater, then the median (Md) falls midway between X_a and X_b:

$$Md_x = \frac{X_a + X_b}{2} = X_a + \frac{X_b - X_a}{2}$$

Table 5-1 shows the calculation of a median for each of three distributions of scores.

Table 5-1 *Calculations for three medians*

Score	Frequency		
	Distribution I	Distribution II	Distribution III
1	1	1	1
2	2	2	2
3	3	3	3
4	1	0	0
5	5	1	0
6	0	5	1
7	0	0	5
Median	3.5	4	4.5

In each distribution, half the scores are three or less, and half the scores are X_b or more, but X_b is 4 for distribution I, 5 for distribution II, and 6 for distribution III.

2. If less than half the scores in the distribution are smaller than some score, X_c, and exactly the same proportion are greater than X_c, then the median is X_c. For example, the median of the set (1, 2, 3, 4, 7) is 3, because there are two smaller scores (1 and 2) and two larger scores (4 and 7).

3. If the median falls in some category or interval X_c but there are unequal proportions of the distribution above and below that interval, then you may wish to interpolate to obtain a more precise estimate of the median.

To interpolate, you assume that scores are distributed evenly throughout the interval in which the median occurs. From this it follows that the proportion of scores below any point in the interval will be the same as the proportion of the interval below that point. From the number of scores in the interval that must be below the median you can find the proportion of scores in the interval that must be below the median. From the latter you can find the proportion of the interval that must be below the median, and by knowing the width of the interval and its lower bound, you can then convert to a score value for the median. An example is given in Table 5-2.

Table 5-2 *Example data for computing interpolated median*

Score	f
11	0
12	1
13	2
14	4
15	1
16	3
N	11
Median	14.1

There are 11 scores in the distribution, so we want the median to separate the upper $5\frac{1}{2}$ scores from the lower $5\frac{1}{2}$ scores. The data are shown in the accompanying graph. Three scores fall below 14 and four above it, so the median interval is 14. We want $2\frac{1}{2}$ of the scores at 14 to fall below the median and $1\frac{1}{2}$ to fall above it. Converting to a proportion of the scores at 14, we want $2\frac{1}{2}/4 = .625$ of those scores to be below the median value. From this it follows that we want .625 of the interval of the score of 14 to be below the median (again, assuming that the true underlying variable is distributed evenly throughout the interval). The interval has real limits (Section 4-1C) of 13.5 and 14.5; thus, the interval width is 1 unit, and we want $(.625)(1) = .625$ score unit to be between the lower bound of the 14 interval and the median. Adding, we find the median to be $13.5 + .625 = 14.125$.

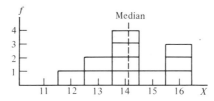

If you prefer to use a formula to find the median, use the formula

$$Md = X_L + p_b(X_U - X_L) \qquad (5\text{-}4)$$

where X_U is the upper bound of the median interval, X_L is the lower bound of the median interval, and p_b is the proportion of scores in the median interval which are below the median. In this example, the median is given by

$$Md = 13.5 + (2.5/4)(14.5 - 13.5)$$
$$= 13.5 + (.625)(1)$$
$$= 14.125$$

Now consider this second example for computing a median, when the interval size is .1 rather than 1.0. Suppose that we want to find a median for the numbers 1.5, 1.6, 1.7, 1.7, 1.7, and 1.8. The median interval is 1.7, having real limits $X_U = 1.75$ and $X_L = 1.65$. There are three subjects in the median interval, only one of which we want to be below the median. Therefore, the computational formula becomes

$$Md = 1.65 + \tfrac{1}{3}(1.75 - 1.65)$$
$$= 1.65 + \tfrac{1}{3}(.1)$$
$$= 1.65 + .033$$
$$= 1.683$$

B. Choosing a Centrality Measure for an Ordinal Scale

With ordinal data, either a median or a mode (the most frequently occurring score or interval) may be used as a measure of centrality. Usually the median is to be preferred, because the mode does not make use of order differences in scores, and therefore is wasteful of information.

The mode is the preferred statistic if you are in a hurry and need only a rough estimate of centrality. It is also preferred when you must select a score to represent all elements in the distribution that will be exactly correct for as many elements as possible.

When you have a complete rank ordering of objects (as many ranks as objects, with no two objects receiving the same rank), the mode is undefined— it is only defined in terms of the number of subjects at a single rank. In this situation, the median is very easily obtained. The median makes use of the order property of ordinal scales and is generally preferable for this reason.

References   The median is discussed briefly in Blommers and Lindquist (1960), 116–118; Freeman (1965), 44–48; and Games and Klare (1967), 90–93.

5-7 Mean

A. Definition of Population Mean

The mean of a distribution of scores is the same as the arithmetic average: to find the mean you simply add all the scores and then divide the result by the number of scores. That is,

$$\text{mean} = \frac{\text{sum of scores}}{\text{number of scores}}$$

We generally compute a mean for a variable that has already been given a label in a prior data table. The population mean is then represented, for each variable, by the Greek letter mu (μ), with the letter representing the variable used as a subscript. If the variable is X, the mean is μ_X; if the variable is Y, its mean is μ_Y; and so on.

The mean can be readily expressed in the statistical notation introduced in Section 5-1. If the subjects are numbered from 1 to N, then there are N scores in the population, going from X_1 to X_N. The sum of the scores can be written $\sum_{i=1}^{N} X_i$ or abbreviated $\sum X$. The final formula for the mean can then be written:

$$\mu_X = \frac{\sum_{i=1}^{N} X_i}{N} \qquad \text{or abbreviated} \qquad \mu_X = \frac{\sum X}{N} \qquad (5\text{-}5)$$

Similar formulas can, of course, be written for variables labeled W, Y, Z, or anything else. The abbreviated formula is preferable when its meaning is unambiguous.

B. Definition of Sample Mean

The sample mean is defined in the same way as the population mean:

$$\text{mean} = \frac{\text{sum of scores}}{\text{number of scores}}$$

It is labeled either with a capital M that has the variable letter as a subscript (M_X, M_Y, $\cdots$) or by the letter itself with a bar over it ($\bar{X}$, $\bar{Y}$, $\cdots$). The formula is identical to the formula for population mean:

$$M_X = \bar{X} = \frac{\sum_{i=1}^{N} X_i}{N} \qquad (5\text{-}6)$$

Note the N is ambiguously defined; it could represent either the size of a population or the size of a sample. You must determine which it is from context, except where the context is confusing. Then the two N's can be distinguished either by giving them subscripts (N_{pop}, N_s) or by making the sample size a lowercase letter (n).

C.   Computation of Means

When data are in the form of a data table, computation involves a direct application of the formulas for mean. An example is given in Table 5-3.

Table 5-3   *Data table for calculating a mean*

i	X_i
1	5
2	5
3	2
4	4
5	6
6	5
7	3
$\sum X$	30
N	7
$\overline{X}$	$\frac{30}{7} = 4.28$

When the data are in the form of a frequency distribution, it is often simpler to do the calculations when a third column of the table is constructed by multiplying each score by its frequency of occurrence. This column is then added to find the sum of scores and the frequencies are added to find N. Then the mean is calculated. This has been done in Table 5-4 for a hypothetical frequency distribution.

Table 5-4   *Calculation of a mean from a frequency distribution*

X	f	fX
6	1	6
5	3	15
4	1	4
3	1	3
2	1	2
1	0	0
$\sum X$		30
N	7	
$\overline{X}$		$\frac{30}{7} = 4.28$

In a grouped frequency distribution, treat all scores in an interval as concentrated at its center, C. Then carry out the calculations as for a frequency distribution, using C instead of X. This has been done for some sample data in Table 5-5.

Table 5-5[†] *Calculation of a mean for a grouped frequency distribution*

Interval	C	f	fC
13–15	14	2	28
10–12	11	1	11
7–9	8	4	32
4–6	5	2	10
1–3	2	1	2
X			83
N		10	
$\overline{X}$		$\frac{83}{10} = 8.3$	

[†] See Section 4-5 for construction of such a table.

In general, the assumption of the grouped-distribution calculations — that all the scores in an interval are concentrated at the center of the interval — is in error. Making this assumption and carrying out the calculations can lead to errors in the results. You should find statistics from grouped distributions only if you aren't very concerned about accuracy of results, if you can't get hold of the raw scores, or if you are convinced that the increased accuracy of using raw scores doesn't justify the extra work of computation.

D. Comparison of Mean, Median, and Mode for Interval Data

Very often distributions of interval-level data are single-peaked and symmetric. When they are, the mean, median (middle value), and mode (most frequent score) are coincident or nearly so, as indicated in Figure 5-9. For skewed distributions, on the other hand, these measures of centrality do not coincide but appear in order: mode, median, and mean from the peak of the distribution to the longer tail. Examples are given in Figure 5-10.

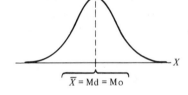

Figure 5-9 *Symmetric single-peaked distribution.*

$\overline{X} = Md = Mo$

A numerical example may help explain why this is so. Suppose that we have a distribution of four scores: 1, 3, 3, and 5. Here the distribution is symmetric

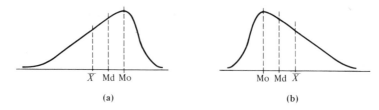

(a) (b)

Figure 5-10 *Skewed distributions.* **(a)** *Negatively skewed.* **(b)** *Positively skewed.*

and the mean, median, and mode are all equal to 3. Now suppose that we form another distribution by replacing the 1 with a 9. This second distribution, with scores 3, 3, 5, and 9, has a mode of 3, but the median has moved to 4 (note that the median would be 4 regardless of what the score was changed to, as long as it was 5 or greater). The mean, however, is changed to 5, because it is affected by the magnitude of the change as well.

In many circumstances you may wish to compute more than one estimate of centrality; in fact, a comparison of the three can give you some indication of the degree of skewness of a distribution. Then, too, you may want to use the same data to investigate more than one problem and need different centrality measures to answer different questions.

Compared to the other centrality measures, the *mean*

1. Is the most reliable (fluctuates least from sample to sample) for symmetric (nonskewed) distributions.
2. Is sensitive to any change in any score in the distribution. This is a particularly important characteristic in experimental work, if you wish to test whether some experimental treatment has any effect at all.
3. Is closest to all scores in the sense that the sum of squared deviations about the mean is smaller than the sum of squared deviations about any other value [$\sum(X - C)^2$ is smallest when $C = \overline{X}$].
4. Has theoretical relationships with other statistics that are very powerful and widely used, e.g., variance, Pearson correlation, and t test.

Compared to other centrality measures, the *median*

1. Divides the distribution into two equal halves for coarse analysis, test construction, pilot research, etc.
2. Is the most reliable centrality measure for use with data that are highly skewed.
3. Is the most representative value in the distribution in the sense of being closest to all the scores ($\sum | X - C |$ is smallest when $C = Md$).*

* $|X - C|$ is the absolute value of the difference between X and C. (See Section 2-4.)

4. Is often the preferred measure of centrality in a skewed distribution, or a grouped distribution with an open-ended category (". . . 10,000 or more") where the mean can't be computed.

5. May be used when there isn't sufficient time to compute the mean.

Compared to the other centrality measures, the *mode*

1. Is a preferred estimate of centrality when a decision based on it will be applied to all elements in the distribution, and you want to be exactly right the greatest proportion of the time.

2. Is a good quick estimate of the mean and median for a symmetric distribution.

To test the effect of different room lighting on test performance, you would probably compare means of subjects operating under different lights, because any change in performance by any subject will be reflected in the mean (but not in the median or mode). To compare the running speed of two groups of rats, the median might be preferred—if some of the rats won't run at all, and you must somehow take account of their behavior as well as the running speeds of others. If you buy shoes for a stranger, you will have the greatest chance of being right if you pick the modal size for people of his height and weight.

References The mean is defined in Blommers and Lindquist (1960), 101-108; Freeman (1965), 54–58; Games and Klare (1967), 93–94; and Hays (1963), 161–165. The mean is compared to other centrality measures in Blommers and Lindquist, 114–115, 119–128; Freeman, 59; Games and Klare, 95–106; and Hays, 165–166.

5-8   Variability measures

A measure of variability or dispersion is an indicator of the degree of spread of scores. The scores of a distribution with greater variability are farther apart from one another than those of a distribution with lesser variability.

A measure of dispersion is often used to indicate just how representative of the distribution of scores the centrality measure is. When the variability is very small, the measure of centrality is very similar to each of the scores, or to most of the scores in the distribution, and so is highly representative of them. When dispersion is great, there are many scores that are different from the measure of centrality, or some scores that are quite different from it, so that it is less representative of all scores in the distribution.

The variability of distributions is an extremely important consideration in comparing their measures of centrality. In general, the greater the variability of the distributions and the less representative their centrality measures are, the less likely that conclusions drawn about their measures of centrality will apply to the distributions as a whole or to the population from which they are drawn. When samples are being compared, the greater the variabilities of the sample scores, the smaller the chances that differences found between a given pair of samples will be found again if these two samples are replaced by two other, similar samples. Chances are smaller that the sample results characterize the populations from which the samples are drawn.

Consider the pairs of distributions pictured in Figure 5-11 (a) and (b).

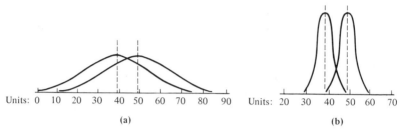

Units: 0 10 20 30 40 50 60 70 80 90   Units: 20 30 40 50 60 70

(a)   (b)

Figure 5-11 *Two pairs of distributions with corresponding centrality measures but different variabilities.*

The measures of centrality are represented by vertical lines in both (a) and (b), and in both cases, they are 10 scale units apart. However, in (a) the variabilities are considerably greater than in (b). Therefore, a difference of 10 units in (b) indicates a much more clear-cut separation of the distributions than in (a).

The distributions in Figure 5-11 appear to be of interval-scaled variables; however, the same principle applies to nominal and ordinal data. For example, the smaller the percentage of cases falling at the mode of a nominal or ordinal distribution, the less useful it will be in coming to conclusions applicable to the distribution as a whole.

5-9   Variation ratio

Relevant earlier sections:
5-5 Mode

It is possible to define a statistic that indicates the extent to which the mode of a distribution is representative of all the scores in a distribution. This statistic is called the *variation ratio* and is defined as

$$v = 1 - \frac{f_{\text{modal}}}{N} \tag{5-7}$$

where f_{modal} is the frequency of scores falling at the mode and N is the total number of scores in the distribution. This amounts to 1 minus the percentage of cases falling in the modal category, or simply the percentage of nonmodal scores.

A greater variation ratio indicates a greater variability and a less representative mode. As examples, consider these three frequency distributions:

A. Eye color	f	B. Eye color	f	C. Eye color	f
Blue	1	Blue	4	Blue	2
Brown	8	Brown	5	Brown	4
Gray	0	Gray	1	Gray	2
Hazel	1	Hazel	0	Hazel	2
N	10	N	10	N	10

In distribution A scores are concentrated at brown, which is reflected in the variation ratio of $v_A = 1 - (8/10) = 0.20$. In both B and C, more scores are nonmodal, giving larger variation ratios: $v_B = 1 - (5/10) = 0.50$, and $v_C = 1 - (4/10) = 0.60$.

Reference Freeman (1965), 40–42, discusses the variation ratio and cites references to other indices of variability for nominally scaled data.

5-10 Range and interquantile ranges

Relevant earlier sections:
4-5 Frequency distribution
5-6 Median

A. Interquantile Ranges

An interquantile range is a preferred measure of variability when the median is the preferred measure of centrality. For an ordinal scale or for an interval variable that is skewed, truncated, and so on (see Section 5-4), it indicates the size of the interval taken up by a given percentage of the scores about the middle of the distribution. Two different interquantile ranges are most often encountered: the *decile range* and the *semi-interquartile range*.

The decile range (D) is found by first finding two points on a distribution: the point below which the bottom 10 percent of the cases fall (C_{10}), and the point above which the top 10 percent of the cases fall (C_{90}). C_{10} and C_{90} are called the tenth and ninetieth centiles, respectively. The decile range is then the difference between these two points: $D = C_{90} - C_{10}$.

The semi-interquartile range is also found by first finding two points on the scale: the point below which the bottom 25 percent of the cases fall (C_{25})

and the point above which the top 25 percent of the cases fall (C_{75}). These points are also called the first and third quartiles (Q_1 and Q_3), respectively. The quartile range (QR) is the difference between Q_1 and Q_3, and the *semi-interquartile range* is half the quartile range: $Q = \frac{1}{2}(Q_3 - Q_1)$. Q is also called the *quartile deviation*.

In using an interquantile range with an interval scale, the result indicates the spread of scores; in general, the greater the value of an interquantile range, the greater the variability of scores. With an ordinal scale, a greater interquantile range does not necessarily indicate a greater spread of scores on some hypothetical underlying interval scale, even when one can be postulated. Instead, it indicates that there are more scores in the distribution or that there are fewer ties in the central portion of the distribution.

When used with interval-level variables, interquantile ranges are unstable relative to, say, the standard deviation; they are often used where a great deal of precision is not required. Interpolating to find the values of centiles used in the computation of interquantile ranges may lead to a false sense of accuracy of measurement. Therefore, it is probably wisest not to interpolate in their calculation unless it is clear that not interpolating would lead to even less precision and be more misleading, for example, when a great number of scores are concentrated at a single value and the needed centile value falls somewhere among them. Interpolation, when it is needed, can be carried out using the methods of Section 5-6.

B. Range

Other interquantile ranges could be defined in a manner analogous to the definition of the semi-interquartile and decile ranges, by specifying the percentages of scores to be cut off at either end of the distribution. However, selecting an unusual interquantile range needs justification to convince others that you have not selected a statistic simply because it supports your hypotheses.

The limiting case of the interquantile range is simply called the range* (see Section 4-5B); it is the total spread of the distribution.

The range (R) may be defined as the difference between the upper limit of the maximum observed score and the lower limit of the minimum observed score, and may be calculated by the following formulas:

$$R = X_{max} - X_{min} \tag{5-8}$$

for a continuous distribution in which there is no rounding or rounding has little effect on score values, and

$$R = X_{max} - X_{min} + 1 \tag{5-9}$$

for a discrete distribution in which scores must take on whole-number values. The $+1$ in formula 5-9 is introduced because the range is inclusive: it takes in

* To be perfectly consistent, we probably should call this the "nil-ile range."

both end scores of the scale. For example, if the scores go from 5 through 9, inclusive, the range includes 5, 6, 7, 8, and 9: five scores. But if you use formula 5-8, you get $R = 9 - 5 = 4$, which is one too few. Another way of thinking of it is that the scores have been rounded; so that the actual range of the example set of scores is from $4\frac{1}{2}$, the lower real limit of the 5 category, to $9\frac{1}{2}$, the upper real limit of the 9 category, $9\frac{1}{2} - 4\frac{1}{2} = 5$. Hence, the range is $(X_{max} + \frac{1}{2}) - (X_{min} - \frac{1}{2}) = X_{max} - X_{min} + 1$.

Often, the range is stated by giving the limiting scores of the distribution rather than the difference between them. In this case one would say that the range is from (X_{min}) to (X_{max}).

Because of its ease of calculation, a range is often obtained for a first look at a distribution; but because of its instability, it is used for little else.

C. Example of the Use of Interquantile Ranges

Consider the data of Table 5-6. The most obvious thing to begin with in characterizing this distribution is the fact that the scores range from 3 to 14. The range can thus be calculated from the formula

Table 5-6 *Scores on an ordinal scale*

Score	Frequency
14	2
13	4
12	3
11	2
10	1
9	1
8	2
7	2
6	1
5	1
4	0
3	1

$$R = X_{max} - X_{min} + 1 = 14 - 3 + 1 = 12$$

or by finding the difference between the upper limit of the highest observed score and the lower limit of the lowest observed score:

$$R = 14.5 - 2.5 = 12$$

An equal number of scores are above and below 11. Therefore, 11 is the median. To use the formula,

$$Md = 10.5 + \tfrac{1}{2}(11.5 - 10.5) = 11.0$$

Ten percent of the scores would be two cases. The lowest two cases are 3 and 5; the highest two are both 14. From the definition of centile scores, $C_{10} = 5.5$ and $C_{90} = 13.5$. Twenty-five percent of the scores are five cases, and five cases are scores of 7 or less; therefore, $C_{25} = 7.5$. For C_{75}, however, we must decide whether we wish to interpolate or not. If not, round C_{75} to 12.5; if so, use method **3** of Section 5-6:

$$C_{75} = 12.5 + \tfrac{1}{4}(13.5 - 12.5) = 12.75$$

Then the decile range, $D = C_{90} - C_{10} = 13.5 - 5.5 = 8$, and the quartile range $QR = Q_3 - Q_1 = 12.75 - 7.5 = 5.25$ if you decide to interpolate and 5.0 if not. From this, the semi-interquartile range is either $Q = \tfrac{1}{2}QR = \tfrac{1}{2}(5.25) = 2.62$ if you interpolated, or $Q = \tfrac{1}{2}QR = \tfrac{1}{2}(5) = 2.5$ if you did not.

Notice that $(Q_3 - Md) \neq (Md - Q_1)$ in this case, indicating that Q does not necessarily mark off equal percentages of scores on both sides of the median and must therefore be interpreted with caution. In this example, the inequality is due to the fact that the distribution is skewed (see Sections 5-2 and 5-13).

References   Discussions of the range and interquantile ranges can be found in Blommers and Lindquist (1960), 136–138; Freeman (1965), 48–53; and Games and Klare (1967), 121–122.

5-11   Variance and standard deviation of a population

If the mean is the preferred measure of centrality, and if a measure of variability is an indicator of how spread out the scores are about the measure of centrality, then it stands to reason that the preferred measure of variability should be an indicator of how far the scores are from the mean.

The difference between any raw score and the mean of the distribution in which it appears is called a *deviation score*. It is represented by the lowercase letter corresponding to the capital letter used to represent the raw scores for the variable. If the raw-score variable is X, deviation scores are represented by $x_i = X_i - \mu_X$; if the raw-score variable is Y, the corresponding deviations are $y_i = Y_i - \mu_Y$.

It would seem reasonable to use the mean of the deviation scores to indicate variability, since each deviation score indicates the distance of its corresponding raw score from the mean. Unfortunately, this value, μ_x, is always equal to zero, as shown below.

$$\mu_x = \frac{\displaystyle\sum_{i=1}^{N} x_i}{N} \qquad \text{(by definition of a mean)}$$

$$= \frac{\sum\limits_{i=1}^{N} (X_i - \mu_X)}{N} \qquad \text{(by definition of } x_i\text{)}$$

Then the summation sign can be distributed according to rule I of Section 5-1 to give

$$\mu_x = \frac{\sum\limits_{i=1}^{N} X_i}{N} - \frac{\sum\limits_{i=1}^{N} \mu_X}{N}$$

Now μ is the same value for all subjects, so it is a constant for this distribution. Then, by rule II, Section 5-1,

$$\mu_x = \frac{\sum\limits_{i=1}^{N} X_i}{N} - \frac{N\mu_X}{N}$$

But

$$\frac{\sum\limits_{i=1}^{N} X_i}{N} = \mu_X$$

so

$$\mu_x = \mu_X - \mu_X = 0$$

The mean of deviations is an average of signed values, so negative values cancel positive ones. The above derivation indicates that the cancellation is complete: the mean is so defined that the sum of negative deviations necessarily equals the sum of positive deviations, and the sum of all deviations is zero.

When the deviations are squared before averaging, this problem does not arise. The square of a negative number is always positive, as is the square of a positive number, so cancellation does not occur. Greater deviations (signs not considered) give greater squared deviations, so a mean of squared deviations is also an indicator of variability. This indicator is called a *variance* and is defined statistically as follows:

$$\text{variance} = \sigma^2 = \frac{\sum\limits_{i=1}^{N} x_i^2}{N} = \frac{\sum\limits_{i=1}^{N} (X_i - \mu_X)^2}{N} \qquad (5\text{-}10)$$

(The value σ^2 is called "sigma squared." This sigma is the lower-case Greek letter whose capital we have used for an entirely different purpose.)

Unfortunately, because all the deviations are squared before they are averaged, the variance is not expressed in the same units as the deviation scores, and the measure cannot serve as a basis for the comparison of scores. To overcome this difficulty, the *standard deviation* is defined as the square root of the variance. The standard deviation is comparable to the original deviations in units and provides a convenient measure for comparing variabilities.

The standard deviation σ is defined as

$$\sigma = \sqrt{\frac{\sum_{i=1}^{N} x_i^2}{N}}$$

or sometimes simply as

$$\sigma = \sqrt{\frac{\sum x^2}{N}} \qquad (5\text{-}11)$$

Reference   A lengthy discussion of variance can be found in Games and Klare (1967), 122–141.

5-12   Variance and standard deviation estimated from a sample

This section assumes that you have already read Section 5-11 and know generally what the variance and standard deviation are. Here we are concerned with computing the best estimates of the population variance σ^2 and standard deviation σ when scores are available only on a small sample taken from the population. In research applications this is almost always the case: you want to know a population variance but have access only to a sample, from which you must estimate the population value.

A. Biased and Unbiased Estimates of the Population Variance

We generally want the sample estimate to be *unbiased*, free from systematic error. We cannot reasonably expect our sample estimate to be free of all error, because for that to happen the sample would have to be an exact replica of the population. To ensure that the sample exactly reflected the population, we would have to examine the population in detail, and that would defeat our very reasons for sampling in the first place: we gather the sample to avoid having to look at the whole population.

However, we can do our best to make the sample estimate free of systematic error; and when we do that, the estimate is said to be unbiased. We can be reasonably sure that the estimate is unbiased if the sample is randomly chosen from the population, and if the statistical formula we use doesn't introduce any systematic errors in the estimate. Then, if we imagine randomly drawing a large number of samples from the same population and calculating the variance estimate for each sample, the formula is said to be unbiased in that the mean of those estimates will be equal to the population variance. The estimates may differ among themselves because of random sampling differences in the samples; but on the average, the formula provides a good guess about the population value. Formula 5-10, used with sample raw scores, would give an unbiased estimate of the population variance:

$$\hat{s}^2 = \frac{\sum (X - \mu)^2}{N}$$

where μ_X is the population mean of X, the X_i are sample scores, and N is the number of subjects in the sample. If we were to draw a large number of samples, as suggested in Figure 5-12, and calculate $\hat{s}^2$ for each, the only differences among the estimates would be the differences in the X's, which would be random. The mean over a large number of such estimates would equal σ^2.

Figure 5-12 *Drawing several samples from the same population.*

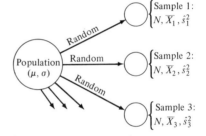

The only difficulty with using this estimate is that we would have to know μ_X, which must be calculated on the entire population. That is generally impractical or impossible; in fact, the reason we want to estimate σ_X^2 is that we don't have access to the whole population.

The natural next step is to replace μ_X by *its* sample estimate, giving formula (5-12):

$$s_X^2 = \frac{\sum_{i=1}^{N} x^2}{N} = \frac{\sum (X - \overline{X})^2}{N} \tag{5-12}$$

Unfortunately, this estimate is *biased*. The sample mean $\overline{X}$ is generally closer to the sample scores (the sum of squared deviations is smaller) than is the population mean μ_X. Thus, the estimate is systematically too small relative to what it would have to be to be unbiased.

We may obtain an unbiased estimate of the population variance by changing formula 5-13 only slightly:

$$\hat{s}_X^2 = \frac{\sum x^2}{N - 1} = \frac{\sum (X - \bar{X})^2}{N - 1} \tag{5-13}$$

Here the denominator has been reduced by 1 to make the variance estimate larger, thus taking into account the fact that our estimate $\bar{X}$ of μ is too good a fit to the sample data. The value $N - 1$ is called the number of *degrees of freedom*; it will be discussed in more general terms in Section 7-8.

Note that a deviation score (x_i) may represent either a deviation from a sample mean $(x_i = X_i - \bar{X})$, as in the present section, or a deviation from a population mean $(x_i = X_i - \mu)$, as in Section 5-11. You will have to determine which meaning is intended by examining the context in which a given deviation score appears.

Notice also that a caret or hat has been placed over the variance estimate to indicate that it is a sample estimate of the population variance. In the case of the variance, the caret distinguishes the unbiased estimate from the biased one. However, when used with other statistics, the caret only indicates an estimate and does not necessarily imply that the estimate is unbiased.

The sample estimate of the standard deviation to be used in this text is the square root of the unbiased variance estimate.

$$\hat{s} = \sqrt{\frac{\sum x^2}{N - 1}} = \sqrt{\frac{\sum (X - \bar{X})^2}{N - 1}} \tag{5-14}$$

(Actually, it is slightly misleading to call $\hat{s}$ an unbiased estimator of σ, although it is less biased than s. In general, a further correction should be applied, as discussed by Dixon and Massey (1957), 77. However, the additional correction usually makes a negligible difference in the magnitude of the estimate. Because $\hat{s}$ is widely called the unbiased estimate, that convention will be followed in this text.)

$\hat{s}^2$ is properly titled the "unbiased sample estimate of the population variance" and $\hat{s}$ the "sample estimate of the population standard deviation." For brevity, however, they will be called the "sample variance" and "sample standard deviation," respectively.

You should also be aware that the caret notation is not universally followed. Many texts use s^2 for the unbiased variance estimate and σ^2 for the biased estimate, even when it is computed on a small sample.

B. Computing Formula

Although it is reasonable to define the variance and standard deviation in terms of deviation measures for some sample, it may be easier to use formulas that require manipulation of raw scores only. Let us therefore find a raw-score equivalent of $\sum x^2$.

$$\sum_{i=1}^{N} x_i^2 = \sum_{i=1}^{N} (X_i - \bar{X})^2 \qquad \text{(by definition of deviation score)}$$

$$= \sum_{i=1}^{N} (X_i^2 - 2X_i\bar{X} + \bar{X}^2) \qquad [\text{by squaring } (X_i - \bar{X})]$$

$$= \sum_{i=1}^{N} X_i^2 - \sum_{i=1}^{N} 2X_i\bar{X} + \sum_{i=1}^{N} \bar{X}^2 \qquad \text{(by rule I, Section 5-1)}$$

$$= \sum_{i=1}^{N} X_i^2 - 2\bar{X} \sum_{i=1}^{N} X_i + N\bar{X}^2 \qquad \text{(by rules II and III, Section 5-1)}$$

Now we can substitute $\sum X/N$ for $\bar{X}$, since this is its formula:

$$\sum_{i=1}^{N} x_i^2 = \sum_{i=1}^{N} X_i^2 - 2\left(\frac{\sum_{i=1}^{N} X_i}{N}\right)\left(\sum_{i=1}^{N} X_i\right) + N\left(\frac{\sum_{i=1}^{N} X_i}{N}\right)^2$$

and simplifying

$$\sum_{i=1}^{N} x_i^2 = \sum_{i=1}^{N} X_i^2 - 2\frac{\left(\sum_{i=1}^{N} X_i\right)^2}{N} + \frac{N\left(\sum_{i=1}^{N} X_i\right)^2}{N^2}$$

$$= \sum_{i=1}^{N} X_i^2 - 2\frac{\left(\sum_{i=1}^{N} X_i\right)^2}{N} + \frac{\left(\sum_{i=1}^{N} X_i\right)^2}{N}$$

$$= \sum_{i=1}^{N} X_i^2 - \frac{\left(\sum_{i=1}^{N} X_i\right)^2}{N}$$

The final computing formula for the variance of a sample can now be written by replacing the expression $\sum_{i=1}^{N} x_i^2$ by the expression that we have shown to be its equivalent, giving

$$\hat{s}_X^2 = \frac{\sum x_i^2}{N-1} = \frac{\sum_{i=1}^{N} X_i^2 - \dfrac{1}{N}\left(\sum_{i=1}^{N} X_i\right)^2}{N-1} \qquad (5\text{-}15)$$

The formula as it stands shows a variance estimate as being equal to a sum of squared differences $\sum x^2$ (often called a "sum of squares"), divided by the number of degrees of freedom, $N - 1$. I like the computational formula found by multiplying numerator and denominator of formula 5-15 by N:

$$\hat{s}_X^2 = \frac{N \sum_{i=1}^{N} X_i^2 - \left(\sum_{i=1}^{N} X_i\right)^2}{N(N-1)} \tag{5-16}$$

The sample estimate of the standard deviation is the square root of this value, as shown in formula 5-17:

$$\hat{s}_X = \sqrt{\frac{\sum X^2 - (\sum X)^2/N}{N-1}} = \sqrt{\frac{N \sum X^2 - (\sum X)^2}{N(N-1)}} \tag{5-17}$$

EXAMPLE   Suppose that we have randomly selected a sample of five subjects from a very large population. We test them on a variable we are interested in and obtain the following scores: 18, 5, 7, 11, and 9. We want to estimate the population mean and variance, so we set up the following table to help with the calculations:

i	X	X^2	x	x^2
1	18	324	8	64
2	5	25	-5	25
3	7	49	-3	9
4	11	121	1	1
5	9	81	-1	1
$\sum$	50	600	0	100

This is a random sample, and we have no way of knowing the value of the population mean. The sample mean, which is an unbiased estimate of the population mean, has the value $\bar{X} = \sum X/N = 50/5 = 10$. The fourth column of the table is found by subtracting the sample mean from each of the scores: $x_i = (X_i - \bar{X})$. Using the deviation-score formula (5-13), we estimate the population variance by

$$\hat{s}_X^2 = \frac{100}{5-1} = \frac{100}{4} = 25$$

The standard deviation is then estimated to be $\sqrt{25} = 5.0$. Using the raw-score formula, (5-15), we can also estimate the population variance

$$\hat{s}_X^2 = \frac{5(600) - (50)^2}{5(4)} = \frac{3000 - 2500}{20} = \frac{500}{20} = 25$$

and obtain the same result. Note, though, that using the biased formula (5-12) would give us a smaller estimate, $\hat{s}_X^2 = 100/5 = 20$. This value would be an underestimate, in general, of the population value.

Although in this example it appeared easier to use the deviation-score formula, it will not always be easier to use it. Remember that it was first necessary to find the mean and then to subtract it from each score. When the mean is not a whole number, this can compound the rounding error of the mean as it is subtracted from each score to get the deviations. Also, when a desk calculator is avilable, both the sum of raw scores and the sum of their squares can be obtained almost as fast as the raw scores can be entered into the machine. When the population is large, use of the raw-score formula can save considerable time over calculating and using deviation scores.

References The biased variance estimate is discussed in Blommers and Lindquist (1960), 139–147, and the unbiased estimate in Games and Klare (1967), 141–146. Edwards (1973) develops the unbiased formula also. The reader is cautioned that none of these texts uses the notation used here, although it is not difficult to translate from one notation to another.

5-13 Should you compute the skewness of a distribution?

As indicated in Figure 5-3, a distribution is said to be skewed when scores tend to be concentrated at one end of the distribution relative to the other. A distribution is positively skewed when the long tail or smaller part is to the right, and negatively skewed when the longer tail or smaller part is to the left.

There are many possible reasons for skewed distributions of scores. For example, an ability test that is too easy or too difficult will give a skewed distribution. If it is too easy, subjects will tend to get high scores, and the distribution will be skewed to the left, or negatively skewed. If it is too hard, they will be positively skewed, because subjects will tend to get low scores. Measures of physiological limits (reaction time, speed of performance, etc.) tend to be skewed; so also are measures that have squared terms or that were formed as the product of two other measures.

Either ordinal or interval data may be skewed. If interval data are skewed, you may be justified in using the median as a measure of centrality and the decile range or semi-interquartile range as a measure of variability. For example, the median is probably a more representative centrality measure for incomes of the employees of a company, because the distribution of pay is certainly skewed positively and use of the mean would probably weigh in the higher paid executives too heavily. An interesting discussion of this problem is found in Huff (1954), Chapter 2.

Many statistical procedures make the assumption that the raw-score distribution of a variable is, at least, symmetric or not skewed. If the distribution is in fact not skewed, the measure of skewness will be zero, and the mean and median will coincide. No statistical test will be presented to show that a distribution is not skewed; however, you can determine how badly asymmetrical it appears to be and decide whether some correction should be made before continuing your analysis.

To make a set of scores more nearly symmetric, one of the many possible nonlinear transformations may be used (see Sections 6-2 and 6-7). Unfortunately, using such a transformation of scores often creates new problems of interpretation which cannot be solved easily.

5-14   Computation of skewness

For a population of subjects and an interval-scaled variable, it is possible to define a measure of skewness (SK):

$$\text{SK} = \frac{\sum x^3/N}{(\sum x^2/N)\sqrt{\sum x^2/N}} = \frac{\sum x^3}{N\sigma^3} \tag{5-18}$$

This measure is positive when the distribution is skewed to the right, negative when it is skewed to the left, and zero when it is not skewed. Such a measure is seldom used with samples, however, because unless the samples are large, it varies a great deal from sample to sample drawn from the same population, and because sufficient precision for most of the decisions that we must make can be attained with much less complicated methods.

A second indicator of the skewness of an interval-scale distribution derives from the fact that the mean of a distribution is more greatly affected by extreme values than is the median, and the median more than the mode. Thus, the greater the difference between mean and median or median and mode, the greater the skewness of a set of scores, with the longer tail and the mean on the same side of the median. This is, of course, more a qualitative indicator than a quantitative measure for small samples, and at the same time that you use this, it might help to look at the graph of the distribution.

In the case of either an ordinal or an interval variable, the median of a distribution will be farther from the quartile in the longer tail (Section 5-10) than from the other quartile.

These relationships can be seen in Figure 5-13. The longer tail of the distribution is to the right, $\overline{X} > Md$, and $(Q_3 - Md) > (Md - Q_1)$. The skewness measure (SK) would be positive for these data.

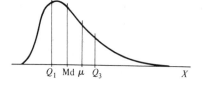

Figure 5-13 *Relations of mean and median to first and third quartiles for a skewed distribution.*

Q_1 Md μ Q_3 X

Homework

1. For the data table, find

(a) N **(b)** X_3 **(c)** Z_4 **(d)** $\sum\limits_{i=1}^{10} X_i$

(e) $\sum\limits_{i=2}^{7} Y_i$ **(f)** $\sum\limits_{i=2}^{8} (Y_i + Z_i)$ **(g)** $\sum\limits_{i=1}^{5} X_i Y_i$

(h) $\sum\limits_{i=1}^{4} (X_i + Y_i - 2Z_i + 3)$ **(i)** $\sum\limits_{i=1}^{6} (814 + 285Z_i)$

(j) $\sum\limits_{i=1}^{4} X_i + \sum\limits_{i=2}^{8} (Y_i/3)$ **(k)** $\sum\limits_{i=1}^{5} (X_i + Z_i - 1)$

i	X_i	Y_i	Z_i
1	4	2	8
2	2	9	1
3	6	3	9
4	8	1	2
5	1	0	0
6	5	4	6
7	6	0	2
8	4	6	3
9	3	2	1
10	3	6	1

2. In which of the following examples is there strong evidence that the sample is nonrandom?
   (a) There are 500 subjects in the population but only 12 in the sample.
   (b) The population has a mean of 45 and a standard deviation of 5, but the sample has a mean of 50 and a standard deviation of 3.
   (c) The sample has 85 percent male subjects and only 15 percent females.
   (d) The proportion of blacks in the sample is twice their proportion in the general population.
   (e) The population is all adult Americans. The sample scores were collected at major intersections in five large cities, by stopping everyone who walked past the intersection during a specified time period.

3. The graph depicts scores on an interval-level variable, X. For this sample, find

(a) N
(b) Mo
(c) Md
(d) $\overline{X}$
(e) Q_1, Q_3, and Q
(f) v

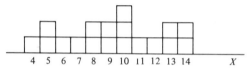

4. Compute the mean and unbiased standard deviation for this sample of cases: 80, 40, 40, 70, 90, 40, 80, 40, 70, 50.

5. In the formula on the first line of the following derivation, all summations are taken from 1 to N. For each subsequent line of the derivation, show how the line was obtained, by

(1) giving the number of the rule used (from Section 5-1);
(2) writing an "A" if the only change was algebraic; or
(3) writing an "E" if an error was made (regardless of whether it was an error in summation or in algebra). Assume that the summation is over a sample of size 200.

$$\sum (X + W)^2 + \sum (Y^2 - 2XY + 5)$$

$$\sum (X^2 + 2XW + W^2) + \sum (Y^2 - 2XY + 5) \underline{\qquad} \square \text{ (a)}$$

$$\sum X^2 + \sum 2XW + \sum W^2 + \sum Y^2 - \sum 2XY + \sum 5 \underline{\qquad} \square \text{ (b)}$$

$$\sum X^2 + 2\sum XW + \sum W^2 + \sum Y^2 - 2\sum XY + \sum 5 \underline{\qquad} \square \text{ (c)}$$

$$\sum X^2 + 2\sum XW + \sum W^2 + \sum Y^2 - 2\sum XY + 1000 \underline{\qquad} \square \text{ (d)}$$

$$\sum X^2 + \sum Y^2 - \sum 2XY + 2\sum XW + \sum W^2 + 1000 \underline{\qquad} \square \text{ (e)}$$

$$\sum (X^2 - 2XY + Y^2) + 2\sum XW + \sum W^2 + 1000 \underline{\qquad} \square \text{ (f)}$$

$$\sum (X^2 - 2XY + Y^2) + 2\sum XW + 200W^2 + 1000 \underline{\qquad} \square \text{ (g)}$$

$$\sum (X - Y)^2 + 2\sum XW + 200W^2 + 1000 \underline{\qquad} \square \text{ (h)}$$

$$\sum (X - Y)^2 + \sum 2(XW + 100W^2) + 1000 \underline{\qquad} \square \text{ (i)}$$

$$\sum [(X - Y)^2 + 2(XW + 100W^2)] + 1000 \underline{\qquad} \square \text{ (j)}$$

6. The sample in the table has been drawn randomly from a large population. For these data, find

X	f
8	2
7	2
6	4
5	2
4	6
3	6
2	6
1	2

(a) The sample size
(b) The sample median
(c) C_{90}
(d) The unbiased estimate of the population mean
(e) The unbiased estimate of the population variance

7. In a psychological experiment 10 subjects were randomly drawn from the population, then 5 were randomly assigned to treatment 1 and 5 to treatment 2. At the end of the experiment the following scores were collected:

Treatment 1	Treatment 2
17	14
4	3
7	3
11	11
11	9

(a) What are the mean and variance of the sample receiving treatment 1?
(b) What are the mean and variance of the sample receiving treatment 2?
(c) What is the best estimate of μ_1? Of σ_1?
(d) Can you legitimately conclude that $\mu_1 > \mu_2$? That $\sigma_1 = \sigma_2$? Why?

8. When there are several samples (j represents sample number) having means $\bar{X}_j$ and sizes N_j, the formula for overall, grand mean is

$$\bar{\bar{X}} = \frac{\sum_j N_j \bar{X}_j}{\sum_j N_j}$$

(a) Find the mean for each of the following samples, then find the grand mean using the formula above.
(b) Show that you obtain the same mean if you treat all scores as appearing in the same sample and calculate a single $\bar{X}$ for that sample.

Data for problem 8.	Sample 1		Sample 2		Sample 3	
	i	*X*	*i*	*X*	*i*	*X*
	1	15	1	8	1	3
	2	8	2	9	2	4
	3	10	3	7	3	1
			4	8	4	7
					5	4

Homework Answers

1. **(a)** 10 **(b)** 6 **(c)** 2 **(d)** 42 **(e)** 17 **(f)** 46
   (g) 52 **(h)** 7 **(i)** 12, 294 **(j)** $27\frac{2}{3}$ **(k)** 36

2. Only **(e)**

3. **(a)** $N = 18$ **(b)** $Mo = 10$ **(c)** $Md = 8.5$ **(d)** $\overline{X} = 9\frac{1}{3}$
   (e) $Q_1 = 7, Q_3 = 12, Q = 2.5$ **(f)** .833

4. $\overline{X} = 60, \hat{s} = 20$

5. **(a)** A **(b)** I **(c)** III **(d)** II **(e)** III **(f)** I
   (g) E **(h)** A **(i)** E **(j)** I

6. **(a)** 30 **(b)** 3.67 **(c)** 7.0 **(d)** 4 **(e)** 4

7. **(a)** $\overline{X}_1 = 10, \hat{s}_1^2 = 24$ **(b)** $\overline{X}_2 = 8, \hat{s}_2^2 = 24$ **(c)** Est.$(\mu_1) = \overline{X}_1 = 10$, est.$(\sigma_1^2)$
   $= \hat{s}_1^2 = 24$ **(d)** No. The sample mean and variance only estimate the corresponding population values. Because of sampling randomness, observed sample differences may or may not reflect corresponding population differences.

8. $\overline{X}_1 = 11, \overline{X}_2 = 8, \overline{X}_3 = 3.8, \overline{\overline{X}} = 7.0$

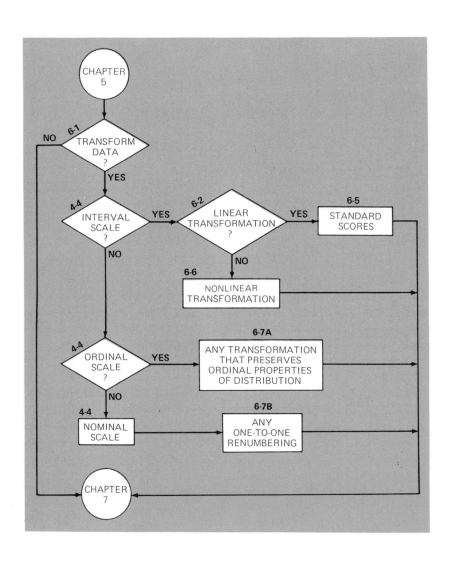

6 Transformations of Scale

6-1 Transform the data?

A transformation of a set of scores is a change in the numbering system used to represent the data: one rule for assigning numbers to responses is replaced by another. Thus, the Fahrenheit temperature scale can easily be transformed to the centigrade scale; the two systems of measurement simply assign different numbers to the same temperature. Inches and centimeters are also transformed from one to the other by a simple mathematical rule. In neither of these examples is a change in the raw observation effected by a transformation of scale; rather, a new set of numbers is used to describe the same set of observations.

The transformation rule is an explicit statement of the transformed scale in terms of the original scale. For Fahrenheit temperature in terms of centigrade temperature, the rule is $°F = \frac{9}{5}°C + 32$; for inches in terms of centimeters, the rule is 1 inch = 2.54 centimeters, or distance in inches = .3937 (distance in centimeters). A transformation rule enables us to change every score of a distribution from one scale to another. Then the measures of centrality and variability of the new distribution can be determined from the old scale and the transformation rule. In other words, if you know the characteristics of the old distribution and the rule of transformation, you can determine what the corresponding characteristics of the new distribution will be.

The transformation of a distribution can be useful for any of three reasons. First, scores can be changed to a form in which computations will be easier to carry out; in this case the results found for the transformed distribution must be converted back into the scale of the original distribution. Alternatively, scores in one distribution can be transformed to give a new distribution with a specified set of characteristics, especially centrality and variability. This may be useful in comparing graphs of two entire distributions, for comparing one person's scores on two tests having different centrality and variability parameters, or for comparing an observed distribution to a standard, theoretical distribution. Finally, a distribution may be transformed in order to change its shape or other characteristics, so that it satisfies assumptions necessary for some statistical test.

The map for this chapter can be divided readily into three regions: linear transformations for interval and ratio scales, nonlinear transformations for interval and ratio data, and transformations for nominal and ordinal scales. Of these, the most important are the linear transformations, since they are a basis for many of the analyses to be presented in the remaining chapters of the text. Accordingly, linear transformations account for the bulk of this chapter.

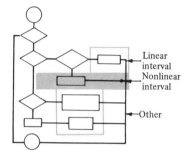

There are two kinds of linear transformations. In an additive transformation, the same number is added to, or subtracted from, each score in a distribution. In a multiplicative transformation, every score in a distribution is multiplied by, or divided by, some number.

Additive and multiplicative transformations do not appear explicitly on the map for this chapter, because for most applications it is just as easy and less confusing to use standard scores for transformations; and standard scores can be used with either interval- or ratio-level raw scores. However, in order for you to understand the section on standard scores, it is important for you first to read the sections on additive and multiplicative transformations.

Nonlinear transformations of interval scales are important mainly in dealing with assumptions of some of the statistics to be covered in later chapters. Although important for some topics, they are properly dealt with in detail in more advanced treatments of statistical analysis. Transformations on nominal and ordinal scales are of minor interest and are included primarily for completeness.

6-2 Linear or nonlinear transformation?

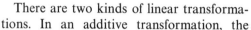

Relevant earlier section:
4-10 Two-dimensional plot

A transformation in which the relative magnitudes of distances between successive scale values are unchanged is called a linear transformation. A linear transformation does not affect the shape of a distribution, and the bivariate plot of the relationship between original and transformed values is a straight line. A nonlinear transformation is one in which the relative magnitudes of distances between scores are altered by the transformation. An example would be to transform each score by squaring it. Here each score in the distribution would be multiplied by a different value (itself). When scores have been transformed in a nonlinear manner, the relationship between the original and the transformed scores is not a straight line.

A plot of X against $10X$ (a linear transformation) is given in Figure 6-1(a) for the range between 0 and 10, and for the same raw scores a plot of X against X^2 (a nonlinear transformation) is given in Figure 6-1(b).

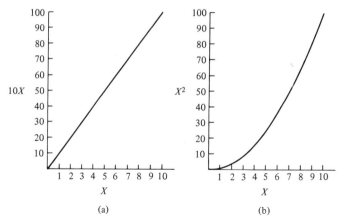

Figure 6-1 *Comparison of linear and nonlinear transformations of scores.* **(a)** *Linear transformation:* $Y = 10X$. **(b)** *Nonlinear transformation:* $Y = X^2$.

Linear transformations are used to put different distributions into the same scale in order to facilitate comparison of their shapes. The comparison of two or more observed distributions may then be done graphically or by the methods of Chapter 10. Alternatively, an observed distribution may be compared to one of the theoretical distributions of Chapter 7, either graphically or statistically. Hypothesis testing (Chapters 9 through 14) involves the transformation of an observed score and its subsequent comparison to a theoretical distribution.

Nonlinear transformations are used to alter the shape of a distribution for some purposes, and usually change the centrality and variability parameters in the process. They are used primarily to assure that assumptions of certain statistical methods are met, especially the assumption of additivity in the analysis of variance (Chapter 13) and of linear regression in correlational analysis (Chapters 15 and 16). If a distribution does not have the shape necessary for the use of some method of analysis, you may transform the distribution nonlinearly so that it does have the proper shape, then carry out the analysis on the transformed scores. The major difficulty with such a procedure lies in trying to make sense of the transformed scores. Interpreting the results of analyses on transformed scores is unfortunately a skill that you will have to develop through familiarity with your own area of interest.

6-3   Additive transformation

One common transformation consists of adding or subtracting the same (constant) number to all scores in the distribution. This is called an additive transformation. You can think of the effect of adding a constant to all scores as either moving the responses up relative to the numbering system, or as moving the numbering scale down relative to the responses of subjects.

A. Additive Transformations with Populations

Consider a population of subjects who have been tested on some behavioral variable and the responses reported as scores, X. Suppose it turns out that the scores have a mean $\mu_X = 50$ and a standard deviation $\sigma_X = 10$. The distribution and scale are shown in Figure 6-2. In this figure, major reference points have been indicated by scale values for the mean and points 1, 2, and 3 standard deviations from the mean.

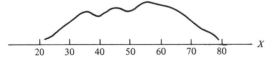

Figure 6-2 *Graph of a distribution of hypothetical population on variable X.*

Now suppose that this numbering system isn't convenient for you, and you decide to add 8 points to each score. This transformation can be represented mathematically by defining a new variable Y, which can be obtained from X using the formula $Y = X + 8$ with each score in the original X distribution. In applying this formula, a score of $X = 20$ converts to $Y = 28$; a score of $X = 43$ transforms to $Y = 51$; and so on. In Figure 6-3, the Y scale has been drawn below the X scale and the scale conversion is shown for the major reference points.

From Figure 6-3 you can see that the mean $\mu_X = 50$ is transformed to $\mu_Y = 58$ for the new set of scores. The effect of the transformation on the standard deviation can be seen by noting that on the X scale 60 was 10 points or one standard deviation above the mean. Under the transformation, $X = 60$ converts to $Y = 68$, which is now 10 points above the Y mean. Thus, the standard deviation on the Y scale is also 10 points.

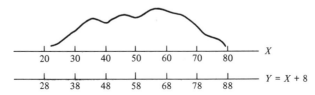

Figure 6-3 *Comparison of scales for X and Y = X + 8.*

This transformation from X to Y is linear, because if you plot values of Y against corresponding values of X from which they are obtained, you get a straight line. (It crosses the Y axis at $X = 0$, $Y = 8$, and makes a 45° angle with both axes.)

Other possible additive transformations include $W = X - 4$, $U = X + 31$, and $V = U - 81 = X - 50$. These are graphed in Figure 6-4. From examination of the graph you can see that the mean μ_X converts to $\mu_W = 46 = \mu_X - 4$; $\mu_U = 81 = \mu_X + 31$; and $\mu_V = 0 = \mu_X - 50$. In each case, the transformation affects the mean in the same way that it affects each score. By contrast, in each case the standard deviation is unaffected by the transformation.

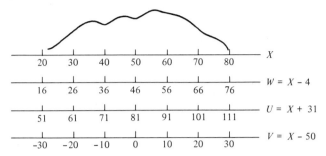

Figure 6-4 *Three additive transformations on X.*

In general, an additive transformation from X to Y is given by the formula $Y = X + K$, where K is a positive or negative constant. Under this transformation, the new mean $\mu_Y = \mu_X + K$, and the standard deviation $\sigma_Y = \sigma_X$.

All the above scales, U, V, W, X, and Y, give equivalent information about the position of a subject's response relative to the responses of other subjects included in the graph of the distribution. They differ only in that they represent the information using different sets of numbers.

You can convert a score on any of these scales to a corresponding score on any other. One way of making the conversion is to set both equal through the X scale. For example, for subject i, if $W_i = 40$, you can find the equivalent on the V scale as follows:

$$W_i = 40 = X_i - 4$$

so

$$X_i = 44$$

Then,

$$V_i = X_i - 50 = 44 - 50 = -6$$

Now you can compare this result to Figure 6-4 to see that -6 on the V scale falls directly below 44 on the X scale and 40 on the W scale. A more general and more often used way of transforming scores will be shown in Section 6-5D using z scores.

It is interesting to note that the V scale of Figure 6-4 is a scale of deviation scores. Recall from Chapter 5 that deviation scores are found by subtracting the mean from each raw score: $x_i = X_i - \mu_X$; and, as in Chapter 5, the mean of deviation scores is always equal to zero.

B. Additive Transformations with Samples

The results shown above for populations also apply to samples: adding a constant to every score in a sample has the effect of adding the same constant to the sample mean, but the sample standard deviation is unaffected. An example is given in Table 6-1 for an original variable X and a linear transformation, $Y = X + 4$, for a sample of five subjects.

Table 6-1 $Y = X + 4$ *for a sample of five subjects*

i	X	x	x^2	$Y = X + 4$	y	y^2
1	3	-3	9	7	-3	9
2	12	6	36	16	6	36
3	9	3	9	13	3	9
4	1	-5	25	5	-5	25
5	5	-1	1	9	-1	1
$\sum$	30	0	80	50	0	80

For the X variable, the mean is given by $\bar{X} = \sum X/N = 30/5 = 6$, and the unbiased variance estimate is $\hat{s}_x^2 = \sum x^2/(N-1) = 80/4 = 20$. Then the estimate of the standard deviation is $\hat{s}_x = \sqrt{20} = 4.4721$, from Table A-2 of the Appendix.

The transformed Y mean is $\bar{Y} = \sum Y/N = 50/5 = 10$. This illustrates the relationship between original and transformed means, as $\bar{Y} = \bar{X} + 4$.

The variance of transformed scores is given by $\hat{s}_y^2 = \sum y^2/(N-1) = 80/4 = 20$, which is the same as the variance of the original X scores. Thus, $\hat{s}_y = \sqrt{20} = \hat{s}_x$. The variance and standard deviation are unaffected by the transformation.

Table 6-2 summarizes the effects of an additive transformation on various statistics.

Table 6-2 *Effects of an additive transformation*

Statistic	Original	$Y = X + K$	$Z = X - K$
Raw score	X_i	$Y_i = X_i + K$	$Z_i = X_i - K$
Mean	$\bar{X}$	$\bar{Y} = \bar{X} + K$	$\bar{Z} = \bar{X} - K$
Median	$Md(X)$	$Md(Y) = Md(X) + K$	$Md(Z) = Md(X) - K$
Mode	$Mo(X)$	$Mo(Y) = Mo(X) + K$	$Mo(Z) = Mo(X) - K$
$\hat{s}$	$\hat{s}_X$	$\hat{s}_Y = \hat{s}_X$	$\hat{s}_Z = \hat{s}_X$
$\hat{s}^2$	$\hat{s}_X^2$	$\hat{s}_Y^2 = \hat{s}_X^2$	$\hat{s}_Z^2 = \hat{s}_X^2$
Q	Q_X	$Q_Y = Q_X$	$Q_Z = Q_X$
v	v_X	$v_Y = v_X$	$v_Z = v_X$

6-4 Multiplicative transformation

A. Multiplicative Transformations with Populations

A second kind of linear transformation occurs when all the scores in a distribution are multiplied by a constant. Consider, for example, a hypothetical population of scores X, having a mean $\mu_X = 50$ and a standard deviation $\sigma_X = 10$. Suppose that you let the constant be 2, and you multiply each score in the X distribution by 2 to get a distribution of scores on Y, using the formula $Y = 2X$. Then $X_i = 20$ becomes $Y_i = 40$; $X_i = 43$ transforms to $Y_i = 86$, and so on. The transformation is graphed in Figure 6-5.

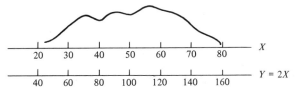

Figure 6-5 *Comparison of scales for X and Y = 2X.*

You can see by examining the two scales that the mean $\mu_X = 50$ transforms to $\mu_Y = 100$, illustrating that whatever is done to each score is also done to the mean. However, in contrast to the additive transformation case, here the standard deviation is affected also. This can be seen by noting that points have been marked on the X scale every 10 points, where the standard deviation of X is 10 points. Corresponding points on the Y scale are 20 units apart, showing that the standard deviation has doubled. Multiplying every score in the X distribution by 2 also multiplies both the mean and the standard deviation by 2.

There is no limit to the number of possible multiplicative transformations; however, Figure 6-6 illustrates one, the transformation $W = 3X$. You can also divide each score in a distribution by a constant. It is convenient to think of division by a constant as multiplication by the reciprocal of the constant. (That way, we at least don't have to come up with any new rules to cover division.) For example, $V = X/2$ is the same as $V = \frac{1}{2}X$; and $U = X/10$ is the same as $U = (1/10)X$. Graphs of these two transformations are also presented in Figure 6-6.

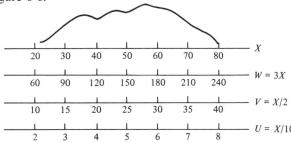

Figure 6-6 *Three multiplicative transformations of X.*

The means on the transformed scales appear directly below the mean of X. By examination you can see that $\mu_W = 3\mu_X$, $\mu_V = \mu_X/2$, and $\mu_U = \mu_X/10$. Looking at the adjacent points along each scale, you can see that on the X scale they are 10 units apart, because $\sigma_X = 10$; on the W scale they are 30 units apart, on the V scale they are 5 units apart, and on the U scale they are 1 unit apart. Thus, the standard deviations have also been affected in the same way as the raw scores.

6-4　Multiplicative transformation

In general, we represent a multiplicative transformation from an original variable X to a new scale Y by the formula $Y_i = KX_i$, where K is a constant for all subjects, and division is represented with a fractional K. Under a multiplicative transformation, the mean $\mu_Y = K\mu_X$, and the standard deviation $\sigma_Y = K\sigma_X$.

As with additive transformations, conversion from one scale to another can be accomplished by making both scores equivalent to values of X. For example, suppose that you want to find the equivalent of $W_i = 192$ on the V scale.

$$W_i = 192 = 3X_i$$

$$X_i = 64$$

Then

$$V_i = \frac{X_i}{2} = \frac{64}{2} = 32$$

From Figure 6-6 you can see that $W = 192$, $X = 64$, and $V = 32$ correspond exactly and are just different ways of numbering the point on the original graph. The more usual way of making the transformation, however, is by use of z scores, as shown in Section 6-5D.

B. Multiplicative Transformations with Samples

The same results apply to samples as to populations: multiplying every score in a sample by a constant K has the effect of multiplying both the mean and standard deviation of the sample by K. This is illustrated by the example of Table 6-3 for an original variable X and a transformed variable $Y = 2.5X$, where $N = 6$.

Table 6-3　*Multiplicative transformation with a small sample*

i	X	x	x^2	$Y = 2.5X$	y	y^2
1	1	-7	49	2.5	-17.5	306.25
2	7	-1	1	17.5	-2.5	6.25
3	19	11	121	47.5	27.5	756.25
4	10	2	4	25.0	5.0	25.0
5	8	0	0	20.0	.0	.0
6	3	-5	25	7.5	-12.5	156.25
$\sum$	48	0	200	120.0	0.0	1250.0

The mean of X is given by $\bar{X} = \sum X/N = 48/6 = 8.0$, and the unbiased variance estimate is $\hat{s}_x^2 = \sum x^2/(N - 1) = 200/5 = 40.0$. Then the estimate of the standard deviation is $\hat{s}_x = \sqrt{40} = 6.3246$.

The mean of Y is given by $\bar{Y} = \sum Y/N = 120/6 = 20$. This 20 is 2.5 times as large as the X mean of 8, showing that the mean has been affected in the same way as each score.

The variance of the transformed scores is found from the formula $\hat{s}_y^2 = \sum y^2/(N-1) = 1250/5 = 250$. Then the standard deviation is $\hat{s}_y = \sqrt{250} = 15.8114$. This is 2.5 times as large as the standard deviation of X values, showing that the standard deviation is also affected in the same way as each score.

A multiplicative transformation affects many sample statistics, some of which are summarized in Table 6-4.

Table 6-4 *Effects of transformation by multiplication or division*

Statistic	Effect of multiplying by K	Effect of dividing by K
Median	$Md(KX) = K \cdot Md(X)$	$Md(X/K) = Md(X)/K$
Mode	$Mo(KX) = K \cdot Mo(X)$	$Mo(X/K) = Mo(X)/K$
Range (discrete data)	$R(KX) = K \cdot R(X) - (K-1)$	$R(X/K) = R(X)/K$
Variance	$\hat{s}^2(KX) = K^2 \hat{s}_X^2$	$\hat{s}^2(X/K) = \hat{s}_X^2/K^2$
Interquantile range	$Q(KX) = K \cdot Q(X)$	$Q(X/K) = Q(X)/K$
Variation ratio	$v(KX) = v(X)$	$v(X/K) = v(X)$

6-5 Standard z scores

Relevant earlier sections:
6-3 Additive transformation
6-4 Multiplicative transformation

It is possible to transform a distribution of scores on an interval- or ratio-scale variable so that the resulting distribution has a mean of zero and a standard deviation equal to 1, regardless of the mean and standard deviation of the original set of raw scores. When this has been done, the resulting scores are called *standard scores* or z *scores*.

A. Computation of Standard (z) Scores for a Population

The obtaining of z scores from a set of raw scores can be thought of as a two-stage process. In the first stage, you convert each raw score X_i to a deviation score x_i as in Chapter 5, by subtracting the population mean: $x_i = X_i - \mu_x$. From Section 6-3, we know that subtracting a constant (in this case, μ_x) from each score in a distribution has the effect of subtracting that constant from the mean of the distribution, while the standard deviation is unaffected. Thus, the deviation scores have a mean of 0 ($\mu_x - \mu_x = 0$) and a standard deviation σ_x.

In the second stage, you divide each of the deviation scores by the standard deviation to obtain the z scores. From Section 6-4, we know that dividing each score in a distribution by a constant has the effect of dividing both the mean and standard deviation by that constant. Thus, the z scores have a mean of

0 $(0/\sigma_x = 0)$ and a standard deviation of 1 $(\sigma_x/\sigma_x = 1)$, regardless of the mean and standard deviation of the original, raw-score distribution. Figure 6-7 represents these transformations graphically, for a hypothetical set of data having a raw-score mean of 43 and standard deviation of 12.

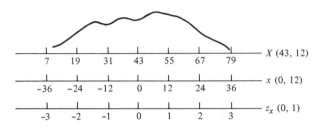

Figure 6-7 *Raw, deviation, and z scales for a distribution.*

In this example, the three numbering scales are shown below the same frequency or percent polygon, with their scales lined up. In going from raw scores to deviation scores, the mean changes from 43 to 0, but the standard deviation remains 12. In the transformation from deviation scores to z scores, the mean remains 0 and the standard deviation changes from 12 to 1. The transformation affects all points on the scale, not just the ones shown. For example, a raw score of 37 would transform to a deviation score of -6, and then to a z score of $-.5$.

The above process can be represented in a formula that is applied to each score in the distribution to make the complete transformation:

$$z_i = \frac{X_i - \mu_X}{\sigma_X} \tag{6-1}$$

B. Computation of z Scores for a Sample

As indicated in Section 5-3, we seldom if ever deal with entire populations; hence the definition of a z score in formula 6-1 is primarily of theoretical interest. In practical situations, z scores are also useful; when dealing with sample values the formula becomes

$$z_i = \frac{X_i - \bar{X}}{\hat{s}_X} \tag{6-2}$$

As an example, consider the following hypothetical data from a sample of six subjects. The raw scores are presented in the X column, the deviation scores in the next column, and the z scores in the last column. Each of the z scores could also be found using formula 6-2. At the bottom of each column is the mean and unbiased standard deviation for that set of scores, again showing the effects of the transformations on those statistics.

Table 6-5 *A z transformation for a small sample*

i	X	$x = X - \mu_X$	$z_X = x/\sigma_X$
1	10	-3	$-\tfrac{3}{4}$
2	20	7	$1\tfrac{3}{4}$
3	12	-1	$-\tfrac{1}{4}$
4	15	2	$\tfrac{1}{2}$
5	9	-4	-1
6	12	-1	$-\tfrac{1}{4}$
Mean	13	0	0
$\hat{s}$	4	4	1

C. Reasons for Obtaining (z) Scores

One advantage of z scores is that they give the position of any score relative to the other scores in a distribution. A z score indicates by its sign whether the corresponding raw score is above or below the mean of the raw-score distribution, and by its magnitude the number of standard deviations the raw score is from the mean. In Figure 6-7, a raw score of 19 converts to a z score of -2; this indicates that 19 is two standard deviations below the raw score mean (of 43). A raw score of 79 transforms to a z score of $+3$, indicating that 79 is 3 standard deviations above the raw score mean. In Table 6-5, the first subject's raw score of 10 is 3/4 or .75 standard deviation below the mean for that set of scores, and the second subject's raw score of 20 is 1.75 standard deviations above the same sample mean.

It is difficult to compare a subject's scores on two distributions if the distributions have different means and/or standard deviations. However, if the distributions have the same shape, converting them to standard scores allows them to be compared directly. Suppose that the little girl next door comes home from school one day and tells you she got a 15 in spelling and a 23 in arithmetic. Should you praise her, or sympathize with her? Or praise her for the arithmetic but not the spelling, or the spelling but not the arithmetic? But if you knew that the spelling test had a mean of 13 and a standard deviation of 2, and that the arithmetic test had a mean of 21 and a standard deviation of 8, then you could calculate her z scores as $+1.0$ and $+0.25$, and you might feel more comfortable about responding to her report. You also would know that she did a little better in spelling than arithmetic, relative to the other children, because her z score for spelling is larger.

D. Using z Scores to Find Transformed Raw Scores

All distributions of z scores, regardless of their raw-score parameters, have exactly the same mean and standard deviation. Because in most behavioral research the means and standard deviations of variables are fortuitous anyway, this does not generally result in loss of information. On the other hand, calculating z transformations takes time and does not alter the shape of the score distribution. No transformation of scores can alter the meaning of the behavior underlying the scores or make two scales more similar conceptually.

Suppose that we have the following transformation from X to Y: $\overline{X} = 50$, $\hat{s}_X = 10$, and $\overline{Y} = 500$, $\hat{s}_Y = 50$. The transformation equation is $Y_i = 5X_i = 250$, and Anna Biotic has scores $X_{12} = 45$ and $Y_{12} = 475$. You can easily verify the consistency of these scores as well as the accuracy of the transformation equation.

Anna's X score can be transformed to a z score.

$$z_{12} = \frac{45 - 50}{10} = \frac{-5}{10} = -.50$$

So can her Y score.

$$z_{12} = \frac{475 - 500}{50} = \frac{-25}{50} = -.50$$

Furthermore, the same applies to anyone else's scores: standard scores are unaffected by linear transformations on the raw-score distribution.

Because a person's z score does not change with linear changes in the raw-score scale, we may equate a person's z scores relative to X and to Y, and use the equation to solve for any term we don't know. Thus, formula 6-3 can be used to solve for any of the six terms it contains:

$$\frac{X_i - \mu_X}{\sigma_X} = \frac{Y_i - \mu_Y}{\sigma_Y} \tag{6-3}$$

if we know the other five. To obtain Anna's Y score from her X score, however, we need an equation based on the sample z-score formula:

$$\frac{X_i - \overline{X}}{\hat{s}_X} = \frac{Y_i - \overline{Y}}{\hat{s}_Y} \tag{6-4}$$

Then given that we know the means and standard deviations for the two distributions, we can write the following equation:

$$\frac{45 - 50}{10} = \frac{Y_{12} - 500}{50}$$

and then solve for Y_{12}.

E. Finding the Equation for a Linear Transformation

Suppose that you have a distribution of scores with mean $\overline{X}$ and standard deviation $\hat{s}_X$, and you want to perform a linear transformation to a new mean

$\overline{Y}$ and standard deviation $\hat{s}_Y$. You would like to know what formula you should apply to each X score to get the new distribution of Y scores.

For sample data, the formula can be obtained by solving equation 6-4 for Y_i. First, multiplying both sides of the equation by $\hat{s}_Y$ gives

$$Y_i - \overline{Y} = \frac{\hat{s}_y}{\hat{s}_x}(X_i - \overline{X})$$

and the final formula is obtained by adding $\overline{Y}$ to both sides:

$$Y_i = \frac{\hat{s}_y}{\hat{s}_x}(X_i - \overline{X}) + \overline{Y} \tag{6-5}$$

For example, suppose that you have a variable X with mean $\overline{X} = 5$ and standard deviation $\hat{s}_X = 3$, and you want to effect a linear transformation to a scale having mean $\overline{Y} = 13$ and standard deviation $\hat{s}_Y = 6$. Applying formula 6-5 you would get

$$Y = \tfrac{6}{3}(X - 5) + 13$$

$$= 2X - 10 + 13$$

$$= 2X + 3$$

This formula, applied to each score in the X distribution, would produce a Y distribution with the desired mean and standard deviation.

F. Using z Scores to Average Results of Different Variables

Suppose that in an introductory statistics course there are two hour examinations and a final, and the instructor wishes to weight all three exams equally. If the three exams have different standard deviations, and if he simply adds raw scores for the three exams, in effect he will be weighting the most variable exam most heavily and the least variable set of scores least heavily in determining the final grade. This is because the greater differences existing among scores on the more variable exam are carried directly into the final total. Using z scores gives all three exams a standard deviation of 1, thus equalizing their contribution to the total score. This procedure can be written

$$z_{total} = \frac{z_1 + z_2 + z_{final}}{3}$$

On the other hand, suppose that the instructor wishes to give more weight to the final-exam mark, say twice as much as to either hour exam, in determining the total grade. This could be done by finding a weighted average of the three scores. He should multiply the z score for the final exam by 2 and divide the total expression by 4 instead of 3:

$$z_{total} = \frac{z_1 + z_2 + 2z_{final}}{4}$$

By extension, any number of variables could be combined using any weights that may seem reasonable.

For each student in the above two formulas, the last operation in computing z_{total} is division by a constant (4, in the last example). Because the final division constitutes a linear transformation on scores, it can be omitted without affecting the relative status of subjects. In this case, the formula for the last example would be

$$4z_{total} = z_1 + z_2 + 2z_{final}$$

References Blommers and Lindquist (1960), 157–176; and Games and Klare (1967), 150–166.

6-6   Some nonlinear transformations

Ordinarily, a nonlinear transformation is used to yield scores that meet the assumptions of statistical analyses more closely than the raw data do. Usually the desired outcome distribution is "normal" (see Section 7-3), which implies that it looks bell-shaped (see Figure 7-6).

Sometimes the variable to be transformed is positively skewed (e.g., when the score is number of seconds to complete a task, a few people will take a very long time, which will yield a positively skewed distribution of scores). If you want the transformed distribution to be nearly normal, use the formula* $Y = \log(X + 1)$ or $Y = 1/X$.

For example, suppose you are running rats in an alley 5 feet long; you can time them for each trial but want to express their running speed in your theory about their behavior. You can transform from time to speed using the formula

$$Y = (1/X) \cdot K$$

where Y is speed in feet/second, X is time in seconds, and $K = 5$ feet. Then if a particular rat completes the alley in 2.5 seconds, its speed is found to be

$$Y = (1/2.5 \text{ seconds})(5 \text{ feet})$$

$$= (5 \text{ feet})/(2.5 \text{ seconds}) = 2 \text{ feet/second}$$

---

*This first formula is a logarithmic transformation. For a discussion of logarithms, see a college algebra text [e.g., Hart (1966)].

Bartlett (1936) suggests another transformation that can be used with a positively skewed distribution (especially when the data are the number of responses in a fixed time period). The formula is $Y = \sqrt{X + .5}$. As an alternative, Freeman and Tukey (1950) recommend the transformation $Y = \sqrt{X} + \sqrt{X + 1}$.

Converting to percentile ranks is a kind of nonlinear transformation, as shown in Section 7-2. It is also possible to *normalize* a distribution by first transforming (nonlinearly) to percentile ranks, then transforming (nonlinearly) the percentile ranks to scores of a normal distribution. This procedure is discussed in Section 7-3.

Many other nonlinear transformations have been used, but they are usually discussed in the context of the statistical methods with which they arc used. A general discussion of the problem of determining a formula for a nonlinear relationship is found in Ezekiel and Fox (1966), Chapter 6.

References Basic references on nonlinear transformations include Bartlett (1937b), 137–138, and (1947), 39–52; Hotelling and Frankel (1938), 87–96; and Mueller (1949), 198–223. Textbook discussions can be found in Edwards (1968), 128–131; Snedecor (1956), 314–328; Walker and Lev (1953), 423–425; and Winer (1962), 218–222.

6-7 Transformations on ordinal and nominal scales

A. Transformations on Ordinal Scales

Any transformation that preserves the ordinal properties of the scores is permissible with ordinal data; that is, any of the linear or nonlinear transformations discussed with respect to interval scales can be used for ordinal data with impunity. Because there are a great number of possible transformations, and because there is less general use for such transformations than for those that involve interval scales, none will bc discussed in detail here. In general, measures of centrality and variability will be affected by changes of scale, but probably the safest procedure for determining their new values would be to recompute these parameters and statistics as the need arises, rather than try to predict mathematically the effects of possible transformations of scale on their values.

B. Transformations on Nominal Scales

The only point to remember in transforming nominal-scaled variables is that if you wish to retain all the information in the original set of scores, the transformation must be one to one; that is, for every scale value of the old scale there must be a scale value of the new scale, and vice versa.

In order to find the mode of a distribution after transformation, simply find the value to which the old mode was transformed. The variation ratio is unaffected by any one-to-one transformation.

Homework

In problems 1 through 5 compute the standard (z) score corresponding to each raw score.

1. $\mu_X = 40$, $\sigma_x = 8$, $X_i = 30$, $z_i = ?$

2. $\mu_X = 28$, $\sigma_x = 2$, $X_i = 21$, $z_i = ?$

3. $\mu_X = 50$, $\sigma_x = 10$, $X_i = 64$, $z_i = ?$

4. $\overline{X} = 100$, $\hat{s}_x = 16$, $X_i = 132$, $z_i = ?$

5. $\overline{X} = 500$, $\hat{s}_x = 100$, $X_i = 320$, $z_i = ?$

In problems 6 through 12, a linear transformation is carried out from X to Y. For each variable, the fourteenth subject is Don D. Drain. The new values of $\overline{Y}$ and $\hat{s}_y$ are indicated. In each case determine what Don's transformed score is.

Problem	From:			To:		
	$\overline{X}$	$\hat{s}_x$	X_{14}	$\overline{Y}$	$\hat{s}_y$	Y_{14}
6.	29	7	18	34	7	
7.	21	12	27	35	20	
8.	8	6	10	17	30	
9.	20	10	15	14	6	
10.	50	6	71	18	2	
11.	41	12	56	128	32	
12.	23	8	1	87	20	

Problems 13 through 17. Go back to problems 6 through 10, and in each case find the equation for transforming from X to Y.

18. An instructor wishes to derive a course grade for each student on the basis of a midterm and a final, with the final counting $\frac{3}{4}$ of the total grade. Let X_i represent a midterm score and Y_i a final-exam score for subject i. If both score distributions have the same shape, what should his computing formula be?

Homework Answers

1. -1.25	**2.** -3.50

3. 1.40   **4.** 2.0

5. -1.80 **6.** 23   **7.** 45   **8.** 27

9. 11   **10.** 25   **11.** 168   **12.** 32

13. (6.) $Y = X + 5$ **14.** (7.) $Y = \frac{5}{3}X$

15. (8.) $Y = 5X - 23$ **16.** (9.) $Y = \frac{3}{5}X + 2$

17. (10.) $Y = \frac{1}{3}X + 34\frac{3}{4}$ **18.** Total $= \dfrac{X_i - \mu_X}{\sigma_X} + \dfrac{3(Y_i - \mu_Y)}{\sigma_Y}$

Theoretical
Distributions

I N this chapter the idea of a decision map breaks down. This is not because there are no connections among the various distributions, nor because they are not used for different kinds of problems—the decision rules for selecting a distribution to solve any of a range of problems will be given in later chapters, but they would only be confusing if presented here. Some of the interrelationships among the distributions presented will be indicated within the sections of this chapter, but they do not in themselves provide a basis for applying the distributions to problems.

The chapter is divided into eight sections, as follows:

If you want an understanding of the theoretical basis on which later chapters build, you can read this chapter in its entirety before going on. The intention of the chapter, however, is to serve to introduce the notion of theoretical distributions and to serve as a reference for later chapters. If you want to use the book in problem solving, or if you find theoretical distributions relatively uninteresting, read only the first three sections of this chapter and go on to Chapter 8. You can always come back to the rest of the chapter as you have a need for it.

7-1 Theoretical distributions

In Chapters 4 and 5 methods were discussed for graphing and characterizing empirical, or observed, distributions. In this chapter, by contrast, the concern is with purely theoretical, or a priori, distributions, which may or may not be approximated by any particular empirical distribution. Several distributions are introduced here preliminary to a discussion of probability, and later, hypothesis testing. The purpose is to indicate something about the characteristics and uses of each distribution, and the similarities and differences among the several distributions.

A major concern of this chapter is with the relationship between scores along a variable and areas under the graph of the distribution of the variable. In each of the distributions it will be indicated how you can use a formula or table to transform from scores to areas and back. Because areas under the graph of a distribution are used to represent proportions, probabilities, or frequencies of cases, these also may be related to scores on variables. The use of a known relationship between scores and areas under the graph of a distribution to effect a change of numbering is often called an *area transformation*. This set of relationships may be represented schematically as in Figure 7-1.

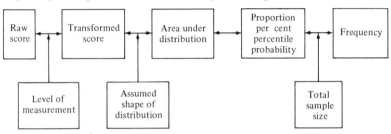

Figure 7-1 *Schematic representation of area transformation using a standard distribution.*

To use an area transformation, you first observe a raw score or set of raw scores. From this a transformed score such as z, t, χ^2, or F may be obtained, provided that the operations performed in obtaining the transformed score appear consistent with the assumed level of measurement (nominal, ordinal, interval, or ratio) of the raw scores. Under conditions to be specified in later chapters, transformed scores may be assumed to have one of the standard distributions discussed in this chapter. When this occurs, a table of the distribution may be used to convert one or more transformed scores into an area under the distribution (the best examples of this appear in Section 7-3A and B for the normal distribution). Areas under graphs may be interpreted as proportions, percents, percentiles, or probabilities, depending on your purpose (probability is defined and explained in Chapter 8). Given a total sample or population size, proportions, percents and probabilities may then be converted to frequencies of cases. Finally, you should note that the horizontal arrows in Figure 7-1 are two-headed, so that by making the appropriate assumption (level of measurement, shape, and sample size), you may work from any kind of information to any other.

This chapter presents standard distributions that represent whole families of distributions. Only one of each family will be discussed, and its properties will be indicated by formulas or tables. When you wish to know the properties of another distribution in the same family, you can perform a linear transformation

(Chapter 6) to put it into this standard form. The relations found through formulas or tables for the standard distribution will necessarily hold also for the distribution that you are interested in.

In most applications (some of which will be indicated later in this text), no attempt will be made to indicate a relationship between an observed distribution and one of the following theoretical distributions; rather, an assumption will be made that, because of such and such conditions, one of these can be used to solve a problem. (The considerations, unfortunately, must be left to later chapters.) However, if you should wish to test the similarity of a given empirical distribution to one of the distributions described here, you may use the methods of Chapter 10.

Under certain conditions, the sampling distribution of a sample characteristic can be specified a priori. Thus, if a population of raw scores is known to be normally distributed, and if a number of samples of size N are randomly and independently drawn from that population, then the means of those samples will tend to be normally distributed, variances will tend to have a chi-square distribution, and ratios of pairs of variances will tend to be distributed as F. Known relationships of this kind are the basis for some of the statistical tests to be considered in Chapters 9 through 18.

7-2 Rectangular distribution

A rectangular distribution occurs whenever there are no subjects with scores below some lower value, there are no subjects with scores above some second value, and subjects are evenly distributed across all scores between those two limits. A rectangular distribution of scores between the values 816 and 877 for some variable might be graphed as shown in Figure 7-2.

Figure 7-2 *Graph of a continuous rectangular distribution with real limits 816 and 877.*

The two most common sources of rectangular distributions are complete rank orders (where each score is given a different rank) and percentiles.

Consider an experiment in which a subject is required to rank five stimuli in order of preference; 1 indicates the stimulus he most prefers and 5 the one he likes least. No ties are to be permitted. The distribution might be represented as a bar graph (see Figure 7-3) where each box represents a stimulus. This distribution is clearly very similar to the previous one. It also would be similar in shape to a distribution in which the subject was required to put two, three, four, or any other number of stimuli in each category, as long as he was required to place the same number in each category.

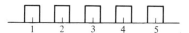

Figure 7-3 *Graph of a discrete rectangular distribution having five scale values.*

A percentile-rank distribution always places subjects according to the same rule: exactly 1 percent of the scores are required to fall between any two adjacent percentile ranks. The distribution is 100 units long (one for each percentile rank) and one percentage unit high, with a total area of 100 percent.

Any of the above distributions could be converted to a standard rectangular distribution, which would run, say, from 0 to 1, and have a height of 1 unit. We could tabulate values of this standard distribution (although it would not seem worth the effort except for didactic purpose) and then refer to the table to find percentages or proportions between specified values.

Figure 7-4 *Graph of a standard rectangular distribution.*

The table might look like Table 7-1, where we will call the score along the standard distribution R. This distribution can be graphed as in Figure 7-4.

Table 7-1

R	Area: 0 to R	Area: R to 1
.01	.01	.99
.02	.02	.98
.03	.03	.97
.04	.04	.96
.05	.05	.95

Values of R might be found that would correspond to any raw score X of a nonstandard, continuous rectangular distribution in the following way. (1) Subtract the lower real limit (X_{min}) of the distribution from each score in the distribution. If the length of the distribution of raw scores is L ($L = X_{max} - X_{min}$, where X_{max} is the upper real limit of the distribution), the transformed scores now run from 0 to L. (2) Divide these transformed scores by L to get R. The formula for R_1 which corresponds to an X_1 of the X distribution is as follows:

$$R_1 = \frac{X_1 - X_{min}}{X_{max} - X_{min}} \tag{7-1}$$

Let us now use the tabulated values of R to find percentages of cases under specified parts of the above standard distribution.

PROBLEM What percentage of subjects in the distribution of Table 7-1 have scores between .44 and .59 inclusive?

ANSWER Each difference of .01 represents one percentile, and the inclusive interval contains both end categories. Hence, if we assume that the distribution is continuous but not rounded, the percentage of subjects is $100\,(.59 - .44) = 15$ percent. If the data have been rounded to the nearest hundredth, then the real limits are $X_{max} = .595$ and $X_{min} = .435$, so the percentage in the inclusive interval is 16 percent.

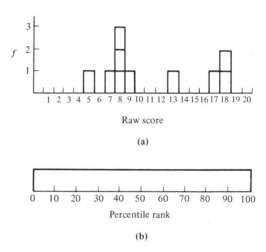

Raw score

(a)

Percentile rank

(b)

Figure 7-5 *A percentile-rank transformation.*

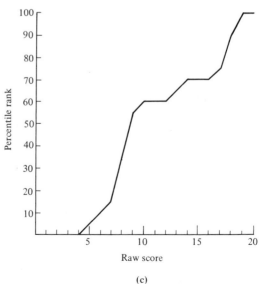

Raw score

(c)

Notice that a conversion from any nonrectangular distribution of scores to percentile ranks is a nonlinear area transformation. Figure 7-5 shows a raw-score distribution, a corresponding percentile-rank* distribution for the same set of scores, and a bivariate plot of the percentile-rank transformation. This third plot indicates that the relationship between the two sets of scores is not linear (Figure 7-5C).

Homework†

If a variable X is continuous and rectangularly distributed between the values $X = 0$ and $X = 80$ for some population, then
1. What proportion of scores in the population are greater than $X = 50$?
2. What proportion of scores fall between $X = 20$ and $X = 44$?
3. What value of X cuts off the lowest 5 percent of this distribution?
4. What two values of X cut off the most extreme 5 percent of this distribution ($2\frac{1}{2}$ percent at each end)?

7-3   Normal distribution

Relevant earlier section:
6-6   Standard scores

A normal distribution is bell-shaped: it has a high point at its center and trails off in a smooth curve in both directions, as indicated in Figure 7-6. The normal distribution is the most important single distribution for this text, as it represents the distribution of random error for continuous data. For example, if you were to measure the width of the room that you are now in 500 times with a 6-inch pocket ruler, and express the results to the nearest $\frac{1}{32}$ inch, you could find that the resultant estimate would not be the same on all tries. The graph of the 500 widths would be very nearly a normal distribution (and, unless you had made some systematic errors as well, the mean of the distribution would be your best estimate of the actual width of the room). Height and weight of same sex subjects who are either adults of the same age tend to be normally distributed, as are many other physical attributes. Intelligence and various special abilities tend

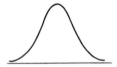

Figure 7-6   *A normal distribution.*

---

* Percentile ranks are calculated according to the method of Section 4-7B.
† Homework answers appear at the end of the chapter.

to be normally distributed; and scores on many personality tests also tend to follow this distribution. When samples are randomly drawn from a population, certain characteristics of the samples will tend to be normally distributed as a result of the randomness of sampling. Among these, the distributions of means and differences between means will be discussed in part C of this section, and other statistics will be discussed later in the book.

The normal distribution is convenient to deal with, because a complete specification of any normal distribution requires only that you specify the mean and standard deviation of the distribution. All normal distributions differ from one another only with respect to these two parameters. Therefore, it is possible to obtain any normal distribution from any other normal distribution by transformations on μ and σ, as indicated in Chapter 6.

The formula for the normal distribution is as follows:

$$y = \frac{1}{\sqrt{2\pi\sigma^2}} \cdot e^{-(1/2)z^2} \tag{7-2}$$

where y is the height of the curve, π is the ratio of the circumference of a circle to its diameter ($\pi = 3.14159265\ldots$), e is the base of the system of natural or Naperian logarithms ($e = 2.71828\ldots$), σ is the standard deviation of the distribution, and

$$z = \frac{X - \mu}{\sigma}$$

is a standard score corresponding to the raw score X, defined as in Section 6-5.

Fortunately, it will probably never be necessary for you to have a deal with this formula directly, because all the information you need to work with normal distributions has already been tabulated by thoughtful statisticians. Two forms of the table of the standard normal distribution appear as Tables A-4 and A-5 of the Appendix.

Notice that these tables are of the *standard* normal distribution, however, and are not directly applicable to every normal distribution. The standard normal distribution is a normal distribution (ND) that has a mean of zero and a standard deviation of 1. This can be abbreviated by saying that it is ND(0, 1), where the first number within the parentheses is taken by convention to be the value of the mean, and the second number is the value of the standard deviation. This table can also be used to determine the characteristics of any other normal distribution by first transforming that distribution to standard form using the z transformation of Section 6-5:

$$z_i = \frac{R_i - \mu_R}{\sigma_R} \tag{7-3}$$

where **R** is any variable. Of course, you could perform this transformation on each score of the original normal distribution with which you are working in order to obtain the standard normal distribution, but this generally would be unnecessarily time-consuming, if possible at all. Instead, you can find particular values of the original distribution that are useful, transform those values, come to some result through the table of the standard normal distribution, and then apply this result to the original distribution. Usually it is not necessary to transform more than one or two values of the original distribution to the standard distribution in order to solve a problem for which the normal distribution is applicable. For example, if you were given that a set of test scores X is ND(500, 100) and asked to find the number of subjects having scores between 545 and 619, you would only have to transform these two values in order to solve the problem:

$$z_{545} = \frac{545 - 500}{100} = \frac{45}{100} = 0.45 \quad \text{and} \quad z_{619} = \frac{619 - 500}{100} = \frac{119}{100} = 1.19$$

These z scores could then be converted to percentages using the table of the standard normal distribution, Table A-4, and the percentage to a frequency by multiplying by N.

A. Using the Normal Table (Table A-4)

Table A-4 lists areas under the normal distribution and the height of the distribution as functions of z scores. Column 1 lists positive z scores from .00 to 3.70. Because the normal distribution is symmetric, negative z's are not listed separately in order to save space. It is therefore up to you to keep track of the sign of z when you use the table and to insert it appropriately.

Columns 2 to 4 of Table A-4 list proportions of the total area under the curve for corresponding values of z. Column 2 lists the area between any given z and the mean of the distribution ($z = 0$). Figure 7-7(a) indicates the area represented by the value in column 2 of Table A-4 when z is negative. Figure 7-7(b) indicates the same value for a numerically identical positive z score. Because the normal distribution is symmetrical, the areas corresponding to positive and negative z's of the same magnitude are themselves of the same size. For example, when $z_1 = -1.26$, the area from z_1 to $z_0 = 0$ is .3962; when $z_2 = +1.26$, the area from z_2 to 0 is also .3962.

(a)   (b)

Figure 7-7 *Areas given by column 2 of Table A-4.* **(a)** *Area tabulated in column 2 for negative z_i (cross hatched).* **(b)** *Area tabulated in column 2 for positive z_i.*

Column 3 lists the area under the larger portion of the normal distribution cut off by any given z; this is diagrammed in Figure 7-8. When $z = +.88$, the larger area under the standard normal distribution is .8106; when $z = -.96$, the larger area under the distribution is .8315; and so on.

(a)   (b)

Figure 7-8 *Areas given by column 3 of Table A-4.* **(a)** *Area when z_i is negative.* **(b)** *Area when z_i is positive.*

Column 4 lists the *smaller* area under the normal distribution cut off by any given z, as indicated in Figure 7-9. When $z = -.62$, the smaller area under the distribution is .2676 (and the larger area is .7324; .2676 + .7324 = 1.00). When $z = +1.00$, the smaller area under the standard normal distribution is .1587.

(a)   (b)

Figure 7-9 *Areas given by column 4 of Table A-4.* **(a)** *Area when z_i is negative.* **(b)** *Area when z_i is positive.*

Finally, column 5 lists the height of the curve for any given z score, as shown by Figure 7-10. This **y** value is the value obtained from the formula for the normal distribution given earlier in this section. Again, because of the symmetry of the distribution, the height of the curve will be the same at corresponding positive and negative values. For example, when $z = +.09$, **y** $= .3973$; when $z = -.09$, **y** $= .3973$ also. More examples follow. Check these examples carefully, using paper and pencil, to make sure that you understand them and can do similar problems.

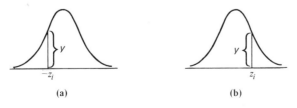

(a)   (b)

Figure 7-10 *Heights of a normal curve given by column 5 of Table A-4.*

1. If a variable is normally distributed, with a mean of 50 and a standard deviation of 10, (a) a raw score of 45 corresponds to a z score of $-.5$; a raw score of 60 corresponds to a z score of 1.0; and a raw score of 68 corresponds to a z score of 1.8. (b) from column 2 of Table A-4 we see that .3413, or 34.13 percent of subjects have scores between 50 and 60 (between z scores of 0 and 1); from column 3, we see that .6915, or 69.15 percent of cases are greater than 45 (a z score of $-.5$); and from column 4 we find that .0359 of cases are greater than 68. (c) In a sample of 200 subjects, we would then expect $200 \times 0.34 = 68$ scores between 50 and 60, $69 \times 200 = 138$ greater than 45, and 7 greater than 68 $[(.0359)(200) = 7.18]$.

2. A distribution of 200 scores is ND(250, 100). What score selects the 66 highest scores from the rest? First, $66/200 = .33$, or 33 percent. From column 4 of Table A-4, the z score corresponding to a smaller area of .33 is .44. Then

$$z = .44 = \frac{X - \mu}{\sigma} = \frac{X - 250}{100}$$

and

$$X = (.44)(100) + 250 = 250 + 44 = 294$$

B. More Complex Problems

Often problems require you to find a percentage or proportion of scores falling between two scores or outside two boundaries. Such problems require you to add or subtract two or more areas from Table A-4 to find a solution. Often such problems can be solved in more than one way, because columns 2, 3, and 4 of that table give equivalent information in only slightly different forms. Because the normal distribution is symmetric, the area to one side of the mean is equal to .5 of the total area under the distribution. Therefore, line by line, the values in column 3 of table A-4 are equal to the value in column 2 plus .5. Because the total area under the distribution is 1.0, the values in column 4 are equal to 1.0 minus the values in column 3 and .5 minus the values in column 2.

Consider a variable that is ND(500, 100). What proportion of cases fall between the values of 563 and 719?

Here again, we can convert from z scores to proportions by Table A-4. The z score corresponding to a raw score of 563 is $z = (563 - 500)/100 = .63$. Similarly, the z score corresponding to a raw score of 719 is $(719 - 500)/100 = 2.19$. Then from column 2 of Table A-4, the problem can be solved by subtraction.

$$p = .4857 - .2357 = .2500$$

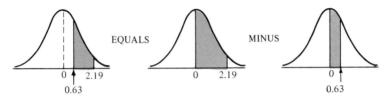

Alternatively, the problem can be solved using column 3.

$$p = .9857 - .7357 = .2500$$

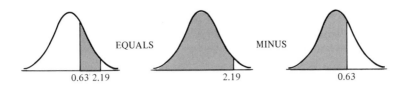

Finally, column 4 can be used to solve the problem.

$$p = .2643 - .0143 = .2500$$

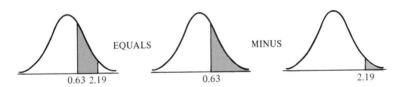

For the same distribution, the proportion of cases between raw scores of 400 and 563 can be found by using the z scores corresponding to the two values and Table A-4. A raw score of 400 corresponds to a z score of -1.0. Hence, the problem can be solved using column 2.

$$p = .3413 + .2357 = .5770$$

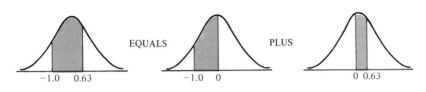

This problem can also be solved using columns 3 and 4 of Table A-4.

$$p = .8413 - .2643 = .5770$$

141

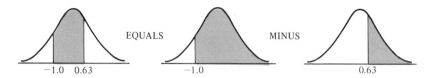

Now consider again the problem of page 138, in which you were asked to find the number of subjects having scores between 545 and 619, when X is ND(500, 100). The z scores obtained were .45 and 1.19. Complete the solution for this problem, showing that the required percentage is .2094 and, if $N = 800$, the expected frequency of scores in that region is 167.52, or approximately 168.

C. Use of the Normal Distribution with Means and Other Statistics

When a large number of samples are drawn from the same population and some statistic is calculated for each of the samples, the magnitude of that statistic will generally vary from sample to sample. We can then make a distribution of the statistic, where each entry in the table is obtained from a different sample. For many statistics, this distribution will be normally distributed, or approximately so, under specifiable conditions.

For example, suppose that we randomly drew 14,000 samples of size $N = 100$ from an infinite population in which X is ND(75, 12), and calculated $\overline{X}$ for each sample. Different samples would have different means, because they contain scores from different subjects; but all 14,000 sample means would be estimates of the same population mean μ. We could record the means in a frequency distribution and graph the distribution with a polygon. The sample means would be approximately normally distributed, with a mean of means $\overline{\overline{X}} \doteq 75$ and a standard deviation (called the *standard error or the mean*) $\hat{s}_{\overline{X}} \doteq 1.20$.

The approximation of our obtained distribution to a normal distribution would be very close but not exact. The following conditions contribute to the closeness of the approximation: (a) the raw scores X are normally distributed in the population; (b) the samples are randomly drawn from the population; and (c) the samples are large. Of these, the most important condition is (b). If it holds, then either (a) or (c) will be sufficient to cause the means of samples to be approximately ND.

When sample means are normally distributed, the mean of sample means is given by the formula

$$\mu_{\overline{X}} \doteq \mu_X$$

and the standard error of the mean by the formula

$$\sigma_{\overline{X}} \doteq \frac{\sigma_X}{\sqrt{N}}$$

where μ_X and σ_X are the mean and standard deviation of the raw scores in the population, and N is the size of each of the samples. Note that in order for you to apply these formulas, all the samples must be of the same size.

This is one of the most important principles underlying many statistical analyses. It is called the *Central Limit Theorem*; and it can be stated, somewhat more formally, as follows. If a population has a mean μ_X and variance σ_X^2, and if random samples of size N are drawn, then as N increases, the distribution of sample means $\overline{X}$ approaches a normal distribution with mean $\mu_{\overline{X}} = \mu_X$ and variance $\sigma_{\overline{X}}^2 = \sigma_X^2/N$.

Now, sample means can be dealt with in the same way as normally distributed raw scores, using the methods of this section. For example, suppose that X is ND(500, 100) and we draw a large number of samples of size $N = 16$ randomly from this population. Then the sample means have as their theoretical mean $\mu_{\overline{X}} = \mu_X = 500$ and as their standard error $\sigma_{\overline{X}} = \sigma_X/\sqrt{N} = 100/\sqrt{16} = 25$. We can specify the proportion of sample means that we expect to fall in any region of the score distribution, using the table of the standard normal distribution. To find the proportion of sample means we can expect to be greater than or equal to 515, we first convert 515 to a z score in the distribution of sample means:

$$z(515) = \frac{515 - 500}{25} = \frac{15}{25} = +.60$$

Then we can look in Table A-4 to find the proportion of the standard normal distribution falling above a z score of .60, which is .2743. Thus, when X is ND(500, 100) and we randomly draw samples of size $N = 16$ from the population, we can expect 27.43 percent of the sample means to be greater than or equal to 515. Alternatively, we can say that the probability of a sample mean being greater than or equal to 515, in this problem, is .2743.

The difference between sample means $(\overline{X}_1 - \overline{X}_2)$ from two samples is also normally distributed, under the same conditions that the sample mean is normally distributed, with a mean

$$\mu_{\overline{X}_1 - \overline{X}_2} = \mu_{X_1} - \mu_{X_2}$$

and standard error

$$\sigma_{\overline{X}_1 - \overline{X}_2} = \sqrt{\frac{\sigma_{X_1}^2}{N_1} + \frac{\sigma_{X_2}^2}{N_2}}$$

where

$\mu_{\overline{X}_1 - \overline{X}_2}$ is the average or expected difference in sample means

μ_{X_1} is the mean of raw scores for population 1 and μ_{X_2} is the mean of raw scores for population 2

$\sigma_{\bar{X}_1 - \bar{X}_2}$ is the standard error of the difference in sample means

σ_{X_1} is the variance of raw scores for population 1 and σ_{X_2} is the variance of raw scores for population 2

N_1 and N_2 are the two sample sizes

EXAMPLE We know that for the general population of adult Americans, scores on the Weltanschauung test are normally distributed, with a mean of 50 and a standard deviation of 10 for both men and women. We randomly draw a sample of $N_m = 32$ men and another of $N_w = 32$ women from the general population of adult Americans, asking, "What is the probability that the women's average will be 5 or more points greater than the men's average?"

SOLUTION Our statistic is $\bar{X}_w - \bar{X}_m$, and we want to know the probability that it is equal to or greater than 5. If we were to randomly draw a large number of pairs of samples, like the pair we actually have drawn, we would expect the average difference between means to be zero:

$$\mu_{\bar{X}_w - \bar{X}_m} = 50 - 50 = 0$$

and the standard error of the difference to be

$$\sigma_{\bar{X}_w - \bar{X}_m} = \sqrt{\frac{100}{32} + \frac{100}{32}} = \sqrt{\frac{200}{32}} = \sqrt{\frac{25}{4}} = \frac{5}{2} = 2.5$$

Now we can form a z score for the statistic of interest:

$$z = \frac{(\bar{X}_w - \bar{X}_m) - \mu_{\bar{X}_w - \bar{X}_m}}{\sigma_{\bar{X}_w - \bar{X}_m}}$$

$$= \frac{5 - 0}{2.5} = 2.0$$

Since $\bar{X}_w - \bar{X}_m$ is normally distributed, this z statistic is also normally distributed, and we can obtain the probability from Table A-4 of the Appendix. The probability is .0228.

A number of other statistics are also normally distributed. However, discussion of these applications will be left to later chapters.

D. Normalizing a Distribution

In the method of normalizing a distribution to be demonstrated here, we assume that the measurement scale is interval level and ask what normal distribution is most closely approximated by the observed distribution. That is, what would the frequency be at each score if the observations were as nearly normal as possible? The result of normalization is a set of expected frequencies, one for each score, which can be used by themselves or compared to the observed frequencies to test for normality of the observed distribution (e.g., see Section 10-3, in which observed and binomial distributions are compared).

The procedure, roughly, is as follows. (1) Find the mean and standard deviation of the observed distribution. (2) Find the boundaries of each score category or interval. (3) Convert these boundaries to z scores using the method of Section 6-6. (4) Use the z scores to find the expected proportion of the distribution in each score interval. (5) Convert the expected proportions to frequencies using the total number of subjects in the sample.

EXAMPLE An observed distribution is given in Table 7-2, with the raw scores X in the first column and the frequency at each raw score in the second column.

The mean and standard deviation of the original distribution were estimated at $\bar{X} = 10.29$ and $\hat{s}_X = 2.66$. The upper real limit of each score interval is given in the third column of Table 7-2 as X_u. It is converted to a z-score estimate using the sample estimates of the mean and standard deviation. Thus, for example,

$$z_u(X = 16) \doteq \frac{16.5 - 10.29}{2.66} \doteq 2.33$$

Table 7-2 Normalization of an observed distribution

X	Observed frequency f, at X	Upper bound of category X		Expected proportion below z_u	Expected proportion at X	Expected frequency at X
		X_u	z_u			
16	1	16.5	2.33	.9901	.0151	1.0
15	3	15.5	1.96	.9750	.0321	2.1
14	5	14.5	1.58	.9429	.0560	3.6
13	6	13.5	1.21	.8869	.0902	5.9
12	6	12.5	.83	.7967	.1195	7.8
11	9	11.5	.46	.6772	.1453	9.5
10	10	10.5	.08	.5319	.1498	9.8
9	7	9.5	− .30	.3821	.1307	8.5
8	8	8.5	− .67	.2514	.1045	6.8
7	4	7.5	−1.05	.1469	.0691	4.5
6	4	6.5	−1.42	.0778	.0419	2.7
5	2	5.5	−1.80	.0359	.0213	1.4
4	0	4.5	−2.18	.0146		

The expected proportion below each z score is found from Table A-4 of the appendix. The expected proportion at a given X score is found by subtracting the expected proportion below its lower real limit (X_u for the next lower category) from the expected proportion below its upper limit, X_u. Thus the expected proportion at $X = 16$ is .9901 − .9750 = .0151.

Finally, the expected proportion at a given score is converted to an expected frequency by multiplying it by the total number of subjects (65 for the distribution of Table 7-2).

The observed and expected frequencies from Table 7-2 have been graphed together in Figure 7-11. We can see from this graph that the distribution of expected frequencies is nore nearly normal in shape, although some slight departures from normality occur because of rounding errors.

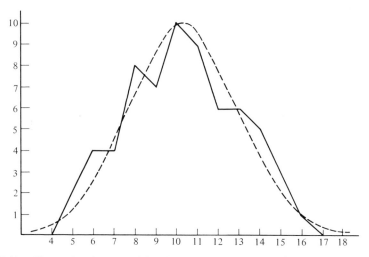

Figure 7-11 *Observed and expected distributions from Table 7-2.*

References   Discussions of the normal distribution can be found in Blommers and Lindquist (1960), 177–193; Dixon and Massey (1969), Chapter 5; Games and Klare (1967), 168–197; and Guilford (1965), 124–132.

Homework

5. The Purdue Test of Mental Acuity is normally distributed, with a mean of 850 and a standard deviation of 100 for college students. Harold O'Dawn gets a score of 894. What is his percentile rank with respect to other college students?

6. A variable is known to be normally distributed in the population, with mean $\mu = 500$ and standard deviation $\sigma = 100$. If a sample of 468 people is randomly drawn from the population, how many people do you expect to have scores between 563 and 719?

7. Find the semi-interquartile range for a normal distribution of scores having a mean of 83 and a standard deviation of 10.

8. A normal distribution has a mean of 4 and a standard deviation of 5. If a large number N of cases is randomly drawn, it is expected that 350 will fall between 1.75 and 5.35 (raw scores). What is the total number N of cases that we are planning to draw from this distribution?

9. Random samples of size $N = 10$ are drawn from a normally distributed population having mean $\mu = 50$ and standard deviation $\sigma = 12$.

(a) What proportion of such samples would you expect to have means greater than 52?

(b) What proportion of such samples should have means less than 47?

(c) What z score cuts off the upper 5 percent of the standard normal distribution? Use this with the formula for standard error of the mean to determine the raw score that separates the 5 percent of largest means from the rest in the present sampling experiment.

7-4 Distribution of Student's t

Relevant earlier section:
7-3 Normal distribution

The t distribution is the distribution by means of which we estimate the position of a score or statistic with respect to other scores or statistics when (1) we are willing to assume that the population is normally distributed, but (2) the sample is small, and (3) we do not know the standard-error parameter (see Section 7-3C) and must estimate it from limited sample information. t has the same general form as z except that the standard error has been replaced by an estimate of it, as indicated in the following:

$$t_i = \frac{\mathbf{R}_i - \mu_{\mathbf{R}}}{\hat{s}_{\mathbf{R}}} \tag{7-4}$$

where $\mathbf{R}$ is a normally distributed variable with mean $\mu_{\mathbf{R}}$ and $\hat{s}_{\mathbf{R}}$ is the sample estimate of its standard error.

It is possible to estimate the standard error with different degrees of precision. Because of this, there is no single distribution of t; rather, it constitutes a whole family of distributions that become more and more similar to a normal distribution as the number of independent observations underlying the estimate $\hat{s}_{\mathbf{R}}$ of $\sigma_{\mathbf{R}}$ gets larger. The number of independent observations has already been called the number of degrees of freedom (Sections 5-12 and 7-8). Thus, there is a different distribution of t for each number of degrees of freedom, but in practice the difference between distributions of t becomes negligible when the degrees of freedom (df) are greater than, say, 30 to 50. In Figure 7-12, a t distribution with 7 df is superimposed on the standard normal distribution for comparison.

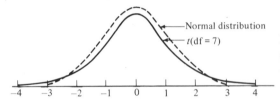

Figure 7-12 *Comparison of a t distribution with the standard normal distribution.*

Probably the most important feature of t to note is that its variance and standard deviation are greater than those of the standard normal distribution. In fact, its variance is equal to $df/(df - 2)$, when $df > 2$; thus the variance ranges from 3.0 when $df = 3$ to 1.0 as $df \to \infty$. Because the t distribution has a greater variance than the standard normal distribution, the same numerical t score and z score may have different meanings. For example, a t score of 1.86 in a distribution for 8 df cuts off a smaller area of .05, whereas a z score of 1.86 cuts off only .0314 of the standard normal distribution. The difference in areas (.0186) is not large compared to the total area of 1.00 under the distribution, but it *is* large compared to the areas under both distributions cut off by the score. We will be very interested in these small areas. The smaller the number of degrees of freedom, the less extreme a given value of t is with respect to its appropriate distribution; hence the greater the need to use a t distribution with the correct degrees of freedom.

A. Use of the t Table

Table A-6 of the Appendix gives t scores corresponding to certain areas under the distributions of t for degrees of freedom ranging from 1 to 30, then for 40, 60, and 120 df. Also given for comparison are z scores that cut off these same areas under the standard normal distribution.

Each row of the table gives values for a different t distribution, except for the last row, which represents the standard normal distribution. Columns represent areas or proportions under the distributions, and entries in the body of the table are scale values of the variables (t or z).

It is clear from an examination of the z line of Table A-6 that it contains much less information about the standard normal distribution than does Table A-4 or Table A-5. It has been necessary to omit information about each distribution in order to tabulate comparable information about a number of different ones.

There are two labels for each column. One gives a value for a "one-tail test" and the other gives a value for a "two-tail test." The column labeled .05 for a one-tail test gives, for each of the distributions tabulated, the score beyond which .05 of the distribution falls. An example is given in Figure 7-13 to show

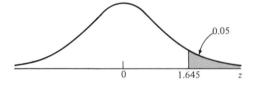

Figure 7-13 *Illustration that a z score of 1,645 cuts off the upper 0.05 of a standard normal distribution.*

that, for the standard normal distribution, .05 of the area falls to the right of a z score of 1.645 (this value can also be confirmed by reference to Table A-3). In Figure 7-14, it is shown that .05 of the area under the same distribution

Figure 7-14 *Illustration that a z score of −1.645 cuts off the lower 0.05 of a standard normal distribution.*

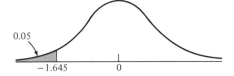

falls to the left of a z score of −1.645. Since the standard normal and t distributions are symmetrical, and if we consider both tails of the distribution simultaneously, we can say that .10 of the distributions fall beyond values of ±1.645 (see Figure 7-15). This is the meaning of the term "two-tail test."

Figure 7-15 *Illustration that z scores of ±1.645 cut off a total of 0.10 of a standard normal distribution.*

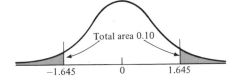

The same considerations apply to the other distributions of t in this table. For example, in a t distribution with 14 df, .025 of the distribution lies to the right of a t score of 2.145, .01 lies to the right of a t score of 2.624, .05 lies to the left of a t score of −1.761, .01 lies outside the values ±2.977, and so on.

B. Two Applications

The t distribution can be used with a single score, to see if it is very unusual, provided an estimate of the raw-score standard deviation is available. Then, in formula 7-4, you can replace **R** by X_i, giving

$$t = \frac{X_i - \mu_X}{\hat{s}_X} \tag{7-4a}$$

and the number of degrees of freedom for the test is equal to the number of degrees of freedom in the estimate $\hat{s}_X$. This is in most cases equal to $N - 1$, as indicated in Section 5-12, where N is the number of scores on which $\hat{s}_X$ is computed.

EXAMPLE Suppose that you have just completed a pilot version of a psychological test, the "Tendency to Project" test, and have obtained scores for 30 randomly chosen undergraduates at your college. The sample mean is calculated to be 25 (which is exactly what you intended it to be) and the sample standard deviation is 5.4 (which you had not predicted). A graph of the score distribution looks about normal, considering how few scores you have. You now want to canvass the college for students with

149

unusually high (upper 5 percent) scores on this test. The critical value of t cutting off the upper 5 percent of the distribution for $N - 1 = 29$ df is 1.699. You can substitute this value in formula 7-4a and solve for X:

$$1.699 = \frac{X_i - 25}{5.4}$$

$$X = (5.4)(1.699) + 25 \doteq 34.17$$

Therefore, in your survey you will look for students with scores greater than 34.

A more common use of t is in comparing the mean of a sample of scores to a hypothetical or population mean, when the population standard deviation is not known. In this case, you replace **R** in formula 7-4 by $\overline{X}$, giving

$$t = \frac{\overline{X} - \mu_{\overline{X}}}{\hat{s}_{\overline{X}}}$$

This formula allows you to compare a particular $\overline{X}$ to the entire distribution of $\overline{X}$'s for samples randomly drawn from some population. The sample means have a mean $\mu_{\overline{X}}$ and standard deviation estimated by $\hat{s}_{\overline{X}} = \hat{s}_X/\sqrt{N}$. The t statistic allows us to indicate how unusual a particular $\overline{X}$ is, compared to the entire distribution of $\overline{X}$'s that we would get by randomly drawing samples of size N from the population. The number of degrees of freedom for the statistic is equal to the number of independent scores used to obtain $\hat{s}_{\overline{X}}$, which in turn is equal to the number of independent scores used to obtain $\hat{s}_X$ (usually $N - 1$). We never actually obtain the distribution of $\overline{X}$'s, and so never calculate $\hat{s}_X$ directly, because the amount of work involved is prohibitive.

EXAMPLE You know from the test manual that fifth graders have an average of 68 on the Standard Test of Mathematics Achievement, but the test author has failed to include the standard deviation of scores. You want to know whether the fifth graders in your class do unusually well on this test: in the upper 1 percent of fifth-grade classes. There are 25 students in your class; their mean on the test is 71.4 and their standard deviation is 6.0. In order for this sample mean to be in the upper 1 percent of sample means, randomly drawn from all fifth-grade classes, it would have to give a t (at 24 degrees of freedom) greater than 2.492 (check this using Table A-6). From your data, $\overline{X} = 71.4$, $\mu_X = 68$, $\hat{s}_{\overline{X}} = 6.0$, $\hat{s}_{\overline{X}} = 6/\sqrt{25} = 6/5 = 1.2$, and so

$$t = \frac{71.4 - 68}{1.2} = \frac{3.4}{1.2} \doteq 2.83$$

This value of t is greater than the value that cuts off the upper 1 percent of the t distribution, so your class is in the upper 1 percent of fifth-grade classes on the test.

Homework

10. What value of t cuts off the lowest 1 percent of a t distribution having 16 degrees of freedom?

11. What value of t cuts off the upper 5 percent of a t distribution having 7 df?

12. What values of t cut off the most extreme 5 percent ($2\frac{1}{2}$ percent in each tail) of a t distribution with 12 df?

13. Suppose that we draw a large number of samples of size $N = 16$ from a population having a raw score mean of 55. One such sample has a mean of 50 and an unbiased standard deviation of 8.0.

    (a) What is our best estimate of the population standard deviation of raw scores, from these data?

    (b) What is our best guess of the standard error (standard deviation) of the sample means?

    (c) What value of t cuts off the lowest 5 percent of sample means in this distribution of sample means?

    (d) Is it reasonable to claim that this sample's mean is in the lowest 5 percent of sample means? Compute t to answer this question.

7-5 Binomial distribution

The binomial distribution, like the t, is very closely related to the normal distribution. Whereas the normal distribution is the theoretical distribution of random error for a continuous variable, the binomial distribution is the theoretical distribution of random error when the variable is discrete (see Section 4-4B), that is, when the variable can only be taken on whole-number values or linear transformations of them.

In a binomial distribution the proportion of the distribution for each value of the variable can be found by expanding the expression $(p + q)^n$, where p and q are themselves proportions or fractions adding to 1 ($p + q = 1.0$). When $n = 2$, $(p + q)^n = (p + q)^2 = p^2 + 2pq + q^2$. In this case, the distribution has three levels; the proportion of the distribution for the first level is given by p^2, for the second level by $2pq$, and for the third by q^2. If $p = .4$, then q must equal $1.0 - .4 = .6$; $p^2 = (.4)^2 = .16$; $2pq = 2(.4)(.6) = .48$; and $q^2 = (.6)^2 = .36$. The graph of the binomial represents the values of these three terms as areas. Thus, for this example, the binomial distribution whose areas correspond to the values of the three terms of the binomial expansion when $p = .4$ is graphed in Figure 7-16. Here the bars have two labels. The first is the algebraic formula for the term of the expansion represented by the bar, and the second is the exponent of q in the corresponding formula, where

151

$q^1 = q$ and $q^0 = 1$. Note that if $p + q = 1.0$, then $(p + q)^n = 1.0$, too, because 1.0 raised to any power is still 1.0. Thus $(p + q)^2 = 1.0$, as exemplified by the above proportions: $.16 + .48 + .36 = 1.00$. This also implies that the total area under the graph of any binomial distribution is equal to 1.0.

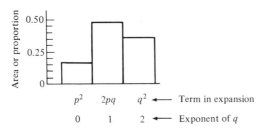

Figure 7-16 *Graph of a binomial distribution when $n = 2$ and $p = 0.4$.*

When $n = 3$, $(p + q)^n = (p + q)^3 = p^3 + 3p^2q + 3pq^2 + q^3$. Then if $p = .5$, $q = .5$, $p^3 = .125$, $3p^2q = .375$, $3pq^2 = .375$, and $q^3 = .125$. As a check we see that $.125 + .375 + .375 + .125 = 1.000$. The distribution may be graphed as in Figure 7-17.

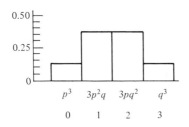

Figure 7-17 *Graph of a binomial distribution when $n = 3$ and $p = 0.5$.*

In general, the expression $(p + q)^n$ may be expanded to

$$(p + q)^n = (p^n) + (np^{n-1}q) + \left[\frac{n(n-1)}{2} p^{n-2}q^2\right]$$
$$+ \left[\frac{n(n-1)(n-2)}{3 \cdot 2} p^{n-3}q^3\right] + \cdots \qquad (7\text{-}5)$$

the last term of which is always q^n. This distribution has a mean $\mu = np$, a standard deviation $\sigma = \sqrt{npq}$, and skewness

$$SK = \frac{p - q}{\sqrt{npq}} = \frac{p - q}{\sigma}$$

Because $q = 1 - p$ and therefore is totally dependent on p, the binomial distribution is completely defined by two parameters, n and p.

The first of the distributions above had $n = 2$, $p = .4$, so its parameters were $\mu = np = 2(.4) = .8$, $\sigma = \sqrt{npq} = \sqrt{2(.4)(.6)} = \sqrt{.48} \doteq .693$, and and $SK = (p - q)/\sigma = (.4 - .6)/.693 \doteq -.289$. The second distribution had

$n = 3$ and $p = q = .5$, so its parameters were $\mu = 3(.5) = 1.5$, $\sigma = \sqrt{3(.5)(.5)} \doteq .866$ and $SK = 0/.866 = 0$.

At the introductory level, the binomial distribution is used almost exclusively with probability problems. For this reason, primary discussion of it will be left for Chapter 8.

Homework

14. What is the magnitude of the third term in the binomial distribution, when $n = 3$ and $p = .40$?

15. What are the mean and standard deviation for a binomial distribution having $n = 16$ and $p = .36$?

7-6 Chi square

Relevant earlier sections:
5-11 Population standard deviation
5-12 Sample standard deviation
6-5 Standard scores
7-3 Normal distribution

A. Definition of Chi Square

The distribution of chi square (χ^2) is also closely related to the standard normal distribution in the following manner. Suppose that you drew a single value from a standard normal distribution at random and squared it. The result would be a single value of chi square:

$$\chi^2 = z^2 \tag{7-6}$$

Of course, the same could be done with any normal distribution, because any normal distribution can be converted to the standard normal distribution by computing z scores (Section 7-3). Therefore, for any normal distribution

$$\chi^2 = z^2 = \frac{(X - \mu_X)^2}{\sigma_X^2}$$

Chi square, as just defined, can be considered to be a curvilinear transformation of z. However, there is a sense in which chi square is a more general statistic that includes z as a particular case. A value of chi square may also be formed by adding together two z^2 scores drawn randomly and independently from the same normal distribution.

$$\chi^2 = z_1^2 + z_2^2 = \frac{(X_1 - \mu_X)^2}{\sigma_X^2} + \frac{(X_2 - \mu_X)^2}{\sigma_X^2}$$

$$\sum_{i=1}^{2} z_i^2 = \sum_{i=1}^{2} \frac{(X_i - \mu_X)^2}{\sigma_X^2} = \frac{1}{\sigma_X^2} \sum_{i=1}^{2} (X_i - \mu_X)^2$$

In general, a value of chi square can be formed from the addition of any number of z^2 scores drawn randomly and independently from the same normal distribution:

$$\sum_{i=1}^{N} z_i^2 = \sum_{i=1}^{N} \frac{(X_i - \mu_X)^2}{\sigma_X^2} \tag{7-7}$$

Values of chi square based on different numbers of z scores are not directly comparable, however, as might be expected from the fact that they are based on different amounts of information about the normal distribution from which they are drawn. Therefore, it is always necessary when discussing chi square to indicate the number of degrees of freedom, or independent observations, on which a particular value of the statistic is based. A chi square based on one z score has one degree of freedom, one based on two independent z scores has two degrees of freedom, and so on. A chi square based on N independent, randomly drawn values from a normal distribution is said to have N degrees of freedom.

An important use of chi square is suggested by a substitution of the formula for z into the chi-square formula.

$$\chi^2[a \text{ df}] = \sum_{i=1}^{a} z_i^2 \qquad \text{where} \qquad z_i = \frac{X_i - \mu}{\sigma}$$

$$\chi^2[a \text{ df}] = \sum_{i=1}^{a} \left(\frac{X_i - \mu}{\sigma}\right)^2$$

$$= \frac{\sum_{i=1}^{a} (X_i - \mu)^2}{\sigma^2}$$

because σ^2 is a constant.

Then dividing both numerator and denominator by a,

$$\chi^2[a \text{ df}] = \frac{\sum_{i=1}^{a} (X_i - \mu)^2/a}{\sigma^2/a}$$

But the numerator of this expression is just the unbiased sample estimate of

σ^2 (it is unbiased because deviations are taken about μ instead of $\overline{X}$; see Section 5-12).

$$\chi^2[a\,\mathrm{df}] = \frac{\hat{s}_X^2}{\sigma_X^2/a} \qquad \text{and} \qquad \chi^2[a\,\mathrm{df}] = a\!\left(\frac{\hat{s}_X^2}{\sigma_X^2}\right) \tag{7-8}$$

This formula suggests that χ^2 can be used to compare an observed variance estimate with a theoretical variance, an application that will be demonstrated in Chapter 11.

Like the normal distribution, the chi-square distribution is a theoretical abstraction that is most accurately defined mathematically. However, a close approximation to a chi-square distribution for 1 degree of freedom can be obtained by randomly drawing a large number of scores from a normal distribution, finding a value of chi square for each score, and then graphing the resulting values of chi square. For each raw score, the corresponding value of chi square is given by the formula

$$\chi_j^2[1\,\mathrm{df}] = z_j^2$$

The theoretically determined distribution is given in Figure 7-18.

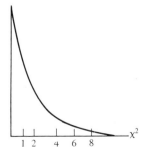

Figure 7-18   *Distribution of chi square for one degree of freedom.*

The equivalence of the standard normal distribution and the chi-square distribution with 1 df can be exemplified by comparison of tabulated values of chi square from Table A-7 with corresponding values of z from Table A-4 (see Table 7-3). When we show the equivalence of the two tables, we must remember that z_j^2 may be the square of either a positive or a negative z score. Thus, the area above some value of chi square is equal to the area above the corresponding positive z score plus the area below the corresponding negative z score. To show equivalence of the tables, we will compare the value of chi square above which some proportion α of the total area appears and the square of the value of z above which half that area ($\alpha/2$) appears in the standard normal distribution. A comparison of the χ^2 and z^2 columns shows them to be identical (except for rounding error).

Table 7-3   *Comparison of chi square and standard normal distributions*

α, upper proportion	$\chi^2[1\ df]$	α/2, upper proportion	$z_{\alpha/2}$	$z^2_{\alpha/2}$
.001	10.828 †	.0005	3.2905	10.827
.01	6.63490	.005	2.5758	6.6347
.05	3.84146	.025	1.9600	3.8416
.10	2.70554 †	.05	1.6449	2.7056
.25	1.32330 †	.125	1.1503	1.3232
.50	.454936 †	.25	.6745	.4550
.75	.1015308 †	.375	.3186	.1015
.90	.0157908 †	.45	.1257	.0158
.95	.0039321	.475	.0627	.003931

† These values were taken from the more complete Pearson and Hartley table (1966), 136–137.

The distribution of chi square for 2 df might be found by drawing a large number of *pairs* of scores from a normal distribution, calculating a value of chi square for each pair of scores, then plotting the values of chi square. For each pair of scores,

$$\chi^2_j\ [2\ df] = z^2_{1j} + z^2_{2j}$$

where j is index for pairs of scores. The resultant graph is given in Figure 7-19.

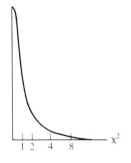

Figure 7-19   *Distribution of chi square for two degrees of freedom.*

Chi square for more than 2 df is found simply by adding together more independent scores from the normal distribution:

$$\chi^2_j[N\ df] = \sum_{i=1}^{N} z^2_{ij}$$

where each value of j refers to a different set of N scores.

Graphs for several chi-square distributions with different degrees of freedom are superimposed in Figure 7-20.

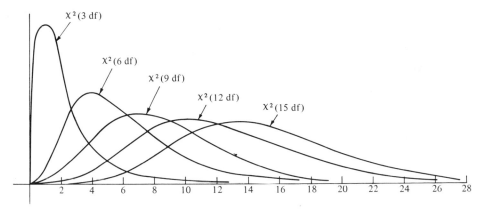

Figure 7-20 *Distributions of chi square for several different degrees of freedom.*

From the distributions of chi square it is, first of all, apparent that chi square is always positive. This is because all the elements of any chi square are squared. Whether a z score is positive or negative, its square will always be positive. Second, as the number of degrees of freedom changes, so do the centrality, variability, and shape of the distribution. The theoretical mean of a chi-square distribution with more than 2 df is equal to its number of degrees of freedom, and the mode is equal to df -2. The fact that the two centrality measures are not equal implies that the distribution must always be skewed. However, the magnitude of skewness decreases with increased degrees of freedom, as can be seen from Figure 7-20. It is therefore necessary to have tabulated values of chi square which differ for different numbers of degrees of freedom.

B. Use of Chi-Square Table for 30 df or Less (Table A-7)

Table A-7 of the Appendix gives values of chi square corresponding to various areas under the distribution for 37 different chi-square distributions, from $df = 1$ to df $= 30$, and for several selected values of degrees of freedom beyond 30. For any distribution, the column value indicates the proportion of the graph of that distribution which is more extreme than the value that appears in the body of the table for that row and column. Because of the most frequent uses for such a table in statistics, only some of the possible areas have been tabulated. When you use the table, degrees of freedom are generally given by the data that you are working with. You usually want to know the value of chi square for the distribution that cuts off a given proportion of the distribution in one or both tails.

Figure 7-21 is the graph of a chi-square distribution for 11 df. Two values are of particular interest on this graph: $\chi^2_{.05}(11$ df, lower tail$) = 4.57481$ is the value that cuts off the lower .05 of the area of the distribution, and $\chi^2_{.05}(df$ 11, upper tail$) = 19.6751$ is the value that cuts off the upper .05 of the distribution. These values were taken from Table A-7 of the Appendix.

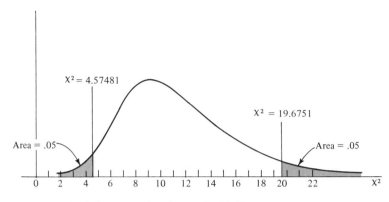

Figure 7-21   *Graph of chi-square distribution for 11 df.*

The left-hand column of Table A-7 gives the number of degrees of freedom. Both of the values of chi square above appear on line 11. In the body of the table are found the values of chi square. The left half of the table has values in the lower tail, and the right half has values in the upper tail. Alpha (α) is the proportion of the distribution to be cut off by the value of chi square found in the table: both values of chi square in Figure 7-21 are for $\alpha = .05$, the one-tail test.

The two values of chi square (4.57481 and 19.6751) considered together cut off .10 of the distribution in both tails (.05 + .05). Therefore, they are the values used for a two-tail test when alpha is .10.

Similarly, you can verify from Table A-7 that $\chi^2 = 13.0905$ cuts off the lower .05 of a chi-square distribution with 23 df; $\chi^2 = 45.5585$ cuts off the upper .005 of a chi-square distribution with 24 df; and $\chi^2 = 1.73493$ and $\chi^2 = 23.5894$ cut off the extreme .01 of a chi-square distribution with 9 df.

C. Normal Approximation When Degrees of Freedom Are Greater Than 30
Although values of chi square are given in Table A-7 for up to 100 df, it is also possible to use a normal approximation to chi square when df > 30. You may get a feeling for the appropriateness of this approximation by looking at Figure 7-20, which shows that as the number of degrees of freedom increases, the distribution of chi square becomes more nearly normal.

To use the normal approximation, transform χ^2 using the formula

$$z = \sqrt{2\chi^2} - \sqrt{2(\mathrm{df}) - 1} \qquad (7\text{-}9)$$

and compare the resultant z score to the standard normal distribution.

EXAMPLE   From Table A-7, we know that $\chi^2 = 90.5312$ cuts off the upper 5 percent of the chi-square distribution for 70 df. Let us see if we can obtain a similar result from the normal approximation. We can do this by taking $\chi^2 = 90.5312$ as obtained with 70 df, transforming it using formula 7-9, and determining what portion of the standard normal distribution it cuts off. The transformed value is

$$z = \sqrt{2(90.5312)} - \sqrt{2(70) - 1}$$

$$= \sqrt{181.0624} - \sqrt{139}$$

$$\doteq \sqrt{181} - \sqrt{139} = 13.4536 - 11.7898$$

$$\doteq 1.6638$$

When we compare this value to Table A-4 of the standard normal distribution, we find that it cuts off between .0485 ($z = 1.66$) and .0475 ($z = 1.67$) of the distribution. Thus, using the normal approximation, we would conclude that with 70 df a χ^2 of 90.5312 cuts off about the upper .048 of the chi-square distribution. This is slightly erroneous, but the approximation is good enough for most purposes. Its value lies in allowing you to use chi square for degrees of freedom and α levels other than those found in Table A-7.

References   Chi square is discussed in Guilford (1965), 227–234; Hays (1963), 336–343, 347–348; and McNemar (1969), 250–255.

Homework

16. What value of chi square cuts off the upper 5 percent of a chi-square distribution with 4 df? With 12 df? With 28 df?
17. What value of chi square cuts off the lowest 1 percent of a chi-square distribution with 3 df? With 18 df? With 60 df?
18. What values of chi square cut off the most extreme 5 percent of a chi-square distribution with 7 df?
19. A chi-square statistic is to be formed from 15 z scores. If there are 8 constraints on the data, what is the lower bound of the 95th percentile for the appropriate theoretical distribution?
20. A population of scores has an unknown mean but is known to have a variance of 11. A large number of samples are randomly drawn and an unbiased sample variance is computed for each. If each sample has 23 subjects, what variance would you expect to cut off the 2 percent of samples having the smallest variances?

7-7   *F* distribution

Relevant earlier section:
7-6   Chi square

The statistic F is defined as the ratio of two chi-square distributions, each divided by its own degrees of freedom:

$$F[a, b \text{ df}] = \frac{\chi^2(a)/a}{\chi^2(b)/b} \tag{7-10}$$

But $\chi^2(a) = a \cdot \hat{s}_a^2/\sigma_a^2$, and similarly for $\chi^2(b)$. Therefore, substituting in formula 7-10 and dividing, we obtain

$$F[a, b \text{ df}] = \frac{\hat{s}_a^2/\sigma_a^2}{\hat{s}_b^2/\sigma_b^2}$$

When the two samples are, in fact, drawn from populations having the same variance, $\sigma_a^2 = \sigma_b^2$, so

$$F[a, b \text{ df}] = \frac{\hat{s}_a^2}{\hat{s}_b^2} \tag{7-11}$$

The statistic is used to test whether two observed variances could have been computed on samples drawn from the same population. Notice that, in contrast to the t and chi-square distributions, F is dependent on two different degrees of freedom: one for the numerator and one for the denominator.

A. Sources and Uses of *F* Scores

The two variance estimates used to calculate the F ratio may be obtained from two different samples. When this is done, it is usually for the purpose of testing whether the two samples are drawn from populations with the same variance. This use will be discussed in detail in Chapter 11.

The variances used in the F ratio may also be obtained from the means and variances of several samples in analysis of variance. This procedure, which is used to determine whether several samples could have come from populations having the same mean, will be discussed in Chapter 13.

B. Graphic Representation of *F* Distribution

Because each F ratio is a function of two independent chi-square distributions, there must be a different distribution of F for each *pair* of degrees of freedom. In general, these distributions are not symmetric, resembling distributions of chi square in shape. In Figure 7-22 two different F distributions have been sketched. Notice that there are no negative values of F; this is because both numerator and denominator are squared terms, and therefore positive.

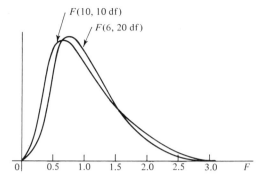

Figure 7-22 *Distributions of F[10, 10 df] and F[6, 20 df].*

C. Tables for the *F* Distribution (Table A-8)

Because of the large number of *F* distributions about which we might need information, only four values are tabulated for each distribution. All four tables are arranged in the same way. The number of degrees of freedom for the numerator of an *F* ratio is found by looking across the top margin of the table, and the number of degrees for the denominator by looking down the left-hand margin. The desired *F* distribution has its entry at the intersection of row and column corresponding to its degrees of freedom. Each of the four tables is used for a certain proportion α of the distribution, and the entry for a given *F* distribution indicates what *F* score cuts off the upper α of that distribution. Table A-8a indicates for each *F* distribution what *F* score cuts off the upper .05 of the area of the distribution, Table A-8b shows which score cuts off the upper .025 of each distribution, and so on.

It is also possible to find the values of *F* cutting off the lower .05, .025, .01, and .005 of each distribution. To find the value of $F[a, b \text{ df}]$ below which α percent of the distribution falls, find the value above which α percent of the distribution of $F[b, a \text{ df}]$ falls, and take its reciprocal. For example, the upper .01 of the distribution $F[9, 10 \text{ df}]$ is cut off by $F_{.01}[9, 10 \text{ df}] = .949$; but the lower .01 of the distribution of $F[9, 10 \text{ df}]$ is cut off by $F_{.99}[9, 10 \text{ df}] = 1/F_{.01}[10, 9 \text{ df}] = 1/5.26 = .19$.

D. Relation of *F* to the Other Theoretical Distributions

Relation of F to chi square It has already been shown (part A, above) that the *F* distribution is formed as the ratio of two chi-square distributions, each divided by its own degrees of freedom. If *b*, the degrees of freedom of the denominator, is infinite, then $\hat{s}_b^2 = \sigma_b^2$. From this it follows that

$$F[a, \infty \text{ df}] = \frac{\hat{s}_a^2/\sigma_a^2}{\sigma_b^2/\sigma_b^2} = \frac{\hat{s}_a^2}{\sigma_a^2}$$

But this is distributed as $\chi^2[a \text{ df}]/a$. Therefore, the line of the *F* table for infinite degrees of freedom in the denominator (the last row of the table) should have the same entries as the corresponding values of chi square, divided by the number of degrees of freedom. Excerpts from these tables are compared in Table 7-4.

Table 7-4   *Comparison of selected values of $\chi^2[a\ df]$ and $F[a, \infty\ df]$, upper area = .05*

a	$\chi^2[a\ df]$	$\chi^2[a\ df]/a$	$F[a, \infty\ df]$
1	3.841	3.84	3.84
2	5.991	2.99	2.99
3	7.815	2.60	2.60
4	9.488	2.37	2.37
⋮	⋮	⋮	⋮
30	43.773	1.46	1.46

Relations of F to t   From the definition of t in Section 7-4, we know that

$$t^2[b\ df] = \frac{x^2}{\hat{s}_X^2}$$

where the b df denote that b independent values are used to estimate σ_X^2 with $\hat{s}_X^2$. Let us now divide both numerator and denominator by σ^2:

$$t^2[b\ df] = \frac{x^2/\sigma^2}{\hat{s}_X^2/\sigma^2}$$

But by definition, the numerator is equal to z^2 (Section 6-5), which is in turn equal to a chi square with 1 df, from Section 7-6.

$$\frac{x^2}{\sigma^2} = \chi^2[1\ df]$$

Also from Section 7-6, we know that the denominator is a chi-square variable:

$$\hat{s}^2/\sigma^2 = \frac{\chi^2[b\ df]}{b}$$

Therefore,

$$t^2[b\ df] = \frac{\chi^2[1\ df]/1}{\chi^2[b\ df]/b}$$

which by definition (formula 7-10) is equal to an F ratio with 1 and b df. An example will be given in Section 13-3, to show that either a t or an F distribution may be used to test the hypothesis that two populations have the same mean value. For the present, it is interesting to note that tabulated values of the F distribution for 1 and b degrees are equal to the squares of corresponding tabulated values in the t distribution (two-tailed, for b df). Some of these values have been given in Table 7-5.

Table 7-5 *Comparison of values of $t[b$ df] and $F[1, b$ df] cutting off the upper 1 percent and 5 percent of their distributions*

b	$t_{.01}[b$ df]	t^2	$F_{.01}[1, b$ df]	$t_{.05}[b$ df]	t^2	$F_{.05}[1, b$ df]
1	63.647	4052	4052	12.706	161.4	161.4
2	9.925	98.50	98.50	4.303	18.51	18.51
3	5.841	34.12	34.12	3.183	10.13	10.13
4	4.604	21.20	21.20	2.776	7.71	7.71
5	4.032	16.26	16.26	2.571	6.61	6.61
⋮	⋮	⋮	⋮	⋮	⋮	⋮
20	2.845	8.10	8.10	2.082	4.35	4.35
⋮	⋮	⋮	⋮	⋮	⋮	⋮
30	2.750	7.56	7.56	2.042	4.17	4.17

References The F distribution is discussed by Hays (1963), 348–351; and McNemar (1969), 283, 286–287.

Homework

21. What value of F cuts off the upper 5 percent of an F distribution with 5 and 7 df? With 7 and 24 df? The upper 1 percent of an F distribution with 9 and 14 df?

22. What value of F cuts off the lower 5 percent of an F distribution with 5 and 8 df? With 7 and 20 df?

23. An F ratio is to be formed with a numerator variance of 7.0 and 10 df, and a denominator variance of 5.0, with 8 df. Does the observed F fall in the upper 5 percent of the appropriate F distribution?

24. In the second example of Section 7-4, a t was calculated for determining whether a sample mean was unusually large. The computed t was 2.87 and there were 24 df. Suppose that you have no t table, and complete the example to find out if the sample is in the upper 1 percent of samples, using the F tables.

7-8 Degrees of freedom

The distributions of t, χ^2, and F are distinguished from the remaining distributions in this chapter by their greater dependence on sample values. In order to make use of these distributions, you must first estimate the parameters on which they depend, on the basis of the sample data.

For example, the formula for t differs from that for z only in that it utilizes an estimate of the standard error in place of the exact parameter value demanded by z. There is greater variability to the t distribution because the standard error is known but must be estimated from sample data: the greater variability comes from the greater uncertainty we have when using t.

The amount of additional error contributed by the estimation of a parameter is inversely related to the size of the sample on which the estimate is based: the larger the sample, the less error there is. Statistics such as t, chi square, and F take into account the amount of additional error through the concept of degrees of freedom.

The number of degrees of freedom (df) is equal to the number of independent scores that are used to estimate a parameter. In the case of t, the number of degrees of freedom is equal to the number of degrees of freedom for the standard error, which depends on the formula used (see Chapter 12). If a standard error is calculated from the scores of N subjects, all of whom are members of the same sample, then $df = N - 1$, the number of independent scores used in calculating the standard deviation (as indicated in Section 5-12). Alternatively, the number of degrees of freedom is equal to the number of scores minus the number of constraints on the scores. An unbiased standard deviation, $\hat{s} = \sqrt{\sum x^2/(N-1)}$, has one constraint on the scores: that the deviations (x) sum to zero (i.e., that the deviations are taken about the sample mean). Then for the t formula

$$t = \frac{X - \mu_0}{\hat{s}_X}$$

there are $N - 1$ df because the unbiased estimate $\hat{s}_X$ used in the standard error is based on $N - 1$ independent measures. If a standard error is estimated from data collected in two separate samples of sizes N_1 and N_2, it is usually obtained by estimating the standard deviation from each sample, then combining the estimates. There are a total of $N_1 + N_2$ observations on which the combined estimate is based, but each sample estimate is taken about the mean of that sample. One degree of freedom is lost for each sample, so the combined estimate is based on $N_1 + N_2 - 2$ df.

The chi-square test for comparing a sample variance to a population variance,

$$\chi^2 = \frac{(N_e - 1)\hat{s}_e^2}{\sigma_0^2} \qquad \text{(Section 11-1)}$$

has $N - 1$ df because the statistic $\hat{s}_e^2$ is based on a sample of size N, but the same constraint, $\sum x = 0$, is needed to compute the variance estimate. When, in Section 12-12, the t distribution is used to compare two independent samples to determine whether their means are equal, the number of degrees of freedom is the total number of independent measures in both samples. When it is used with difference scores (Section 12-9), the number of degrees of freedom is the number of independent difference scores. When chi square is used to compare distributions, the basic data are frequencies of subjects at each score in the distribution. In this application, the degrees of freedom are the number of independent frequencies (for a more detailed discussion, see Section 10-6). At this point it is impossible to give more specific directions to allow you to determine the degrees of freedom for all applications of these statistics. Instead, the appropriate number of degrees of freedom will be indicated in conjunction with each application.

Homework Answers

1. $3/8 = .375$ **2.** $.30$ **3.** 4.0 **4.** 2.0 and 78.0 **5.** 67 **6.** 117

7. 6.7 **8.** 1250 **9.** **(a)** $.30$ **(b)** $.21$ **(c)** 1.64; 56.2

10. -2.583 **11.** 1.895 **12.** 2.179; -2.179

13. **(a)** 8.0 **(b)** $8/\sqrt{16} = 2.0$ **(c)** $t_{.05}$(lower tail) $= -1.753$

   (d) $t = \dfrac{50 - 55}{2} = -2.5$; yes, it is reasonable.

14. $.432$ **15.** $\mu = 5.76$; $\sigma = 1.92$ **16.** 9.48773; 21.0261; 41.3371

17. $.114832$; 7.01491; 37.4849 **18.** 1.68987; 16.0128 **19.** 14.0671

20. 5.49 **21.** 3.97; 2.42; 4.03 **22.** $.2075$; $.2907$

23. $F_{obs} = 1.4$; $F_{.05} = 3.35$; no.

24. $F = t^2 \doteq 8.23$; $F_{.01}(1, 24 \text{ df}) = 7.82$; yes, it is.

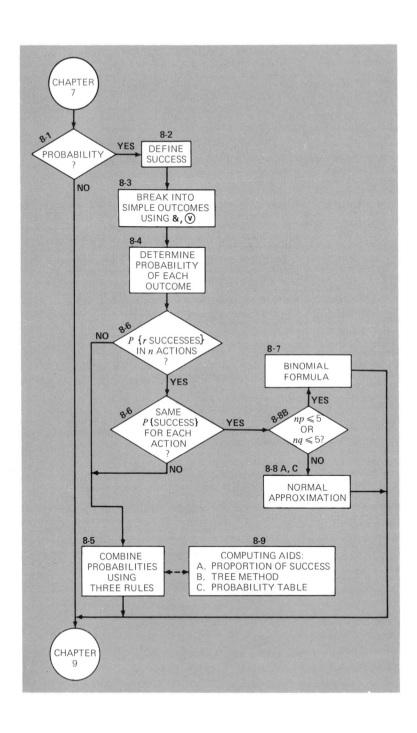

8 Probability

PROBABILITY is a concept used in the prediction of future observations. The methods for studying probability presented in this chapter may either be used directly to predict subject performance, or indirectly, through various inferential techniques that will be examined in subsequent chapters.

An understanding of the conception of probability is necessary background for the chapters on hypothesis testing (Chapters 9 through 14). Chapter 8 defines probability and relates it to areas under the graph of a distribution. In Chapter 9, probability will be related to inferences about populations; in subsequent chapters, probabilities will be associated with most of the distributions discussed in Chapter 7, in order to make the needed inferences.

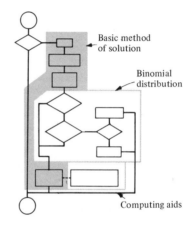

Basic method of solution

Binomial distribution

Computing aids

Sections 8-2 through 8-5 (the left-hand column of the decision map) indicate the sequence of steps to be followed in solving the most elementary problems in probability. Section 8-9 (computing aids) provides shortcut solutions to these same problems.

For more complicated problems, the binomial distribution may provide a solution. Two approaches to the binomial distribution are possible: an exact solution, and an approximation that can be used under conditions specified in Section 8-8B.

8-1 Probability?

Elementary probability theory is used to estimate the relative frequency with which an as yet unobserved result can be expected to occur, or to indicate the degree of confidence one can have that on a single future occasion, the effect will be observed. It is used, in other words, as a basis for extrapolating from what is already known about events and their occurrence and for making statements about our degree of confidence in the occurrence of specified but

167

presently unobservable effects. For example, from our knowledge of the relative frequencies of heads and tails on a single coin in the past, we can predict, with a great deal of accuracy, the results of tossing this same coin in the future, or on occasions in the past that we have not been able to observe directly.

There are some problems in which you wish to know the probability of a particular result as an aid to decision making. For example, you can figure betting odds as the ratio of the probability that a result will occur to the probability that it won't occur; if you like to gamble, this knowledge might be very useful. Or you may wish to estimate the probability of an accident with a certain piece of equipment to determine whether additional safety devices should be developed. Problems in the direct calculation of probabilities will be considered in this chapter.

Another use of probability is as a background against which to evaluate particular observations. For example, a score of 40 on a 100-item test has no meaning in itself; but if the test is composed of true–false items, then the score represents performance below chance. The probability of getting each item right by guessing is .50, and we would expect a person who only guessed on the test to get about 50 items right. On the other hand, a score of 40 would be twice as good as chance, if the items were five-option multiple-choice items. In the latter case, the probability of getting each item right by guessing would be .20; and we would expect a person to get at least 20 items right by guessing alone.

In most behavioral research, we wish to make statements about population parameters; for example, "Low-anxiety subjects obtain higher scores on test X than high-anxiety subjects." This refers to the population of all low-anxiety subjects and the population of all high-anxiety subjects. However, we can never observe the entire populations, so we must infer their relative status on the basis of sample data. Further, we cannot expect sample results to be identical to population results, because chance factors enter into the selection of sample subjects. We cannot make any statements with certainty about population values on the basis of observations of samples; however, we can determine the probability with which our sample results are in agreement with the results that we would observe if we could test the entire populations. This application of probability notions in hypothesis testing is introduced in Chapter 9 and appears throughout the remaining chapters.

References See Blommers and Lindquist (1960), 193–198; and Games and Klare (1967), 198–203.

8-2   Defining success

In order to solve problems in elementary probability, you will need to be able to state each one and to break it down into components that can be solved.

Then you must recombine the results that you have obtained to arrive at a final result or conclusion.

In many problems, there will be several levels of analysis, and you will need to organize your results at each level in order to arrive at a final solution. Suppose that you are asked to find the probability of getting three or more heads in six tosses of an unbiased coin. The simplest level of analysis would be to determine the probability of heads in a single tossing of a coin; the next level, to determine the probability of getting, say, three heads in six tosses; and at the broadest level, to determine the probability of getting either 3 or 4 or 5 or 6 heads in tossing the coin six times. Although we could define "success" on any of these levels, in this chapter it will be reserved for the broadest level only. This use of "success" has the advantage of reminding you to recombine results from subanalyses and to complete the problem.

Success (S) is therefore defined arbitrarily, by the problem, as that which you wish to find the probability of. If you want to know the probability of getting a 7 on a given roll of a pair of dice, success equals getting a 7 on that roll of those dice; if you wish to figure the probability that a given nag will come in last in a horse race, success equals having that nag come in last (even though the horse's owner might have a different view of the matter).

8-3   Breaking the problem into actions and outcomes

A. Some Definitions

Most of the problems in elementary probability theory cannot be solved by simple inspection, so a general technique has been devised for solving such problems. The problem is first analyzed into components, called *actions*. Each action must have two or more possible *outcomes*, or results. Next, the outcomes that contribute to success are identified, and if necessary, each outcome may be analyzed into more elementary actions and outcomes. At some point in the analysis, the actions and outcomes may be simple enough so that we can estimate their probabilities of occurrence. At this point, it is possible to combine the separate probabilities of the various simple outcomes, using the formulas of probability theory, to find the probability of the complex outcome for the original problem.

As defined here, an *action* is something that is done; an outcome is a possible consequence of an action. Actions and outcomes can be either simple or complex, depending on the problem and the stage of the analysis at which you are focussing. In the example of Section 8-2, we wanted the probability of obtaining 3 or more heads in 6 tosses of an unbiased coin. At the broadest level of analysis, the action in this problem consists in tossing the coin 6 times, and there are 2 possible outcomes: getting 3 or more heads, or getting fewer than 3 heads. At the next level of analysis, there are 7 possible outcomes: no heads, 1 head, 2 heads, 3 heads, and so on.

At a yet finer level of analysis, the outcome 3 heads can be seen to consist of 20 different possible outcomes, one of which is heads-tails-heads-heads-tails-tails.

We can also focus on single tosses of the coin as the actions, for each of which the outcome is either a head or a tail. Through our analysis, we can derive the probability of 3 or more heads in 6 tosses from the probabilities of combinations of individual heads and tails. We can also deduce the probability of heads and tails on each toss of the coin from the fact that it is unbiased. Then, inserting the probabilities of individual heads and tails into the formula we have found, we can compute the overall probability we want.

In order for elementary probability theory to be applicable to a problem, the outcomes of each action must be *mutually exclusive* and *exhaustive*, and the tions must be *independent*.

Consider an action A with outcomes a_1 or a_2; the two outcomes are *mutually exclusive* if and only if the occurrence of either a_1 or a_2 necessarily precludes the occurrence of the other. In tossing a coin (A), the outcomes "head" (a_1) and "tail" (a_2) are mutually exclusive because it is impossible for both sides to be up at the same time. In rolling a single die (B), the outcome "3 or less" (b_1) and the outcome "4 or more" (b_2) are mutually exclusive, also; however, the outcomes "3 or less" and "3 or more" are not, since the occurrence of a 3 could be classified as an instance of either outcome.

The outcomes of an action are *exhaustive* if and only if any observed result can be classified as an instance of one of them. When you roll a single die, the outcomes "3 or less" and "4 or more" are exhaustive. The outcomes "3 or less" and "5 or more" are not exhaustive, however, because a 4 could not be counted as an instance of either and is therefore unclassifiable. On a questionnaire that has true–false questions, the outcomes "true" and "false" to a question are exhaustive only if each respondent must give an answer to the question.

Two actions are *independent* if and only if the outcomes of one action in no way whatsoever influence the probabilities of the outcomes of the other. If two people toss two coins at the same time, the actions can probably be considered to be independent, but it seems highly unlikely, for example, that a young man's choice of a date for a Friday night is entirely independent of the choice he made the previous week.

B. Formal Method for the Solution of Probability Problems

Once a statement of success has been made, the next step in the formal method is to analyze it into an exhaustive set of mutually exclusive outcomes of one or more independent actions. In describing this method, uppercase letters will be used to designate actions and corresponding lowercase letters, subscripted, will be used to represent the simple outcomes of those actions. For example, if the actions are called A and B, their outcomes might be a_1, a_2, and a_3 of A and b_1 and b_2 of B. Two logical operators, & and ⓥ, will be used: a_1 & b_1 is read "both a_1 and b_1"; a_1 ⓥ a_2 is read "either a_1 or a_2 but not both."

When specific actions are contemplated, labeling of outcomes may use mnemonic aids; for example, the outcomes of tossing a coin are h and t, the outcomes of running a rat in a T maze are r and l turns, and so on.

In order to simplify the writing of probability statements, a special notation has been developed. The words "the probability of" are replaced by a letter P with brackets. The outcome for which the probability is being stated is placed, in abbreviated form, in the brackets. For example, the statement "The probability of heads in a single toss of this coin is $\frac{1}{2}$" can be written $P\{h\} = \frac{1}{2}$. "The probability of getting four heads in six tosses of this coin is .234" can be written $P\{4h|6 \text{ tosses}\} = .234$. "The probability that the rat will turn right at a choice point is .5" can be written $P\{r\} = .50$.

As an example of simplifying probability statements, consider tossing two coins, where success (S) is defined as getting one head and one tail. Success can be broken down into either getting a head on the first coin and a tail on the second, or getting a tail on the first and a head on the second. This can be written

$$\text{Success} = (h_1 \ \& \ t_2) \ \textcircled{v} \ (t_1 \ \& \ h_2)$$

and

$$P\{S\} = P\{(h_1 \ \& \ t_2) \ \textcircled{v} \ (t_1 \ \& \ h_2)\}$$

This may be further abbreviated by omitting the subscripts and the &'s:

$$P\{S\} = P\{ht \ \textcircled{v} \ th\}$$

Consider a second example: "Find the probability of getting a 5 on a single roll of a pair of dice." Success can be achieved in any one of several ways. You can get a 1 on the first die and a 4 on the second, a 2 on the first and a 3 on the second, a 3 on the first and a 2 on the second, or a 4 on the first and a 1 on the second. The probability of success can thus be written

$$P\{S\} = P\{1_1 \ \& \ 4_2) \ \textcircled{v} \ (2_1 \ \& \ 3_2) \ \textcircled{v} \ (3_1 \ \& \ 2_2) \ \textcircled{v} \ (4_1 \ \& \ 1_2)\}$$

or, simplifying,

$$P\{S\} = P\{1, 4 \ \textcircled{v} \ 2, 3 \ \textcircled{v} \ 3, 2 \ \textcircled{v} \ 4, 1\}$$

where in this case each pair of scores has been divided by a comma so that it will not appear to be a single, two-digit number.

8-4 Determining the probability of each outcome

To be able to assign a probability value to an outcome that is the combination of several simple outcomes, it is necessary to be able first to assign probability values to each of the simple outcomes.

Probability values are expressed similarly to proportions. If something cannot possibly happen, it is assigned the probability 0; if an outcome cannot fail to occur, it is assigned a probability of occurrence of 1.0. All other probability values fall within the range from zero to 1.

You are always free to estimate probabilities by blind guessing or astrology or any other method you wish, but usually one of two methods is followed: the a priori rational method or the empirical method.

In the a priori method, some simplifying assumption is made and the probabilities are derived on the basis of the simplifying assumption and the number and nature of the possible outcomes. Examples of such simplifying assumptions are "The coin is unbiased," "It is a fair die," and "The animal's behavior is random." These three assumptions are different ways of expressing the hypothesis that all outcomes of a particular action are equally probable. An unbiased coin has two sides, each of which is equally likely to come up on a given toss of the coin. Therefore, we can assign the probability of $\frac{1}{2}$ or .50 to both "heads" and "tails". A fair die has six sides, each of which is equally likely to come up on a given roll of the die. Therefore, the a priori probability of each side is $\frac{1}{6}$. If the behavior of a rat in a maze is random, then at each choice point he will have a .50 probability of turning right and a .50 probability of turning left.

In the empirical method of estimating probabilities, past experience is used as a basis for estimating the probability of a future or otherwise unknown outcome. To use this method to find the probability of heads for a given coin, you would toss the coin a large number of times and use the proportion of heads on those trials as the best estimate of the probability of heads on any subsequent toss of the same coin. If a rat has turned right in 65 of the last 100 choice points in a maze, then if all known independent variables remain constant, your best estimate of his probability of a right turn at the next choice point is .65.

Here a word of caution is in order. When an action is repeated, the probabilities of its outcomes may vary from trial to trial, and you should be wary lest you use an inappropriate estimate of the probability of an outcome for a successive trial. For example, if you draw cards from a well-shuffled deck but do not replace the cards already drawn before selecting additional ones, probabilities shift. Consider the problem of drawing three spades in three successive draws. For the first draw, there will be 13 spades out of 52 cards; for the second draw if the first draw was successful there will be only 12 spades and 51 cards; while for the third draw there will be 11 spades and 50 cards in the deck. Therefore, the probability of a spade on each draw is different: $\frac{13}{52} = .25$ on the first draw, $\frac{12}{51} = .235$ for the second, and $\frac{11}{50} = .22$ for the third.

The same problem can occur in many different forms. In a learning experiment involving a rat in a T maze or a monkey with two doors or a college student with two buttons, behavior will be changed by the reinforcement schedule followed, and the probabilities of the responses may shift from trial to trial.

8-5 Three rules for combining probabilities

The following three rules comprise a basis for solving all the probability problems that you will encounter in this text, although for many problems other methods of solution may prove more efficient. These three rules are stated and illustrated below.

Rule I If a_1 and a_2 are mutually exclusive outcomes of action A, then $P\{a_1 \text{ ⓥ } a_2\} = P\{a_1\} + P\{a_2\}$.

EXAMPLE If a die is fair, and if the probability of each side coming up is $\frac{1}{6}$ (as indicated in Section 8-4), then the probability of either a 3 or a 4 coming up is $\frac{1}{6} + \frac{1}{6} = \frac{1}{3}$. Note that the 3 and 4 sides are mutually exclusive, because it is not possible for both to come up on top at the same time. Action A in this case is rolling the die once; the outcomes are $a_1 =$ the 3 comes up and $a_2 =$ the 4 comes up. The 3 and 4 are mutually exclusive, because it is not possible for both to come up on top at the same time.

Rule II If $a_1, a_2, \ldots, a_n$ are all the mutually exclusive outcomes of action A (i.e., they comprise an exhaustive set of outcomes), then

$$P\{a_1 \text{ ⓥ } a_2 \text{ ⓥ} \cdots \text{ ⓥ } a_n\} = 1.0$$

Since a probability of 1.0 represents certainty, this rule says that you are certain that one of the outcomes listed is going to occur as the result of the action. For example, when you toss a coin, you ordinarily are certain that it will fall with either the head or the tail side up. Thus you say: "The probability of getting either a head or a tail on tossing this coin once is 1." The action A is tossing the coin once, and the outcomes are $a_1 =$ head, $a_2 =$ tail. They are mutually exclusive, because you can't get both heads and tails at the same time. The probability statement can be written $P\{h \text{ ⓥ } t\} = 1.0$. Rules I and II can be used to find the probability of heads on a single toss of an unbiased coin. By rule II, $P\{h \text{ ⓥ } t\} = 1.0$. Then, by rule I, $P\{h\} + P\{t\} = P\{h \text{ ⓥ } t\} = 1.0$. But the coin is assumed to be unbiased; hence $P\{h\} = P\{t\}$. Substituting equals for equals in the previous equation gives $P\{h\} + P\{h\} = 1.0$. Therefore, $2P\{h\} = 1.0$ and $P\{h\} = \frac{1}{2}$.

Similar reasoning can be used to find the probability of a 2 on a single roll of a fair die. There are six sides to the die, so $P\{1\} + P\{2\} + P\{3\} + P\{4\} + P\{5\} + P\{6\} = 1.0$. Because the die is assumed fair, all six sides have an equal probability of showing on any throw: $P\{1\} = P\{2\} = \cdots = P\{6\}$, from which it follows that $6 \cdot P\{2\} = 1.0$ and $P\{2\} = \frac{1}{6}$.

This general principle holds any time that all outcomes of an action are mutually exclusive and equally probable. If there are n such outcomes, the probability that any one of them will occur is $\frac{1}{n}$.

Rules I and II are applicable only if the outcomes are mutually exclusive. However, in some cases where the outcomes of an action are not mutually exclusive, the problem can be restated so that an equivalent action with mutually exclusive outcomes is used. For example, we might be tempted to answer the question, "What is the probability of getting a 3 on the first roll or a 4 on the second roll of a fair die?" in the following *wrong* way: $P\{3 \text{ ⓥ } 4\} = P\{3\} + P\{4\} = \frac{1}{6} + \frac{1}{6} = \frac{1}{3}$. This would be wrong, because getting a 3 on one roll of the die is not mutually exclusive with getting a 4 on the next roll: you could get both the 3 and the 4. (Actually there is another error here, too, because the 3 and the 4 are not outcomes of the same action A: they are outcomes of different rolls of the die.) However, if we think of our action A as rolling the die *twice*, we can restate the problem so that it can be solved. In this case, we can define four mutually exclusive outcomes: (a) a 3 on the first roll and a 4 on the second; (b) a 3 on the first roll and something other than a 4 on the second; (c) something other than a 3 on the first roll and a 4 on the second; and (d) neither a 3 on the first roll nor a 4 on the second. Either (b) or (c) would be a solution, and they are mutually exclusive; so we can use rule I: $P\{b \text{ ⓥ } c\} = P\{b\} + P\{c\}$. This problem is still not solved, however, because we haven't yet found $P\{b\}$ or $P\{c\}$. To do so, we need another rule.

Rule III If A and B are independent actions and if a_1 is an outcome of A and b_1 is an outcome of B, then $P\{a_1 \text{ \& } b_1\} = P\{a_1\} \cdot P\{b_1\}$.

Rule III is sometimes called the *multiplicative rule for combining probabilities*.

EXAMPLE Let us go back to the previous example and find the probability of getting a 3 on the first roll and a 4 on the second roll of a fair die. We can interpret rule III as follows: $A =$ the first roll; $a_1 =$ getting a 3; $B =$ the second roll; and $b_1 =$ getting a 4. Then P 3(first roll) & 4(second roll) $= P\{3(\text{first roll})\} \cdot P\{4(\text{second roll})\} = (\frac{1}{6})(\frac{1}{6}) = \frac{1}{36}$. Rule III applies because it is reasonable to assume that neither roll of the die influences the other: they are independent. Now let us simplify the notation and complete the earlier problem. Let's call something-other-than-3, not-3, and similarly for not-4; and let's use subscripts to indicate the first and second rolls. From rule II, $P\{\text{not-3}\} = 1.0 - \frac{1}{6} = \frac{5}{6}$, and in the same way, $P\{\text{not-4}\} = \frac{5}{6}$. Now, applying rule III to outcome (b) of the previous example, we get

$$P\{3_1 \text{ \& not-}4_2\} = P\{3_1\} \cdot \{P \text{ not-}4_2\}$$
$$= \frac{1}{6}(\frac{5}{6})$$
$$= \frac{5}{36}$$

In the same way, applying rule III to outcome (c) gives a probability of $\frac{5}{36}$; and for outcome (d) we get

$$P\{\text{not-}3_1 \text{ \& not-}4_2\} = P\{\text{not-}3_1\} \cdot P\{\text{not-}4_2\}$$
$$= \tfrac{5}{6}(\tfrac{5}{6}) = \tfrac{25}{36}$$

Now we can verify that the four outcomes are exhaustive, using rule II, because their probabilities add to 1: $\frac{1}{36} + \frac{5}{36} + \frac{5}{36} + \frac{25}{36} = \frac{36}{36} = 1.00$. Finally, we can complete the example by adding the probabilities for parts (b) and (c):

$$P\{b \text{ ⓥ } c\} = P\{b\} + P\{c\}$$
$$= \tfrac{5}{36} + \tfrac{5}{36}$$
$$= \tfrac{10}{36} \doteq .278$$

As another example of the use of rule III, consider the following problem: An unbiased coin is tossed and a fair die is rolled at the same time. What is the probability of getting a tail on the coin and a 5 on the die? The solution is as follows:

$$P\{t \text{ \& } 5\} = P\{t\} \cdot P\{5\} \quad \text{(by rule III)}$$
$$= \tfrac{1}{2}(\tfrac{1}{6}) \qquad \text{(since both are unbiased)}$$
$$= \tfrac{1}{12}$$

8-6   Finding the probability of r occurrences of an outcome in n repetitions of an action

Relevant earlier section:
7-5   Binomial distribution

In many of the most interesting and important problems in elementary probability, it is theoretically possible but a long and tedious job to analyze "success" into all the component elements that can go into making it up. For example, if you are asked to find the probability that exactly four out of eight rats will turn right in a T maze when the probability of a right turn is .60 for each rat, you will find yourself writing out 70 different outcomes, each of which is a compound of eight simple outcomes. Not only would the task be tedious; your chances of making a mistake in carrying it out would also be great. And this is a relatively simple problem compared to many that you may face.

Fortunately, there is a way of solving some problems of this kind which is economical and provides an exact solution. Under some conditions, you can use the *binomial formula*.

First, it must be possible to state the problem in the form, "What is the probability of getting r occurrences of the same outcome in n independent repetitions of the same action, where the probability of the chosen outcome is p for each trial?" Here p is a probability of occurrence for the chosen outcome, and n and r are frequencies, r being less than or equal to n. In the problem of the rats in the T maze, $n = 8$, $r = 4$, and $p = .60$; and the problem can be restated as "What is the probability of exactly 4 right turns in 8 trials, when the probability of a right turn is .60 for each trial?"

Some other examples are: (1) "What is the probability of 3 or more heads in 5 tosses of an unbiased coin?" (2) "What is the probability of a score below 5 on an 8-item true–false test, by guessing only? In this case, $n = 8$, $p = .50$, and r varies from 0 to 5 (we have to compute the binomial formula for each of these 6 values of r and then add). (3) "What is the probability of drawing exactly 4 spades in 6 draws from a deck of cards, when each card is replaced after it is drawn and the deck shuffled well?" Here, $n = 6$, $r = 4$, and $p = .25$ (the proportion of spades in the deck at each draw = the probability of drawing a spade each time).

8-7   Binomial formula

A. Calculating Probabilities

If an experiment consists of n actions, all independent of one another, and if $p = P\{\text{outcome } a_1 \text{ occurs}\}$ and $q = (1 - p) = P\{\text{outcome } a_1 \text{ does not occur}\}$ do not change from action to action, then $P\{r \text{ occurrences of outcome } a_1 \text{ in } n \text{ actions}\}$ is given by the binomial formula

$$P\{r|n\} = {}_nC_r p^r q^{n-r} \qquad (8\text{-}1)$$

where ${}_nC_r$ is the number of combinations of n things taken r at a time, defined in formula 8-2.

Consider the following problem. A biased coin has a probability of heads $P\{h\} = .60$. What is the probability of getting exactly 3 heads in 5 tosses of this coin? The formal or algebraic solution (see Sections 8-3 and 8-5) is as follows:

$$P\{3h|5 \text{ tosses}\} = P\{(hhhtt)\text{ⓥ } (hhtht)\text{ⓥ } (hhtth)\text{ⓥ } (hthht)\text{ⓥ}$$
$$(hthth)\text{ⓥ } (htthh)\text{ⓥ } (thhht)\text{ⓥ } (thhth)\text{ⓥ}$$
$$(ththh)\text{ⓥ } (tthhh)\}$$

where, for example, $(hhhtt)$ is an abbreviation for $(h_1 \,\&\, h_2 \,\&\, h_3 \,\&\, t_4 \,\&\, t_5)$, in which the &'s and subscripts have been omitted for simplicity.

From rule II we know that $P\{t\} = .40$. Then, using rules I and III, we find that

$$P\{3h|5 \text{ tosses}\} = (.6)(.6)(.6)(.4)(.4) + (.6)(.6)(.4)(.6)(.4) +$$
$$(.6)(.6)(.4)(.4)(.6) + (.6)(.4)(.6)(.6)(.4) +$$

$$(.6)(.4)(.6)(.4)(.6) + (.6)(.4)(.4)(.6)(.6) +$$

$$(.4)(.6)(.6)(.6)(.4) + (.4)(.6)(.6)(.4)(.6) +$$

$$(.4)(.6)(.4)(.6)(.6) + (.4)(.4)(.6)(.6)(.6)$$

Because the order of multiplication does not affect the result, each of the above products has the same value:

$$(.6)(.6)(.6)(.4)(.4) = .03456$$

There are 10 of these outcomes; therefore,

$$P\{3h \mid 5 \text{ tosses}\} = 10(.03456) = .3456$$

The binomial formula gives this same exact solution with considerably less effort. We let $p = P\{h\} = .60$ and $q = P\{t\} = 1 - P\{h\} = .40$. We seek a compound outcome in which there are $r = 3$ heads and $n - r = 5 - 3 = 2$ tails. In this case, the portion $p^r q^{n-r}$ of formula 8-1 is equal to $(.6)^3(.4)^2 = (.6)(.6)(.6)(.4)(.4)$, which is the probability of each of the 10 successful outcomes as obtained by the algebraic solution. The other expression in the binomial formula, $_nC_r$, is in this case equal to 10; it tells the number of ways success can be achieved.

The distinct advantage of using the binomial formula in solving problems for which it is appropriate is that it is not necessary to list all the successful outcomes, but simply to know how many of them there are and what the probability of each is. Application of formula 8-1 would thus have been much easier than using the formal method of solution:

$$P\{3h \mid 5 \text{ tosses}\} = {}_5C_3(.6)^3(.4)^2$$

$$= 10(.216)(.16)$$

$$= .3456$$

The value of $_nC_r$ can be found in either of two different ways. The easiest thing to do, if n is 20 or less, is to look up the value in Table 8-1, which is a variation of Pascal's triangle.

In Table 8-1, n is listed down the left-hand margin and r across the top. In the body of the table are values of $_nC_r$, the binomial coefficient, for specified values of n and r. In the problem just completed, $n = 5$ and $r = 3$, so the value of $_5C_3$ appears at the intersection of row 5 and column 3 and is equal to 10. When $n = 8$ and $r = 2$, $_nC_r = 28$, and so on. Values of $_nC_r$ missing from the lower right-hand corner of the table can be determined from the relationship $_nC_r = {}_nC_{n-r}$, because the coefficients form a symmetric distribution. For example, $_{20}C_{12} = {}_{20}C_8 = 125970$.

Table 8-1 *Binomial coefficients*

n	r										
	0	1	2	3	4	5	6	7	8	9	10
1	1	1									
2	1	2	1								
3	1	3	3	1							
4	1	4	6	4	1						
5	1	5	10	10	5	1					
6	1	6	15	20	15	6	1				
7	1	7	21	35	35	21	7	1			
8	1	8	28	56	70	56	28	8	1		
9	1	9	36	84	126	126	84	36	9	1	
10	1	10	45	120	210	252	210	120	45	10	1
11	1	11	55	165	330	462	462	330	165	55	11
12	1	12	66	220	495	792	924	792	495	220	66
13	1	13	78	286	715	1287	1716	1716	1287	715	286
14	1	14	91	364	1001	2002	3003	3432	3003	2002	1001
15	1	15	105	455	1365	3003	5005	6435	6435	5005	3003
16	1	16	120	560	1820	4368	8008	11440	12870	11440	8008
17	1	17	136	680	2380	6188	12376	19448	24310	24310	19448
18	1	18	153	816	3060	8568	18564	31824	43758	48620	43758
19	1	19	171	969	3876	11628	27132	50388	75582	92378	92378
20	1	20	190	1140	4845	15504	38760	77520	125970	167960	184756

The value of $_nC_r$ can also be calculated, using formula 8-2:

$$_nC_r = \frac{n!}{r!(n-r)!} = \frac{n \text{ factorial}}{(r \text{ factorial})(n\text{-minus-}r\text{-factorial})} \tag{8-2}$$

where $n! = n(n-1)(n-2)(n-3)\cdots 3\cdot 2\cdot 1$, $r! = r(r-1)(r-2)(r-3)$ $\cdots 3\cdot 2\cdot 1$, and $(n-r)! = (n-r)(n-r-1)(n-r-2)\cdots 3\cdot 2\cdot 1$. In other words, to find the factorial of a number, multiply that number by every smaller integer (i.e., whole number) down to 1.* When $n = 5$ and $r = 3$,

$$_nC_r = {_5C_3} = \frac{5!}{3!2!} = \frac{5\cdot 4\cdot 3\cdot 2\cdot 1}{(3\cdot 2\cdot 1)(2\cdot 1)} = \frac{5\cdot 4\cdot 3\cdot 2\cdot 1}{(3\cdot 2\cdot 1)(2\cdot 1)}$$

$$= \frac{5\cdot 4}{2} = 10$$

This is the same number found in Table 8-1 for $_5C_3$. In fact, every entry in

* Exception: $0! = 1$, by definition.

Table 8-1 can be found by applying formula 8-2 to the values of n and r appearing in the margins of the table for that entry. When, for example, $n = 8$ and $r = 2$,

$$_8C_2 = \frac{8!}{2!6!} = \frac{8 \cdot 7 \cdot 6 \cdot 5 \cdot 4 \cdot 3 \cdot 2 \cdot 1}{(6 \cdot 5 \cdot 4 \cdot 3 \cdot 2 \cdot 1)(2 \cdot 1)} = \frac{8 \cdot 7 \cdot 6!}{6! \cdot 2 \cdot 1}$$

$$= \frac{8 \cdot 7}{2} = 4 \cdot 7 = 28 \qquad \text{(as listed in Table 8-1)}$$

Note the manipulation I used in the numerator of the above formula: $8! = 8 \cdot 7 \cdot 6!$ This is legitimate, because the products can be grouped in various ways, such that the last few numbers form a factorial. For example,

$$8! = 8 \cdot 7 \cdot 6 \cdot 5 \cdot 4 \cdot 3 \cdot 2 \cdot 1$$

$$= 8(7 \cdot 6 \cdot 5 \cdot 4 \cdot 3 \cdot 2 \cdot 1) = 8 \cdot 7!$$

$$= 8 \cdot 7 \cdot 6(5 \cdot 4 \cdot 3 \cdot 2 \cdot 1) = 8 \cdot 7 \cdot 6 \cdot 5! \qquad \text{and so on.}$$

I used $8 \cdot 7 \cdot 6!$ in the numerator of the $_nC_r$ formula, because 6! was the largest factorial in the denominator. Thus I could divide the 6!'s and simplify the expression very easily.

The expression $_nC_r$ is very often read by reading off the letters: "$n - C - r$," or by the phrase "the number of combinations of n things taken r at a time." Two other common notations for this same expression are $\binom{n}{r}$ and C_r^n. However it is written, there is no algebraic manipulation to be performed on the expression itself: n and r serve as labels, not as multipliers, divisors, or anything else.

B. Special Case: When $p = \frac{1}{2}$

Probably the most common binomial distributions in research are ones in which, for each action, the two outcomes are equally probable. In this case, the binomial formula simplifies somewhat, because when $p = \frac{1}{2}$, so does q:

$$_nC_r p^r q^{n-r} = {}_nC_r(\tfrac{1}{2})^r(\tfrac{1}{2})^{n-r}$$

Because we have the same term $(\frac{1}{2})$ raised to two different powers and then multiplied, the result is equal to the same term $(\frac{1}{2})$ raised to the sum of the two powers. But $r + (n - r) = n$; so the formula reduces to

$$_nC_r(\tfrac{1}{2})^n = \frac{{}_nC_r}{2^n}$$

Values of $_nC_r$ are given in Table 8-1, and values of 2^n are easily calculated, as shown by the following short table:

n	0	1	2	3	4	5	6	7	8
2^n	1	2	4	8	16	32	64	128	256

EXAMPLE What is the probability of 3, 4, or 5 heads in 5 tosses of an unbiased coin? For an unbiased coin, the probability of heads $p = \frac{1}{2}$, the same as the probability of tails. Then for 3 heads, $P\{3h\} = {}_5C_3/2^5 = 10/32$. For 4 heads, $P\{4h\} = {}_5C_4/2^5 = 5/32$; and for 5 heads, $P\{5h\} = {}_5C_4/2^5 = 1/32$. Thus, from rule I, $P\{3, 4 \text{ or } 5h\} = 10/32 + 5/32 + 1/32 = 16/32$.

C. Relation of the Binomial Formula to the Binomial Distribution

The binomial distribution was discussed in Section 7-5, where terms of the expansion of $(p + q)^n$ were taken as proportions and related to the graph of the distribution. The binomial formula, $_nC_r\,p^r q^{n-r}$, can be used to find each term in the binomial expansion, and thus the entire distribution for any values of n and p.

For example, when $n = 3$, $(p + q)^n = p^3 + 3p^2q + 3pq^2 + q^3$. Then if $p = 0.5$, $q = 0.5$, $p^3 = 0.125$, $3p^2q = 0.375$, $3pq^2 = 0.375$, and $q^3 = 0.125$. Now, when we compute formula 8-1 letting $r = 3$, we get $_3C_3(.5)^3(.5)^0 = .125$, the value of the first term in the expansion; then letting $r = 2$, we get $_3C_2(.5)^2(.5) = 3(.125) = .375$, the value of the second term in the expansion; and similarly for the other two terms. The graph of this distribution is presented in Figure 7-17. In the present section the areas under the bars are interpreted as probabilities, rather than proportions, which has no effect on the distribution itself.

References The binomial distribution is discussed in Guilford (1965), 118–123; and Hays (1963), 131–155.

8-8 The normal approximation to the binomial distribution

Relevant earlier sections:
8-7 Binomial distribution
7-3 Normal distribution

A. Smoothing the Binomial Distribution

The binomial distribution can be graphed by listing values of r along the horizontal axis and probabilities along the vertical axis. For simplicity, consider the case where $p = q = \frac{1}{2}$. Then when $n = 3$, the histogram of Figure 8-1 can be drawn as shown. Because the bars are all of the same width, the height of each bar is proportional to the probability it represents, as is the area of the bar. The total area under the distribution is 1.0.

Figure 8-1 *Graph of a binomial distribution when n = 3 and p = q = ½.*

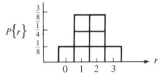

Values of r are integers (whole numbers) only, so that it really makes no sense to speak of fractional values of r. However, let us imagine another, continuous scale corresponding to the scale of r. On this scale, the bar for $r = 0$ extends from $-\frac{1}{2}$ to $+\frac{1}{2}$; the bar for $r = 1$ extends from $\frac{1}{2}$ to $1\frac{1}{2}$; the bar for 2 goes from $1\frac{1}{2}$ to $2\frac{1}{2}$; and so on. The probability distribution consists of all the bars. Therefore, we can say that the probability of $r = 1$ is given by the area under the distribution between $r = \frac{1}{2}$ and $r = 1\frac{1}{2}$; the probability of $r = 3$ is given by the area under the distribution between $r = 2\frac{1}{2}$ and $r = 3\frac{1}{2}$; and so on. In general, the probability of a value of r is given by the area under the distribution of $P\{r\}$ between $(r - \frac{1}{2})$ and $(r + \frac{1}{2})$.

Figure 8-2 *Graph of a binomial distribution when n = 6 and p = q = ½.*

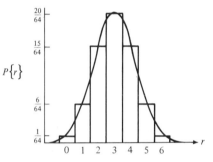

Now compare Figure 8-1 to Figures 8-2 and 8-3. Imagine in each distribution drawing a smooth curve through the centers of the bars, as has been done in Figures 8-2 and 8-3.

Figure 8-3 *Graph of a binomial distribution when n = 9 and p = q = ½.*

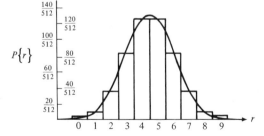

As n increases from 3 to 6 to 9 and more, the histogram that represents the binomial distribution becomes more and more like the smooth curve. Therefore, the areas under corresponding portions of the binomial distribution and the smooth curve become more and more nearly the same.

Consider the areas under two distributions for some value of r when n is large. This may be graphed as in Figure 8-4.

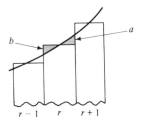

Figure 8-4 *Comparison of part of a binomial distribution with part of a normal distribution.*

The shaded area (a) is under the continuous distribution but not under the binomial distribution, and the shaded area (b) is under the binomial distribution but not the continuous one. As n increases, the bars get narrower by comparison to the width of the entire distribution, and both a and b get smaller. The difference in areas under the distributions for the single value of r is equal to $|a - b|$. As n gets larger, this difference gets smaller as well. Therefore, as n gets larger, the area under the smooth curve between $(r - \frac{1}{2})$ and $(r + \frac{1}{2})$ is a better and better approximation to the area under the binomial distribution for the value r.

That the smooth curve approximating the binomial distribution is the normal distribution should not be surprising. The binomial distribution is a distribution of chance, or randomness, or error, when the outcomes are restricted to whole-number values, whereas the normal distribution is a distribution of randomness or error, when the measurement scale is continuous. As the number of discrete-scale values increases, differences between successive values become smaller and smaller relative to the entire distribution, and the scale effectively approaches a continuous scale. Since the two distributions relate to the same concept, random error, for two different kinds of scales, it seems reasonable that as those scales become more similar, their respective distributions should also approach one another.

B. Criteria for Approximating the Binomial Distribution

The area under the normal distribution is not always an equally accurate estimate of the corresponding area under the binomial distribution. The discrepancy between the two distributions may occur as a function of the size of n, as indicated above, or of the inequality of p and q. The above examples were taken for the case where $p = q = \frac{1}{2}$, in which instance the binomial distribution is symmetric. If $p \neq q$, the binomial distribution becomes skewed; however, the normal distribution is always symmetric. Therefore, as the difference between p and q increases, the binomial is less and less accurately approximated by the normal distribution, assuming that n is fixed. This is especially true when n is small: values in the shorter tail tend to have greater probabilities than values in the longer tail. On the other hand, a highly skewed binomial distribution may be approximated very closely by a normal distribution if n is large enough.

A simple rule of thumb takes into account both possible sources of discrepancies between the two distributions: use the normal to estimate areas under the binomial only when both $n \cdot p$ and $n \cdot q$ are greater than 5. Thus, the approximation is reasonably good if $n = 20$ and $p = .3$, because then $n \cdot p = 20(.3) = 6$ and $n \cdot q = 20(.7) = 14$, both of which are greater than 5. The approximation would not be good enough, according to this rule of thumb, if $n = 9$ and $p = .6$, or if $n = 30$ and $p = .15$. In the latter two cases, use of the approximation would lead to more error than most statisticians would be comfortable with.

C. Applying the Normal Approximation

The normal distribution is determined completely by specification of two parameters, μ and σ. Therefore, the problem is to estimate these parameters for the normal distribution that best fits a given binomial distribution from the parameters of the binomial distribution. It is known that the following formulas provide the values of the parameters for such a best-fitting normal approximation to any binomial distribution.

$$\mu = np \tag{8-3}$$

$$\sigma = \sqrt{npq} \tag{8-4}$$

Once you have found the mean and standard deviation for the normal distribution which best fits the binomial for a given problem, you must next find the limits of the area under the normal distribution which corresponds to the area under the binomial that interests you. You do this by calculating z scores; for a single value of r, the z scores desired are those corresponding to $(r - \frac{1}{2})$ and $(r + \frac{1}{2})$. These are the real limits of the category, corresponding to the bounds of the histogram bar, as shown in Figure 8-4. The area under the normal distribution between these two values is approximately equal to the area under the binomial distribution for r.

$$z(r - \tfrac{1}{2}) = \frac{(r - \tfrac{1}{2}) - \mu}{\sigma} = \frac{r - \tfrac{1}{2} - np}{\sqrt{npq}} \tag{8-5}$$

$$z(r + \tfrac{1}{2}) = \frac{(r + \tfrac{1}{2}) - \mu}{\sigma} = \frac{r + \tfrac{1}{2} - np}{\sqrt{npq}} \tag{8-6}$$

The final step is to determine the area between these two limits by using the table of the standard normal distribution in accordance with the procedures described in Section 7-3.

EXAMPLE   To determine the probability of 15 heads in 36 tosses of an unbiased coin, first test to see whether the normal approximation can be used: $np = nq = (36)(\frac{1}{2}) = 18$, and since $18 > 5$, the normal approximation is appropriate. Next, find $\mu = np = 18$; $\sigma = \sqrt{npq} = \sqrt{(36)(\frac{1}{2})(\frac{1}{2})} = \sqrt{9} = 3$. Then, from formulas 8-5 and 8-6,

$$z(r - \tfrac{1}{2}) = z(14\tfrac{1}{2}) = \frac{14\tfrac{1}{2} - 18}{3} = \frac{-3\tfrac{1}{2}}{3} = \frac{-7}{6} = -1.17$$

$$z(r + \tfrac{1}{2}) = z(15\tfrac{1}{2}) = \frac{15\tfrac{1}{2} - 18}{3} = \frac{-2\tfrac{1}{2}}{3} = \frac{-5}{6} = -.83$$

Therefore, approximate the area under the binomial distribution for $n = 36$, $r = 15$ by the area under the normal distribution, $\mu = 18$, $\sigma = 3$, between the values $z = -1.17$ and $z = -.83$. From column 2 of Table A-4, this area is equal to $.3790 - .2967 = .0823$; hence, $P\{15h|36 \text{ tosses}\} \doteq .0823$.

This result can be compared to the true value given by the binomial distribution:

$$P\{15h|36 \text{ tosses}\} = \frac{n!}{r!(n-r)!} p^r q^{n-r} = \frac{36!}{15!21!} (\tfrac{1}{2})^{15}(\tfrac{1}{2})^{21}$$

$$= \frac{3.7199 \times 10^{41}}{(1.3077 \times 10^{12})(5.1091 \times 10^{19})} \left(\frac{1}{68,719,476,736} \right)$$

$$\doteq .08103$$

Clearly the normal approximation is reasonably accurate and involves less computational drudgery than direct application of the binomial formula to this problem.

When the problem consists in finding the probability for two or more adjacent values of r, it is not necessary to carry out a separate set of calculations for each value of r. Finding the probabilities for r_1, r_2, and r_3 is the same as finding the area from $(r_1 - \tfrac{1}{2})$ to $(r_1 + \tfrac{1}{2})$, from $(r_2 - \tfrac{1}{2})$ to $(r_2 + \tfrac{1}{2})$, and from $(r_3 - \tfrac{1}{2})$ to $(r_3 + \tfrac{1}{2})$, and then adding these three areas (see Figure 8-5). But when r_1, r_2, and r_3 are successive values, $r_1 + \tfrac{1}{2} = r_2 - \tfrac{1}{2}$, and $r_2 + \tfrac{1}{2} = r_3 - \tfrac{1}{2}$.

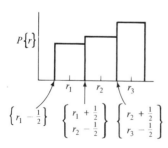

Figure 8-5   *Part of a binomial distribution.*

Hence the area under the normal distribution from $r_1 - \frac{1}{2}$ to $r_2 + \frac{1}{2}$ is a good approximation of the area under the binomial distribution for r_1 and r_2, and the area under the normal distribution from $r_1 - \frac{1}{2}$ to $r_3 + \frac{1}{2}$ is a good approximation of the area under the binomial distribution for $r_1 + r_2 + r_3$. In general, the area under the binomial distribution for $r_i + r_{i+1} + r_{i+2} + \cdots + r_j$ is approximated by the area under the corresponding normal distribution from $r_i - \frac{1}{2}$ to $r_j + \frac{1}{2}$.

EXAMPLE Consider the problem of finding the probability of getting between 25 and 34 heads, inclusive, in 100 tosses of an unbiased coin. Here the simple outcome that we are counting is a head on one toss of the coin, and $P\{h\} = p$ is the same for all tosses. Therefore, the binomial distribution is appropriate, and because $np = 50$ and $nq = 50$ are both greater than 5, the normal approximation can be used. We first find

$\mu = np = 50$ and $\sigma = \sqrt{npq} = \sqrt{(100)(\frac{1}{2})(\frac{1}{2})} = \sqrt{25} = 5$. Then the problem can be graphed as shown in the illustration. The desired probability can be approximated by the area under the normal distribution between $z(25 - \frac{1}{2})$ and $z(34 + \frac{1}{2})$.

$$z(25 - \tfrac{1}{2}) = z(24\tfrac{1}{2}) = \frac{24\tfrac{1}{2} - 50}{5} = \frac{-25\tfrac{1}{2}}{5} = -5.1$$

$$z(34 + \tfrac{1}{2}) = z(34\tfrac{1}{2}) = \frac{34\tfrac{1}{2} - 50}{5} = \frac{-15\tfrac{1}{2}}{5} = -3.1$$

To four decimal places, the area less than $z = -5.1$ is about .0000, and the area less than $z = -3.1$ is about .0010 (from Table A-4). The probability of finding between 25 and 34 heads is the area between the two points, or $.0010 - .0000 = .0010$.

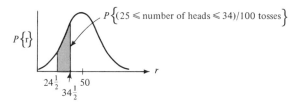

$P\{(25 \leqslant \text{number of heads} \leqslant 34)/100 \text{ tosses}\}$

EXAMPLE Let us rework the previous problem to find the probability of getting between 25 and 34 heads, inclusive, in 100 tosses of a *biased* coin, which has $P\{h\} = .2$ and $P\{t\} = .8$. The normal approximation is still appropriate, since $np = 100(.20) = 20$ and $nq = 100(.80) = 80$ are both greater than 5. The mean of the best-fitting normal distribution is

$np = 20$, and the standard deviation is $\sqrt{100(.20)(.80)} = \sqrt{16} = 4$. The desired area under the normal distribution is bounded by $z(24\frac{1}{2})$ and $z(34\frac{1}{2})$, which have different values from those of the previous example:

$$z(24\tfrac{1}{2}) = \frac{24\tfrac{1}{2} - 20}{4} = \frac{4\tfrac{1}{2}}{4} = +1.12$$

$$z(34\tfrac{1}{2}) = \frac{34\tfrac{1}{2} - 20}{4} = \frac{14\tfrac{1}{2}}{4} = +3.63$$

From Table A-4 we find the areas corresponding to these two z scores and, by subtraction, the area under the standard normal distribution between them:

$$P\{25 \leq \text{number of heads} \leq 34 \text{ in } 100 \text{ tosses}\} = .4998 - .3686$$

$$= .1312$$

Note that even the graph for this problem is different from that for the previous one:

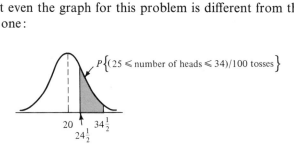

8-9   Computing aids in the solution of probability problems

Many elementary problems in probability can be solved with the use of special techniques for combining probabilities. Three useful techniques will be discussed in this section. The first technique is used when the outcomes can all be enumerated and when all outcomes have the same probability of occurrence. The second, the tree method, is used when all the outcomes and their probabilities can be listed, but when all outcomes are not necessarily equally probable. The third method, the probability table, is a simplification of the tree method which examines only the successful outcomes, thus simplifying the computations but omitting some computational checks.

A. Proportion of Successful Outcomes

This technique can be used to solve problems in which there are a number of final (simple or compound) outcomes that are all equally probable. The formula used is as follows:

$$P\{\text{success}\} = \frac{\text{number of successful outcomes}}{\text{total number of outcomes}} = \frac{N_s}{N_T} \qquad (8\text{-}7)$$

As an example, consider a three-item true–false test in which the subject has no information about any of the items. Suppose (arbitrarily) that the correct answer sequence is *t-f-f*. Then there are several possible answer sequences that the subject might give, that is, *ttt, ttf, tft, ftt, tff, ftf, fft,* and *fff*, a total of eight possible outcomes in all. Of these eight outcomes, only one leads to success, so that $P\{\text{success}\} = \frac{1}{8}$.

Fortunately, it is not necessary to enumerate all the compound outcomes in order to know how many there are. We have defined the problem so that both responses to each item are compatible with both responses to each of the others. Therefore, there are $2 \cdot 2 = 4$ response combinations to two items, and $2 \cdot 2 \cdot 2 = 8$ response combinations to three items. Because each of the simple outcomes (t, f) to each item has the same probability, and because the same number of simple outcomes is combined in the same way to produce each compound outcome, all the compound outcomes are necessarily equally probable.

The formal method of solution gives the same results:

$$P\{S\} = P\{t_1 \ \& \ f_2 \ \& \ f_3\}$$
$$= P\{t_1\}P\{f_2\}P\{f_3\} \qquad \text{(by rule III)}$$
$$= \tfrac{1}{2} \cdot \tfrac{1}{2} \cdot \tfrac{1}{2} = \tfrac{1}{8}$$

As a second example of this method, consider a two-item multiple-choice test in which there are five responses to each item. What is the probability of getting one or both items right by chance alone?

You might be tempted to say that the answer is $\frac{1}{5} + \frac{1}{5}$, but this would be wrong. You would be implicitly applying rule I, which doesn't apply here: a correct answer on item 1 and a correct answer on item 2 are *not* mutually exclusive outcomes of the same action. It is quite possible for someone to get both items right.

To solve this problem, make a table listing all the responses to item 1 down the left-hand side and all the responses to item 2 across the top. Assume arbitrarily that B is the correct answer to item 1 and D is the correct answer to item 2. Place a check mark in each cell of the table which represents a correct answer.

		Item 2			
Item 1	A	B	C	D	E
A				✓	
B	✓	✓	✓	✓	✓
C				✓	
D				✓	
E				✓	

There are $5 \cdot 5 = 25$ possible response combinations, 9 of which are successful. The successful combinations have been indicated by check marks. The probability of a successful response combination is therefore $P\{S\} = N_S/N_T = 9/25 = .36$. Can you solve this problem using the formal method? (For the solution, see the homework answers.)

B. The Tree Method

For the tree method, a computing diagram is used to solve probability problems as an aid to the application of the three rules. This method can be used for any problem for which the proportion-of-successes method can be used. It has the additional advantage of being appropriate for problems in which the compound outcomes are not all equally probable, and of providing some checks on the computations.

As an example of this method, consider the following problem. A rat is placed on an electrified Lashley jumping stand, so that he is forced to jump to one of two doors in a vertical wall placed in front of him. In previous trials he has been found to jump to the right-hand door 60 percent of the time. If no learning occurs on the next two trials, and if his jumps can be considered to be independent, what is the probability that he will jump exactly once to each door?

Here the proportion-of-success method cannot be used because the outcomes are not equally probable. Instead, consider the following illustration:

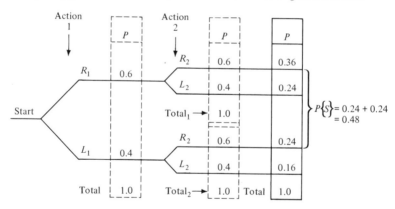

Begin reading the diagram at the left-hand side of the page; indicate the first jump by splitting the line in two, since the rat can jump either to the right or to the left on his first trial. Label each alternative and indicate its probability of occurrence. Then no matter which way he jumped the first time, the rat will still have to jump again. Therefore, divide both lines at action 2, label each response on the second jump, and indicate its probability.

Rule II is verified by the totals in the dashed boxes, and the values in the final column are found for each row by multiplying together all probabilities found along the path to that row, in accordance with rule III. Finally, since success is defined as the two middle rows, their probabilities are added, in accordance with rule I.

The second example in part A of this section asked the probability of getting one or both of two five-response, multiple-choice items right by chance alone. This problem can also be solved by the tree method.

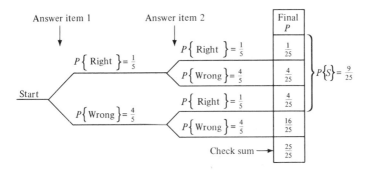

C. Probability Table

This is yet another device that can be used to combine the three rules into a standard technique for solving elementary probability problems. It shares with the formal method the potential weakness that it does not include the rule II checks built into the tree method, but it does allow for a saving of computation since it does not consider the unsuccessful outcomes.

Consider the following problem. What is the probability of getting a 7 on a single roll of a pair of dice? This problem can be solved as follows:

$$P\{7\} = P\{(1_1 \& 6_2) \textcircled{v} (2_1 \& 5_2) \textcircled{v} (3_1 \& 4_2) \textcircled{v} (4_1 \& 3_2) \textcircled{v} (5_1 \& 2_2) \textcircled{v} (6_1 \& 1_2)\}$$

$$= P\{1_1 \& 6_2\} + P\{2_1 \& 5_2\} + \cdots + P\{6_1 \& 1_2\} \quad \text{(by rule I)}$$

$$= P\{1_1\}P\{6_2\} + P\{2_1\}P\{5_2\} + \cdots + P\{6_1\}P\{1_2\} \quad \text{(by rule II)}$$

$$= \tfrac{1}{6} \cdot \tfrac{1}{6} + \tfrac{1}{6} \cdot \tfrac{1}{6} + \tfrac{1}{6} \cdot \tfrac{1}{6} + \tfrac{1}{6} \cdot \tfrac{1}{6} + \tfrac{1}{6} \cdot \tfrac{1}{6} + \tfrac{1}{6} \cdot \tfrac{1}{6} \quad \text{(since the dice are fair)}$$

$$= \tfrac{6}{36} = \tfrac{1}{6}$$

The same problem can be solved by table:

A: First die	B: Second die	$P\{A\}$	$P\{B\}$	$P\{A\}P\{B\}$
1	6	$\tfrac{1}{6}$	$\tfrac{1}{6}$	$\tfrac{1}{36}$
2	5	$\tfrac{1}{6}$	$\tfrac{1}{6}$	$\tfrac{1}{36}$
3	4	$\tfrac{1}{6}$	$\tfrac{1}{6}$	$\tfrac{1}{36}$
4	3	$\tfrac{1}{6}$	$\tfrac{1}{6}$	$\tfrac{1}{36}$
5	2	$\tfrac{1}{6}$	$\tfrac{1}{6}$	$\tfrac{1}{36}$
6	1	$\tfrac{1}{6}$	$\tfrac{1}{6}$	$\tfrac{1}{36}$

Here each row of the table represents a successful combination of a result on the first die and a result on the second die, so that the sum of both dice is 7. The first two columns of the table state what the combination is, the next two columns give the separate probabilities of getting these scores, and the final column gives the joint probability as found from rule III. Rule I is then used to find the total probability of success by adding down the last column.

Homework

1. Use the three rules to solve each of the following:
   (a) $P\{$at least one head in two tosses of an unbiased coin$\} = ?$
   (b) You have two biased coins, one with $P\{h_1\} = .40$, the other with $P\{h_2\} = .30$. If you toss both coins, what is the probability of getting one head and one tail?
   (c) You draw two cards from a deck without replacement. What is the probability that both are aces?
2. Solve (a), (b), and (c) of problem 1 with the use of the computing aids.
3. A biased coin has $P\{h\} = .58$. How many tails would you expect in the next 50 tosses of this coin?
4. Two thirds of all women passing a particular mirror in a hotel lobby are observed to adjust their hair. What is the probability that exactly two of the next five women passing the mirror will adjust their hair?
5. What is the probability of getting 212 or more heads in 400 tosses of an unbiased coin?
6. A bent coin is observed to fall heads up $\frac{3}{4}$ of the time. What is the probability that it will show tails in three of the next four tosses?
7. What is the probability of getting a score of 5 on a given roll of a pair of dice?
8. It is 3:00 A.M., you are driving 30 mph down Main Street and there are four traffic lights between here and home. Each light is red 50 percent of the time, and green and yellow 50 percent; they are independently set, and there is no traffic. You want to get home without changing speed, stopping, or running a red light.
   (a) Define success in terms of traffic-light colors.
   (b) What is your probability of success?
   (c) Solve the problem another way.
9. Naive flatworms have an equal probability of turning either way in a T maze. Using glycerine as a reinforcer and electric shock for punishment, you train a group of flatworms always to turn toward the lighted arm of the maze and away from the darkened arm. You then grind them up and feed them to a second batch of flatworms. If your theory is correct, each new flatworm thus gastronomically enriched should have a $\frac{3}{4}$ probability of turning toward the lighted arm of the maze. You randomly select four of

the cannibal worms (after lunch) and run them through the T maze. What is the probability, if your theory is correct, that

(a) All four will turn toward the lighted arm?

(b) Exactly three will turn toward the lighted arm?

(c) Three or more will turn toward the lighted arm?

10. You are training naive rats to make a right turn in a T maze, with food reinforcement for each correct trial. Each rat is given three trials. You predict that on each correct trial the probability of error will be reduced by $\frac{1}{5}$ (on the next trial it will be $\frac{4}{5}$ of what it was on the trial in which the correct response was made), but that a wrong response will leave the probabilities unaffected. Find the probabilities for 0, 1, 2, and 3 right turns.

11. Only one of the two previous problems can be solved using the binomial distribution. Indicate which one and explain the difference.

12. Eighty percent of graduate students at a large metropolitan university have full-time jobs in addition to their schoolwork. If a sample of 25 students is randomly drawn from this graduate population, what is the probability that 23 or more will have full-time jobs?

13. At any given time, 3 of every 10 citizens of Transcendentia are high on pot. In a random national sampling of four subjects, what is the probability that exactly half will be in the stated condition?

14. A small Eastern state sends 6 representatives to the House every two years. This year there are no major issues, there are equal numbers of registered Democrats and Republicans, and all candidates are trying to find the political center in their campaigns; therefore, the electoral process can be described as random with .50 probability of election for each candidate. What is the probability that four of the six representatives elected will be Republican?

Answer to Problem, Section 8-9A

You can get one or both items right by getting the first right and the second wrong, the first wrong and the second right, or both right. Call the items A and B and use subscript 1 for the correct response and 2 for the wrong answer.

$$
\begin{array}{ll}
a_1 = \text{A right} & P\{a_1\} = \frac{1}{5} \\
a_2 = \text{A wrong} & P\{a_2\} = \frac{4}{5} \\
b_1 = \text{B right} & P\{b_1\} = \frac{1}{5} \\
b_2 = \text{B wrong} & P\{b_2\} = \frac{4}{5}
\end{array}
$$

Then, $S = a_1 b_2 \; \textcircled{v} \; a_2 b_1 \; \textcircled{v} \; a_1 b_1$ and

$$
\begin{aligned}
P\{S\}(\tfrac{1}{5})(\tfrac{4}{5}) &+ (\tfrac{4}{5})(\tfrac{1}{5}) + (\tfrac{1}{5})(\tfrac{1}{5}) \\
&= \tfrac{4}{25} + \tfrac{4}{25} + \tfrac{1}{25} \\
&= \tfrac{9}{25}
\end{aligned}
$$

Homework Answers

1. (a) $\frac{3}{4} = .75$　**(b)** .46　　　　　**(c)** $\frac{1}{221} \doteq .0045$　**3.** 21

4. $\frac{40}{243} \doteq .165$　**5.** .1251　　　　**6.** $\frac{3}{64}$　　　**7.** $\frac{1}{9}$

8. $\frac{1}{16}$　　　　**9. (a)** $\frac{81}{256} \doteq .32$　**(b)** $\frac{27}{64} \doteq .42$　　**(c)** $\frac{189}{256} \doteq .74$

10. _____　**11.** 10 cannot

Number of right turns	P
0	.125
1	.305
2	.366
3	.204
$\sum$	1.000

12. Use the normal approximation:

$$P\{23 \text{ or more}\} = .4970 - .3944 = .1026$$

13. .2646　　　**14.** $\frac{15}{64} \doteq .23$

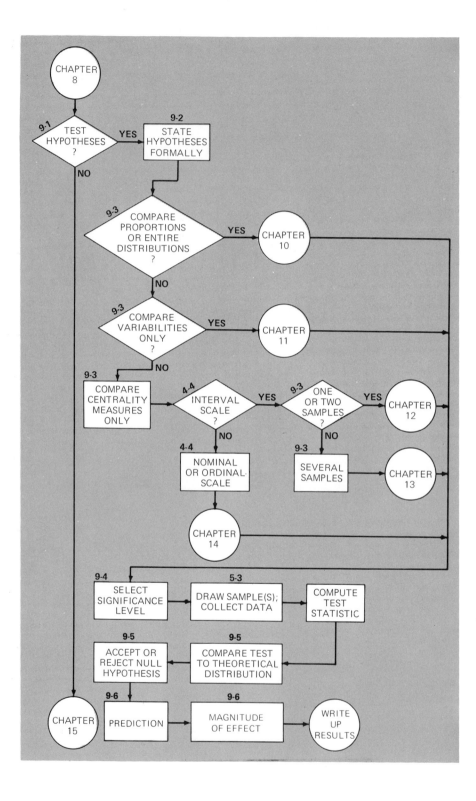

9 Hypothesis Testing

HYPOTHESIS testing, as outlined in the map for this chapter, has three broad stages: (1) assuming that you wish to test a hypothesis, you (2) define your problem and make a general selection of a method for solving it. You then (3) to go one of Chapters 10 through 14 for more detailed considerations in selecting a statistic, (4) collect your data and test your hypothesis; (5) make any predictions you need to make, and (6) establish the effectiveness of the experimental effects. At the end of the research you generally write up the findings.

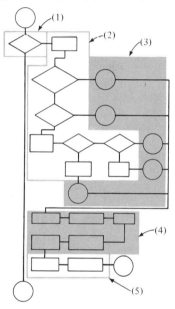

Hypothesis testing is not so amenable to a mapped representation, as might be hoped, because the considerations developed in Sections 9-4 and 9-5 are necessary to an understanding of Chapters 10 through 14. Therefore, it will be wisest to read these two sections along with the sections of later chapters which are most relevant to your problem.

9-1 Hypothesis testing

A. Basic Problem: Use of Hypothesis Testing

Hypothesis testing may be regarded as a strategy for obtaining information in usable form from data. This methodology lies at the juncture of the theoretical and the practical, by means of which data are evaluated and decisions are made about what the data imply. It is theoretical because the questions asked are always about populations, and they may relate to the "purest" of "pure science" issues (as well as to "applied science" issues). It is practical because it furnishes a method for arriving at a decision about population parameters that is reasonable, on the basis of limited sample information. Through hypothesis testing we satisfy our need to know answers to our problems without exhausting our resources in doing so.

A hypothesis is a statement about some characteristic of a population that is capable of being supported or not supported by the examination of empirical evidence. Examples might include: "Frustration produces aggression"; "Children from the cities are more aggressive than children from the suburbs"; and "In the learning of a foreign language, individual differences are reduced by oral drills."

As shown in the decision map for this chapter, it is possible to formulate and test hypotheses about many different kinds of characteristics of populations: measures of centrality and variability, shapes of distributions, and proportions and percentages. These kinds of hypotheses will be discussed in Section 9-3 and elaborated further in Chapters 10 through 14. Later chapters on degree and form of relationships will also include discussions of methods for testing hypotheses about these characteristics. Finally, there are many kinds of hypotheses for which the tests are too advanced or too little used to be included in this text. For these you will have to refer to other sources.

B. Logic of Hypothesis Testing

Hypothesis testing is based on the logical form of denying the consequent. This form can be stated as follows:

If A is true, then B must be true	(first premise)
B is not true	(second premise)
Therefore, A cannot be true	(conclusion)

The conclusion follows in a straightforward manner because if A were true, then B would have been true, from the first premise; and for B to be true would contradict the second premise.

Statistical hypothesis testing puts a slight twist on this logic, however, changing the deduction to read:

If A is true, then B will probably be true
B is not true

Therefore, A is probably not true

Or, to give a more concrete example: (1) If this sample is drawn from a population with a mean of 40, the probability is .95 that the sample mean will fall between 30 and 50. (2) This sample has a mean of 58.23. (3) Therefore, the probability that this sample came from a population with a mean of 40 is .05 or less.

There is a second twist to keep in mind: remembering to make A something that you want to disprove. The statement that you want to disprove is called the *null hypothesis* and is labeled H_0. The twist amounts to the fact that you usually don't even mention the hypothesis that you are interested in testing, here to be called the *motivated hypothesis, H_m*. It is often not necessary to mention H_m, because you arrange your hypotheses so that H_0 and H_m are mutually exclusive and exhaustive (for definitions of these terms, see Section 8-3). Then in disproving H_0 you automatically demonstrate H_m, with the same degree of certainty.

C. Outline of Procedure

Broadly speaking, the procedure for hypothesis testing follows the sequence indicated in the map for this chapter.

1. State your hypotheses. First state formally the motivated hypotheses (also called the "alternative hypothesis" or "experimental hypothesis,") which is the one you want to substantiate. Then, state the null hypothesis, which is simply that there is no such effect as the one you have postulated (see Section 9-2).
2. Select a statistic for making the comparison in which you are interested (see Section 9-3). Select an appropriate sample of subjects, carry out the necessary experiment, and compute the magnitude of your selected statistic.
3. Imagine running the same experiment a large number of times when the null hypothesis was known to be true, and each time computing the same statistic as used in the real experiment. The values of the statistic would have a distribution as a result of the sampling error and the nature of the statistic. This distribution can be specified theoretically.
4. Compare the value of the statistic computed on your sample to the theoretical distribution of the same statistic obtained when the null hypothesis is assumed true. If the probability of obtaining the observed result is sufficiently small under the null hypothesis, reject the null hypothesis and accept the motivated hypothesis.

This outline is necessarily highly abbreviated. However, it does emphasize the central notion of hypothesis testing: if a set of data has been collected under a null hypothesis, it should be similar to other sets of data collected under the same null hypothesis. Therefore, we decide to accept the null hypothesis when our data are similar to what we expect under the null hypothesis and to reject it when they are not. We realize that we may erroneously accept or reject the null hypothesis, but we learn to live with error, knowing that we can make the probability of its occurrence very small even if we can't eliminate it entirely.

D. Three Example Problems

This part is strictly optional and should be omitted if you already have a problem of your own. If you do not have a problem, this part serves as a device for pulling together and introducing Chapters 9 through 14. You should not expect to gain much from reading the following problems unless you refer ahead to the indicated sections for more detailed explanations of each phase of each analysis.

PROBLEM 1 During the past three years, 100 rats have been run a total of 400 trials through the same maze without reinforcement. The mean number of errors per trial is 2.5, with a standard deviation of .50, and these values are unaffected by the number of previous trials an animal has had. We wish to determine whether feeding animals on completion of the maze will reduce their error rate.

Our hypotheses (Sections 9-2 and 9-3) must compare the population of nonreinforced (NR) rats to the population of reinforced (R) animals with respect to number of errors. We wish to know whether there is an effect on the average animal or, equivalently, whether the population means are the same or different. It seems obvious that the reinforced animals should have fewer errors, but let's be open about the direction of effect by using a two-tail test. Therefore, our hypotheses are

$$H_0 : \mu_R = \mu_{NR} \qquad H_m : \mu_R \neq \mu_{NR}$$

where μ_R is the mean number of errors in the reinforced population and μ_{NR} is the mean number of errors made by the nonreinforced population.

The research design is diagrammed in Figure 9-2. We randomly sample a small number of rats from the null population of nonreinforced animals, give them several trials with reinforcement, then find their mean number of errors after treatment. We assume that they are then representative of all reinforced rats, and compare the sample mean to the mean of the null population in order to test the hypotheses.

We are comparing a sample mean to a population mean, so we turn to Chapter 12. From the statement of the null hypothesis, we know that we must use a two-tail test. We have a standard deviation based on 400 observations, so a z test can be used. From Section 12-4 the z formula is found to be

$$z = \frac{\overline{X} - \mu_0}{\sigma_X / \sqrt{N}}$$

We arbitrarily set $\alpha = .01$ (Section 9-4) and note that critical values of the standard normal distribution are ± 2.58, as indicated in Table A-4.

Next, 16 animals are randomly selected from the population of rats on hand, which have been run through the maze without reinforcement. Each animal is run through the maze again, once on each of six successive days, with reinforcement. The number of errors made by each animal on the sixth day is recorded and the mean computed to be 1.125.

Inserting values into the computational formula for z gives

$$z = \frac{1.125 - 2.50}{.50/\sqrt{16}} = \frac{-1.375}{.50/4} = \frac{-1.375}{.125}$$

$$= -11.0$$

The computed value $z = -11.0$ is less than the smaller critical value of -2.58, so the null hypothesis is rejected. The reinforcement condition used reduces the number of errors that rats make in this maze, as compared to their being run without reinforcement.

PROBLEM 2   You wish to determine the effect of combat on soldier's attitudes toward war. You are able to obtain a random sample of five enlisted men just before they leave for the front lines and another of six enlisted men just returning from combat. You administer to both groups a scale of favorability toward the war, hypothesizing that the returning soldiers will have greater variability of attitudes.

Your independent or treatment variable is combat experience; your dependent variable is attitude toward the war. You have two independent samples which represent the same population before and after treatment. The closest design given in Section 9-3 is in Figure 9-6. You wish to compare the population of soldiers with combat experience to the population without it, by comparing samples randomly drawn from them.

If the scale can be assumed to give interval-level scores, the variabilities of the two groups can be compared. The hypotheses (Section 9-2) are

$$H_0 : \sigma^2_{\text{before}} = \sigma^2_{\text{after}} \qquad H_m : \sigma^2_{\text{before}} \neq \sigma^2_{\text{after}}$$

where again, a nondirectional test is carried out so that it will be possible to identify a difference that is opposite from the one predicted, if it should occur in the data.

Choose an α level of .05 (Section 9-4) and decide to use the test of Section 11-2:

$$F = \frac{\hat{s}^2_{\text{larger}}}{\hat{s}^2_{\text{smaller}}} \qquad df = N_L - 1, N_S - 1$$

You collect the following data:

Before	13, 8, 11, 9, 9
After	14, 3, 4, 9, 9, 9

Using the methods of Sections 5-7 and 5-12, you find out that $\overline{X}_b = 10$, $\hat{s}^2_b = 4$, $\overline{X}_a = 8$, and $\hat{s}^2_a = 16$. Clearly the sample variance after treatment is greater than the sample variance before treatment; the question is whether the observed difference in sample variances is sufficient to allow you to infer a difference in the corresponding population variances.

Your test statistic is

$$F = \frac{\hat{s}_a^2}{\hat{s}_b^2} = \frac{16}{4} = 4.0$$

to be compared to the critical value $F_{.05}$(two tail, 5, 4 df) = 9.36. The observed F of 4.0 is less than the critical value, so the null hypothesis is not rejected: there is insufficient evidence for a difference in population variances. You conclude that combat experience does not increase the differences among soldiers in their favorability toward the war.

PROBLEM 3 In a large university over 2000 students take a one-semester introductory psychology course every year. You wish to know whether there is any sex difference at all in grade distribution. Your hypotheses (Section 9-2) are

$$H_0 : \text{Pop. Dist. (m)} = \text{Pop. Dist. (f)}$$
$$H_m : \text{Pop. Dist. (m)} = \text{Pop. Dist. (f)}$$

The appropriate design is given in Figure 9-6, with the two populations "male students who are taking introductory psychology" and "female students who are taking introductory psychology." You set α at .05 and collect data on 75 randomly drawn students. Graphing the data, you see that males appear to have a more even grade distribution than females. Can you legitimately generalize to the two populations?

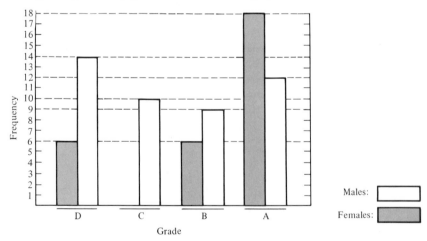

Since you are comparing entire distributions, Chapter 10 applies. The observed frequencies may be tabulated as follows:

	Grade				
	D	C	B	A	Σ
Males	14	10	9	12	45
Females	6	0	6	18	30
Σ	20	10	15	30	75

The expected frequencies are found from the marginal totals in accordance with Section 10-1:

	Grade				
	D	C	B	A	Σ
Males	12	6	9	18	45
Females	8	4	6	12	30
Σ	20	10	15	30	75

No expected frequency is less than 1.0, and $\frac{7}{8} = .875$ of the expected frequencies are greater than 5, so the table can be kept as it is. Its size is 2×4, so Section 10-3 is used for solution. The general formula 10-2 is

$$\chi^2 = \Sigma \frac{(O - E)^2}{E}$$

which is equal to 12.50 in this case. The degrees of freedom are $(2 - 1) \times (4 - 1) = 3$, and $\chi^2_{.05}(3 \text{ df}) = 7.81473$. The observed chi square is greater than the critical value, so you conclude that the two populations (males and females) have different grade distributions in this course.

References General introductions to hypothesis testing can be found in Blommers and Lindquist (1960), 264–281; Games and Klare (1967), 271–275; and Hays (1963), 245–253.

9-2 State your hypotheses formally

The great value of hypothesis testing lies in your ability, using this system, to infer characteristics of populations from information gathered on very small samples. There is never any interest in a sample per se, but only for what information it can give you about the population from which it is drawn.

Therefore, hypotheses are always stated formally in terms of population characteristics. Because any value can only be interpreted by comparison with other values, statistical tests involve comparison of two or more parameters.

Suppose that we wish to infer the value of the mean in some population. We may have two competing theories about what it is: theory 1 says $\mu = 45$ and theory 2 says $\mu = 52$. We set up two mutually exclusive and exhaustive *exact hypotheses:*

$$H_1 : \mu = 45 \qquad H_2 : \mu = 52$$

H_1 and H_2 are mutually exclusive because the population mean can take on only one value for a given population. They are exhaustive in a theoretical, not mathematical sense—we do not have a third theory that predicts μ to be other than 45 or 52. If we set things up carefully, any observed extimate $\overline{X}_j$ of μ should allow us to decide between the two theories, because $\overline{X}_j$ must have been drawn under one or the other.

In actual practice this approach is not often used, because we can seldom find theories that specify parameters precisely enough to permit such a test. Therefore, it will only be remarked parenthetically that we could also compare three or more specific hypotheses in this way, using a single sample estimate of μ, provided that the hypotheses were all mutually exclusive and exhaustive.

Often it is possible to specify some parameter value that represents a base line against which a special population or treatment effect may be compared. Suppose that the verbal portion of the Scholastic Aptitude Test has a mean $\mu = 500$ and standard deviation $\sigma = 100$ for the population of college applicants required to take the test. If you design a college preparatory program intended in part to raise verbal proficiency in this population, you could use the population mean of 500 as a base line to test the effectiveness of the program in raising test scores.

If the entire population were given your program, it would obtain a mean μ_e (*e* for experimental). You hope that μ_e will be greater than 500, so make this the motivated hypothesis:

$$H_m : \mu_e > 500$$

The null hypothesis is then that the program is not effective in raising scores, so it is

$$H_0 : \mu_e \leqq 500$$

These last two are called *inexact hypotheses*. Like exact hypotheses, they are mutually exclusive and exhaustive, but here each hypothesis represents a whole range of possible parameters.

Perhaps this is a good place to introduce some convenient abbreviations. In this example, we call the population of college applicants required to take the test (with $\mu = 500$ and $\sigma = 100$) the *null population* and its mean the *null mean*. We call the hypothetical population formed from the null population by giving every subject the experimental treatment (the college preparatory program) the *experimental population* or *treatment population* and its mean the *experimental mean*. Similar abbreviations may be used for other parameters.

Inexact hypotheses may be either *directional* or *nondirectional*. The hypotheses

$$H_0 : \mu_e \leq 500 \qquad H_m : \mu_e > 500$$

are directional, or one-tail, because they specify a direction for the desired effect of the treatment. We are only interested in the experimental conditions if they raise the average test performance.

Suppose, on the other hand, that we had wanted to be able to conclude something if the program either raised or lowered scores. In this case, we could have stated nondirectional, or two-tail, hypotheses:

$$H_0 : \mu_e = 500 \qquad H_m : \mu_e \neq 500; \quad \text{that is, } H_m : (\mu_e > 500) \textcircled{v} (\mu_e < 500)$$

In this case, we could conclude that H_0 is false if μ_e is different from 500 in either direction.

The same distinctions among kinds of hypotheses may also be made when two different experimental treatments are compared. Suppose that theory A predicts that treatment 1 will raise scores on some outcome measure by 10 points more than treatment 2 will, and suppose that theory B predicts that the two treatments will be equally effective in altering scores. The exact hypotheses for this pair of theories would be

$$H_A : (\mu_1 - \mu_2) = 10 \qquad \text{(theory A)}$$
$$H_B : (\mu_1 - \mu_2) = 0 \qquad \text{(theory B)}$$

With directional hypotheses, you might predict that treatment 1 will be more effective than treatment 2:

$$H_0 : (\mu_1 - \mu_2) \leq 0 \qquad H_m : (\mu_1 - \mu_2) > 0$$

With nondirectional hypotheses, you might simply try to determine whether the two treatments affected means differently:

$$H_0 : (\mu_1 - \mu_2) = 0 \qquad H_m : (\mu_1 - \mu_2) \neq 0$$

With any single set of data, you could only test one pair of hypotheses. However, you do have a choice about which pair you test. If you wanted to run the experiment three times with three independent sets of data, you could test all three pairs of hypotheses.

Formal hypotheses about other population characteristics can also be stated and tested. You can specify exact hypotheses about the variance of scores, for example,

$$H_1 : \sigma_e^2 = 100 \qquad H_2 : \sigma_e^2 = 110$$

You might have inexact hypotheses about the relative magnitude of variances for male (m) and female (f) subjects, predicting that males' scores will be more variable:

$$H_0 : \sigma_f^2 \geqq \sigma_m^2$$
$$H_m : \sigma_f^2 < \sigma_m^2$$

You might hypothesize that the shapes of test-score distributions in the populations of males and females would be different. This could be stated as conditional, nondirectional hypotheses:

$$H_0 : \text{Pop. Dist. (m)} = \text{Pop. Dist. (f) in shape}$$
$$H_m : \text{Pop. Dist. (m)} \neq \text{Pop. Dist. (f) in shape}$$

(Here a linear transformation of one or both distributions could be used to equate means and variances before testing the hypothesis of interest.)

You might hypothesize that the proportion of males in some particular range (e.g., greater than 140) is different from the proportion of females in the same range. If you let π (pi) represent a population proportion, then you might state the nondirectional hypotheses as follows:

$$H_0 : (\pi_m > 140) = (\pi_f > 140) \qquad H_m : (\pi_m > 140) \neq (\pi_f > 140)$$

Of course, this is not intended to be an exhaustive list of kinds of hypotheses, merely a few illustrations of some of the possibilities open to you. Many additional hypotheses will be stated and tested in the remaining chapters of the book.

References   Formal stating of hypotheses is discussed in Dixon and Massey (1969), 75–77; and Games and Klare (1967), 275–278.

9-3   Selecting the parameter to be compared and the design for making the comparison

A. The Parameter To Be Compared

Once you have decided that your purposes can best be met by a hypotheses-testing design and you are committed to comparing two or more populations with respect to some attribute, you must then ask yourself, "What measure will be most sensitive to the kind of effect I wish to observe in this experiment?"

The map for this chapter indicates this issue by a series of three separate questions: Do you want to compare centrality measures only, variabilities only, or proportions or entire distributions? Your choice at this point will affect the entire direction of subsequent research, and your perspicacity in finding the most appropriate basis for comparison will greatly affect your ability to observe whatever differences may or may not be present.

1. *Questions about centrality* Centrality measures are compared whenever the hypotheses to be tested specify that members of one population have more or less of some attribute than members of some other population. For example: "Are Jews more intelligent than Christians?"; "Are French women sexier than American women?"; or (perhaps less realistically) "Do cats have more lives than other animals?" Corresponding null hypotheses can be stated formally as follows, where the dependent variable has been indicated in parentheses in each case:

(i) $H_0 : \mu_J \leqq \mu_C$ (intelligence)
(ii) $H_0 : \mu_F \leqq \mu_A$ (sexiness)
(iii) $H_0 : \mu_C \leqq \mu_{OA}$ (number of lives)

(Here hypotheses have been stated in terms of population means, but other centrality measures could have been used if more appropriate. In general, one's choice of statistic will be determined both by the appropriateness of the statistic, as indicated in Chapter 5, and by the availability of a procedure for testing hypotheses about it. Many of the available procedures appear in the following chapters: if you can't find one you need there, get the help of a statistician about other possible tests. In any case, it's generally a good idea to plan out your entire research, including your choice of statistical test and ability to draw reasonable conclusions, before collecting any data.)

Centrality measures are also compared when it is expected that the effect of one or more treatments will be either to raise or lower scores on some dependent measure. For example, "Is intergroup competition or cooperation more effective in reducing intergroup hostility?"; "Can an understanding of psychology (as measured by some test) be best gained through concentrating on general principles or through studying a large number of specific examples?"; or "Which of four new tranquilizers is the most effective in reducing anxiety in a population of neurotic subjects?" Hypotheses comparing centrality measures are tested with the methods described in Chapters 12, 13, and 14.

2. *Questions about variability* Variability measures are compared whenever your hypotheses relate to the within-group differences in two populations; for example, "Is there greater diversity in intelligence test scores for schizophrenic patients than for pure paranoid patients?"; "Are there greater differences among subjects in speed of reaction to taboo words than to nontaboo

words, in a word-association task?"; or "Is there greater variability of adult income for children from broken homes than for children who are not from broken homes?"

Corresponding null hypotheses about population variances can be stated formally as follows, where the dependent variable has been indicated in parentheses in each case:

(i) $H_0 : \sigma_S^2 \leq \sigma_P^2$ (intelligence)

(ii) $H_0 : \sigma_T^2 \leq \sigma_{NT}^2$ (reaction time)

(iii) $H_0 : \sigma_{BH}^2 \leq \sigma_{NBH}^2$ (adult income)

Hypotheses about variability are tested using the methods found in Chapter 11.

3. *Questions about entire distributions* Entire distributions are compared when a comparison of centrality or variability measures will be either inappropriate or insufficient. Here you have your choice of allowing centrality and/or variability to be considered in comparing the distributions or not, depending on your hypotheses. For example, "Are horses from all starting positions equally likely to win a race?" Here, an empirical distribution is compared to a rectangular one (implying no differences), and differences in centrality or variability, as well as differences in shape, are allowed to affect the outcome. "Are the creativity test scores of mongaloid children normally distributed (in the statistical sense)?" Here again, an empirical distribution is compared to a theoretical (normal) one; but here the mean and variance are not considered, because the question of statistical normality is a question of shape alone. "Is the distribution of eye colors different for males and females?" Here, two empirical distributions are compared on a nominal scale, and centrality and variability differences would both contribute toward an affirmative response to the question. Methods for comparing entire distributions are presented in Chapter 10.

Proportions, percentages, and probabilities are compared whenever the rate of occurrence of some attribute is at issue. For example, "Do public-school children have a greater probability of acceptance at the University of Michigan than private-school children?" "Is handedness a sex-related characteristic? (i.e., is the proportion of right-handed males different from the proportion of right-handed females?)" or "Is the proportion of subjects agreeing to item 27 of the Psychic Syndrome Test different from .5?"

Percentages, proportions, and probabilities can be converted to frequencies using the number of subjects on which they were computed. These frequencies are then amenable to analysis with the methods of Chapter 10.

B. Comparing a Statistic to a Parameter

This kind of design uses a characteristic of a single sample either to compare two populations with respect to their corresponding parameters or to decide whether an experimental treatment has produced an effect on the population characteristic of interest. Two populations can be compared by the use of a single sample if either (a) the variable has been examined for so many subjects of one population that for all practical purposes the entire population has been examined, or (b) the one population is a theoretical abstraction whose parameters can be specified without recourse to actual observation.

In the following discussions,

ρ_ϵ is a parameter of the population we are examining, a mean, a variance, or some other characteristic

ρ_0 is the value we expect that parameter to have when the null hypothesis is true

R is an estimate of ρ_ϵ obtained from a random sample drawn from the population we are examining

1. *Naturally existing populations*   The first of these uses can be diagrammed as in Figure 9-1. Here two populations are compared: a null population for which the parameter ρ_0 is known, and a comparison population for which the parameter ρ_ϵ is not known. A sample of N subjects is randomly drawn from the comparison population and the computed value of its characteristic **R** is compared to the null population parameter. From this test, it can be inferred whether the comparison population differs from the null population in its characteristic on the attribute compared.

EXAMPLE PROBLEMS   **1** Means. Replace ρ_0 by μ_0, and ρ_ϵ by μ_ϵ, and **R** by $\bar{X}$ in Figure 9-1. (a) A test of masculine interests has been developed on a population of thin 10-year-old boys. Do fat 10-year-old boys differ from the thin population on the average? Here, the null hypothesis is $H_0 : \mu_T = \mu_F$, where μ_T is the mean of the null (thin) population. (b) A test of creativity has been given to children in urban schools for a number of years. It is now administered to a sample of rural children. Is the population of rural children less creative as scored by this test? Here, the null hypothesis is given by $H_0 : \mu_R \geq \mu_U$. **2** Variabilities. Replace ρ_0 by σ_0^2, ρ_ϵ by σ_ϵ^2, and **R** by $\hat{s}_\epsilon^2$ in Figure 9-1. Then (a) Do 10-year-old fat boys have less variability of masculine interests than their thin counterparts? Here, the null hypothesis would be given by $H_0 : \sigma_F^2 \geq \sigma_T^2$. (b) Do the rural children have greater variability in creativity test scores than the urban chil-

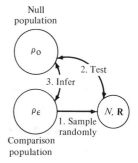

Null
population

2. Test

3. Infer

1. Sample
randomly

N, **R**

ρ_0

ρ_ϵ

Comparison
population

Figure 9-1   *Comparison of two extant populations using a sample drawn from one.*

dren? Here the null hypothesis is given by $H_0 : \sigma_R^2 \leq \sigma_U^2$. **3** Entire distributions. (a) Are the fat boys' masculinity scores normally distributed? (b) Is the distribution of creativity scores for rural children different in any respect from the distribution of scores for the urban children? $[H_0 : \text{Pop. Dist. (rural)} = \text{Pop. Dist. (urban).}]$

2. *One population produced experimentally* The second design can be diagrammed as indicated in Figure 9-2.

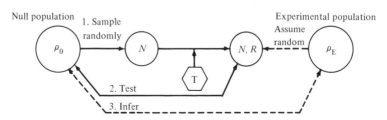

Figure 9-2 *Using a single sample to test for the effectiveness of an experimental treatment.*

A sample of size N is drawn randomly from a null population and a treatment (T) is applied. After the treatment, data are collected and the statistic of interest **(R)** is found for the sample. This sample is now assumed to be randomly drawn from the experimental population, which differs from the null population only in that all its members have received the treatment. The sample statistic is compared to the corresponding parameter, ρ_0, of the null population, and from this comparison it is inferred whether the parameters of the two populations (ρ_0, ρ_e) are the same or different (i.e., whether the treatment has any effect when applied to the entire population). The statistical methods used to analyze the results of this second design are the same as those used to analyze data produced by the design of Figure 9-1. The only difference is that the experimental population is not given; it is produced (hypothetically) by the experimental treatment.

EXAMPLE A test of hypochondriasis has a known distribution for hypochondriacs. A small sample of hypochondriacs is given a new kind of short-term psychotherapy. Possible questions to be tested might include: (1) Does the new treatment reduce scores on the test? $(H_0 : \mu_T \geq \mu_0.)$ (2) Does the treatment reduce scores more for more severely disturbed patients and less for patients who are less hypochondriacal? $(H_0 : \sigma_T^2 \geq \sigma_0^2$ and $H_m : \sigma_T^2 < \sigma_0^2$, because a reduction of high scores relative to low ones will tend to compress the distribution and thus to decrease its variability.) (3) Is there any effect at all on the score distribution as a result of treatment? $[H_0 : \text{Dist. (treatment)} = \text{Dist. (no treatment).}]$

C. Comparing Statistics from Two Different Samples
The designs considered in parts 1 and 2 below begin with a single null population, from which two samples are drawn in some manner. The designs differ

in the ways in which the two samples are obtained. If the parameter of the null population is known, this information is not used. In part 3, two different naturally existing populations are compared by drawing samples from each and comparing the sample values.

1. *Independent samples from the same population*   Here two samples are randomly and independently drawn from the null population. Each sample is given a different treatment, and after the treatment scores are obtained and the statistics of the two samples are compared. It is assumed that the two post-treatment samples are randomly drawn from two hypothetical post-treatment populations: one population differs from the null population only in that all members have received treatment 1; the other population differs from the null population only in that all members have received treatment 2 (see Figure 9-3).

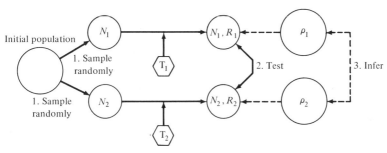

Figure 9-3   *Using two samples to test the relative effectiveness of two experimental treatments.*

EXAMPLES   Suppose that two methods for teaching reading skills are to be compared on separate samples for the same population. If you wish to select the method that has the greater overall effect, compare centrality values. ($H_0 : \mu_1 = \mu_2$.) To discover whether method 1 is relatively more helpful to better initial readers than is method 2, compare variabilities. ($H_0 : \sigma_1^2 \leq \sigma_2^2$.) (They would tend to "catch up" to good initial readers, thus reducing overall variability of performance.) To see if there is any difference at all in the effect of the two methods, compare the entire distributions. [$H_0 :$ Pop. Dist. (1) = Pop. Dist. (2).]

This design may also be used to test for the effectiveness of a single treatment, by making T_2 a control condition. The control group is treated in the same way as the experimental group, except for the part thought to produce the experimental change. Such a design is particularly useful when you suspect that the mere fact of experimenting with subjects will affect their responses on the outcome measure. A prototype for this design occurs in drug research, where control subjects may be given a sugar pill known to be chemically ineffective.

2. *Correlated data*   Correlated data occur when two samples are not drawn independently of one another; instead, there is a one-to-one relationship be-

tween the members of one sample and the members of the other. This can occur either when pretesting and post-testing are carried out on the same sample or when the members are assigned to the two samples in matched pairs.

The test–retest design is diagrammed in Figure 9-4. A sample of size N is randomly drawn from the population of subjects who have not had the treatment, the sample is tested, and the magnitude of the statistic of interest is calculated. The treatment is administered, the sample is tested again, and its post-treatment value is computed. The post-treatment sample is assumed to be randomly drawn from the population of all subjects who have had the treatment. The pretreatment statistic is compared to the post-treatment statistic for the same sample, and an inference is made as to whether the corresponding parameters are the same or different. The two samples are correlated because the same subject appears on the same line of the data table for the two samples. Measurement error is reduced and precision increased by reducing between-subject differences.

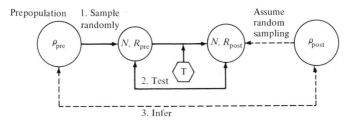

Figure 9-4 *Test-retest design with a single subject sample.*

This design can be used whenever it is feasible to test both before and after the treatment is administered. It might be used, for example, to evaluate the effectiveness of teaching test-taking strategies on subjects' intelligence-test scores. On the other hand, there are many cases for which this design is particularly inappropriate. These occur when the combination of pretest and experimental treatment allow subjects to develop hypotheses about the nature of the research and thus to affect the results of the post-test. For example, some sensory-deprivation experiments have been criticized on the grounds that giving intelligence tests before and after told subjects what should be affected; that removal of stimulation for a period of time suggested that scores should be lowered; and that subjects could lower their test scores easily, even if there were no "real" effect of the treatment.

Matched-pairs sampling can be diagrammed as indicated in Figure 9-5. Subjects are drawn randomly in pairs from the original population and assigned either randomly or systematically to two samples, depending on the type of hypothesis being tested. If the hypothesis has to do with the effects of treatments that are applied after the samples are drawn, then the subjects are assigned randomly from each pair to the two samples. For example, we might take same-sex twins and randomly assign them to two treatment groups to study,

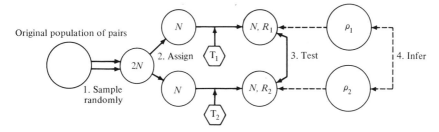

Figure 9-5 *Use of matched samples to compare the effectiveness of two experimental treatments.*

say, the effects of different kinds of verbal reinforcement on the learning of a finger maze. We might then test to see if one group learns the maze faster (H_0: $\mu_1 = \mu_2$), whether one kind of reinforcement has greater diversity of effect (H_0: $\sigma_1^2 = \sigma_2^2$), or whether there is any difference in the effects produced by the two treatments at all (H_0: Pop. Dist.$_1$ = Pop. Dist.$_2$).

If the hypothesis deals with differences among subjects that are already present in the population of untreated subjects, assignment of subjects to samples will be systematic, and no treatment will be administered. This might occur, for example, in comparing the verbal fluency of older and younger sisters, tested at the same age. The older girls would naturally be assigned to one sample, the younger to the other sample, and the treatment, whatever it might be, would be assumed already to have occurred as a result of birth order. The matching occurs by virtue of the girls in each pair being sisters: we expect two girls with the same parents to be more similar in verbal fluency than two girls picked at random. Thus, this procedure eliminates some random variability and leads to a more sensitive test of the hypothesis.

3. *Naturally existing populations, independent samples*   Here two samples are drawn from their respective populations for the purpose of comparing the population values (Figure 9-6). The characteristics of the two samples are compared, and from this comparison it is inferred whether the two populations have the same value for the parameter of interest.

EXAMPLES   (1) You wish to compare 20 legends from each of two primitive societies, to determine which society has the greater average amount of hostile imagery in its legends. (2) You ask, "Is there greater variability

Figure 9-6 *Comparison of two extant populations using samples drawn from each.*

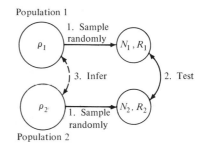

of pay for house carpenters where unions are well established than in areas where there are no unions?'' (3) In a ring-toss game, subjects are allowed to toss from any distance they wish, but the distances they choose are covertly recorded, as measures of level of aspiration. You compare effective and ineffective factory foremen to see if there are differences in the distributions of the dependent variable (level of aspiration), as indicated by differences in distances chosen. From this you make an inference as to whether the two populations of foremen are the same or different in level of aspiration.

D. Comparing Statistics from Several Samples

Often we may wish to ask the same kinds of questions already discussed, but with respect to several populations simultaneously. Here, the populations compared may either be naturally existing, or they may be produced experimentally. The two kinds of designs may be diagrammed differently, but the hypotheses and methods of analysis are identical.

1. *Naturally existing populations* In this situation, we draw one sample from each of three or more populations already in existence, then calculate some statistic on each of the samples, compare the several statistics simultaneously, and from the results infer relationships among the corresponding parameters.

EXAMPLE   Suppose that we wish to determine whether there are social-class differences in knowledge of family-planning techniques. We use a widely accepted formula for determining social class which distinguishes five classes. We obtain a random sample of married adults from each class so defined. (1) To compare the five social classes on overall amount of knowledge, we choose a test of knowledge of family-planning technique, then compare the means for the five groups. Our null hypothesis would be $H_0: \mu_1 = \mu_2 = \mu_3 = \mu_4 = \mu_5$, and we would reject it if there were any differences at all among the averages. (2) To compare the distributions of knowledge within classes, we might compare the variances of the five samples. Our null hypothesis would be $H_0: \sigma_1^2 = \sigma_2^2 = \sigma_3^2 = \sigma_4^2 = \sigma_5^2$; rejection of this hypothesis would indicate that there are greater differences in amount of knowledge within some class than within some other class. (3) To determine whether people in different classes know about different kinds of techniques, we might indicate for each subject the technique he knows most about, then obtain a distribution of techniques for each class. We could then compare these distributions for the five classes.

2. *Experimentally produced populations* In this design, we begin with a single population, draw three or more samples from it, and give each sample a different experimental treatment (see Figure 9-7). We assume that each of the treatment samples is representative of a treatment population consisting of all subjects in the original population after they have been given that treat-

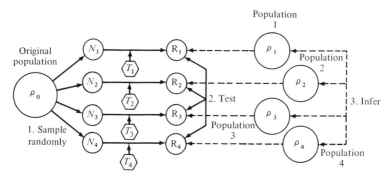

Figure 9-7 *Use of several samples to test for the relative effectiveness of different treatments.*

ment. Comparison of the statistics from the samples allows us to infer relationships among the corresponding parameters and, therefore, to compare the hypothetical populations corresponding to the various samples. There are two possible sources of differences among the samples: sampling differences, which we try to ensure are as nearly random as possible, and treatment effects. If we can infer population differences on the variable tested, we can also infer differences among the treatments, because there are, in theory, no sampling differences among the populations (they represent the same subject population under different treatment conditions).

    EXAMPLE Suppose that, in order to find out more about the neural centers critical to higher mental processes, you wish to study the effects of six different brain lesions on the problem-solving ability of cats. You obtain a sample of 30 cats and randomly assign 5 to each treatment group, then make the necessary lesions. (1) To test the hypothesis that some lesion will produce a greater handicap than some other, obtain a measure of degree of problem-solving ability and compare post-treatment centrality measures of the several groups on this measure. (2) To determine if some lesion is more difficult to perform than some other, compare variabilities of the six groups. (3) If you are concerned with performance on a single problem, and if you expect some treatment to produce total inability to solve the problem a greater proportion of the time than some other treatment, you can compare proportions passing the problem in the several groups.

    3. *Multiple classifications* Subjects may be classified on two or more independent variables at once, so that you can test for effects of each independent variable singly and in combination with others. For the two-way design, you can construct a table that has as many rows as there are levels of one independent variable and as many columns as there are levels of the other. Then randomly draw a sample from your initial population and randomly assign it to a cell of the table. Continue this procedure until all the cells are filled, as indicated in Figure 9-8.

213

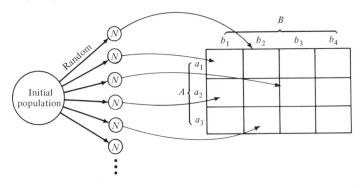

Figure 9-8 *Sampling for a two-way analysis.*

In the figure the two independent variables are labeled A and B. In contrast to previous figures in this section, Figure 9-8 omits the administration of treatments, collection of post-treatment data, or inference to post-treatment populations. The subjects in each cell receive a combination of treatments which differs from row to row and from column to column. You can test to see whether A or B or the combination of the two produces any effect on the dependent variable.

EXAMPLE You feel that a certain amount of cognitive dissonance will be generated when a subject who has just participated in a boring experiment for very little pay has to convince another subject to participate in the same experiment. As a result, you predict that such subjects will, on the average, be more persuasive than subjects who have either participated in an interesting experiment or received more pay. There are six treatment combinations, in the following table, of amount paid and degree of interest in the experiment:

	Amount paid		
	$1	$5	$10
Boring experiment	I	II	III
Interesting experiment	IV	V	VI

Equal numbers of introductory-psychology students are assigned randomly to each of the six treatment combinations, and afterward each student is asked to convince another student (a confederate) to participate in the same experiment. The outcome variable, indicating the effectiveness of the treatments, is amount of persuasiveness as rated by a hidden observer.

In a multiple-classification experiment, the effects need not be produced experimentally. Both variables can be characteristics already present in the population (e.g., age, sex, race, height, etc.), or one a status variable and the other a manipulated variable (such as reinforcement schedule, differences in instructions, different drugs, different amounts received, etc.).

Extremely complicated experimental designs are possible but cannot be dealt with in this text. Some are mentioned in Chapters 10 (proportions and entire distributions) and 13 (means), but for detailed treatment you will have to refer to more advanced texts.

Reference Experimental designs are discussed in Games and Klare (1967), 313–321.

9-4   Select a significance level

It is usually impractical or impossible to compare all the members of one population with all the members of another in order to answer a research question. Therefore, you must usually infer the relationships among population characteristics on the basis of the observed characteristics of samples drawn from them. Whenever you randomly sample from a population, however, you run the risk that your observed sample characteristic will differ from the population characteristic that you are trying to estimate. The mere fact of sampling introduces the possibility of sampling errors; you may, by chance alone, select a sample whose mean is quite different from the mean of the population, whose variance is quite different from the population variance, and so on.

Let us assume that you are comparing two populations by comparing samples drawn from them, and that you observe a difference in some characteristic of the two samples. This difference may correctly reflect a difference in population characteristics, or it may be due to error in selecting the samples from their respective populations, or both. Similarly, if you do not observe a difference in sample characteristics between the two samples, this equality may correctly reflect the equality of the two populations with respect to the characteristic, or a real difference between populations may be masked by an equal but opposite sampling error. Because you do not observe the populations directly, you cannot tell whether your sample results reflect population differences or sampling error or some combination of the two.

Your hypotheses are stated in terms of population characteristics, and your decision as a result of hypothesis testing will be made in terms of population characteristics. Therefore, in making your decision to accept or reject the null hypothesis, you may be correct in either of two ways or incorrect in either of two ways. These possible outcomes of hypothesis testing are given in Table 9-1.

Table 9-1   *Outcomes of hypothesis testing*

	Decision	
	Accept H_0	Reject H_0
H_0 is true	Correct	Type I error
H_0 is false	Type II error	Correct

You may either accept or reject the null hypothesis when it is true, and you may either accept or reject it when it is false. Although you can't avoid the possibility of making an erroneous decision on the basis of sample information, you do have control over the probability of making either a type I or a type II error. The probability of making a type I error is called the *level of significance* and is represented by the Greek letter α (alpha). The probability of a type II error is represented by the Greek letter β (beta) but has no other designation. People prefer to talk about the probability of correctly rejecting the null hypothesis rather than incorrectly accepting it. The probability of rejecting the null hypothesis when it is false is called *power*; it is equal to $1 - \beta$.

In testing hypotheses, you must learn to live with uncertainty about whether you have made a correct decision and accepted the correct hypothesis. On the other hand, it is possible to control the uncertainty. One way of doing this is to decide on the α and β levels that you are willing to tolerate; you can set α and β arbitrarily small, but the smaller you set them, the more subjects you must test and the greater expenditure you must incur in terms of testing time and money. Or, for a fixed sample size, you may be able to choose α or β; but the smaller you set α, the greater β must necessarily become, and vice versa. Careful use of hypothesis-testing strategies in the planning of research allows you to answer research questions while allocating limited experimental resources efficiently.

You must be prepared to justify your selection of α and sample size to other researchers, who will not be sympathetic to a selection that is both arbitrary and unusual. In most research the attempt is made to keep α low, so that the results of the research will be conservative. This is reasonable because the null hypothesis generally represents the status quo, while the motivated or alternative hypothesis is a formal statement of your quarrel with the commonly held viewpoint. If you set α too high, you will have a greater probability of rejecting a generally held hypothesis when it is correct, and of building your own theory on error.

In order to reduce the probability of type I error, a convention has developed in using α of .05 or .01 or smaller. No one insists that you use an α level of .05 or smaller, but if you use one greater than .10, it would be a good idea to justify your choice. The most convincing arguments for using $\alpha > .05$ might be either that it is impossible to set it at .05 or less within the context of your experiment (for practical reasons) or that, for your purpose, a type I error is less critical than some other consideration (e.g., type II error).

Most research reports of hypothesis testing indicate that levels of significance are explicitly considered, but most apparently do not consider power in planning of experiments. This is unfortunate, because in neglecting power, an experimenter may be left with nothing to say, in the event that the results of his experiment are consistent with the null hypothesis. If, through neglect, an experimenter has permitted his β level to be high, then a result consistent with the null hypothesis may also be consistent with the alternative hypothesis; and unless he can reject the null hypothesis, he can say nothing. This is indicated in Table 9-1 if you remember that you never know whether the null hypothesis is true or false. You merely make a decision to accept or reject it. If you reject the null hypothesis, either the null hypothesis is false or you have made type I error. You weigh the balance in favor of being correct by making the type I error rate, α, very small. Similarly, in order to be relatively confident that you are correct when you accept the null hypothesis, you must make the type II error rate, β, very small. This is equivalent to making the power $(1 - \beta)$ large.

References   Discussions of type I and type II errors can be found in Blommers and Lindquist (1960), 281–287; Dixon and Massey (1969), 83–84; Games and Klare (1967), 278–284; Guilford (1965), 205–207; and Hays (1963), 280–286.

9-5   Reaching a decision about the null hypothesis

At this point we have stated our hypotheses, selected our test statistic, decided on a level of significance, and collected our data. We now compute the value of the appropriate statistic from the observed data and determine whether the magnitude of the statistic is such that we ought to accept or reject the null hypothesis.

From our significance level we can derive a decision rule that will permit an answer to our question. We have already decided that we will be willing to reject the null hypothesis with probability α when it is true, provided that this α level would permit us to correctly reject the null hypothesis when it is false. We need a method for converting from an α level to a rejection region —a range of values of the test statistic such that, if our observed statistic falls in this range, we will decide to reject the null hypothesis. Thus, if our null hypothesis is

$$H_0 : \mu \geq 100$$

there will be some value of a sample mean so small that if our observed mean is that small or smaller, we will feel compelled to reject the null. That is, at some point we must come to a sample mean so small that the sample could not reasonably have been drawn from a population with mean greater than or equal to 100, and our α level tells us how small the mean has to be so that the null hypothesis is not reasonable.

In Chapters 7 and 8 it was indicated that a probability value could be represented as an area under a distribution and that area transformations could be effected between probability values and values of some characteristic or statistic. The statistics used in Chapters 10 through 14 for testing hypotheses have the distributions discussed in Chapters 7 and 8. Therefore, it is possible to use an area transformation from the α level to a scale value of the test statistic. This *critical value* of the test statistic then serves as a boundary value for a region of scores in which the null hypothesis is rejected: the *rejection region*. Any test statistic at or beyond the critical value is said to be so extreme compared to the values expressed in the null hypothesis that we are willing to decide that the null hypothesis isn't true. When a value of a test statistic is in the rejection region, it is said to be *significant*. Any test statistic at or beyond the critical value is said to be so extreme compared to the values expressed in the null hypothesis that we are willing to decide against the null.

In effect, we imagine replicating our experiment a very large number of times. Each time we calculate a value of our test statistic, just as we did in the real experiment. The difference lies in running our imaginary experiments *when the null hypothesis is known to be true*. This hypothetical distribution of observed values of the statistic under the null hypothesis is what we are approximating when we use a tabled distribution. The critical value we obtain from the tabled distribution is the one we would have obtained by replicating the experiment a great many times when the null was true, and marking off the extreme proportion of the distribution given by the α level. Values of the test statistic in the rejection region are extremely improbable when the null hypothesis is true. Therefore, when such a value occurs, it seems inconsistent with maintaining confidence in the null, and the null is rejected.

There are at least three parts to stating the conclusions from a hypothesis test. First, state the conclusion in statistical terms, having to do with the probability of the findings when the null hypothesis is presumed true. Thus, you might say, "The results are significant (or nonsignificant) at the .05 level (or at an α of .05)"; or, "The null hypothesis was rejected (or accepted." In example problem 1 of Section 9-1, the conclusion was stated: "The computed value $z = -11.0$ is less than the smaller critical value of -2.58, so the null hypothesis is rejected." Second, state your findings conceptually, in terms of the variables you studied, to indicate whether or not there was an effect of the independent variable (or variables) on the dependent variable. In example problem 1, the conclusion was stated: "The reinforcement condition used reduces the number of errors that rats make in this maze, as compared to their being run without reinforcement."

The above conceptual conclusion is very loosely put, but it contains the components of a reasonable conclusion. It states the two levels of the independent variable: (1) the reinforcement condition used, and (2) without reinforcement. It states the dependent variable, number of errors, and the effect of the treatment condition on it: "reduces." And it states what population is being generalized to: rats (in other words, all laboratory rats). Finally, the statement tells the direction of the findings without saying anything about how large or important the effect is. Magnitude of effects is discussed in Section 9-6 and in Chapters 15 through 18.

The third part of most conclusions ties the findings back to previous research, or to a broader theoretical position, or to the practical problem that the research was designed to help solve. This is what is done in most experimental articles in journals, but because of space and time limitations, it will be done very little in this text.

Incidentally, there are a couple of locutions widely used in discussing nonsignificant results that are meaningless within the context of the research methods indicated here. One is the conclusion, ". . . the results *approached* significance," or ". . . . the results were *nearly* significant." Freely translated, this means, "The null hypothesis is nearly false," or "I am *almost* concluding that the null hypothesis is false"; in either case, however, the conclusion is absurd.

The other meaningless expression is "The results, although not significant, were in the predicted direction." What this really means is that the experimenter has selected an α level of .05, .01, or something (which is fine) and then changed to an α level of .50* (which is questionable) *after* looking at his results (which is cheating).

9-6 Prediction and magnitude of effect

Hypothesis testing allows us to generalize beyond the particular sample or samples we are working with, to the populations from which they were drawn. It allows us to make inferences, based on a very small number of subjects, about large numbers of additional subjects that we have not and may never observe directly.

But hypothesis testing has its limitations. When we reject a null hypothesis we generally conclude that there is a nonzero difference between two population parameters. Now, a nonzero difference can be very small indeed; and particularly with large samples, it is often possible to reject a null hypothesis when the difference between parameters is, for practical purposes, negligible. No one, presumably, wants to build a scientific theory or institute a training program or in any other way carry out the work of a psychologist as a result of having been impressed by significant but negligible findings. Unless we examine our results further, we could become victims of our own methodology.

*There are only two directions in a directional test, so the a priori probability of picking the direction that the results will be in is $\frac{1}{2} = .50$.

The fact that a null hypothesis has been rejected gives us little or no information about the size of the population parameters, or the magnitude of the differences between them. Further, it doesn't tell us how much it helps to know the populations that subjects belong to, in predicting their status on the dependent variable. All it tells us is that there is a difference between population parameters and, sometimes, the direction of the difference.

Suppose, for example, you read in a journal that men have been found to have significantly higher scores than women on the Friedman Test of Agressive Tendencies, at the .05 level. There are a number of additional questions you might want to ask about these findings, some of them conceptual, some statistical. Among the conceptual questions, you might ask, (a) "What men, and what women?" (The population is vaguely stated: nobody really samples *all* men and *all* women . . .) (b) "What is meant by 'agressive tendency'?" (It could refer to verbal agressiveness, physical aggressiveness, energy level and drive, or maybe something else.) (c) The validity question: "How do test scores relate to actual behavior?" (The test itself might have no practical value.) Among the statistical questions, you might ask, (a) "What was the average score for men, and how great was the variability among scores?" (b) "What was the average score for women, and how great was their variability?" (c) "What was the magnitude of the difference between the men's and the women's averages?" (d) "How important is the difference? What proportion of tne error of predicting subjects' scores on the agressiveness test is eliminated by knowing subjects' sexes?"

Among the statistical questions, the first three are prediction questions, which are discussed in Chapter 15; the last question has to do with degree of effectiveness of prediction, or degree of relationship between independent and dependent variables, or magnitude of effect; it is discussed in Chapters 16 and 17.

Homework

For problems 1 through 3, give the best response to each.
  1. A one-tail test of a hypothesis is appropriate
    (a) When one wishes to be certain that his data will turn out to be significant.
    (b) When the null hypothesis is some number other than zero.
    (c) When the result is known before the test is run.
    (d) When the underlying distribution of scores is not normal.
    (e) When one wishes to perform power computations.
    (f) When one wishes to be very conservative with respect to significance level.
    (g) When there is good reason to specify an alternative hypothesis that is directional.

    (h) When a sample value is compared directly to a population value rather than to another sample value.
2. You set $\alpha = .01$. This means that
    (a) If the null hypothesis is true, you will have a .01 probability of rejecting it.
    (b) If the null hypothesis is false, you will have a .01 probability of rejecting it.
    (c) If the null hypothesis is true, you will accept it 1 percent of the time.
    (d) If the null hypothesis is false, you will accept it 1 percent of the time.
    (e) The total area under the probability curve is .01.
    (f) The variance of the mean is 1 percent of the variance of the raw scores.
    (g) The variance of the raw scores is 1 percent of the variance of the means.
    (h) The test is accurate to within 1 percent of the variance.
    (i) 1 percent of your experiments will work out.
    (j) There is a .01 probability that your result will be right.
3. What can you conclude if your result is significant?
    (a) The result is quite important for psychological research.
    (b) You were wrong by a considerable amount.
    (c) The motivated hypothesis is probably correct.
    (d) The null hypothesis is probably correct.
    (e) The sample was probably drawn from the experimental population.
    (f) The sample was probably drawn from the null population.
    (g) The sample mean is different from the null population.
    (h) The sample mean is different from the experimental population.
    (i) The sample is clearly different from the null population.
    (j) The sample is clearly different from the experimental population.
4. An experimental second-grade class of 10 children (group 1) is given a special reading course, while a control group (group 2) of 10 children receives the usual Dick and Jane. At the end of the school year, both groups are tested with a standard reading test.
    (a) What characteristic of a distribution will be most sensitive to an overall difference between the two groups, if you assume that the reading test gives approximately interval-level measurement?
    (b) What are your formal hypotheses?
    (c) What difference in sample characteristics must you observe to conclude that the experimental course is more effective than the control?
    (d) What figure from Section 9-3 best illustrates the appropriate experimental design?
    (e) What section of Chapters 10 through 14 should be followed to complete the analysis if it turns out that the two groups have roughly similar distributions of test scores.
5. Suppose that, instead of the two groups of problem 4, you had three treatment groups and one control group, each group having 10 subjects, and you again tested at the end of the year. Suppose that you want to know

whether there are any differences among the means of the four groups of subjects at the end of the year. Again assume that the reading test gives approximately interval-level measurement.

(a) What are your formal hypotheses?

(b) What figure of Section 9-3 best illustrates the experimental design?

(c) What section of Chapters 10 through 14 should be used to complete the analysis?

6. Suppose that we have a single class of 40 second graders, all of whom receive regular reading instruction by their classroom teacher. The three experimental and one control treatments occur 1 hour per week, during which time the class is always divided into the same four groups of 10 students each. The dependent variable is letter grade in reading (A+, A, A−, B+, . . .) given by the regular teacher at the end of the year.

(a) What figure of Section 9-3 best illustrates the experimental design?

(b) What section of Chapters 10 through 14 should be followed to complete the analysis?

(c) Look at the section chosen under (b). What are your safest (most conservative) hypotheses?

(d) What kind of result would you have to observe to reject your null hypothesis?

7. The author of an intelligence test claims that forms A and B of the test give equivalent scores. Among other things, this implies that the scores on the two forms should have equal variances. You doubt the claim. You therefore randomly and independently draw two samples of 121 subjects each, giving one sample form A and the other form B.

(a) What are your formal hypotheses?

(b) What figure from Section 9-3 best describes the design? How do you interpret T_1 and T_2 in this figure for the present problem?

(c) What section from Chapters 10 through 14 should be used to solve this problem?

8. Suppose that in problem 7 you had given both forms of the test to one group of 122 randomly chosen subjects and compared their scores on the two forms.

(a) There is no figure in Section 9-3 which accurately diagrams this experimental design. Which of those available is most similar?

(b) Can you produce a diagram that accurately reflects the present design?

(c) What section from Chapters 10 through 14 should be used to sove this problem?

(d) What is/are the critical value(s) of the statistic you chose in part (c) of this problem, if α is set at .01?

(e) If $s_1^2 = 10$ and $s_2^2 = 12$, what else must you calculate to solve this problem?

9. In a public opinion poll, single men were asked, "If all other things were equal, would you prefer to marry a girl who (I) had never held a job, (II) had held a job and was moderately successful at it, or (III) had held a job

and was extremely successful at it?" The men responding were classified into three groups: high income, middle income, and low income. The problem was to determine whether there were any differences at all in the responses of the three groups of subjects.

(a) What were the formal hypotheses?

(b) What figure from Section 9-3 best describes this design?

(c) What section from Chapters 10 through 14 should be used to solve this problem?

(d) What is the critical value of the statistic you have chosen, if α is set at .05?

10. Suppose, in problem 9, that you consider the three responses to the question as relating to a prospective wife's "work potential," which would be low for option (I) and high for answer (II). Then suppose that the research question is whether there are any differences in the work potential demanded by the average high-income subject, the average middle-income subject, and the average low-income subject.

(a) What figure from Section 9-3 best describes this experimental design?

(b) What section of Chapters 10 through 14 should be used to solve this problem?

(c) Can your conclusions be used to make inferences about corresponding population means? Why or why not?

Homework Answers

1. g   2. a   3. c

4. (a) Mean   (b) $H_0: \mu_1 \leq \mu_2$, $H_m: \mu_1 > \mu_2$. (The special course is presumably of interest only if it raises reading achievement scores.)
   (c) $\bar{X}_1 > \bar{X}_2$ by a significant amount   (d) Figure 9-3: T_1 = experimental course, T_2 = control   (e) Section 12-12

5. $H_0: \mu_1 = \mu_2 = \mu_3 = \mu_4$; H_m: not $(\mu_1 = \mu_2 = \mu_3 = \mu_4)$ (b) Figure 9-7
   (c) Section 13-4

6. (a) Figure 9-7   (b) Section 14-6   (c) $H_0: \mu_{\mathcal{R}_1} = \mu_{\mathcal{R}_2} = \mu_{\mathcal{R}_3} = \mu_{\mathcal{R}_4}$;
   H_m: not $(\mu_{\mathcal{R}_1} = \mu_{\mathcal{R}_2} = \mu_{\mathcal{R}_3} = \mu_{\mathcal{R}_4})$ (d) The average of ranks for some group(s) greatly differed from the average for some other(s).

7. (a) $H_0: \sigma_1^2 = \sigma_2^2$; $H_m: \sigma_1^2 \neq \sigma_2^2$ (b) Figure 9-5; T_1 = form A, T_2 = form B   (c) Section 11-2

8. (a) Figure 9-5   (c) Section 11-3   (d) ± 2.617 (e) r_{AB}

9. (a) H_0: Dist.(Hi) = Dist.(Mid) = Dist.(Lo); H_m: not [Dist.(Hi) = Dist. (Mid) = Dist.(Lo)]   (b) Figure 9-8   (c) Section 10-3
   (d) 9.48773

10. (a) Figure 9-7   (b) Section 14-6   (c) Not unless you are willing to assume that the three underlying distributions of "work potential desired in a wife" have similar shapes, which seems unlikely.

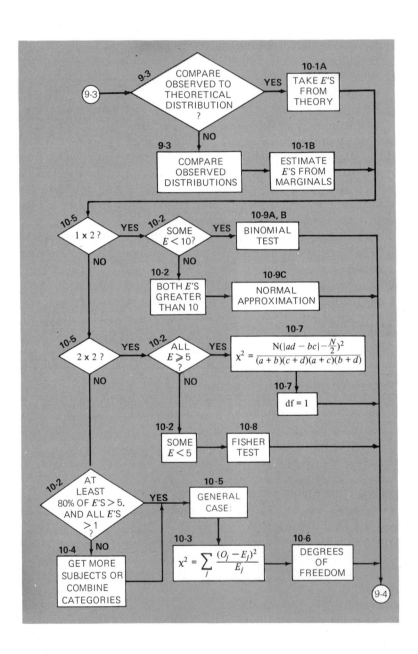

10

Comparing Proportions or Entire Distributions

PROPORTIONS or distributions are compared by comparing frequencies, rather than by comparing means, variances, or proportions of samples directly.

A univariate, bivariate, or multivariate frequency distribution is constructed as a preliminary step in the analysis (see Sections 4-5 and 4-13). In the first region of the map for this chapter, expected frequencies are calculated for each cell of the frequency distribution on the basis of observed frequencies and/or theoretical considerations. The remaining three regions contain different analyses, depending on the size of the data matrix to be analyzed.

It should be noted that whichever test is used, the general hypothesis-testing format of Chapter 9 is to be followed in carrying out the analysis.

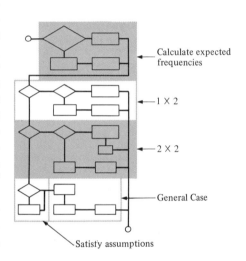

Calculate expected frequencies

1 X 2

2 X 2

General Case

Satisfy assumptions

10-1   Finding expected frequencies

Each of the tests discussed in this chapter requires the calculation of expected frequencies, either preliminary to selecting the test or in making the comparisons required by the test. In every case, an observed distribution is explicitly or implicitly compared to a theoretical one to determine whether the observed distribution could have been obtained by chance under the null hypothesis.

Two approaches are used to determine the expected frequencies; the approach depends on the hypothesis tested. The expected frequencies may be given by theory, when the characteristics to be tested for are known a priori (e.g., when testing a distribution for normality or rectangularity). When two or more distributions are compared to one another, no form of distribution is specified ahead of time; instead, it is asked whether the observed distributions could have been randomly obtained from the same population. In this case, the marginal values of the multiple-frequency distribution are used to obtain the expected frequencies.

A. Comparing an Observed to a Theoretical Distribution

In this case, theory or prior knowledge is used to specify the proportions of observations in each category. These proportions are multiplied by the total number of cases observed in the distribution to obtain expected frequencies for each category. The null hypothesis to be tested is that the sample distribution could have been obtained randomly from a population that had the theoretical distribution specified. Designs for this kind of comparison are given in Section 9-3B.

EXAMPLE In a five-item, true–false test, given that subjects have no information at all, the distribution of total scores can be predicted on the basis of the binomial distribution to be as shown. (These proportions were calculated using the binomial formula, Section 8-7, with $n = 5$ and $p = \frac{1}{2}$.)

Score	Proportion of subjects
0	$\frac{1}{32}$
1	$\frac{5}{32}$
2	$\frac{10}{32}$
3	$\frac{10}{32}$
4	$\frac{5}{32}$
5	$\frac{1}{32}$

To get the expected frequencies, we must know how many subjects we wish to predict for. If, for example, the observed distribution is as follows,

Score	Observed frequency
0	0
1	2
2	10
3	15
4	14
5	7
N_t	48

then the expected frequencies can be calculated by multiplying each expected proportion by N_t.

$$E_0 = 48\left(\frac{1}{32}\right) = 1.5$$

$$E_1 = 48\left(\frac{5}{32}\right) = 7.5$$

$$E_2 = 48\left(\frac{10}{32}\right) = 15 \quad \text{and so on}$$

You can also easily compare an observed distribution to either a rectangular or a normal distribution. In the case of a normal a priori distribution, the expected frequencies can be calculated using Table A-4 of the Appendix, as indicated in Section 7-3D. For a rectangular distribution, divide the total number of subjects by the number of categories to find the expected frequency in each category, as indicated in Section 7-2.

B. Comparing Two or More Observed Distributions

In this kind of problem, there is no a priori distribution to which the observed distributions can be compared. Instead, the null hypothesis is that all the distributions compared could have been drawn randomly from the same population. Designs for these comparisons are given in Section 9-3C and D.

The data are first placed in a multiple-frequency distribution, as illustrated in Table 10-1, in which grade-point averages of field-independent (FI) and field-dependent (FD) subjects are compared. In Tables 10-1 and 10-2 the rows and columns are numbered, as well as being labeled, to make the subscripting of the cell entries clear. In Table 10-1, N_1 is the total number of field-independent subjects, and N_2 is the total number of field-dependent subjects (i.e., these are the row sums). S_1, S_2, etc., are the column sums, and N_t is the total number of subjects tested.

Table 10-1 *Form for comparing grade-point averages of two groups of subjects*

| Column: | 1 | 2 | 3 | 4 | 5 | |
G.P.A.:	0–0.49	0.50–1.49	1.50–2.49	2.50–3.49	3.50–4.00	Σ
1. FI	O_{11}	O_{12}	O_{13}	O_{14}	O_{15}	N_1
2. FD	O_{21}	O_{22}	O_{23}	O_{24}	O_{25}	N_2
Σ	S_1	S_2	S_3	S_4	S_5	N_t

The expected frequencies are the frequencies that would occur under the null hypothesis, given the same marginal values (Table 10-2).

Table 10-2 *Expected frequencies for field-independent* (FI) *and field-dependent* (FD) *subjects of Table 10-1*

Column: G.P.A.:	1 0–0.49	2 0.50–1.49	3 1.50–2.49	4 2.50–3.49	5 3.50–4.00	Σ
1. FI 2. FD	E_{11} E_{21}	E_{12} E_{22}	E_{13} E_{23}	E_{14} E_{24}	E_{15} E_{25}	N_1 N_2
Σ	S_1	S_2	S_3	S_4	S_5	N_t

The null hypothesis says that the two distributions (FI and FD) are the same; thus we expect the same proportion of the FI and FD distributions to be in column 1 of the table. The expected proportion of FI in column 1 is defined as E_{11}/N_1, using the notation of Table 10-2; similarly, the expected proportion of FD in column 1 is E_{21}/N_2. If the null hypothesis were known to be true, we would expect these two proportions to be the same:

$$\frac{E_{11}}{N_1} = \frac{E_{21}}{N_2}$$

Let us now use this expectation to calculate values for the expected frequencies. We will show that $E_{11} = N_1 S_1/N_T$. First, it is always true that

$$S_1 = E_{11} + E_{21}$$

When the null hypothesis is that two or more distributions (rows) will have the same shape, then we expect the same proportion of every row to appear in the first column. Then,

$$\frac{E_{21}}{N_2} = \frac{E_{11}}{N_1} \quad \text{so} \quad E_{21} = \frac{E_{11} N_2}{N_1}$$

Replacing E_{21} by its equivalent in the formula for S_1 gives

$$S_1 = E_{11} + \frac{E_{11} N_2}{N_1}$$

and, multiplying both sides by N_1,

$$N_1 S_1 = N_1 E_{11} + N_2 E_{11}$$

Collecting terms on the right side of this equation,

$$N_1 S_1 = (N_1 + N_2)E_{11}$$

but because $N_1 + N_2 = N_t$,

$$N_1 S_1 = N_t E_{11}$$

Then dividing both sides by N_t, we get

$$E_{11} = \frac{N_1 S_1}{N_t}$$

A similar derivation can be carried out for each of the other cells of the table. In general, it can be said that

$$E_{ij} = \frac{N_i S_j}{N_t} \tag{10-1}$$

That is, any expected frequency is equal to the product of its marginal values divided by the total N for the table. This is the computing formula for expected frequencies.

EXAMPLE Following is a table of hypothetical observations corresponding to Table 10-1. Here grade-point-average values have been replaced by letters to simplify subscripting.

	Grade					Σ
	F	D	C	B	A	
FI	10	20	30	18	12	90
FD	18	36	28	12	6	100
Σ	28	56	58	30	18	190

The calculations for some of the expected frequencies are as follows:

$$E_{\text{FI,F}} = \frac{90(28)}{190} = \frac{252}{19} \doteq 13.27$$

$$E_{\text{FI,D}} = \frac{90(56)}{190} = \frac{504}{19} \doteq 26.53$$

$$E_{\text{FI,C}} = \frac{90(58)}{190} = 27.47$$

$$E_{\text{FI,B}} = \frac{90(30)}{190} = 14.21$$

The remaining expected frequencies can be found by subtraction, because the sum of expected frequencies for any column must add to the column total, and similarly for rows. Thus, $E_{FD,F} = S_F - E_{FI,F} = 28 - 13.27 = 14.73$. The expected frequency for the A column, FI row can be found by subtracting all the other expected frequencies in that row from the row total, 90, giving 8.53. The completed table of expected frequencies is as follows:

	Grade					
	F	D	C	B	A	Σ
FI	13.27	26.53	27.47	14.21	8.53	90
FD	14.73	29.47	30.53	15.79	9.47	100
Σ	28	56	58	30	18	190

Note that the null hypothesis of similar shapes among the distributions leads to no tests of the marginal sums or totals: these are taken as givens and used in calculating the expected frequencies. You can test hypotheses about the marginal frequencies using these same data, if you wish, although it is seldom done. For example, you could test the null hypothesis that there are equal proportions of field-independent and field-dependent subjects in the population; or that the population distribution of grades was normally distributed, using part A of this section, provided that you had randomly sampled subjects with respect to the hypothesis being tested.

10-2 Decisions based on sample size and expected frequencies

Relevant earlier sections:
7-6 Chi-square distribution
8-8 Normal approximation to binomial distribution

In Section 8-8, the area under a continuous distribution (the normal) was used to approximate the area under a corresponding discrete one (the binomial). The approximation was indicated to be poor when samples were small or the parameters p and q were discrepant. A rule of thumb was indicated to determine when the normal distribution could give answers reasonably close to those given by the binomial. This approximation is used in Section 10-9.

Similarly, chi square is used in Sections 10-3 and 10-7 as a continuous-variable approximation to a discrete distribution, the multinomial. The formula for the multinomial distribution is a generalization of the binomial formula (formula 8-1). If an event is repeated n times, and on each occasion there are J mutually exclusive and exhaustive possible outcomes (instead of only two possible outcomes, as in the binomial case), then the probability of obtaining r_1 ocurrences of outcome 1 and r_2 occurrences of outcome 2, and so on, and r_J occurrences of outcome J is

$$\frac{n!}{r_1!\, r_2! \cdots r_J!} p_1^{r1} p_2^{r2} \cdots p_J^{rJ}$$

where $p_1, p_2, \cdots, p_J$ are the probabilities of the outcomes, with $p_1 + p + \cdots + p_J = 1.00$ for each of the n trials.

As in the case of the binomial distribution, the approximation of the multinomial distribution by chi square is none too good unless certain parameters (the E's in this case) are large enough. The generally accepted criteria for the approximation are given in the map. However, many authorities feel that these criteria are not cautious enough; for example, Tate and Hyer (1973) recommend that all E's be greater that 20, based on a comparison of multinomial probabilities with chi-square approximations. I have decided not to use their recommendations here, mainly to avoid presenting the complexities of the multinomial distribution in an introductory course. For further discussion of the chi-square approximation, see Tate and Hyer; Hays (1963), 580–581; or Li (1964), Chapter 22.

10-3   Use of chi square to compare proportions or entire distributions

The general form of the chi-square approximation used for this purpose is quite unlike its definition as given in Section 7-6. However, its use in this form is more widespread and familiar than the earlier formula.

Three kinds of hypotheses may be tested using the chi-square approximation. Several sample proportions may be compared and an inference made to corresponding population proportions:

$$H_0 : \pi_1 = \pi_2 = \cdots = \pi_J$$

$$H_m : \text{It is not true that } (\pi_1 = \pi_2 = \cdots = \pi_J)$$

You may hypothesize that the population from which a sample was actually drawn (say, an experimental population) differs in some way from a null population:

$$H_0 : \text{Pop. Dist. (expt'l.)} = \text{Pop. Dist. (null)}$$

$$H_m : \text{Pop. Dist. (expt'l.)} \neq \text{Pop. Dist. (null)}$$

Finally, two or more entire population distributions may be compared:

$$H_0 : \text{Dist. (1)} = \text{Dist. (2)} = \cdots = \text{Dist. } (J)$$

$$H_m : \text{not } [\text{Dist. (1)} = \text{Dist. (2)} = \cdots = \text{Dist. } (J)]$$

I like to think of this application of chi square as asking the question, "Do the percent histograms of the population distributions differ in any respect at all?" This allows me to visualize a graph for each distribution, and to see the differences between distributions as differences in the heights of corresponding bars.

For any table of frequencies, you will have an observed frequency in each cell; on some basis you should be able also to calculate an expected frequency for each cell. Then for each cell of the table, you can obtain an index of the magnitude of discrepancy between what you expected under the null hypothesis and what you actually found by calculating the ratio

$$\frac{(O - E)^2}{E}$$

where O is the observed frequency and E is the expected frequency for that cell Note (1) that this ratio is always positive, since the denominator is a frequency and the numerator is squared, and (2) the bigger the difference between observed and expected frequencies for any cell, the bigger this ratio is.

The value of chi square is then found by adding these ratios across the cells of the table. For a single variable of classification, the formula is

$$\chi^2 = \sum_j \frac{(O_j - E_j)^2}{E_j} \tag{10-2a}$$

and for two variables of classification, the formula becomes

$$\chi^2 = \sum_j \sum_k \frac{(O_{jk} - E_{jk})^2}{E_{jk}} \tag{10-2b}$$

In the latter case, one variable of classification takes the j subscript and the other the k subscript, and you simply add up the ratios over all combinations of j and k.

The calculated value of chi square must always be positive, as it is the sum of positive terms (the ratios); and the bigger the total amount of differences in the table, the bigger chi square will be. The magnitude of χ^2 is an index of the overall dissimilarity between the observed frequencies and the expected frequencies. At some point, the observed χ^2 gets so large that we no longer accept the null hypothesis on which it is based. This occurs when the probability of obtaining a value of χ^2 that large or larger, by chance, is less than our chosen

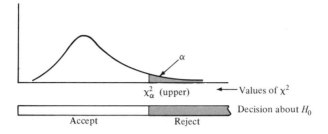

Figure 10-1 *Theoretical distribution of χ^2 and decision regions for the null hypothesis.*

α. We find that the observed χ^2 falls in the rejection region, so we reject the null hypothesis, as shown in Figure 10-1.

EXAMPLE Let us consider the example of Section 10-1A in which an observed distribution was to be compared to a theoretical one. Our null hypothesis is that the sample is drawn from a population of subjects who have no information at all and, therefore, that the population score distribution is binomial with $p = q = .50$. This may be stated formally as follows:

$$H_0 : \text{Pop. Dist. } (X) = \text{Bin. Dist. } (n = 5, p = .50)$$
$$H_m : \text{Pop. Dist. } (X) \neq \text{Bin. Dist. } (n = 5, p = .50)$$

The design is represented by Figure 9-1, where the null population is a hypothetical population of subjects who have no information at all.

We decide ahead of time to use an α level of .05. Then the completed data table is given as follows: (scores of 0 and 1 have been combined in accordance with Section 10-4):

Score	Observed frequency O	Expected frequency E	$O - E$	$(O - E)^2$	$(O - E)^2/E$
0–1	2	9	−7.0	49.00	5.44
2	10	15	−5.0	25.00	1.66
3	15	15	.0	.00	.00
4	14	7.5	6.5	42.25	5.63
5	7	1.5	5.5	30.25	20.17
Σ	48	48			$\chi^2 = 32.90$

The test is completed by comparing the observed chi square of 32.90 to the critical value of chi square at α of .05 for $K - 1 = 5 - 1 = 4$ df (one less than the number of categories; for explanation, see Section 10-7).

$\chi^2_{.05}$ (4 df) = 9.488. The observed chi square is greater than the critical value, hence the null hypothesis is rejected.*

EXAMPLE Let us now test the example of Section 10-1B, in which two observed distributions were compared: the grade-point averages of field-independent and field-dependent subjects. The null hypothesis equates the population distributions:

$$H_0 : \text{Pop. Dist. (FI)} = \text{Pop. Dist. (FD)}$$

The design is given in Figure 9-6. Let us arbitrarily set α at .05 and decide to collect a total sample of 190 subjects. Then the critical value of chi square will be the value cutting off the upper .05 of the chi-square distribution for $(J - 1)(K - 1) = (2 - 1)(5 - 1) = 4$ df. As in the previous example, this value is 9.488.

The frequency table is given below, taken from Section 10-1. In each cell of the table, the observed frequency is given in the upper part of the cell and the expected frequency below it.

| | Grade | | | | | |
	F	D	C	B	A	Σ
FI	$O = 10$ $E = 13.27$	$O = 20$ $E = 26.53$	$O = 30$ $E = 27.47$	$O = 18$ $E = 14.21$	$O = 12$ $E = 8.53$	90
FD	$O = 18$ $E = 14.73$	$O = 36$ $E = 29.47$	$O = 28$ $E = 30.53$	$O = 12$ $E = 15.79$	$O = 6$ $E = 9.47$	100
Σ	28	56	58	30	18	190

Then,

$$\chi^2 = \sum_j \sum_k \frac{(O_{jk} - E_{jk})^2}{E_{jk}}$$

$$= \frac{(10 - 13.27)^2}{13.27} + \frac{(20 - 26.53)^2}{26.53} + \frac{(30 - 27.47)^2}{27.47} + \cdots + \frac{(6 - 9.47)^2}{9.47}$$

$$= \frac{(-3.27)^2}{13.27} + \frac{(-6.53)^2}{26.53} + \cdots + \frac{(-3.47)^2}{9.47}$$

* One could also test for no knowledge by assuming that scores are approximately normally distributed and comparing the observed mean of 3.29 to the theoretical mean of 2.50 for these data, using the method of Section 12-5.

$$= \frac{10.69}{13.27} + \frac{42.64}{26.53} + \cdots + \frac{12.04}{9.47}$$

$$= 0.80 + 1.61 + 0.23 + \cdots + 1.27$$

$$= 9.63$$

This calculated value of chi square is greater than the critical value of 9.488; thus, the probability of obtaining a χ^2 this big when the null hypothesis is true is less than .05, and the null hypothesis is rejected.

Note that we could test either set of marginal totals, if we wanted, to see whether they conformed to a predicted distribution. If we were checking to see whether the grade distribution was rectangular, we would expect $190/5 = 38$ students to get each grade; or we could compare the observed distribution to any other set of grade proportions that were of interest.

Two other questions can be asked of these data. The first is, "What prediction shall we make about a subject's status on one of the variables?" In the field dependence–independence example, we might be interested in students' grades. Then we might want to predict grades for the entire sample, or just for the FI subjects, or just for the FD subjects. For whichever group we chose, we could use either the median or the mode to make the prediction, as indicated in Chapter 5 for ordinal data. This issue is discussed further in Section 15-2.

A second issue is, "How well can we predict a subject's status on one of the variables from his or her status on the other?" This question is equivalent to asking how big the difference is between the two distributions, or, what the correlation or degree of relationship is between them. The way you answer the question depends on the level of measurement you attribute to the two variables; and the best way to approach the problem is to have a look at the map for Chapter 17. In the example of the field independent–dependent subjects, if I were really interested in any differences between the two distributions (and not just which group got higher grades), I would probably treat both variables of classification as nominal and use λ (Section 17-2) as a measure of the degree of relationship between them.

References This application of chi square is discussed in Edwards (1960), 63–69, and (1973), 135–138; Hays (1963), 578–595; McNemar (1969), 266–267; and Siegel (1965), 42–47, 104–111, 175–179.

10-4 Combining categories and other strategies for dealing with small expected frequencies

As indicated in Section 10-2, chi square is a continuous approximation to a discrete distribution, which is reasonably accurate when the expected frequencies are large. When expected frequencies are too small, however, use of the general chi-square formula (Section 10-3) leads to a distressingly large error rate.

There are three ways in which small expected frequencies can be handled. The most generally applicable and satisfying solution is probably to draw a larger sample. With greater total sample size, the expected frequencies in all the cells of the table should increase until, eventually, the criteria of large E's are satisfied. A second-best solution for the case of a 1×2 or 2×2 table is to use one of the special tests provided in this chapter. With large matrices, the usual way of handling small expected frequencies is to combine categories for one or the other variable until the criteria are met. This is less than ideal, because in combining categories you are grouping things together that you originally chose to keep separate. By doing so, you lose some of the information you originally wanted, and you may distort the results so much that you can no longer answer your original question.

EXAMPLE In a pilot study for a questionnaire on marital happiness, subjects have been asked the question, "Do you consider yourself generally happy or unhappy with your present marital status?" Subjects had already indicated their status to be single, married, widowed, or divorced. The problem is to determine whether some groups report themselves happier than others. These are the data:

Observed	Single	Married	Widowed	Divorced	Σ
Happy	7	11	0	1	19
Unhappy	9	10	7	5	31
Σ	16	21	7	6	50

The most appropriate analysis would seem to be a general chi square, in accordance with Section 10-3. However, as a preliminary step you calculate the expected frequencies according to the method of Section 10-1B.

$$E_{hs} = \frac{19(16)}{50} = 6.08$$

$$E_{hm} = \frac{19(21)}{50} = 7.98$$

and so on. This is the completed table of expected frequencies:

Expected	Single	Married	Widowed	Divorced	Σ
Happy	6.08	7.98	2.66	2.28	19
Unhappy	9.92	13.02	4.34	3.72	31
Σ	16	21	7	6	50

In this table of expected frequencies, four of the eight, or 50 percent, are smaller than 5. Therefore, the general chi square will probably be greatly in error as a test of the hypothesis that all four marital groups have equal proportions of persons who describe themselves as happy.

There is no feasible exact procedure for such a case (they exist for 1×2 and 2×2 tables, but the multinomial distribution for a 2×4 table is too messy to calculate). Therefore, two possibilities are open: (1) collect data on more subjects, or (2) collapse the data table by combining categories.

If you decide to collect more data, continue sampling subjects as before until the criterion for application of chi square is met. In the present example, we can afford to have one E smaller than 5 (one cell out of eight is 12.5 percent of the E's; two cells would be 25 percent, which would be greater than the 20 percent permitted). If we continue sampling in the same manner, all cell frequencies should increase in about the same ratio. We will permit the smallest E to be less than 5, so let us concentrate on the second smallest, which must be 5 or more. Because all cells will probably increase at about the same rate, we find the new sample size using the formula

$$\frac{E'_{hw}}{E_{hw}} = \frac{N'}{N}$$

where E'_{hw} is the expected frequency (5) needed in the second-smallest cell, and N' is a best guess of the total number of subjects needed. Inserting values into this equation gives

$$\frac{5}{2.66} = \frac{N'}{50}$$

Then

$$N' = \frac{50(5)}{2.66} = \frac{250}{2.66} = 94$$

Therefore, we should probably draw at least 94 subjects in all, or 44 in addition to those already drawn, in order to test this hypothesis as it now stands.

Alternatively, we might test the hypothesis using the data already collected by combining cells of the frequency matrix. Let us collapse the last two columns of the table to get a reduced frequency matrix, as follows:

Observations	Single	Married	Widowed or divorced	Σ
Happy	7	11	1	19
Unhappy	9	10	12	31
Σ	16	21	13	50

The expected frequencies are calculated as before, giving

Expected	Single	Married	Widowed or divorced	Σ
Happy	6.08	7.98	4.94	19
Unhappy	9.92	13.02	8.06	31
Σ	16	21	13	50

In this reduced frequency matrix, 5 out of 6, or 83 percent, of the E's are greater than 5, and no E is less than 1, so the general formula for chi square can now be used. As suggested earlier, however, we now have one messy category: "Widowed or divorced." Before going on with the analysis, we must decide whether we can get enough information out of a table containing this category to make the study meaningful, or whether we have distorted the data too much by the combination. This is a conceptual decision, based on our knowledge of widowed and divorced people and on the kind of results we want to have.

If we decide to complete the analysis, we will find that for the reduced table, $\chi_{.05} = 5.99146$. Also for the reduced table, the calculated χ^2 is 7.14, so the results are significant at the .05 level. It is easy to see from the table the modal response of each of the three groups; to show the degree of relationship between marital status and happiness, we could calculate lambda for these same data (Section 17-2).

10-5   Size of matrix

The data to be used in the methods of this chapter are in the form of frequencies, with the frequencies appearing in matrices. For convenience in selecting a method of analysis, the observed levels of the variables are listed at the decision points of the flow chart. In a 1 × 2 table, there is one observed distribution that has two levels, for example, when you ask whether more than 50 percent of right-handed men have dominant right eyes. There is only one distribution, that of right-handed men. The variable of interest is eye dominance, which has two levels, "right" and "left." In this example, expected proportions are specified a priori to be .5 and .5.

In a 2 × 2 table, two distributions are compared along a variable that has two levels; for example, "Is the proportion of men with dominant right eyes greater for right-handed or for left-handed men?" Here you may decide that the variable to be compared is eye dominance, and that the distributions are right-handed and left-handed men, or you may prefer to think of men in general as classified along two variables: handedness and eye dominance. Your interpretation of the problem may affect the manner in which you gather your samples. If you think of having two distinct distributions, you will probably try to collect about equal numbers of right-handed and left-handed men. On the other hand, if you think of your reference population as men in general, you will probably collect your subjects randomly, which will affect the proportions of left-handed and right-handed men in the total sample.

In the general $J \times K$ case, there are J distributions and K levels of the comparison variable; or if you prefer, J levels of one variable and K of the other. For example, you might compare subjects in different marital categories on the type of physical symptoms that they find most troublesome. The marital categories might be single, married, widowed, and divorced. The symptoms might be poor appetite, headaches, feelings of weakness, stomach ulcer, general restlessness, and allergies. There would be four marital categories and six symptoms, giving a 4 × 6 matrix of frequencies.

If you wished to compare married men and women on this same set of symptoms, you would have a 2 × 6 matrix. A 1 × 6 matrix would be in order if you considered only unmarried males and predicted a priori that there would be equal frequencies in each of the six symptom categories.

In all the above examples, both variables of classification have been nominally scaled. The procedure works equally well with ordinal and interval scales, however. The only problem is that sometimes it is difficult to decide how to break up a continuous distribution into discrete categories for the purpose of doing the analysis. You might, for example, wish to determine whether scores on the Semantic Aberration Test differed significantly from normal. Once you decided how to divide the distribution of scores into J categories, you would have a 1 × J matrix of frequencies, which could be tested for normality like any other 1 × J matrix.

10-6   Determining the number of degrees of freedom in the chi-square approximation

A. General Rule
The general rule to be applied in determining the number of degrees of free-

dom for a given chi-square problem is found from the number of observation cells minus the number of separate values needed to estimate all the expected frequencies. This general rule can be exemplified by considering two special applications of it.

B. Application to $1 \times K$ Frequency Tables for Comparing Observed and Theoretical Distributions

When you are comparing an observed and a theoretical distribution, with one variable of classification, the number of degrees of freedom is $K - a - 1$, where K is the number of levels of the variable of classification and a is the number of independent parameters of the theoretical distribution that must be estimated from the observed data.

The a parameters allow us to specify the shape and/or location of the chosen theoretical distribution that best fits the observed data (i.e., the percentages of subjects in each of the specified categories of the theoretical distribution). If we are fitting a rectangular distribution, $a = 0$, because we can specify in advance of looking at the data that the frequencies in all categories of the theoretical distribution must be equal. To find out how closely a set of data fits a binomial distribution, we need to know two parameters of the binomial distribution, n and p, and we lose 2 df in estimating them from the data. In this case, $a = 2$ for the two bits of information taken from the data.

The one additional degree of freedom is lost due to the fact that the chi-square distribution deals with frequencies, whereas theoretical distributions are generally described by specifying the proportions of the distribution within each scale value or region of the distribution. To use the chi-square test as given in this chapter, we must first multiply each of the theoretical proportions specified by the theoretical distribution by N, the number of observations in our sample: this gives us the expected frequencies for the theoretical distribution, against which the observed frequencies of our actual distribution can be compared. We lose one additional degree of freedom for using the value of N, our total sample size, to convert the proportions to frequencies.

EXAMPLE Suppose that you toss a coin a large number of times to determine whether it is biased. You are comparing it to a theoretical distribution of results in which there are equal numbers of heads and tails. The variable of classification is "side," which has two levels: heads and tails. The theoretical distribution is rectangular, for an unbiased coin by definition comes up heads and tails equally often. Thus we can specify the proportions of observations to be expected in each category (heads, tails) without reference to the observed distribution, and $a = 0$. However, on the basis of Section 10-3 we know that one marginal value (the total number of subjects, N) is necessary to obtain expected frequencies from the theoretical distribution. Hence the number of degrees of freedom given by the general rule is the number of observation cells minus the number of values needed to apply the theoretical distribution: $df = (K - a - 1) = (2 - 0 - 1) = 1$.

EXAMPLE Suppose that we wish to determine whether scores on a 30-item multiple-choice test depart significantly from normality. Let $n = 30$, the number of items. Then there are $n + 1 = 31$ possible scores, ranging from 0 to 30, hence $K = 31$ levels of classification along the variable "score." From Section 8-8, we know that we may specify for the theoretical distribution a mean $np = 30p$ and standard deviation $\sqrt{npq} = \sqrt{30pq}$, under the assumption that each item is responded to independently from each other, where p is the a priori probability of passing each item and $q = 1 - p$. (We must assume that p is the same for every item, i.e., that all items are equally difficult.) From the normal distribution so specified, we may obtain an expected frequency for each possible score, using Table A-4 and N, the number of subjects. If we assume that subjects guess randomly to all items and that there are four response alternatives to each item, then the probability of guessing the correct response to an item is $p = .25$ (Section 8-9A) and $q = .75$. In this case, $\mu = np = 30(.25) = 7.5$ and $\sigma = \sqrt{npq} = \sqrt{(.30)(.25)(.75)} \doteq 2.37$. Only N is taken from observations, so $df = K - a - 1 = 31 - 0 - 1 = 30$.

If, on the other hand, we do not assume complete randomness of response, we cannot reasonably specify μ and σ of the best-fitting normal distribution a priori. In this case, we may choose to estimate the two parameters on the basis of sample statistics, and merely test the distribution for normality of shape. If so, $a = 2$ and the degrees of freedom for the test are $df = K - a - 1 = 31 - 2 - 1 = 28$.

In addition, it may be necessary to reduce K by combining categories because of small expected frequencies (see Section 10-4), unless N is extremely large. With combined categories the rule would remain the same; the number of degrees of freedom would be equal to the number of categories in the revised data table minus the number of parameters estimated, minus 1.

C. Application to the Comparison of Two or More Observed Distributions

The rule usually given to compute degrees of freedom in this situation is $df = (J - 1)(K - 1)$, where J and K are the numbers of rows and columns in the data matrix. Let us consider the case of a 3×4 matrix to show that this is a special instance of the general rule. A frequency table of observations is given schematically as Table 10-3. In this matrix, the O's are observed frequencies; T_1, T_2, and T_3 are row totals; and S_1, S_2, S_3, and S_4 are column sums.

Table 10-3 *Schematic representation of a 3 × 5 frequency matrix of observations*

Variable A	Variable B					Σ
	1	2	3	4	5	
1	O_{11}	O_{12}	O_{13}	O_{14}	O_{15}	T_1
2	O_{21}	O_{22}	O_{23}	O_{24}	O_{25}	T_2
3	O_{31}	O_{32}	O_{33}	O_{34}	O_{35}	T_3
Σ	S_1	S_2	S_3	S_4	S_5	N

From the special rule we would find df $= (3 - 1)(5 - 1) = 8$. Under the null hypothesis, all the expected frequencies are determined from marginal totals (see Section 10-1B), but the marginal totals are determined by the observed frequencies. Therefore, we lose a degree of freedom for each independent marginal value.

For this problem, we must know N and the totals for the three rows, T_1, T_2, and T_3. But if we know any two row totals and N, we can get the third total by subtraction (e.g., $T_2 = N - T_1 - T_3$). We also need to know five column sums; but if we know any four and N, we know the fifth. Hence we need to know N, two row totals, and four column sums, for a total of seven independent marginal values, in order to calculate expected frequencies for the cells of the table. There are $5 \times 3 = 15$ cells, so the number of degrees of freedom by the general rule is $15 - 6 - 1 = 8$. This is the same number of degrees of freedom as given by the formula.

In general, for J rows and K columns, there are JK cells to the table. We need the value of N, plus $J - 1$ row totals, plus $K - 1$ column totals, to calculate the expected frequencies. Therefore, the number of degrees of freedom is given by

$$\begin{aligned} df &= JK - 1 - (J - 1) - (K - 1) \\ &= JK - 1 - J + 1 - K + 1 \\ &= JK - J - K + 1 \\ &= (J - 1)(K - 1) \end{aligned}$$

Thus, the special formula can be derived from the general rule and the nature of the $J \times K$ table.

10-7 Application of chi square to a 2 × 2 frequency matrix

Relevant earlier section:
10-3 Chi-square test

In this section, the chi-square statistic is applied to comparison of two distributions, where each distribution has only two possible scores, or equivalently, to the comparison of two proportions. The design is given by Figure 9-3 or 9-6. The statistic used is a special case of the general chi-square statistic for comparing distributions (Section 10-3), which in the general case is always nondirectional. For 2 × 2 tables, however, a directional test can be developed from it. This section has three parts: part A shows how the general formula can be simplified for the special case of 2 × 2 tables; part B modifies the formula to produce one that is more accurate; and part C extends the application to one-tailed tests. For parts A and B, the null hypothesis can be written either

$$H_0 : \pi_1 = \pi_2 \quad \text{or} \quad H_0 : \text{Pop. Dist. (1)} = \text{Pop. Dist. (2)}$$

where π is a population proportion of some attribute and the subscripts distinguish two populations.

A. A First Approximation

The chi-square statistic as defined in Section 10-3 may be applied directly to this problem; however, the approximation to the exact multinominal distribution is rather poor. When the frequency table is labeled like Table 10-4, the chi-square formula (10-2)

$$\chi^2 = \sum_j \sum_k \frac{(O_{jk} - E_{jk})^2}{E_{jk}}$$

can be shown to simplify to formula 10-3:

$$\chi^2 = \frac{N(ad - bc)^2}{S_1 S_2 T_1 T_2} \tag{10-3}$$

Table 10-4 *General form of a 2 × 2 frequency table*

Distribution	Comparison variable		$\sum$
	Level 1	Level 1	
1	a	b	$T_1 = a + b$
2	c	d	$T_2 = c + d$
$\sum$	$S_1 = a + c$	$S_2 = b + d$	N

B. A Better Approximation: Yates' Correction

The chi-square formula used in the present context is only an approximation to the exact solution, which is given by the multinominal distribution. For a

2 × 2 table, the approximation is not very good, but it can be improved by making a small correction in the formula. The corrected formula is

$$\chi^2 = \frac{N[|ad - bc| - (N/2)]^2}{S_1 S_2 T_1 T_2} \tag{10-4}$$

This is called *Yates' correction*. [If $|ad - bc| < (N/2)$, set chi square equal to zero. The bars on both sides of the expression $ad - bc$ indicate that you always consider this difference to be positive (even if bc is larger than ad).]

EXAMPLE You wish to compare right-handed and left-handed males on eye dominance. To do this, you select 15 males from each population and observe which eye is dominant for each. The hypotheses can be stated in terms of expected proportions of right-handed and left-handed males who have right-eye dominance:

$$H_0 : \pi \begin{pmatrix} \text{right-eye dominant} \\ \text{given right handed} \end{pmatrix} = \pi \begin{pmatrix} \text{right-eye dominant} \\ \text{given left handed} \end{pmatrix}$$

$$H_m : \pi \begin{pmatrix} \text{right-eye dominant} \\ \text{given right handed} \end{pmatrix} \neq \pi \begin{pmatrix} \text{right-eye dominant} \\ \text{given left handed} \end{pmatrix}$$

The hypotheses can also be stated, perhaps even more clearly, in terms of the cells of the 2 × 2 matrix:

$$\left. \begin{aligned} H_0 &: \text{Pop.}\left(\frac{a}{a+b} = \frac{c}{c+d}\right) \\ H_m &: \text{Pop.}\left(\frac{a}{a+b} \neq \frac{c}{c+d}\right) \end{aligned} \right\} \quad \text{or} \quad \left\{ \begin{aligned} H_0 &: \text{Pop.}\left(\frac{a}{a+c} = \frac{b}{b+d}\right) \\ H_m &: \text{Pop.}\left(\frac{a}{a+c} \neq \frac{b}{b+d}\right) \end{aligned} \right.$$

and it doesn't matter whether you state the hypotheses in terms of row proportions of column proportions. The statistic, critical value, observed value, and conclusions will be the same both ways.

The design is given in Figure 9-6. Alpha is chosen to be .05, and degrees of freedom are $(J - 1)(K - 1) = (1)(1) = 1$. The critical value of chi square is 3.84146. The data appear in the following table, with expected frequencies calculated in accordance with Section 10-1B:

Handedness	Dominant eye Right	Left	Σ
	Right	Left	Σ
Right	$O = 9$ $E = 6$	$O = 6$ $E = 9$	15
Left	$O = 3$ $E = 6$	$O = 12$ $E = 9$	15
Σ	12	18	30

We could apply either of the formulas of part A to these data, obtaining the same value of chi square. From formula 10-2 we would get

$$\chi^2 = \frac{(9-6)^2}{6} + \frac{(6-9)^2}{9} + \frac{(3-6)^2}{6} + \frac{(12-9)^2}{9}$$

$$= 1.5 + 1 + 1.5 + 1$$

$$= 5$$

From formula 10-3 we also obtain

$$\chi^2 = \frac{30(9 \cdot 12 - 3 \cdot 6)^2}{15 \cdot 15 \cdot 12 \cdot 18} = \frac{2(108 - 18)^2}{15 \cdot 12 \cdot 18} = \frac{8100}{15 \cdot 18 \cdot 6} = \frac{900}{15 \cdot 12} = \frac{75}{15} = 5$$

So much for establishing the equivalence of the two formulas, since they both overestimate the magnitude of chi square. The corrected formula applied to these same data gives

$$\chi^2 = \frac{30(|108 - 18| - 15)^2}{15 \cdot 15 \cdot 12 \cdot 18} = \frac{30(75)^2}{(15)^2 \cdot 12 \cdot 18} = \frac{5(5)^2}{2 \cdot 18} = \frac{125}{36} \doteq 3.47$$

The corrected value of 3.47 is less than the critical value of 3.84146; therefore, the null hypothesis is not rejected. Yates' correction always gives a more conservative value of chi square and should always be used when a chi-square solution can be found for a 2 × 2 table.

C. Prediction and Correlation

If we wanted to use these data to predict eye dominance of some new subject on the basis of his or her handedness, we would predict right-eye dominance if he or she was right-handed and left-eye dominance if left-handed (12 of 15 subjects in the latter group were left-eye dominant). Almost any correlation coefficient could be used to show the degree of relationship between the two variables, inasmuch as both are dichotomous and can be treated as nominal, ordinal, or interval level. In context, I would choose a coefficient that would be the same as others I might want to compare these results to. Otherwise, using the map for Chapters 16 and 17, I would probably use phi.

D. Directional Test

In this case, the motivated hypothesis is that one proportion is greater than the other in the 2 × 2 table:

$$H_m : \text{Pop.}\left(\frac{a}{a+b} > \frac{c}{c+d}\right) \quad \text{or} \quad H_m : \text{Pop.}\left(\frac{a}{a+c} > \frac{b}{b+d}\right)$$

where the letters correspond to frequencies in Table 10-4.

The first step after collecting the data is to see whether the sample proportions are unequal in the direction predicted by the motivated hypothesis. If not, you can accept the null hypothesis and stop. If the proportions are unequal as predicted, then the next step is to find out if they are significantly unequal.

In this case, the formula is 10-4, the same as used in the nondirectional case, and the computational procedures are the same. The only difference is that the chi-square table (Table A-7) is inappropriate, as formula 10-4 is inherently a nondirectional statistic, even when it is being compared to only one tail of the chi-square distribution. We could compare our computed value of χ^2 to $\chi^2_{.05, \text{upper tail}}$, but this value would be too large: it is the value to be used when checking whether either proportion in the table is larger than the other. The correct values are also in the upper tail of the χ^2 distribution but at double the α level that we use for a nondirectional test. The correct values don't appear in Table A-7 because of space limitations, so they are presented in Table 10-5.

Table 10-5 *Critical values of χ^2 for directional hypotheses about proportions*

α for directional test	Corresponding α for nondirectional test	Critical value
.05	.10	2.706
.025[†]	.05[†]	3.841[†]
.01	.02	5.412
.005[†]	.01[†]	6.635[†]

[†] These values are included for comparison only, as we usually don't test at α's of .025 or .005.

At the .05 level, the benefit to be gained from using directional rather than nondirectional hypotheses can be seen in the difference between .05 critical values for the two cases: when we use directional hypotheses, our χ^2 only has to achieve 2.706 to be significant, whereas for a directional hypothesis it must be at least 3.841; and similarly for the .01 level.

References   Yates' correction is discussed in Edwards (1960), 69–71, 73, and (1973), 138–139; Hays (1963), 585–586, 595–596; and Siegel (1956), 107–109. McNemar (1969), 263 discusses the directional test presented here.

10-8   Fisher exact probability test

Relevant earlier section:
8-7   Binomial distribution (for factorials)

A. Use of the Fisher Test

The Fisher test can be used with a 2×2 matrix of frequencies to determine the probability of obtaining a given set of cell frequencies by chance when the marginal frequencies are assumed to be fixed. You might, for example, compare two subject populations for the proportion of each having a given characteristic. Thus, you might entertain the null hypothesis that males and females do not differ with respect to handedness, or that as great a proportion of males as females is right-handed. In both of these examples, the design would be given by Figure 9-6; but Figure 9-3 could also be used with this section. The data are cast into a 2×2 table such as the following, which is the same as Table 10-4:

	Variable 2		
Variable 1	First level	Second level	Σ
First level	a	b	$a + b$
Second level	c	d	$c + d$
Σ	$a + c$	$b + d$	N

In this case, variable 1 could be sex and variable 2 handedness; then each subject could be uniquely assigned to one cell of the table based on his or her status on these variables. The letters in the cells of the table are frequencies of subjects at each combination of levels on the two variables.

For the first of the example hypotheses – that males and females do not differ with respect to handedness – the null hypothesis could be written

$$H_0 : \pi_m = \pi_f \quad \text{or} \quad H_0 : \text{Pop.}\left(\frac{a}{a + b} = \frac{c}{c + d}\right)$$

where π is a population proportion of right-handed persons. For the second example, the null hypothesis – that as great a proportion of males as females is right-handed – could be written

$$H_0 : \pi_m \geq \pi_f \quad \text{or} \quad H_0 : \text{Pop.}\left(\frac{a}{a+b} \geq \frac{c}{c+d}\right)$$

The test is appropriate for any 2 × 2 table, but the calculations become unnecessarily burdensome when N is large. In the latter case, use chi square with Yates' correction, as shown in Section 10-7. The criteria for choice of method are given in the map for the chapter.

B. Calculations for a Directional Test

The first step in applying Fisher's test is to place the data in a 2 × 2 frequency table and to calculate the marginal totals by adding across each row and column, as indicated in the table.

We can tell at the outset whether the data are in the predicted direction or not, and if they are not, stop the analysis at that point. Suppose, for example, that we predicted a greater proportion of right-handed males than females $(H_m : \pi_m > \pi_f)$ and observed the following distribution of frequencies:

	Right-handed	Left-handed	Σ
Males	3	6	9
Females	5	1	6
Σ	8	7	15

Here it is obvious that the proportion of right-handed males $(3/9 = 1/3)$ is less than the proportion of right-handed females $(5/6)$, so there is no point in going on: the results are in the wrong direction.

If the results are in the direction predicted by the motivated hypothesis, then we ask the statistical question, "What is the probability of obtaining data that extreme or more so, in the predicted direction, when the null hypothesis is true?" If the probability is very small, (smaller than our α level), then we reject the null hypothesis.

Formula 10-5 gives the exact probability for a particular set of cell frequencies, given the marginal totals.

$$p = \frac{(a+b)!(c+d)!(a+c)!(b+d)!}{N!\,a!\,b!\,c!\,d!} \tag{10-5}$$

We can use it to calculate the probability of the one set of data that we actually obtained. However, it doesn't give the probability of our obtained set of data or any more extreme set: to find out that probability, we must (1) determine what cell-frequency configurations would have been more extreme, given the current marginal totals; (2) calculate the probability for each of the more extreme configurations; and (3) add those probabilities to the probability for the observed data set.

EXAMPLE Suppose that our null hypothesis for a given problem is

$$H_m: \text{Pop.}\left(\frac{a}{a+b} < \frac{c}{c+d}\right)$$

and our observed data matrix turns out to be $\begin{array}{|cc|}\hline 2 & 6 \\ 6 & 2 \\\hline\end{array}$. First we add row

and column frequencies to obtain the marginal totals: $\begin{array}{|cc|c|}\hline 2 & 6 & 8 \\ 6 & 2 & 8 \\\hline 8 & 8 & 16 \\\hline\end{array}$. The

data matrix most consistent with no difference between the populations

is $\begin{array}{|cc|}\hline 4 & 4 \\ 4 & 4 \\\hline\end{array}$. We want to find out what the probability is of obtaining a

matrix as extreme as the one we observed, or more so, when the null hypothesis is true and there is no difference in the populations. Given the above marginal totals, the only more extreme matrices in the direction

of the motivated hypothesis are $\begin{array}{|cc|}\hline 1 & 7 \\ 7 & 1 \\\hline\end{array}$ and $\begin{array}{|cc|}\hline 0 & 8 \\ 8 & 0 \\\hline\end{array}$. Therefore, if

$$P\left\{\begin{array}{|cc|}\hline 2 & 6 \\ 6 & 2 \\\hline\end{array}\right\} + P\left\{\begin{array}{|cc|}\hline 1 & 7 \\ 7 & 1 \\\hline\end{array}\right\} + P\left\{\begin{array}{|cc|}\hline 0 & 8 \\ 8 & 0 \\\hline\end{array}\right\}$$

is less than our α level, we will reject the null hypothesis: the data are unusual in the predicted direction. If this sum of probabilities is greater than α, we will accept H_0.

Now suppose that we set α at .05. Then

$$P\left\{\begin{array}{|cc|}\hline 2 & 6 \\ 6 & 2 \\\hline\end{array}\right\} = \frac{8!\,8!\,8!\,8!}{16!\,2!\,6!\,6!\,2!} \doteq .061$$

$$P\left\{\begin{array}{|cc|}\hline 1 & 7 \\ 7 & 1 \\\hline\end{array}\right\} = \frac{8!\,8!\,8!\,8!}{16!\,1!\,7!\,7!\,1!} \doteq .005$$

$$P\left\{\begin{array}{|cc|}\hline 0 & 8 \\ 8 & 0 \\\hline\end{array}\right\} = \frac{8!\,8!\,8!\,8!}{16!\,0!\,8!\,8!\,0!} \doteq .00008$$

and

$$P\left\{\begin{array}{|cc|}\hline 2 & 6 \\ 6 & 2 \\\hline\end{array}\right\} + P\left\{\begin{array}{|cc|}\hline 1 & 7 \\ 7 & 1 \\\hline\end{array}\right\} + P\left\{\begin{array}{|cc|}\hline 0 & 8 \\ 8 & 0 \\\hline\end{array}\right\} = .066$$

This is greater than our chosen α level of .05, so we accept the null hypothesis.

If our observed data matrix had been $\begin{array}{|cc|} 1 & 5 \\ 6 & 3 \end{array}$, we would first find the

marginal totals: $\begin{array}{|cc|c} \hline 1 & 5 & 6 \\ 6 & 3 & 9 \\ \hline 7 & 8 & 15 \\ \hline \end{array}$. In this case there would be only one more

extreme matrix in the same direction: $\begin{array}{|cc|} 0 & 6 \\ 7 & 2 \end{array}$. The easiest way to find

the next more extreme matrix in each case is to subtract 1 from the frequency in the smallest cell, then see how the other frequencies must be distributed to give the same marginal totals. Here there can be only one matrix more extreme than the observed, because you can't have negative frequencies: so when you come to a matrix with a zero cell entry, it's the most extreme one. The calculations are shown below, where each matrix is labeled by the frequency in the smallest cell.

$$P(0) = \frac{9!\,6!\,8!\,7!}{7!\,2!\,0!\,6!\,15!}$$

$$= \frac{9!\,8!}{15!\,2!\,0!} = \frac{9!\,8 \cdot 7 \cdot 6 \cdot 5 \cdot 4 \cdot 3 \cdot 2 \cdot 1}{15 \cdot 14 \cdot 13 \cdot 12 \cdot 11 \cdot 10 \cdot 9!\,2 \cdot 1}$$

$$= \frac{4}{13 \cdot 55} = \frac{4}{715} = .0056$$

$$P(1) = \frac{9!\,6!\,8!\,7!}{15!\,3!\,6!\,5!\,1!}$$

$$= \frac{9!\,8!\,7 \cdot 6 \cdot 5!}{15 \cdot 14 \cdot 13 \cdot 12 \cdot 11 \cdot 10 \cdot 9!\,3!\,5!} = \frac{8!\,7}{15 \cdot 14 \cdot 13 \cdot 12 \cdot 11 \cdot 10}$$

$$= \frac{8 \cdot 7 \cdot 6 \cdot 5 \cdot 4 \cdot 3 \cdot 2 \cdot 7}{15 \cdot 14 \cdot 13 \cdot 12 \cdot 11 \cdot 10} = \frac{8 \cdot 7}{13 \cdot 11 \cdot 5} = \frac{56}{705} = .079$$

Then the probability of obtaining a distribution as extreme as the observed one in the specified direction is $P(0) + P(1) = .0056 + .079 = .086$. The observed probability, .086, is greater than the α level of .05, so the null hypothesis is not rejected. The result is not unlikely enough when H_0 is true to cause us to reject H_0.

C.   Calculations for a Nondirectional Test

For a nondirectional test, the motivated hypothesis is

$$H_m : \text{Pop.} \left(\frac{a}{a+b} \neq \frac{c}{c+d} \right),$$

which is the same as

$$H_m:\text{Pop.}\left(\left[\frac{a}{a+b}>\frac{c}{c+d}\right]\textcircled{v}\left[\frac{a}{a+b}<\frac{c}{c+d}\right]\right)$$

Once we get the observed data set, we can calculate the probabilities for that matrix and any more extreme in the same direction; but it is often hard to see what matrices could be more extreme in the other direction. (Try it for the second example of part B.) So we don't sum all the probabilities in both directions and compare with α.

Instead, we go through all the calculations exactly as for a directional test, finding the sum of probabilities for the observed distribution and all matrices more extreme in the same direction. Then we compare this sum of probabilities to $\alpha/2$, and reject the null hypothesis if the sum is less.

D. Predictions and Correlation
The methods used to make predictions and to correlate variables are not affected by the number of cases. Therefore, the recommendations here are the same as for the 2 × 2 chi square with Yates' correction: see Section 10-7C.

References The Fisher test can be found in McNemar (1969), 272–275; and Siegel (1956), 96–104.

10-9   Binomial test

Relevant earlier section:
8-9   Binomial distribution

A. Nature of the Test and Choice of Distribution
The binomial test is used when the observations can fall into either of two mutually exclusive and exhaustive classes, and when you want to determine whether a population falls into the two classes in specific proportions. The data can be collected using designs like Figures 9-1 and 9-2. Let us call the population that we are interested in the experimental population, and let us assume that subjects in the population can be divided into two classifications. Then a table for this population would appear as shown:

	Classification 1	Classification 2
Experimental population	π_ϵ	$1-\pi_\epsilon$

where π_ϵ and $1-\pi_\epsilon$ are population proportions of subjects. If we then randomly draw subjects from this population, they should fall into the two classifications in roughly the same proportions as does the population. If O_1 and O_2 are the observed frequencies of subjects in the sample, and if $O_1 + O_2 = N$, the sample size, then O_1/N is an estimate of π_ϵ, and O_2/N is an estimate of $(1 - \pi_\epsilon)$.

If we do not know π_ϵ but have a hypothesis about its value, we can test that hypothesis using the sample values O_1 and O_2. If we specify a null proportion π_0 against which we wish to compare our experimental proportion, there are three possible relationships between the two proportions:

(a) $\pi_\epsilon = \pi_0$
(b) $\pi_\epsilon > \pi_0$
(c) $\pi_\epsilon < \pi_0$

Our null hypothesis is formed from (a), either standing alone, in the case of a two-tail hypothesis, or in combination with either (b) or (c) in a one-tail test. In any case, the motivated hypothesis is formed from the remaining relationship(s). For a two-tail test,

$$H_0 : \pi_\epsilon = \pi_0 \qquad H_m : \pi_\epsilon \neq \pi_0 ; \quad \text{that is, } (\pi_\epsilon > \pi_0) \, \textcircled{V} \, (\pi_\epsilon < \pi_0)$$

while for a one-tail test, either

$$\left. \begin{array}{l} H_0 : \pi_\epsilon \leq \pi_0 \\ H_m : \pi_\epsilon > \pi_0 \end{array} \right\} \qquad \text{or} \qquad \left\{ \begin{array}{l} H_0 : \pi_\epsilon \geq \pi_0 \\ H_m : \pi_\epsilon < \pi_0 \end{array} \right.$$

Whatever the null hypothesis may be, it will be rejected if the sample proportion O_1/N is greatly different from the proportion specified by the null hypothesis. Given a total sample size N, we can say that the null hypothesis will be rejected if the frequency O_1 is greatly different from that predicted under the null for that category.

The test is carried out by means of the binomial distribution, with N for this problem corresponding to the n of Chapter 8, π_0 corresponding to p, and $(1 - \pi_0)$ corresponding to q. If we were to follow the decision rule of Section 8-8B, we would use the binomial distribution if either $E_1 = N\pi_0$ or $E_2 = N(1 - \pi_0)$ were less than or equal to 5, and the normal approximation if both terms were greater than 5. However, the binomial test is carried out using the tails of the distribution, where even a small error is large proportional to the area of interest. Therefore, we reduce the probability of error by using the normal approximation only if $E_1 = N\pi_0$ and $E_2 = N(1 - \pi_0)$ are both greater than 10.

B. Use of Binomial Distribution

We begin with a significance level α, and for a one-tail test find the α proportion of the binomial distribution at the extreme of the tail specified by the motivated hypothesis, counting in from the most extreme proportion possible until α of the distribution is divided off from the rest. We then draw a sample and observe the frequency O_1; if it appears in the rejection region, we reject the null hypothesis, otherwise we do not. Perhaps this can best be clarified by an example.

EXAMPLE Suppose for a problem that our hypotheses relate the proportion of subjects (π) in some experimental population (ϵ) having a given characteristic.

$$H_0: \pi_\epsilon \geq .3 \qquad H_m: \pi_\epsilon < .3$$

We decide to draw 10 subjects and to reject the null hypothesis if the observed proportion p_ϵ is much less than .30. The design is given in Figure 9-1. The data appear as follows:

Characteristic	1	2	Σ
Frequency	O_1	O_2	10

where the column labeled 1 indicates the characteristic of interest. We would like to know what O_1 must be in order for us to reject the null hypothesis. Under H_0, O_1 will be 3 or greater, since $3/10 = .30$, as given by π_0, and under H_0, $\pi_\epsilon \geq \pi_0$. Under H_m, O_1 will be less than 3.

The smallest O_1 can be is 0. We can find $P(O_1 = 0)$ from the binomial distribution:

$$P(O_1 = 0) = {}_{10}C_0(.3)^0(.7)^{10}$$

$$= \frac{10!}{10!\,0!}(.7)^{10} = (.7)^{10}$$

$$\doteq .028$$

which is less than .05. Therefore, if $O_1 = 0$, we will reject H_0. But we haven't exhausted our rejection region: it may also be possible to reject H_0 when $O_1 = 1$. To find out, we must do some additional computations.

$$P(O_1 = 1) = {}_{10}C_1(.3)(.7)^9$$

$$= \frac{10!}{9!\,1!}(.3)(.7)^9$$

$$\doteq .1211$$

But $.121 + .028 > .05$; therefore, we restrict our decision rule: reject only when $O_1 = 0$.

A two-tail test of a hypothesis is carried out in the same way, except that $\alpha/2$ of the distribution must be cut off at each end.

EXAMPLE Suppose that our hypotheses are

$$H_0 : \pi_\epsilon = .4 \qquad H_m : \pi_\epsilon \neq .4; \quad \text{that is, } \pi_\epsilon > .4 \text{ⓥ} \pi_\epsilon < .4$$

Suppose that we decide to draw 9 subjects and to use a .05 significance level. Then we will want an $\alpha/2 = .025$ probability of rejecting H_0 at either end of the distribution. At the upper end,

$$P(O_1 = 9) = {}_9C_9(.4)^9(.6)^0$$

$$\doteq .0002$$

$$P(O_1 = 8) = {}_9C_8(.4)^8(.6)$$

$$\doteq 0.0035$$

$$P(O_1 = 7) = {}_9C_7(.4)^7(.6)^2$$

$$\doteq .0208$$

Then, $P(O_1 = 9) + P(O_1 = 8) + P(O_1 = 7) \doteq .0245$, and clearly $P(O_1 = 6)$ when added to this would make the sum greater than .025. Working from the other end,

$$P(O_1 = 0) = {}_9C_0(.4)^0(.6)^9$$

$$= .0101 \quad \text{(which is less than .025)}$$

$$P(O_1 = 1) = {}_9C_1(.4)^1(.6)^8$$

$$= .0605$$

But $.0605 + .0101 = .0706$, which is greater than .025; therefore only $(O_1 = 0)$ is in the rejection region for this tail. For the problem, then, H_0 will be rejected if $O_1 = \{0, 7, 8, \text{ or } 9\}$ and accepted otherwise. Note that in this problem it is not possible to reject H_0 with exactly equal probabilities (.0245 and .0101) in both tails.

C. Normal Approximation

The normal approximation is carried out as indicated in Section 8-8. Both $N\pi_0$ and $N(1 - \pi_0)$ must be greater than 10 for sufficient accuracy of the approximation. Then the best-fitting normal distribution to the appropriate binomial distribution has parameters

$$\mu_0 = N\pi_0$$

$$\sigma_0 = \sqrt{N\pi_0(1 - \pi_0)}$$

If the α level is specified ahead of time, the probability of obtaining the observed frequency O_1, or one which is more extreme in the same direction, can be calculated and compared to it. If $O_1 > N\pi_0$, use

$$z = \frac{(O_1 - \mu_0) - .50}{\sigma_0} \tag{10-6a}$$

and for $O_1 < N\pi_0$, use

$$z = \frac{(O_1 - \mu_0) + .50}{\sigma_0} \tag{10-6b}$$

In either case, find $P\{O_1$ or more extreme$\}$, which is the smaller area under the standard normal distribution corresponding to the calculated z score. When you use formulas 10-6, you will always include the entire area under the histogram for O_1 in the extreme portion of the distribution, because the correction of .50 always subtracts from the difference between O_1 and μ_0.

For a one-tail test, reject if $P\{O_1$ or more extreme$\} < \alpha$, for a two-tail test, reject if $P\{O_1$ or more extreme$\} < \alpha/2$.

EXAMPLE Suppose that, in the two-tail example of part B, we had drawn 30 subjects instead of 9. Hypotheses are $H_0: \pi_\epsilon = .4$ and H_m: $\pi_\epsilon \neq .4$; $\alpha = .05$. Again, the design is given by Figure 9-1. And suppose that O_1, the number of subjects in the sample having the characteristic of interest, is 18. Then $N\pi_0 = 30(.4) = 12$ and $N(1 - \pi_0) = 30(.6) = 18$, both of which are greater than 10; so the normal approximation is appropriate.

$$\mu_0 = N\pi_0 = 12$$

$$\sigma_0 = \sqrt{N\pi_0(1 - \pi_0)} = \sqrt{7.2} \doteq 2.68$$

$O_1 > N\pi_0$, so use formula 10-7a:

$$z = \frac{(18 - 12) - .5}{2.68} = \frac{5.50}{2.68} \doteq 2.05$$

The probability of a z score of 2.05 or greater is given in the smaller area under the standard normal distribution from Table A-4; it is equal to .0202. This is less than .025 (for a two-tail test at $\alpha = .05$), so H_0 is rejected: $\pi_\epsilon \neq .40$.

Reference   Siegel (1956), 36–52.

Homework

1. In the following data table, subjects in each of three groups have been classified as A or B. Cell entries are numbers of subjects. You wish to know whether the proportion of A's varies from group to group.

Group	A	B	Σ
1	33	7	40
2	9	11	20
3	33	7	40
Σ	75	25	100

   (a) What is the value of chi square?
   (b) What are the degrees of freedom?
   (c) Let α be .05; what is the critical value?
   (d) What is your conclusion?

2. A large introductory-psychology class contains 200 freshmen, sophomores, and juniors. The following data were collected to test the hypothesis that the three groups had the same grade distribution at the end of the semester.

Grade	Freshmen	Sophomores	Juniors
A	10	12	8
B	24	18	18
C	46	24	10
D and F	20	6	4

   (a) What are the null and motivated hypotheses?
   (b) What is the critical value of chi square for these data? ($\alpha = .05$.)
   (c) What is the computed value of chi square?
   (d) What is your conclusion?

3. In a test of pressures to masculine conformity on sons of a certain group of immigrant parents, first-born sons were compared to later-born with respect to the number enlisting in the army before the age of 21. The following frequency distribution was obtained:

	First-born	Later-born	Σ
Enlisting	6	1	7
Not enlisting	1	5	6
Σ	7	6	13

Is the difference between the two groups of subjects significant at the .05 level?

4. Test the hypothesis of problem 3 with the following data, gathered from the same populations. As before, use the .05 level of significance.

	First-born	Later-born	Σ
Enlisting	13	17	30
Not enlisting	3	15	18
Σ	16	32	48

5. You believe that, when the first two children in a family are of the same sex, the probability that the third will be of the opposite sex is less than .5. You have just completed a nationwide sampling of families in which you gathered the ages and sexes of all children in each family; in your sample there were eight families having three or more children and with the first two of the same sex. In only two of the eight families was the third child of the opposite sex. State your null hypothesis and test at the .05 level.

6. The following data collected to determine whether Republicans and Democrats in a particular city differed in their "religious persuasion":

Observations	Catholic	Protestant	Jewish	Muslim
Republicans	2	7	1	0
Democrats	8	5	5	2

   (a) Indicate how you would solve this problem by combining categories (but do not carry out the calculations).
   (b) Criticize the method of part (a).
   (c) Indicate how many additional subjects you would probably need to test in order to solve this problem without combining categories.

7. A colleague has a theory that alcoholism is inherited according to Mendelian genetic laws. He assumes that each parent has two genes, either of which can be transmitted to a child. He further theorizes that alcoholism is a recessive gene: a person becomes an alcoholic only if both his genes are "alcoholic" genes. From parents both of whom have one "normal" and one "alcoholic" gene, you would expect on the basis of the theory to find alcoholic children $\frac{1}{4}$ of the time. In an attempt to disprove the theory, you test 200 children of parents of this type and find 135 normals and 65 alcoholics.
   (a) What are your hypotheses?
   (b) Let α be .05. What is your conclusion?

8. You decide to test whether a new sampling method (collecting subjects as they go in and out of the Poppy College Bookstore) gives a truly random

sample of PC students. You know that in the entire undergraduate population of PC there are 25 percent freshmen, 20 percent sophomores, 30 percent juniors, and 25 percent seniors. You draw your sample by selecting every fourth student and asking for his or her class year. In 100 subjects drawn this way, you find 35 freshmen, 30 sophomores, 20 juniors, and 15 seniors.

(a) Complete the analysis and test at the .01 level.

(b) Criticize this method for testing the randomness of a sample.

Homework Answers

1. (a) $\chi^2_{obs} = 12$ (b) df $= 2$ (c) $\chi^2_{.05} = 5.99146$ (d) Reject the hypothesis that the proportion of A's is the same for all three populations.

2. (a) H_0: the population distributions of grades are the same for all three classes.   (b) $\chi^2_{.05}(6$ df$) = 12.5916$ (c) $\chi^2_{obs} = 14.02$ (d) Reject H_0.

3. Use the Fisher test, $P = .0501$; reject H_0.

4. Use χ^2 with Yates' correction, $\chi^2_{.05}(1$ df$) = 3.84146$, $\chi^2_{obs} = 2.50$; do not reject H_0.

5. Let π_0 be the probability that the third child will not be the same sex, then $H_0: \pi_0 \geq .5$; use the binomial distribution, $P\{0 \text{\textcircled{v}} 1 \text{\textcircled{v}} 2$ of the opposite sex$\} = 37/256 = .144 > .05$. Do not reject H_0.

6. (a) You might combine Catholic, Jewish, and Muslim in a rough New Deal coalition category, and then do a Fisher test.

   (b) You lose detail in the religion classification, and you make use of test data to decide which test to use.

   (c) 83

7. (a) Let π be the proportion alcoholic. Then $H_0: \pi = .25$, $H_m: \pi \neq .25$.

   (b) Use the normal approximation to the binomial test, $z = 2.37$; reject H_0.

8. (a) $\chi^2_{obs} = 16.33$; $\chi^2_{.01}(3$ df$) = 11.3449$; reject H_0.

   (b) Randomness cannot be tested by the results of sampling but only by examining the method of sampling.

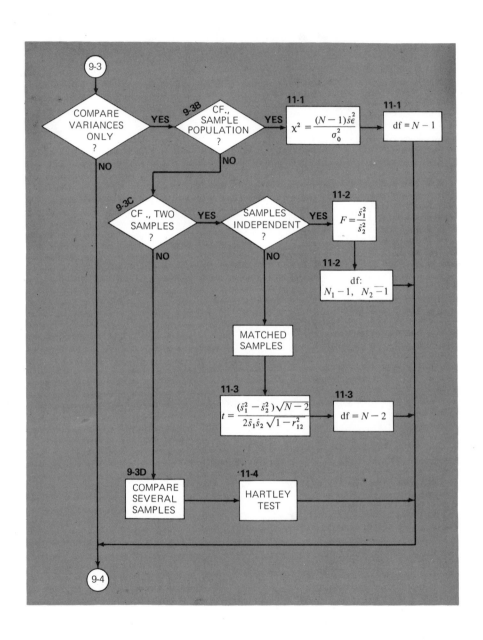

11 Comparing Variances

In this chapter we will consider several ways for comparing population variances. These differ in the number of variances that can be compared, and in the kinds of information that they require. The regional map indicates that one sensible way of breaking up the methods might be by number of samples collected. The chi-square method of Section 11-1 requires only one sample but also requires that you first know one of the population variances. In Section 11-2 and 11-3 you collect data on two samples; Section 11-3 is more sensitive because it demands that the samples be matched on a person-to person basis (i.e., you permit a more exact test because you have done more experimental controlling of your data). The method of Section 11-4 allows you to compare several sample variances at once and decide whether there are any differences among the corresponding population variances.

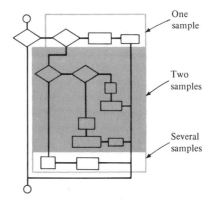

One sample

Two samples

Several samples

The general hypothesis-testing format of Chapter 9 is to be followed, no matter which method of analysis is chosen.

11-1 Comparing a sample variance to a population variance

Relevant earlier sections:
5-12 Sample variance
7-6 Chi-square distribution
Chapter 9 Hypothesis testing

In this section we are concerned with comparing the variance (and standard deviation) of an experimental population to the variance (and standard deviation) of a specified (null) population. Designs for this kind of comparison are given in Section 9-3B. We assume that we already know the null population variance, but that we must obtain a sample estimate of the variance of the experimental population.

Three relationships are possible between the two population variances:

(a) $\sigma_\epsilon^2 = \sigma_0^2$
(b) $\sigma_\epsilon^2 > \sigma_0^2$
(c) $\sigma_\epsilon^2 < \sigma_0^2$

where σ_ϵ^2 is the variance of the experimental population and σ_0^2 is the variance of the null or comparison population. These relationships may be combined to give either directional or nondirectional hypotheses.

A. Directional Hypotheses

Directional tests occur when either (b) or (c) is taken as the motivated hypothesis and the other is combined with (a) to form the null. For example:

$$H_0: \sigma_\epsilon^2 \leqq \sigma_0^2 \qquad H_m: \sigma_\epsilon^2 > \sigma_0^2$$

Here you are hoping to find the experimental variance to be greater than the variance of the null population. Decide on your α level and sample size, then look up the critical value for chi square, with $N - 1$ df, in Table A-5. Obtain an experimental sample of N subjects, and calculate $\hat{s}_\epsilon^2$ as an estimate of σ_ϵ^2. Then compute formula 11-1, which is a variant of formula 7-8.

$$\chi_{obs}^2 = \frac{(N - 1)\hat{s}_\epsilon^2}{\sigma_0^2} \tag{11-1}$$

Select an α level, look up a critical value of chi square for that α, with $N - 1$ df in Table A-7, and compare χ_{obs}^2 to χ_α^2. To reject H_0, $\hat{s}_\epsilon^2$ must be much greater than σ_0^2, and since σ_0^2 is a constant, the greater $\hat{s}_\epsilon^2$, the greater χ_{obs}^2 for fixed degrees of freedom. In other words, χ_{obs}^2 amounts to a linear transformation on observed values of $\hat{s}_\epsilon^2$, using $(N - 1)$ and σ_0^2 as transformation constants. The situation is represented graphically in Figure 11-1. Thus, χ_α^2 fixes a lower bound to the rejection region, and H_0 is rejected if $\chi_{obs}^2 > \chi_\alpha^2$.

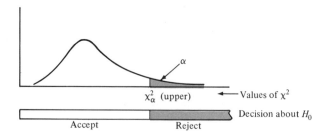

Figure 11-1 *Theoretical distribution of χ^2 for $N - 1$ df and decision regions for $H_0: \sigma_\epsilon^2 \leqq \sigma_0^2$.*

When relationships (a) and (b) are combined to give the null and (c) is used for the motivated hypothesis, all is the same except that rejection occurs at the opposite end of the chi-square distribution.

$$H_0 : \sigma_\epsilon^2 \geq \sigma_0^2 \qquad H_m : \sigma_\epsilon^2 < \sigma_0^2$$

Here the null hypothesis will be rejected only when the observed sample variance is much smaller than the null variance. The distribution and decision regions are represented in Figure 11-2.

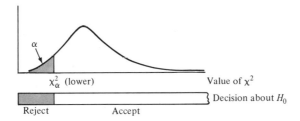

Figure 11-2 *Theoretical distribution of χ^2 and decision regions for $H_0 : \sigma_\epsilon^2 \geq \sigma_0^2$.*

EXAMPLE A test of creativity has been developed and standardized on urban children, and is known to have a mean $\mu_0 = 80$ and a standard deviation $\sigma_0 = 10$ for this population. It is now administered to a sample of 25 rural children to test the hypothesis that there will be greater variability of scores for the rural population. The hypotheses are $H_0 : \sigma_R^2 \leq \sigma_0$ and $H_m : \sigma_R^2 > \sigma_0$. For the sample, the mean $\overline{X} = 78$ and the standard deviation $\hat{s} = 16$. The design is presented in Figure 9-1. Next, α is set at .05; then the critical value of χ^2 in the upper tail is 36.4151, for a one-tail test with 24 df. From formula 11-1,

$$\chi_{obs}^2 = \frac{(25 - 1)(16^2)}{10^2} = \frac{(24)(256)}{100} = (24)(2.56) = 61.44$$

This value lies in the rejection region, so the null hypothesis is rejected (see Figure 11-1).

EXAMPLE Let us now imagine that we are administering the same test to Puerto Rican immigrant children, whose test scores we hypothesize to be constrained because of a language handicap. Let us again select 25 children and use an α of .05. Here our hypotheses are $H_0 : \sigma_\epsilon^2 \geq \sigma_0^2$ and $H_m : \sigma_\epsilon^2 < \sigma_0^2$. Suppose that we observe a sample mean $\overline{X} = 50$ and a standard deviation $\hat{s}$ of 8. The design is the same as for the previous example (Figure 9-1). The critical value is at the other end of the distribution, because we will reject the null hypothesis only when the observed variance is small, and chi square is a linear transformation on the observed variance. The critical value of χ^2 is thus 13.8484, for a one-tail test with 24 df. The observed value is calculated to be

$$\chi^2_{obs} = \frac{(25 - 1)(8^2)}{10^2} = \frac{24(64)}{100} = 15.36$$

The situation is represented graphically in Figure 11-2. We only reject the null hypothesis when the observed χ^2 is less than the critical value. The value we observe for our sample is greater than the tabulated value, so we do not reject the null hypothesis.

B. Nondirectional Hypotheses

Nondirectional tests occur when relationship (a) is taken as the null hypothesis and (b) and (c) are combined to form the motivated hypothesis. Then the hypotheses can be stated:

$$H_0 : \sigma^2_\epsilon = \sigma^2_0 \qquad H_m : \sigma^2_\epsilon \neq \sigma^2_0$$

As in the case of directional hypotheses, the sample is drawn and its variance calculated; then formula 11-1,

$$\chi^2 = \frac{(N - 1)\hat{s}^2_\epsilon}{\sigma^2_0}$$

is computed. A level of significance α is selected and critical values are found in both tails of the χ^2 distribution. Because we cannot say that a type I error that occurs at one end of the distribution is more troublesome than a type I error occurring at the other, we divide α equally between the two tails of the distribution, as indicated in Figure 11-3. Table A-7 lists the proportion of the χ^2 distribution falling above the tabled value; so our critical values correspond to areas of $\alpha/2$ (the upper value) and $\alpha/2$ (the lower value). When $\alpha = .05$, for example, the areas are $.05/2 = .025$ in each tail of the chi-square distribution with $N - 1$ df. Table A-7 is set up so that you can look up your α level directly, for a two-tail test, in the upper margin. Then for any number of degrees of freedom, you take critical values of χ^2 in both tails that fall under the chosen

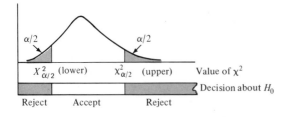

Figure 11-3 *Chi-square distribution and decision regions for two-tail tests.*

α level. The null hypothesis is rejected if the observed χ^2 is either larger than the critical value in the upper tail or smaller than the critical value in the lower tail.

EXAMPLE Suppose that the above test of creativity was developed on Northern children, and we wish to compare a Southern population to it before releasing the test for nationwide distribution. We decide to select 25 children randomly from major Southern cities, and to use α of .05. The critical values of χ^2 are thus 12.4012 and 39.3641, for 24 df. We select our sample and find its standard deviation to be 7. The observed value of chi square is thus

$$\chi^2_{obs} = \frac{24(7)^2}{10^2} = \frac{24(49)}{100} = 11.76$$

This situation is represented graphically in Figure 11-3. Here, we reject the null hypothesis if the observed chi square is either too great or too small, so there are rejection regions at both ends of the chi-square distribution. Our observed value, 11.76, falls in the lower of these rejection regions; therefore, we reject the null hypothesis and conclude that children from Southern cities have less variability in scores on this test.

References The use of chi square to test hypotheses about variances is discussed in Dixon and Massey (1969), 101–106; Hays (1963), 344–345; and McNemar (1969), 279–282.

11-2   Using the *F* test to compare the variances of two experimental populations

Relevant earlier sections:
5-12 Sample variance
7-7   *F* distribution
Chapter 9 Hypothesis testing

There are three mutually exclusive and exhaustive relationships between two variances, σ^2_a and σ^2_b:

(a) $\sigma^2_a = \sigma^2_b$
(b) $\sigma^2_a > \sigma^2_b$
(c) $\sigma^2_a < \sigma^2_b$

From these possible relationships it is necessary to construct two mutually exclusive and exhaustive hypotheses.

11-2 *Using the F test to compare the variances of two experimental populations*

A. Directional Hypotheses

One set of hypotheses that can be constructed from these pairs (a) and (b) against (c):

$$H_0 : \sigma_a^2 \geqq \sigma_b^2 \qquad H_m : \sigma_a^2 < \sigma_b^2$$

Here you have reason to expect the variance of the first population to be smaller than the variance of the second. Hence, compute formula 11-2:

$$F = \frac{\hat{s}_b^2}{\hat{s}_a^2} \qquad (11-2)$$

This is the same as formula 7-11, with the larger expected variance (when H_m is true) in the numerator. Look up the critical value $F_\alpha(N_b - 1, N_a - 1 \text{ df})$ in Table A-8. If the computed value F_{obs} is less than or equal to F_α, do not reject the null hypothesis; if F_{obs} is greater than F_α, reject H_0. (Note that if the calculated F is less than 1.0, you need not even bother looking up a critical value, as the difference is surely not significant in the predicted direction.) The rejection region is represented graphically in Figure 11-4. Notice that the null hypothesis is only rejected in the upper tail of the distribution (i.e., when $\hat{s}_b^2$ is much greater than $\hat{s}_a^2$, as predicted by H_m).

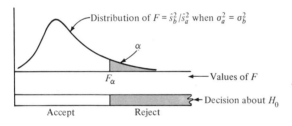

Distribution of $F = \hat{s}_b^2/\hat{s}_a^2$ when $\sigma_a^2 = \sigma_b^2$

α

F_α ← Values of F

← Decision about H_0

Accept   Reject

Figure 11-4 *Rejection region for a one-tail F test.*

EXAMPLE We hypothesize that a new method of teaching reading will favor poor readers over good readers. Two samples are drawn, each sample containing both good and poor readers. One sample receives the new method, and one the usual method of teaching. We compare variances, predicting that the new method will reduce variances in reading-test scores (poor readers "catch up"). Our hypotheses are $H_0 : \sigma_{New}^2 \geqq \sigma_{Usual}^2$ and $H_m : \sigma_N^2 < \sigma_U^2$. The experimental design is diagrammed in Figure 9-6. Next, we set α at .05, and we decide to make $\hat{s}_U^2$ the numerator of the F ratio, since we expect it to be the greater variance. If $N_N = 15$ and $N_U = 20$, then the degrees of freedom are 19 and 14, so we use $F_{.05}(15, 14) = 2.46$ as a conservative critical value (since there is no column for 19 df in the numerator, and since $F_{.05}(15, 14)$ is greater than $F_{.05}(20, 14)$ and thus is less likely to be reached by chance differences between the two samples. If standard deviations are $\hat{s}_N = 10$ and $\hat{s}_U = 15$, then formula 11-2 gives

$$F_{obs} = \frac{(15)^2}{(10)^2} = \frac{225}{100} = 2.25$$

This is less than the critical value, so we cannot reject H_0.

B. Nondirectional Hypotheses

The second pair of hypotheses that can be constructed from the three possible relationships among population variances is

$$H_0 : \sigma_a^2 = \sigma_b^2 \qquad H_m : \sigma_a^2 \neq \sigma_b^2 ; \qquad \text{that is, } (\sigma_a^2 > \sigma_b^2) \, Ⓥ \, (\sigma_a^2 < \sigma_b^2)$$

This pair of hypotheses would be used whenever you had no prior expectation as to which variance ought to be greater. Then collect your data and compute both $\hat{s}_a^2$ and $\hat{s}_b^2$.

Compute $F = (larger\ \hat{s}^2 / smaller\ \hat{s}^2)$ and look up F_α in Table A-8 for a two-tail test (you will reject if either $\hat{s}_a^2 > \hat{s}_b^2$ or $\hat{s}_b^2 > \hat{s}_a^2$). If the calculated value of F is greater than the tabulated value, reject the null hypothesis.

Representation of the rejection region requires two graphs, as indicated in Figure 11-5. However, use of the upper tail of one distribution gives identical information to using the lower tail of the other. Hence, this procedure is really the equivalent to finding an observed F (with the numerator variance chosen a priori, regardless of its magnitude) and comparing it to critical values in both tails of the corresponding theoretical distribution of F.

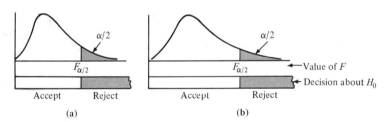

Figure 11-5 *Rejection regions for a two-tail F test.*

The latter procedure is illustrated in Figure 11-6 for a distribution having a and b df (i.e., where $\hat{s}_a^2$ is always placed in the numerator). Here, the critical value in the lower tail of the distribution is found as in Section 7-7C:

$$F_{lower\ \alpha/2}(a, b\ df) = \frac{1}{F_{upper\ \alpha/2}(b, a\ df)}$$

where $F_{lower\ \alpha/2}$ is the critical value in the lower tail of the original distribution and $F_{upper\ \alpha/2}$ is the critical value in the upper tail of an F distribution with df reversed. Unless specified otherwise, critical values will always be taken from the upper tails of F distributions.

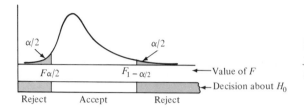

Figure 11-6 *Rejection regions for a two-tail F test in terms of a single distribution.*

EXAMPLE Suppose that you question whether children from broken homes will have the same variability of adult income as children who are not from broken homes. Your hypotheses are $H_0: \sigma^2_{BH} = \sigma^2_{NBH}$ and $H_m: \sigma^2_{BH} \neq \sigma^2_{NBH}$. You select an α level of .05 and draw two samples of children: 33 from broken homes and 21 not from broken homes. You follow all 54 through to adulthood and find that the standard deviation of incomes of the children from broken homes is \$2500, whereas that for children not from broken homes is \$1500. From the statement of the hypotheses, the test must be two-tailed. Because α is .05, each critical value must cut off an extreme .025 of the F distribution. Because 32 df is not tabled for F in Table A-8, let us use a conservative df (30) for children from broken homes (we know that $F_{.05}$ for 32 df would be smaller than this). There are two ways of solving the problem at this point. (a) Find $F_{.05}(20, 30) = 2.20$ and $F_{.05}(30, 20) = 2.35$; compute $F_{obs} = (larger \; \hat{s}^2/ smaller \; \hat{s}^2)$, which is equal to $(\$2500)^2/(\$1500)^2 = 2.78$. Here the numerator has 30 df and the denominator has 20, so $F_{crit} = F_{.05}(30, 20) = 2.35$. Then $F_{obs} > F_{crit}$, so reject H_0. (b) Decide to place one of the variances in the numerator (e.g., the variance for children from intact homes). Then the critical values are $F_{.05}(20, 30) = 2.20$ for the upper tail and $F_{lower\,.05}(20, 30) = 1/F_{.05}(30,20) = 1/2.35 = .426$. The observed value is $F = (\$1500)^2/(\$2500)^2 = 2{,}250{,}000/6{,}250{,}000 = .36$. Thus, the observed value of F is smaller than the lower critical value, so you reject the null hypothesis.

References The F test for equality of variances is discussed in Dixon and Massey (1969), 111–112; Guilford (1965), 191–193; and Hays (1963), 351–352.

11-3 Comparing two variances when the samples are matched or correlated

Relevant earlier sections:
5-12 Sample variance
7-4 t Distribution
Chapter 9 Hypothesis testing

In this section we will consider a method for comparing two population variances with the use of sample information, when the subjects in the samples have been matched on a one-to-one basis. The experimental designs used in correlated sample problems are given in Figures 9-4 and 9-5.

The problem is the same as that for Section 11-2, except for the a priori pairing of subjects, which is used to reduce the error caused by subject sampling differences between the two samples. Thus, there is a possibility of discovering a relationship that might otherwise be obscured.

Hypotheses may be either directional or nondirectional, as indicated in Section 11-2. In either case, the statistic used to test for differences in variance is

$$t = \frac{(\hat{s}_1^2 - \hat{s}_2^2)\sqrt{N - 2}}{2\hat{s}_1\hat{s}_2\sqrt{1 - r_{12}^2}} \qquad (11\text{-}3)$$

with $N - 2$ *df*, where N is the number of pairs of subjects (i.e., the number in each sample), and r_{12}^2 is the square of the correlation between the two samples on the variable tested, when the correlation has been obtained in accordance with Section 16-2.

The squared correlation coefficient indicates the extent to which one variable can be predicted linearly from knowledge of the other. It ranges from zero, when knowledge of scores on one variable is no help in predicting scores on the second, to 1.0, when knowledge of a subject's score on one variable allows you to predict his score on the second without error.

Suppose in the above formula that $\hat{s}_1^2$, $\hat{s}_2^2$, and N are held constant; what happens to t as r_{12}^2 varies? When there is no relationship between variables 1 and 2, $r_{12}^2 = 0$, $\sqrt{1 - r_{12}^2} = \sqrt{1} = 1$, and the formula is as though this square-root term were not included. But as the relationship between variables 1 and 2 increases, r_{12}^2 increases and $\sqrt{1 - r_{12}^2}$ decreases. But $\sqrt{1 - r_{12}^2}$ is in the denominator, so as it gets smaller, the entire fraction gets larger; thus t increases as r_{12}^2 increases. With an increased t you are more likely to call a given difference between $\hat{s}^2$'s significant; thus, the larger the correlation between variables, the more likely you are to find a difference between their variances to be signficant.

If you want a computing formula that you can use without having to refer to Chapter 16, or without having to compute a correlation coefficient, the above t formula is equivalent to

$$t = \frac{(\sum x_1^2 - \sum x_2^2)\sqrt{N - 2}}{2\sqrt{\sum x_1^2 \sum x_2^2 - (\sum x_1 x_2)^2}} \qquad (11\text{-}4)$$

where, as usual, the x's are deviation scores, and the subscripts on the x's are variable labels.

As an example of the construction of the data table, suppose that we are comparing the incomes of two matched samples of subjects to see if population 1 incomes are more variable. Subjects are paired on some variable related to income, that might otherwise confound the results, such as age, education, occupation, etc. Alternatively, the subjects in each pair may be related (e.g., brothers), or two sets of scores may be taken from the same subject under different conditions. The data are organized as shown in Table 11-1. The X's are the incomes of the N subjects in sample 1; the Y's are incomes of the N subjects in sample 2. Then the formula becomes

$$t = \frac{(\hat{s}_X^2 - \hat{s}_Y^2)\sqrt{N-2}}{2\hat{s}_X\hat{s}_Y\sqrt{1-r_{XY}^2}} \qquad df = N - 2$$

Table 11-1   *General form of data table*

Subject pair	Subject score from sample 1	Subject score from sample 2
1	X_1	Y_1
2	X_2	Y_2
3	X_3	Y_3
⋮	⋮	⋮
N	X_N	Y_N

For further discussions of the use of the t distribution in hypothesis testing, see the first three sections of Chapter 12.

EXAMPLE We wish to know whether hallucinogenic drugs produce personality changes in the persons who take them. We select 13 twin pairs of male subjects and randomly assign one of each pair to the experimental group and the other to the control group. The experimental group receives a specified amount of LSD, the controls, a placebo.* Three years later we administer a creativity scale and find the following:

	Experimental	Control
Mean	51	42
Variance	165	36
N	13	13
Correlation	.47	

Has the drug had an apparent effect on the variance of test scores?

This is a matched-pairs experiment, with hypotheses $H_0: \sigma_E^2 = \sigma_C^2$ and $H_m: \sigma_E^2 \neq \sigma_C^2$. Use the t test for comparing variances at $\alpha = .01$.

---

* (Dropping acid is not recommended as an undergraduate research project.)

$$t = \frac{(\hat{s}_E^2 - \hat{s}_C^2)\sqrt{N-2}}{2\hat{s}_E\hat{s}_C\sqrt{1 - r_{EC}^2}}$$

$$= \frac{(165 - 36)\sqrt{13 - 2}}{2\sqrt{36}\sqrt{158}\sqrt{1 - (.47)^2}} = \frac{129\sqrt{11}}{2(6)(12.56981)\sqrt{1 - .2209}}$$

$$= \frac{129(3.316625)}{(150.84)(.8827)} = \frac{427.84}{133.14} = 3.21$$

The critical value of t at $\alpha = .01$ and for 11 df is found from Table A-6 to be ± 3.106. The observed t of 3.21 is greater than the positive critical value, so the null hypothesis is rejected. There is some effect in the population which increases the variance of test scores following use of the drug.

It is interesting to note that in this instance some additional sensitivity has been gained by use of the information in the correlation coefficient. If the data were treated as though unmatched (Section 11-2), the difference in variances would appear nonsignificant, and the null hypothesis would not be rejected:

$$F_{.01}(\text{two-tail}; 12, 12 \text{ df}) = 4.91$$

$$F_{obs} = \frac{\hat{s}_{greater}^2}{\hat{s}_{smaller}^2} = \frac{165}{36} = 4.58$$

Thus $F_{obs} < F_{crit}$, and H_0 is accepted under the assumption that samples are independent. By matching samples we have eliminated some error and have therefore been able to observe an effect that is otherwise masked.

References The t test for comparing variances when data are correlated can be found in Guilford (1965), 193–194; and McNemar (1969); 282.

11-4 Comparing the variances from three or more independent samples

Relevant earlier sections:
5-12 Sample variance
7-6 Chi-square distribution
Chapter 9 Hypothesis testing

In this section we will consider the problem of comparing the variances of three or more populations on the basis of sample data drawn from them. The hypotheses to be tested are

$$H_0: \sigma_1^2 = \sigma_2^2 = \cdots = \sigma_J^2$$
$$H_m: \text{not } (\sigma_1^2 = \sigma_2^2 = \cdots = \sigma_J^2)$$

where each of the J variances is for a different population.

A. Equal Sample Sizes

In order to test these hypotheses, we draw a sample of subjects from each of the populations and compare the variances of the samples. If the differences among the sample variances are very large, we reject the null hypothesis that the population variances are identical. There are J samples (one from each population) with N subjects each.

For each sample we calculate the unbiased variance estimate $\hat{s}_j^2$, using formula 5-14. Then we form the ratio F_{max} by dividing the largest variance estimate $(\hat{s}_{max}^2)$ by the smallest $(\hat{s}_{min}^2)$:

$$F_{max} = \frac{\hat{s}_{max}^2}{\hat{s}_{min}^2} \tag{11-5}$$

This ratio cannot, however, be compared to the standard F table to test for significance, because it is systematically inflated. We did not compare the variances of two randomly chosen samples, but instead looked at the sample variances to decide on the pair to use. In general, the more samples there are, the larger the F_{max} ratio will probably be. Therefore, we use Table A-12 to find the critical value of F_{max}.

The sample sizes are listed down the left-hand column of Table A-12 and the number of samples across the top. Entries in the body of the table are critical values. If the observed value of F_{max} is greater than the tabulated value, the null hypothesis is rejected.

EXAMPLE We randomly assign 10 animals to each of three treatment groups, then implant an electrode in the brain of each animal, according to its treatment condition. We let α be .05; then from Table A-12 the critical value of F_{max} for three groups with 10 subjects per group is 4.85. After treatment, performance data are collected and the following variances are computed:

	$\hat{s}^2$
Group 1	9.56
Group 2	6.00
Group 3	6.78

The largest of these variances is 9.56 and the smallest is 6.00, so $F_{max} =$ 9.56/6.00 = 1.59. This value is less than the critical value of 4.85, so we conclude that there is no evidence for a difference in the variability of performance for the three treatment conditions.

References The basic paper on this test is by Hartley (1950). Textbook discussions can be found in Peatman (1963), 330–331; Walker and Lev (1953), 192; and Winer (1962), 92–94.

B. Unequal Sample Sizes
The most commonly used method for comparing variances is the Bartlett test (Bartlett, 1937a, 268). However, it is complicated to use, and it is no longer recommended as a preliminary test in analysis of variance, which was its principal application. Discussion of the test can be found in Li (1964), 552–556; McNemar (1969), 285–286; Walker and Lev (1953), 192–195; and Winer (1962), 95–96.

Homework

For each of problems 1 through 4, select the best response from the answers given.

(a) $\hat{s}_A^2 = \hat{s}_B^2$ (b) $\hat{s}_A^2 > \hat{s}_B^2$ (c) $\hat{s}_A^2 \neq \hat{s}_B^2$ (d) $\hat{s}_A^2 \leq \hat{s}_B^2$
(e) $\sigma_A^2 = \sigma_B^2$ (f) $\sigma_A^2 > \sigma_B^2$ (g) $\sigma_A^2 \neq \sigma_B^2$ (h) $\sigma_A^2 \geq \sigma_B^2$

1. Two new methods of teaching reading are compared on separate samples from the same initial population. You want to know whether method A is relatively more helpful to better initial readers than is method B. Your motivated hypothesis is
2. We are comparing two kinds of verbal reinforcement on the learning of a finger maze. We expect that method B will be less consistent or predictable in its effect on subjects than will method A. Our null hypothesis is
3. Our null hypothesis states that there is no greater variability of pay for house carpenters (a) where unions are well established than (b) where there are no unions. To reject it we would have to observe
4. Generate an item similar to the above in a research context of your own choosing.

For each of the problems 5 through 10, (a) state the decision rule for rejecting the null hypothesis, in terms of the statistic to be computed; then (b) compute the statistic from the data provided; and (c) accept or reject the null hypothesis. An example answer might be, "Use Section 11-3; reject H_0 if t_{obs} is greater than 2.776 or less than -2.776; $t_{\text{obs}} = 1.86$; accept H_0.*

5. $H_0: \sigma_\epsilon^2 \geq 4.95$; $H_m: \sigma_\epsilon^2 < 4.95$; $N = 15$; $\hat{s}^2 = 4.26$; $\alpha = .01$.

---

* Parts (b) and (c) are messy unless you have access to a slide rule or a calculator.

6. $H_0: \sigma_\epsilon^2 = \sigma_c^2$; $H_m: \sigma_\epsilon^2 \neq \sigma_c^2$; $N_\epsilon = 10$; $N_c = 17$; $\hat{s}_\epsilon^2 = 52$; $\hat{s}_c^2 = 40$; $\alpha = .05$.

7. $H_0: \sigma_1^2 = \sigma_2^2 = \sigma_3^2$; $H_m:$ not $(\sigma_1^2 = \sigma_2^2 = \sigma_3^2)$; $N_1 = N_2 = N_3 = 12$; $\hat{s}_1^2 = 4$; $\hat{s}_2^2 = 7.28$; $\hat{s}_3^2 = 12.42$; $\alpha = .05$.

8. $H_0: \sigma_\epsilon^2 = 16.25$; $H_m: \sigma_\epsilon^2 \neq 16.25$; $N = 20$; $\hat{s}^2 = 15.08$; $\alpha = .05$.

9. $H_0: \sigma_{pre}^2 \leq \sigma_{post}^2$; $H_m: \sigma_{pre}^2 > \sigma_{post}^2$; $N = 25$; $\hat{s}_{pre}^2 = 28$; $\hat{s}_{post}^2 = 18$; $\alpha = .05$; $r_{pre-post} = .45$.

10. Generate and solve an item similar to the above.

11. Our null hypothesis is that the variance of a particular population is exactly equal to 1.50. We randomly draw a sample of subjects from the population and find their raw scores to be 5, 6, 5, 5, and 4.

    (a) State both hypotheses formally.

    (b) What figure from Section 9-3 most closely represents this experimental design?

    (c) What formula should be used?

    (d) What is/are the critical value(s), if α is .05?

    (e) What is the observed value of the test statistic?

    (f) What is your conclusion?

12. A scientist is testing for the equality of two population variances. If his sample sizes are $N_1 = 10$, $N_2 = 12$, and the sample standard deviations are $\hat{s}_1 = 5$, $\hat{s}_2 = 7$, complete the analysis by answering questions (a) to (f) of problem 11 for this problem.

13. We wish to test the hypothesis that Americans have greater variability in height than do Syrians. We therefore wait on a street corner in Chicago and stop and measure the heights of the first 12 Americans who will allow us to. We then go to Syria and do the same until we have a sample of 5 Syrians (the smaller sample size is due to the language barrier). The Americans have an average height of 70 inches and an unbiased variance estimate of 36; the Syrians have an average height of 66 inches and a variance estimate of 9. Let $\alpha = .05$.

    (a) What are the null and motivated hypotheses?

    (b) What statistic should be used?

    (c) What is the critical value of the statistic?

    (d) What is the computed value?

    (e) What is your conclusion?

14. You predict that alcohol will have a greater effect on the coordination of poor drivers than on the coordination of good drivers. You obtain 18 subjects and test the driving accuracy of each subject both before and $\frac{1}{2}$ hour after having two drinks. The results are as follows:

	Before	After
Mean number of errors	8	11
SD	4	5
$r_{pre-post} = .80$.		

Complete the analysis as for problem 11; answer questions (a) to (f) for this problem.

15. You hypothesize that showing a movie each day in a statistics class will decrease the variability of examination grades, by motivating the poorer students and boring the better ones. You carry out this policy in a class of six students, between the midterm and the final exam, predicting that the final exam grades will be less variable. You find the means for the two exams both to be 18, with standard deviations of 5 and 3 for the midterm and final, respectively. To increase your precision, you correlate the two sets of grades, finding an r of .60. If you test your hypothesis at the .05 level, and if you can reasonably attribute any differences to changes in subjects (rather than to differences in the exams), what is your conclusion?

16. When a psychological test is used in personnel selection, the scores of the persons selected are generally less variable than the scores of the entire group from which they were drawn. An intelligence test having a mean of 100 and a standard deviation of 15 is used to select 19 employees having IQ's between 80 and 95 to be line workers in an automobile assembly plant. After one year on the job the employees are retested and found to have a variance of scores of 200. Determine whether this is a statistically significant result in the predicted direction at the .01 level.

Homework Answers

1. f   **2.** h   **3.** b

5. Section 11-1; reject if $\chi^2_{obs} < 4.6603$; $\chi^2_{obs} = 12.05$; accept H_0.

6. Section 11-2; reject if $F > 3.05$; $F_{obs} = 1.30$; accept H_0.

7. Section 11-4; reject if $F_{max} > 3.28$; $F_{max} = 3.10$; accept H_0.

8. Section 11-1; reject if $\chi^2 < 8.90655$ or $\chi^2 > 32.8523$; $\chi^2 = 17.63$; accept H_0.

9. Section 11-3; reject if $t > 1.714$; $t = 1.196$; accept H_0.

11. (a) $H_0: \sigma^2 = 1.50$, $H_m: \sigma^2 \neq 1.50$ (b) Figure 9-1   (c) Formula 11-1   (d) $\chi^2_{lower} = 0.484419$; $\chi^2_{upper} = 11.1433$ (e) $\chi^2_{obs} = 1.33$ (f) Accept H_0.

12. (a) $H_0: \sigma^2_1 = \sigma^2_2$, $H_m: \sigma^2_1 \neq \sigma^2_2$ (b) Figure 9-6   (c) Formula 11-2   (d) $F_{.05} = 3.96$ (e) $F_{obs} = 1.96$ (f) Accept H_0.

13. $H_0: \sigma^2_A \leq \sigma^2_S$, $H_m: \sigma^2_A > \sigma^2_S$, where A = American, S = Syrian. (b) Use a one-tail F test, formula 11-2. (c) $F_{.05} > 5.9117$ (d) $F_{obs} = 4.0$ (e) Do not reject H_0.

14. (a) $H_0: \sigma^2_{post} \leq \sigma^2_{pre}$; $H_m: \sigma^2_{post} > \sigma^2_{pre}$, where X = driving performance score (b) Figure 9-2 (c) Formula 11-3 (d) $t_{.05} = 1.746$ (e) $t_{obs} = 1.50$ (f) Do not reject H_0.

15. Use Section 11-3, with null hypothesis $H_0: \sigma^2_{final} \geq \sigma^2_{mid\ term}$. $t_{.05} = 2.132$, $t_{obs} = 1.33$, so do not reject H_0.

16. Use Section 11-1, with null hypothesis $H_0: \sigma^2_S \geq \sigma^2_U$, where S is the hypothetical selected population, U the unselected population. $\chi^2_{.01} = 7.01491$, $\chi^2_{obs} = 16.0$, so do not reject H_0.

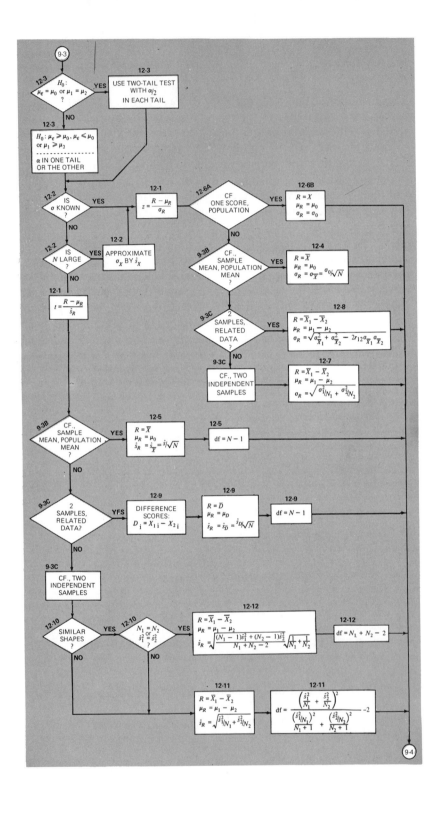

12

Comparing Means: One or Two Samples

THE map for this chapter may be divided into conceptual regions in at least two different ways, as shown in the regional maps. In the first region you determine whether to use a one-tail or a two-tail test. Next you decide between t and z distributions for testing your hypotheses. The remainder of the first regional map is divided between t and z tests. In the second regional map the remaining regions pertain to type of comparison being made, rather than to the distribution used in making the comparison. Similar comparisons, based on different

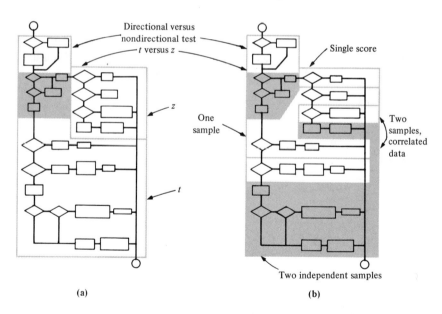

assumptions, are discussed in Chapter 14. The major distinction is that Chapter 12 assumes interval-level raw data, whereas Chapter 14 assumes ordinal-level data. Whichever distribution and type of comparison is used, the overall approach is that indicated in Chapter 9.

12-1   *z* and *t* tests

Relevant earlier sections:
6-5   Standard scores
7-3   Normal distribution
7-4   *t* distribution

All *z* and *t* tests have the same rationale. We begin by calculating some statistic on a sample or a pair of samples — a mean, a difference between two means, a mean of change scores or difference scores, etc. Call this statistic **R** in the general case. This is what has been done in the decision map: the formulas for *t* and *z* are in terms of **R**, and in order to find out what the formulas are for a particular kind of problem, you replace the **R**, $\mu_\mathbf{R}$, and $\sigma_\mathbf{R}$ of the general formula by what they are equal to for that formula. Thus, when $\mathbf{R} = \overline{X}$, $\mu_\mathbf{R} = \mu_{\overline{X}} = \mu_0$ and $\sigma_\mathbf{R} = \sigma_{\overline{X}}$, the general formula for *z* is replaced by the formula specific to a statistic $\overline{X}$:

$$ z = \frac{\mathbf{R} - \mu_\mathbf{R}}{\sigma_\mathbf{R}} = \frac{\overline{X} - \mu_0}{\sigma_{\overline{X}}} $$

We learn what the theoretical distribution of **R** would be if the null hypothesis were true, and we compare the calculated value of **R** to this theoretical distribution. The comparison is done by imagining the drawing of an infinite number of samples, of the same size as the one we actually drew, from the null population (see Figure 12-1).

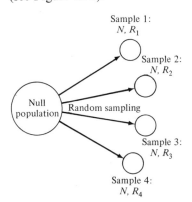

Figure 12-1   *Hypothetical drawing of samples from a null population.*

Because the selection of subjects to form the several samples is random, we cannot expect a perfect match between the scores of the subjects in any sample and the scores of the subjects in every other sample. Therefore, we do not expect all the samples to give the same value of the statistic **R**. Let us therefore imagine forming a distribution of the values of **R** calculated from the samples drawn under the null hypothesis, as in Figure 12-2.

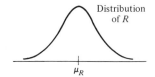

Figure 12-2 *Hypothetical distribution of sample statistic* **R** *when H_0 is true.*

This distribution is known as the sampling distribution of the statistic **R**: $\mu_{\mathbf{R}}$ is the mean of the sampling distribution of **R** when H_0 is true, and $\sigma_{\mathbf{R}}$ is the standard error of the statistic **R**. Notice that when **R** is a raw score, the standard error of **R** is called the standard deviation. Perhaps this would best be expressed the other way around: the standard error of a statistic is the standard deviation of its sampling distribution.

Suppose that we have done an experiment, and at the end of it we collect data and calculate a statistic, $\mathbf{R}_{\epsilon}$, that ought to be sensitive to the experimental treatment. We say that, if our experiment has produced the effect we were looking for, $\mathbf{R}_{\epsilon}$ will have been raised or lowered, relative to what it would otherwise have been. To find out whether the experiment has been successful, we compare the experimental $\mathbf{R}_{\epsilon}$ to the distribution of **R**'s obtained when the null hypothesis is true. If $\mathbf{R}_{\epsilon}$ is unusually large or small (depending on our hypotheses), we say there has been an effect of the experimental treatment. To carry out the test, we do a linear transformation on **R** to a standard distribution for which we can easily identify unusual values. The sections of this chapter specify these linear transformations and the ways of identifying unusual values.

z and *t* tests are often used for testing hypotheses about means, differences between pairs of means, and so on, because these statistics tend to be normally distributed even when the raw-score variables from which they are derived are not. This is true because, no matter what the shape of the raw-score distribution, differences among sample means depend on the subjects who are chosen for the samples. If the subjects are truly randomly chosen, then sample means will tend to differ randomly, and for these kinds of statistics, random error is normally distributed. This phenomenon was discussed earlier, in Section 7-3C. The shapes of the raw-score distributions are relatively unimportant when comparing sample means or differences between means, but it is extremely important that the sampling process be as nearly random as possible. For a discussion of random sampling, see Section 5-3.

We use a *z* test if the parameters $\mu_{\mathbf{R}}$ and $\sigma_{\mathbf{R}}$ can both be specified from our knowledge of the null population. If it is necessary to estimate the value of $\sigma_{\mathbf{R}}$ from the sample on which our value of **R** has been calculated, then we must test by comparison with the *t* distribution.

If the value of $\sigma_{\mathbf{R}}$ is known from the null population and we therefore can use the *z* test, we do the following. We imagine transforming the distribution of **R** (which must be normally distributed with mean $\mu_{\mathbf{R}}$ and standard deviation $\sigma_{\mathbf{R}}$) to the standard normal distribution, using the *z* transformation (Section 6-5) for each of the hypothetical sample values of **R**, as indicated in Figure 12-3. This distribution is found in Tables A-4 and A-5 of the Appendix. We make the

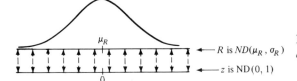

Figure 12-3 *z transformation of a normal distribution.*

same transformation on the value of **R** calculated from our real sample, which gives it a *z* score comparable to this same standard distribution.

If the value of σ_R is not known, we estimate it from the sample and use a *t* test. We imagine transforming the **R** distribution into the distribution of *t* (see Section 7-4) for each of the hypothetical samples and the real sample. Critical values of the *t* distribution are found in Table A-6 of the Appendix for comparison.

Because *t* and *z* are linear transformations of the statistic **R**, any decision arrived at relating a *t* or *z* score to areas, percentages, or probabilities also applies to the value **R** corresponding to that *t* or *z* score.

Whichever distribution is used, one or two critical values are found from the table on the basis of the hypotheses and the level of significance (and the number of degrees of freedom, in the case of the *t* test). The *t* or *z* score corresponding to the calculated value of **R** is compared to the critical value(s), and if it appears in the critical region, the null hypothesis is rejected.

Reference   A general comparison of *t* and *z* can be found in Games and Klare (1967), 298–302.

12-2   Deciding between *t* and *z*

Relevant earlier sections:
6-5   Standard scores
7-3   Normal distribution
7-4   *t* distribution

Although the *t* and *z* tests have the same general form, they make different assumptions. In the *z* test, the statistic (**R**) is transformed directly to the standard normal distribution, so that it must be assumed that both the mean μ_R and the standard deviation σ_R of the statistic are known exactly.

How might we come to know both μ_R and σ_R? The first of these, μ_R is easy: we hypothesize a value for it and therefore always know it exactly. The only error that might occur here is in deciding on our hypothesis. For example, consider the question, "Are the intelligence test scores of college students generally greater than 100?" Here μ_R is the population mean of scores μ_0 against which students are compared, and is exactly 100 because that is the value you have specified in asking the question. (Of course, if your question is in error, your answer will be in error relative to the question you ought to have asked.) For a second example, when we compare two population means by comparing the means of samples drawn from them, our null hypothesis often specifies that $\mu_R = \mu_1 - \mu_2$ is equal to zero (i.e., that the means of the two populations are equal).

σ_R is not always known exactly, however. You might be given the population value in a textbook problem, or you might actually know it from testing all members of a population. However, in general neither of these conditions applies, and σ_R must be estimated from a sample.

The t test is used when both the statistic (R) and the estimate of its standard error ($\hat{s}_R$) are obtained from sample data. In order for the obtained t to be compared to the tabulated values to test hypotheses, it is necessary that the estimates R and $\hat{s}_R$ be independent. If the sample is small, and if both R and $\hat{s}_R$ are estimated from the same sample, independence will be assured only if R is normally distributed, and this will occur only if the raw scores on which R is based are normally distributed in the population. On the other hand, as sample sizes increase, the values of R computed on the samples become more and more nearly normally distributed, regardless of the shape of the raw-score distribution on which R is based, if raw scores are randomly chosen. Therefore, the assumptions of the t test are either that the population of raw scores is normally distributed or that sample sizes are large.

Strictly speaking, whenever σ_R must be estimated, a t test ought to be used, because a z test will allow you to reject the null hypothesis a greater proportion of the time than you have specified, when the null is true (see Section 9-4). If you want an idea of the amount of error caused by using the z test where the standard error must be estimated, compare the critical value of z to the critical value of t at the appropriate number of degrees of freedom in Table A-6.

As the number of degrees of freedom increases, the t distribution approaches the standard normal distribution, so that for large degrees of freedom it really makes very little difference whether you use a t or a z test. Most authorities feel that the two tests are virtually identical by the time the number of degrees of freedom (which is approximately equal to sample size) reaches 30 or 40.

12-3   Directional or nondirectional test?

This section should be read in connection with Sections 9-4 and 9-5 and the section of this chapter which best fits your problem.

In Section 9-2, it was pointed out that any of three relationships is possible between two parameters, ρ_1 and ρ_2:

(a) $\rho_1 = \rho_2$
(b) $\rho_1 > \rho_2$
(c) $\rho_1 < \rho_2$

For the present chapter, each of these ρ's is replaced by either a population mean μ or a difference between means ($\mu_a - \mu_b$). From these relationships, two mutually exclusive and exhaustive hypotheses will be constructed.

When we take either (b) or (c) to be the alternative or motivated hypothesis, the other two combine to form the null. In this case we say that the hypotheses are directional, because only an inequality in the direction specified by the motivated hypothesis can be used to reject the null hypothesis.

Suppose, for example, that we have a motivated hypothesis that the mean of our experimental population, μ_ϵ, will be greater than the mean μ_0 of some specified (null) population. Our null hypothesis would be $H_0: \mu_\epsilon \leq \mu_0$. Suppose that we now draw a sample and calculate its mean. The only way we can obtain sufficient evidence from the sample to reject the null hypothesis is if the sample mean is much greater than the mean of the null population. If the sample mean is much greater than the mean of the null population, this is evidence that the population mean of which it is an estimate is also greater than the null mean.

Because of randomness in sampling, our sample means will not all be identical to the null population mean, even when the null hypothesis is true; there is a distribution of sample means. Theoretically, this distribution extends indefinitely in both directions along the score variable. Therefore, in order to be able to reject the null hypothesis when it is false, we must mark off part of the range of scores as being so extreme that, when an observation falls in that region, we find the null hypothesis very hard to believe.

If we cut off any portion of the range of possible sample means under the null hypothesis, we are guaranteeing a probability of a type I error; to control this error, we specify its magnitude a priori. Because we can reject the null only, in the case of the example, when the sample mean is large, we make our rejection region at the upper end of the distribution of sample means under the null hypothesis. This is graphed in Figure 12-4.

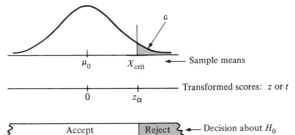

Figure 12-4 *Sampling distribution of means and decisions about the null hypothesis when $H_0: \mu_\epsilon \leq \mu_0$ is true.*

On the other hand, if our null hypothesis is $H_0: \mu_\epsilon \geq \mu_0$, we will be able to reject it only if our sample mean is much smaller than the null mean. Hence, we make our rejection region in the lower tail of the distribution of sample means. This is diagrammed in Figure 12-5.

Finally, when our null hypothesis is $H_0: \mu_\epsilon \geq \mu_0$, we will have evidence that our experimental population is not the same as the null population if our sample mean is either too great or too small. In this case, we split the significance

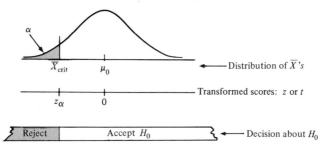

Figure 12-5 *Sampling distribution of means and decisions about the null hypothesis when* $H_0: \mu_\epsilon \geqq \mu_0$ *is true.*

level α into two equal parts, and draw critical values so that $\alpha/2$ of the distribution is cut off at either end. This situation is represented in Figure 12.6.

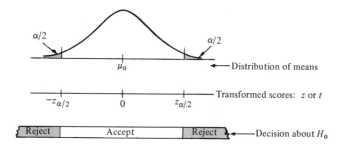

Figure 12-6 *Sampling distribution of means and decisions about the null hypothesis when* $H_0: \mu_\epsilon = \mu_0$ *is true.*

In actuality, of course, we do not obtain a distribution of means under the null hypothesis; we only obtain one mean, that for our observed sample. Our comparison distribution is a theoretical one—the tabulated distribution of either t or z, which depends on whether we know the population variance or not. We actually transform our sample mean by the t or z formula to make it comparable to the appropriate distribution, and make our decision on the basis of the results using the transformed values.

For clarity of presentation, the discussion of this section has been in terms of means; however, the results are more general than that. What we actually get is a distribution of some statistic **R**, which may be a mean, a difference between means, a difference score, or other statistic. We hypothesize a value of the parameter $\mu_\mathbf{R}$, which serves as the mean of the sampling distribution of **R** under the null hypothesis. For example, suppose we predict that the mean of population 1 is 53 points greater than the mean of population 2. Our statistic **R** $=$ $\bar{X}_1 - \bar{X}_2$, and our population parameter $\mu_\mathbf{R} = (\mu_1 - \mu_2) = 53$. For this problem, the test appears as in Figure 12-7.

283

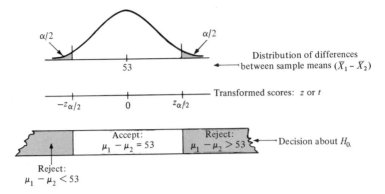

Figure 12-7 *Sampling distribution of differences between pairs of means and decisions about the null hypothesis when* $H_0: \mu_1 - \mu_2 = 53$ *is true.*

References Comparisons of directional and nondirectional tests can be found in Edwards (1973), Chapter 6; and McNemar (1969), 63–65.

12-4 Comparing a sample mean to a population mean when the population standard deviation is known

Here the problem is to decide whether two populations have the same mean, when the mean of one population is known but the mean of the other must be estimated on the basis of sample information. It is assumed that both populations have the same standard deviations and that it is known. Any of the three sets of formal hypothesis can be tested, where μ_ϵ is the mean of the experimental population and μ_0 is the mean of the null population.

1. $H_0: \mu_\epsilon \leqq \mu_0$ $H_m: \mu_\epsilon > \mu_0$
2. $H_0: \mu_\epsilon \geqq \mu_0$ $H_m: \mu_\epsilon < \mu_0$
3. $H_0: \mu_\epsilon = \mu_0$ $H_m: \mu_\epsilon \neq \mu_0$

Some relevant designs are given in Section 9-3B.

Let us call the population for which we know both the mean and standard deviation the null population. Imagine randomly drawing an infinite number of samples of size N from the null population, computing the mean for each, and plotting the distribution of these $\bar{X}$'s. Figure 12-8 diagrams this way of obtaining the distribution of $\bar{X}$'s.

Under the null hypothesis, $\bar{X}$ is normally distributed with a mean μ_0 and a standard deviation (the standard error of the mean) of $\sigma_0/\sqrt{N}$. Use the z table and your hypotheses to find the critical value(s) corresponding to your

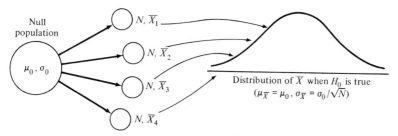

Figure 12-8 *Hypothetical sampling distribution of means.*

chosen signficance level α. Your observed $\overline{X}$ can also be transformed into the standard normal distribution by subtracting the mean of $\overline{X}$'s under the null hypothesis and dividing by the standard error:

$$z_{\overline{X}(\text{obs})} = \frac{R - \rho_0}{\sigma_R} = \frac{\overline{X}_{\text{obs}} - \mu_{\overline{X}}}{\sigma_{\overline{X}}} = \frac{\overline{X}_{\text{obs}} - \mu_0}{\sigma_0 / \sqrt{N}} \qquad (12\text{-}1)$$

The z score corresponding to the observed $\overline{X}$ is then compared to the critical values of z. If it appears in the critical region, the null hypothesis is rejected; if not, the null hypothesis is not rejected. The null hypotheses and decision regions have already been related to one another, in Figures 12-4, 12-5, and 12-6.

EXAMPLE   Freshmen at Hotstuff University get a mean of 130 and a standard deviation of 16 on a standardized test of mathematical reasoning. 25 randomly chosen freshmen are given a special experimental course in algebra in hopes it will improve their standing on the test. As a safeguard it is decided also to see if the course lowers their test scores. Alpha is selected to be .05. The hypotheses are

$$H_0 : \mu_\epsilon = \mu_f = 130 \qquad H_m : \mu_\epsilon \neq \mu_f$$

where μ_ϵ is the mean of the experimental population and μ_f is the mean for all freshmen. Assume that the only effect of the experimental course will be on the average performance, not on the variability or shape of the score distribution. Then $\sigma_\epsilon = \sigma_f = 16$. Use formula 12-1.

The rejection region is the area on both sides of μ_0 beyond which $\alpha/2$ of sample means fall when H_0 is true, as indicated in Figure 12-6. Critical values are given by Table A-4 to be ± 1.96; we reject H_0 if the observed value of z is less than -1.96 or greater than $+1.96$.

Suppose that we observe a sample mean of 138 for students in the experimental course. The calculated value of z is then

$$z = \frac{138 - 130}{16/\sqrt{25}} = \frac{8}{16/5} = \frac{40}{16} = \frac{5}{2} = 2.5$$

The calculated value of z, 2.5, is greater than the positive critical value, 1.96. Therefore, reject the null hypothesis: the sample is not representative of the null population. It is drawn from a population the mean of which is greater than 130, and we conclude that the experimental course raises test scores.

References This normal test is discussed in Dixon and Massey (1969), 95–98; and Winer (1962), 14–20.

12-5 Comparing a sample mean to a population mean when the population standard deviation is not known

The problem for this section is the same as that for Section 12-4: we want to know whether two populations have the same mean value when one of those means is known and one must be estimated from sample values. The hypotheses and sampling procedure are presented in Section 12-4, and relevant experimental designs are indicated in Section 9-3B. By contrast with Section 12-4, the population standard deviation is also unknown and must be estimated from sample values. We specify a null population with a mean μ_0 and unknown standard deviation σ_0. Then we imagine drawing a large number of samples of size N from that population. These will be distributed as in Figure 12-8. However, because σ_0 is not known, we cannot use the z test of Section 12-4. In fact, we only draw one sample, as indicated in Section 12-4, and compute both mean $\overline{X}$ and standard deviation $\hat{s}_X$ for the sample. We assume that, if the population from which our sample is drawn differs from the null population, it differs only in mean and not in standard deviation: $\sigma_e = \sigma_0$. Then our sample standard deviation $\hat{s}_X$ is an unbiased estimate of the standard deviation of the null population.

Under these assumptions, $\overline{X}$ is normally distributed for large N and approximately so for small N, with mean $\mu_{\overline{X}} = \mu_0$ and standard deviation $\sigma_{\overline{X}} = \sigma_X/\sqrt{N}$, when H_0 is true. If we knew σ_0, we could use formula 12-1, but because of sampling error in our estimate $\hat{s}_X$ of σ_0, we must test using the t distribution.

$$t = \frac{\mathbf{R} - \mu_R}{\hat{s}_{\mathbf{R}}} = \frac{\overline{X} - \mu_0}{\hat{s}_{\overline{X}}} = \frac{\overline{X} - \mu_0}{\hat{s}_X/\sqrt{N}} \tag{12-2}$$

Calculation of the sample mean and standard deviation are carried out in accordance with Sections 5-7 and 5-12, and then entered into the formula for t. If the computed value of t is more extreme than the tabulated value for the appropriate number of degrees of freedom, the null hypothesis is rejected.

EXAMPLE   Children in a "disadvantaged" neighborhood usually do poorly on a standardized aptitude test in mathematics—obtaining a mean score of 50. Nine of these children, drawn at random, are given a sequence of four academic-success experiences. Test for the effectiveness of the treatment in raising performance, at $\alpha = .01$.

SOLUTION   The design is given in Figure 9-2. We will reject the hypothesis of no treatment effect only if the sample mean is much larger than the null population mean. We thus have a directional test, with hypotheses

$$H_0: \mu_\epsilon \leq 50 \qquad H_m: \mu_\epsilon > 50$$

Suppose that we collect the following scores after treatment: 50, 56, 48, 48, 53, 52, 55, 55, and 51. $N = 9$, hence df $= 8$, and for a directional test, critical value of t is 2.896. The computations are as follows, where the data have been transformed to simplify the calculations; the arithmetic has been done on $Y = X - 48$ and the results transformed back to the original distribution later:

Subject	X	$Y = X - 48$	Y^2
1	50	2	4
2	56	8	64
3	48	0	0
4	48	0	0
5	53	5	25
6	52	4	16
7	55	7	49
8	55	7	49
9	51	3	9
Σ		36	216

$$\bar{Y} = \frac{36}{9} = 4; \qquad \text{hence,} \quad \bar{X} = 4 + 48 = 52$$

$$\hat{s}_Y^2 = \frac{\sum Y^2 - (1/N)(\sum Y)^2}{N - 1} = \frac{216 - (36)^2/9}{8} = \frac{216 - 4(36)}{8} = \frac{72}{8} = 9$$

$$\hat{s}_Y = \sqrt{\hat{s}_Y^2} = \sqrt{9} = 3 = \hat{s}_X$$

Then

$$t = \frac{\bar{X} - \mu_0}{\hat{s}_X/\sqrt{N}} = \frac{52 - 50}{3/\sqrt{9}} = 2.0$$

But $t_{.01} = 2.896$, which is greater than 2.0. Hence, do not reject H_0; we do not have statistical evidence that the treatment improved performance on the mathematics test.

References See Dixon and Massey (1969), 98–100; Edwards (1973), 62–63; Games and Klare (1967), 303–306; and Winer (1962), 20–24.

12-6 Comparing one score to a population mean when the population variance is known

A. The Decision

This design is included primarily for comparison with the descriptive use of the z distribution, where a z score permits a standardized comparison of an individual with the other members of the group to which he is being compared.

In the present case, the exact same value of z is an indication, not of the individual's personal standing with respect to some specified population, but rather of whether the population from which he was drawn has the same mean value as the null population to which his score is compared. This is therefore a limiting case of the comparison of Section 12-4 of a sample mean to a population mean—limiting in that a sample size of 1 is the smallest possible.

Because only one case is being used to test a hypothesis, it must be assumed that both the null population and the population from which the subject was drawn are normally distributed, and that the variances of the two populations are the same. Inasmuch as it is impossible to obtain a variance estimate from a single subject, there is no t test corresponding to the present use of z.

B. The Computations

This problem constitutes the limiting case of the comparison of a sample mean to a population mean. It is limiting in that the sample size of 1 is as small as samples can be. Relevant designs are given in Section 9-3B. The formula for z, derived as a special case of formula 12-1, is identical to the z-score formula of Section 6-5.

$$z = \frac{\mathbf{R} - \mu_R}{\sigma_\mathbf{R}} = \frac{\overline{X} - \mu_{\overline{X}}}{\sigma_X / \sqrt{N}}$$

But $\overline{X} = X$ and $N = 1$, so

$$z = \frac{X - \mu_0}{\sigma_0} \qquad (12\text{-}3)$$

One- and two-tail hypotheses can be tested with formula 12-3, using the figures of Section 12-3 and the table of the standard normal distribution (Table A-4). Hypotheses are indicated in Section 12-4.

In most research, you would be better advised to draw a sample of more than one subject in order to test a hypothesis. However, there is no theoretical reason why an inference cannot be made about a parameter on the basis of a single subject drawn randomly from the population. The major difficulties with such a design would appear to lie in observing a large-enough discrepancy from the null value to permit rejection, and in convincing your colleagues that the subject used was, in fact, randomly drawn from a normally distributed population.

12-7 Comparing the means of two independent samples when both population variances are known

A. Hypotheses and the Sampling Design

In this section we wish to compare the means of two populations by comparing the means of samples drawn randomly from them. The usual experimental designs are given in Figures 9-3 and 9-6. The population variances may be "known" either because we have a great deal of data on one of the populations and assume that the other has the same variance, or because our sample sizes are large enough that we can assume that both sample variances are fairly accurate estimates of their respective population values (see Section 12-2).

If we label the populations 1 and 2 by subscripting their parameters, we can state our hypotheses formally by comparing the actual (but unobservable) difference between the two means $(\mu_1 - \mu_2)$ to some hypothesized, numerical difference $(\mu_1 - \mu_2)_0$, $(\mu_1 - \mu_2)_a$, or $(\mu_1 - \mu_2)_b$, as indicated in Table 12-1,

Table 12-1 *Hypotheses comparing two population means*

Exact hypotheses	Inexact hypotheses	
	Directional	Nondirectional
$H_a : (\mu_1 - \mu_2) = (\mu_1 - \mu_2)_a$ $H_b : (\mu_1 - \mu_2) = (\mu_1 - \mu_2)_b$	$H_0 : (\mu_1 - \mu_2) \geq (\mu_1 - \mu_2)_0$ $H_m : (\mu_1 - \mu_2) < (\mu_1 - \mu_2)_0$	$H_0 : (\mu_1 - \mu_2) = (\mu_1 - \mu_2)_0$ $H_m : (\mu_1 - \mu_2) \neq (\mu_1 - \mu_2)_0$

where the subscripted differences are specified a priori. $(\mu_1 - \mu_2)_0$ is usually but not necessarily specified to be zero.

The sampling design for this problem can be represented as indicated in Figure 12-9. Here we identify two populations; from each we draw a sample and compute its mean, $\overline{X}$. We then find the difference between these means and test to see whether this difference is significant.

The test is effected by imagining drawing an infinite number of pairs of samples, when H_0 is true, finding the difference between means for each pair of samples, and plotting these differences. If the observed difference between sample means would be a very rare occurrence under the null hypothesis, then

12-7 *Comparing the means of two independent samples*

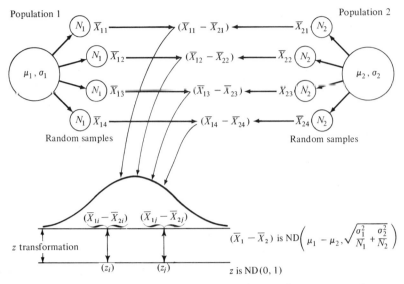

Figure 12-9 *Hypothetical sampling distribution of differences between pairs of means.*

this difference will appear in one of the tails of the distribution of hypothetical differences. The distribution of hypothetical differences is normal with a mean $\mu_1 - \mu_2$ and a standard deviation $\sigma_{\bar{X}_1 - \bar{X}_2}$ so that we only have to compare the observed difference to the mean theoretical difference by the z distribution. The test statistic is given by formula 12-4:

$$z = \frac{R - \mu_R}{\sigma_R} = \frac{(\bar{X}_1 - \bar{X}_2) - (\mu_{\bar{X}_1 - \bar{X}_2})_0}{\sigma_{\bar{X}_1 - \bar{X}_2}} = \frac{(\bar{X}_1 - \bar{X}_2) - (\mu_1 - \mu_2)_0}{\sqrt{\sigma_1^2/N_1 + \sigma_2^2/N_2}} \quad (12\text{-}4)$$

The critical value(s) at the chosen level of significance can be looked up in the table of the standard normal distribution. Then transform the observed difference into a z score and compare it to the critical z values obtained from the standard normal distribution.

B. Example When $\sigma_1 \neq \sigma_2$
An intelligence test has been administered to a large number of high-school seniors in the past, with these results:

	Mean	Sigma
Males	100	20
Females	100	15

We wish to know whether a new method of training in test-taking skills will affect male and female average performance differently on this test. We have no hypothesis as to which group should be more affected. Therefore, we use a nondirectional test. We assume that training had no effect on the σ's for the two groups and that we can therefore use the known values for these parameters in formula 12-4. Next, we arbitrarily set α at .05; thus, we can look up the critical value of z, which is ± 1.96. Twenty-five males and 25 females are given the special training. The results are:

	Mean
Males	115
Females	105

The hypotheses are $H_0: \mu_1 = \mu_2$ and $H_m: \mu_1 \neq \mu_2$. Therefore, under H_0, $(\mu_1 - \mu_2)_0 = 0$, and the z formula becomes

$$z = \frac{(\bar{X}_1 - \bar{X}_2) - 0}{\sqrt{\sigma_1^2/N_1 + \sigma_2^2/N_2}}$$

The method of solution involves converting the observed difference into a z score and comparing that z score to the critical z score(s). Here,

$$z_{\text{obs}} = \frac{115 - 105}{\sqrt{(20)^2/25 + (15)^2/25}} = \frac{10}{\sqrt{(400 + 225)/25}} = \frac{10}{5} = 2.0$$

This observed value of 2.0 for z is greater than the positive critical value of z; hence, it is in the rejection region, and the null hypothesis is rejected. We conclude that the null hypothesis was incorrect: there is a difference between the effect of the training on the males and the females. Now we can look back at the post-training means and be more explicit about our conclusions: the training raises the scores of males ($\bar{X} = 115$) more than females ($\bar{X} = 105$).

C. Example When $\sigma_1 = \sigma_2$

When $\sigma_1^2 = \sigma_2^2$, the denominator of formula 12-4 can be simplified by removing from under the radical the common term σ^2:

$$\sqrt{\frac{\sigma_1^2}{N_1} + \frac{\sigma_2^2}{N_2}} = \sqrt{\frac{\sigma^2}{N_1} + \frac{\sigma^2}{N_2}} = \sqrt{\sigma^2\left(\frac{1}{N_1} + \frac{1}{N_2}\right)} = \sigma\sqrt{\frac{1}{N_1} + \frac{1}{N_2}}$$

Suppose, in the example for part B, that population standard deviations had been 20 for both males and females, and both initial population means had been 100. Our formula would then be

$$z = \frac{(\overline{X}_1 - \overline{X}_2) - (\mu_1 - \mu_2)_0}{\sigma \sqrt{1/N_1 + 1/N_2}} \qquad (12\text{-}5)$$

The critical values of z would still be ± 1.96. $(\mu_1 - \mu_2)_0$ would still be zero, under the null hypothesis. The problem would be solved by converting the ob-served differences to a z score and comparing that result to the critical value(s) of z:

$$z_{obs} = \frac{115 - 105}{20\sqrt{1/25 + 1/25}} = \frac{10}{(20/5)\sqrt{2}} = \frac{10\sqrt{2}}{4(2)} = \frac{14.14}{8} = 1.77$$

This observed z score would lie between the critical values of z; hence, we would accept the null hypothesis of no difference between accept the null hypothesis that there is no difference in the effectiveness of the training for males and females.

D. Prediction and Correlation

When comparing the effects of two different treatments on some dependent variable, your best prediction for subjects in the population for each treatment condition is the sample mean for that condition. From the map for Chapter 16, you can see that the most reasonable correlation coefficient is the point bi-serial, because the independent variable is dichotomous (there are two treat-ment groups) and the dependent variable is continuous and interval-level. The point biserial is a special case of the Pearson coefficient, and its square tells the proportion of variance of the dependent variable that can be accounted for by knowledge of the independent variable.

References   Discussion of this test can be found in Dixon and Massey (1969), 114–116, 119; and Guilford (1965), 173–177.

12-8   Comparing the means of two related samples when the population variances are known

In this section we wish to make an inference about the relative sizes of two population means on the basis of the means of samples drawn from them, when there is a known correlation between the scores in one sample and the scores in the other. The scores in the two samples are correlated if there is a reason to pair each score in one sample with a score in the other, on a priori grounds, and if larger scores in one sample tend to be paired with larger scores in the other, and smaller scores in the one sample tend to be paired with smaller scores in the other. This might occur, for example, when the twin of each mem-ber of one sample appears in the other sample, or where the same subject is tested twice, say before and after some treatment. Some designs by means of which correlated data are obtained are given in Section 9-3C2.

When the hypotheses are directional, they may be stated

$$H_0: \mu_1 \geq \mu_2 \quad \text{and} \quad H_m: \mu_1 < \mu_2$$

When the hypotheses are nondirectional, they may be written

$$H_0: \mu_1 = \mu_2 \quad \text{and} \quad H_m: \mu_1 \neq \mu_2$$

With the design of this section, there is a possibility that we might be able to remove some of the differences between subjects from consideration as error and thus to increase the sensitivity of the test to differences between population means. Subjects are drawn in matched pairs from some population and then randomly or systematically assigned to one of two treatments. Then the data table can be described as in Table 12-2.

Table 12-2 *Raw scores and difference scores*

Subject or subject pair	X_1	X_2	Difference $D = X_1 - X_2$
1	X_{11}	X_{12}	D_1
2	X_{21}	X_{22}	D_2
$\vdots$	$\vdots$	$\vdots$	$\vdots$
N	X_{N1}	X_{N2}	D_N
Mean	$\bar{X}_1$	$\bar{X}_2$	$\bar{D}$

In this table, X_{i1} might be the score of subject i in group 1 and X_{i2} might be the score of the corresponding subject in group 2; or X_{i1} might be the post-test score of subject i and X_{i2} might be the pretest score of the same subject.

A. Difference-Score Formula

It can easily be shown that if $D_i = X_{i1} - X_{i2}$, then $\bar{D} = \bar{X}_1 - \bar{X}_2$:

$$\bar{D} = \frac{1}{N} \sum D_i$$

$$= \frac{1}{N} \sum (X_{i1} - X_{i2}) \qquad \text{(by substitution)}$$

$$= \frac{1}{N} \sum X_{i1} - \frac{1}{N} \sum X_{i2} \qquad \text{(by rule I, Section 5-1)}$$

$$= \bar{X}_1 - \bar{X}_2 \qquad \text{(by definition of the mean)}$$

12-8 Comparing the means of two related samples

By a similar derivation it can be shown that $\mu_D = \mu_1 - \mu_2$. Thus, the hypotheses can be restated as follows. For a nondirectional test,

$$H_0: \mu_1 = \mu_2 \text{ is equivalent to } H_0: \mu_D = 0$$

$$H_m: \mu_1 \neq \mu_2 \text{ is equivalent to } H_m: \mu_D \neq 0$$

Similarly, directional hypotheses can be written

$$\left.\begin{array}{l} H_0: \mu_D \geq 0 \\ H_m: \mu_D < 0 \end{array}\right\} \quad \text{or} \quad \left\{\begin{array}{l} H_0: \mu_D \leq 0 \\ H_m: \mu_D > 0 \end{array}\right.$$

It is now relatively easy to represent sampling under the null hypothesis by the drawing of matched pairs of subjects from a population of matched pairs to constitute samples similar to the one actually drawn (see Figure 12-10).

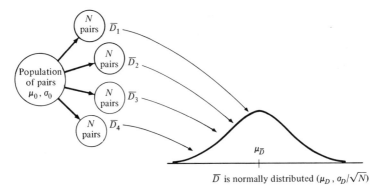

$\bar{D}$ is normally distributed $(\mu_D, \sigma_D/\sqrt{N})$

Figure 12-10 *Drawing of samples in related pairs from a single null population.*

Now we can complete the analysis for this section by applying the methods of Section 12-4 to the difference-score column of Table 12-2: we now have a one-sample normal test using difference scores instead of raw scores. In formula 12-1 we replace $\bar{X}$ by $\bar{D}$, μ_X by μ_D, and $\sigma_{\bar{X}} = \sigma_X/\sqrt{N}$ by $\sigma_{\bar{D}} = \sigma_D/\sqrt{N}$, where N is now the number of difference scores, and the standard deviation of differences σ_D is given by using D's instead of X's in formula 5-10:

$$\sigma_D = \sqrt{\frac{\sum (D_i - \mu_D)^2}{N}}$$

Thus, the appropriate z formula is

$$z = \frac{\mathbf{R} - \mu_R}{\sigma_{\mathbf{R}}} = \frac{\bar{D} - \mu_D}{\sigma_D/\sqrt{N}} \tag{12-6}$$

This is simply the formula of Section 12-4 applied to the last column of Table 12-2. For this column, you find the mean $\bar{D}$ and the standard deviation σ_D, then apply formula 12-6.

Formula 12-6 would seem to embody a contradiction. σ_D is to be calculated from data, without reference to prior knowledge about σ_1 or σ_2; this is strictly possible only if both populations are available, since σ_D is a population standard deviation. But if both populations are available for examination, you can simply calculate μ_1 and μ_2, and see if they are the same or different without resort to a statistical test.

In fact, the formula is useful, if you do not know σ_1 or σ_2 but have large enough samples so that σ_D may be closely approximated by $\hat{s}_D$ calculated on sample values. This is better written as formula 12-7:

$$z = \frac{\bar{D} - \mu_D}{\hat{s}_D / \sqrt{N}}, \qquad N \text{ large} \tag{12-7}$$

This formula is identical to formula 12-9 of Section 12-9 and is here compared to the z distribution only if the number of pairs of observations is large enough to make the approximation very good. Any time formula 12-7 is appropriate (because samples are large), the t test of Section 12-9 is also appropriate. Because the two formulas are identical, and because the t test can also be used with small samples, an example of the application of these statistics is deferred to Section 12-9.

B. Raw-Score Formula

A second formula equivalent to formula 12-6 can be derived from it. It was shown above that $\bar{D} = \bar{X}_1 - \bar{X}_2$ and indicated that $\mu_D = \mu_1 - \mu_2$. With a little more algebra it could have been shown that

$$\sigma_{\bar{D}} = \sqrt{\frac{\sigma_1^2}{N} + \frac{\sigma_2^2}{N} - 2r_{12}\frac{\sigma_1\sigma_2}{N}} = \sqrt{\frac{\sigma_1^2}{N} + \frac{\sigma_2^2}{N} - 2\left(\frac{\sum x_{1i}x_{2i}}{N^2}\right)}$$

where r_{12} is the correlation between X_1 and X_2 (see Chapter 16 for a discussion of correlation). The z formula can then be written

$$z = \frac{(\bar{X}_1 - \bar{X}_2) - (\mu_1 - \mu_2)}{\sqrt{\sigma_1^2/N + \sigma_2^2/N - 2r_{12}(\sigma_1\sigma_2/N)}} = \frac{(\bar{X}_1 - \bar{X}_2) - (\mu_1 - \mu_2)}{\sqrt{\sigma_1^2/N + \sigma_2^2/N - 2(\sum x_{1i}x_{2i}/N^2)}} \tag{12-8}$$

This formula might be preferred over the first difference-score formula in two cases: (1) when the population variances σ_1^2 and σ_2^2 are known, some sampling error can be reduced if you use them in the second formula and rely on sample data only for $\bar{X}_1, \bar{X}_2$, and r_{12}; and (2) when data have been analyzed by a computer correlation program. Some common programs yield values for

$\overline{X}_1, \overline{X}_2, \sigma_1, \sigma_2$, and r_{12}, which can then be entered into the second formula and the calculations easily completed.

Notice in the second formula that when the two groups tested have a zero correlation, the third term of the denominator goes to zero and the formula reduces to formula 12-4 (which is indeed the formula for uncorrelated groups). A zero correlation could result either if the subjects in the two groups were independently drawn (as in Figures 9-3 and 9-6) or if subjects were matched on some variable that was unrelated to the dependent variable.

When there is a positive correlation between matched subjects (high scores in one group tend to be paired with high scores in the other, and similarly for low scores), the method of the present section is more sensitive than formula 12-4, because the standard error is reduced by a term proportional to the magnitude of the correlation. Thus the denominator is smaller and the overall value of z larger. The greater the correlation between the two samples, the more the standard error will be reduced and the more likely that a consistent difference between the two populations will be identified on the basis of the sample data, when all other things ($N, \overline{X}_1, \overline{X}_2, \sigma_1$, and σ_2) remain constant.

EXAMPLE   We want to know whether electroshock therapy increases the tested intelligence of patients receiving the treatment. The dependent variable is score on an intelligence test known to have a mean of 100 and a standard deviation of 16 for the general adult population. The design chosen is that of Figure 9-4, under the assumption that $\sigma_{pre} = \sigma_{post} = 16$. Alpha is chosen to be .05, and the hypotheses are $H_0 : \mu_{post} \leq \mu_{pre}$ and $H_m : \mu_{post} > \mu_{pre}$. The critical value of z is $z_{.05}$ (upper tail) $= 1.645$, and the null hypothesis will be rejected if the obtained value is greater than that critical value.

A group of 64 patients is administered the test, obtaining a mean score of 85. After treatment, they are administered the test again, obtaining a mean of 88, and the correlation between scores on the two administrations is .50. Is this increment significant at the .05 level?

SOLUTION   Substituting in formula 12-8, we obtain

$$z = \frac{88 - 85}{\sqrt{256/64 + 256/64 - 2(.5)(16 \cdot 16/64)}}$$

$$= \frac{3}{\sqrt{(256/64)[1 + 1 - 2(.5)]}} = \frac{3}{\sqrt{4(1)}} = \frac{3}{2} = 1.50 < 1.645$$

Therefore, we do not reject the null hypothesis. The conclusion is that electroshock therapy does not significantly raise intelligence-test scores.

Reference   This test is discussed in Guilford (1965), 177–181.

12-9 Comparing the means of two related samples when the population variances are not known

Relevant earlier section:
12-8 Related samples, variances known

A. Defiinition and Application

The problem of this section is the same as that of Section 12-8: we wish to test a hypothesis about the relative magnitudes of two population means μ_1 and μ_2 when subjects in the two samples have been paired. Relevant designs are given in Section 9-3C2 and hypotheses are stated in Section 12-8. A *t* test is used because the population standard deviations are not known and the sample size is small (less than 30 or 40 *pairs* of observations). The appropriate form of the *t* formula is

$$t = \frac{\mathbf{R} - \mu_{\mathbf{R}}}{\hat{s}_{\mathbf{R}}} = \frac{\bar{D} - \mu_D}{\hat{s}_{\bar{D}}} = \frac{\bar{D} - \mu_D}{\hat{s}_D/\sqrt{N}} \tag{12-9}$$

where terms are as defined in the earlier section: $D_i = X_{i1} - X_{i2}$, a difference between scores of two matched subjects, one subject on two occasions, etc.; $\bar{D}$ is the mean of difference scores for the sample; μ_D is the hypothesized population mean of difference scores; $\hat{s}_D$ is the unbiased standard deviation of differences; and $\hat{s}_{\bar{D}}$ is the standard error of the mean of difference scores.

EXAMPLE Consider the following data table, in which subjects are administered the same test before and after an experimental treatment. Let $H_0: \mu_D = 0$ and $D = X_{\text{after}} - X_{\text{before}}$. The design is given in Figure 9-4. Let $\alpha = .05$; then if $N = 9$, $t_\alpha = \pm 2.306$. Then

$$t = \frac{\bar{D} - \mu_D}{\hat{s}_{\bar{D}}} = \frac{5 - 0}{\sqrt{6}/\sqrt{9}} = \frac{5(3)}{\sqrt{6}} = \frac{5\sqrt{6}}{2} \doteq 6.12$$

Subject	Before	After	D	$d = D - \bar{D}$	d^2
1	23	29	6	1	1
2	18	22	4	-1	1
3	21	29	8	3	9
4	22	30	8	3	9
5	19	20	1	-4	16
6	18	24	6	1	1
7	16	20	4	-1	1
8	19	25	6	1	1
9	21	23	2	-3	9
Σ			45	0	48
			$\bar{D} = 5$		$\hat{s}_D^2 = \frac{48}{8} = 6$

297

The critical values of t at the .05 level for a nondirectional test with 8 df are ± 2.306. The computed value of 6.12 is greater than $+2.306$. Therefore, reject the null hypothesis; scores are raised by the experimental treatment.

B. Example That Shows the Advantages of Using a Correlated Design When Data Are Related

Suppose that we have a problem in which pre- and post-treatment data are collected on the same subjects. We wish to know whether the treatment has affected their scores. We use four subjects.

1. Suppose that we ignore the fact that the data are correlated. Then,

$$H_0 : \mu_{pre} = \mu_{post}$$

$$H_m : \mu_{pre} \neq \mu_{post}$$

Set $\alpha = .01$. Collect the data in Figure 12-11 and graph.

i	X_{pre}	X_{post}
1	3	4
2	6	7
3	9	10
4	12	13

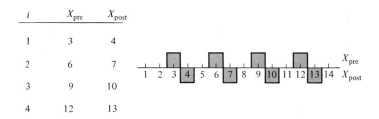

Figure 12-11

The shapes of the two distributions are identical, as are the N's; therefore, we use Section 12-12. Thus, $df = N_1 + N_2 - 2 = 6$ and $t_\alpha = \pm 3.707$.

i	X_{pre}	X_{post}	X_{pre}^2	X_{post}^2
1	3	4	9	16
2	6	7	36	49
3	9	10	81	100
4	12	13	144	169
Σ	30	34	270	334
Mean	7.5	8.5		

$$\sum x_{\text{pre}}^2 = \sum X_{\text{pre}}^2 - \frac{(\sum X_{\text{pre}})^2}{N}$$

$$= 270 - \frac{900}{4} = 270 - 225$$

$$= 45$$

$$\sum x_{\text{post}}^2 = 334 - \frac{1156}{4} = 334 - 289$$

$$= 45$$

Then

$$\hat{s}_{\bar{X}_{\text{post}} - \bar{X}_{\text{pre}}} = \sqrt{\frac{\sum x_{\text{pre}}^2 + \sum x_{\text{post}}^2}{N_1 + N_2 - 2}} \sqrt{\frac{1}{N_1} + \frac{1}{N_2}} = \sqrt{\frac{45 + 45}{6}} \sqrt{\frac{1}{4} + \frac{1}{4}}$$

$$= \sqrt{\frac{90}{12}} = \sqrt{\frac{30}{4}} = \frac{1}{2}\sqrt{30} = \frac{5.477}{2} = 2.738$$

and

$$t = \frac{(\bar{X}_{\text{post}} - \bar{X}_{\text{pre}}) - 0}{\hat{s}_{\bar{X}_{\text{post}} - \bar{X}_{\text{pre}}}} = \frac{8.5 - 7.5}{2.738} = \frac{1}{2.738} = .366$$

which is nonsignificant $(.366 < 3.707)$.

2. Suppose now that we take into account the correlation in the data due to the fact that the same subjects are tested on the same variable twice.

$$H_0: \mu_{\text{pre}} = \mu_{\text{post}} \leftrightarrow H_0: \mu_D = 0 \qquad \text{(where } D \text{ is a difference score)}$$

$$H_m: \mu_{\text{pre}} \neq \mu_{\text{post}} \leftrightarrow H_0: \mu_D \neq 0$$

Set $\alpha = .01$. Use formula 12-9; then $t_\alpha = \pm 5.841$.

i	X_{pre}	X_{post}	$D = X_{\text{post}} - X_{\text{pre}}$	D^2
1	3	4	1	1
2	6	7	1	1
3	9	10	1	1
4	12	13	1	1
$\sum$			4	4
Mean			1	

$$\hat{s}_D^2 = \frac{N \sum D^2 - (\sum D)^2}{N(N-1)} = \frac{4(4) - (4)^2}{4(3)} = \frac{16 - 16}{12} = 0$$

$$\hat{s}_{\bar{D}}^2 = \frac{\hat{s}_D^2}{N} = \frac{0}{4} = 0; \qquad \hat{s}_{\bar{D}} = 0$$

$$t = \frac{\bar{D} - \mu_D}{\hat{s}_{\bar{D}}} = \frac{1 - 0}{0} = +\infty$$

Since t observed $> t$ critical, reject H_0, in contrast to your action when you did not assume the scores correlated. What is the meaning of t of $+\infty$? How can you obtain a t that large when the distributions overlap so much?

C. Prediction and Correlation

Relative to Sections 12-7, 12-11, and 12-12, prediction and correlation are more complicated for the present section, because the data are related. However, the recommendations of those sections can be followed here also, by ignoring the difference scores and working only with the raw scores: this solution is conservative but reasonable. In addition, you can predict the average change in the population, using $\bar{D}$ as the estimator.

References The use of t with correlated data is found in Dixon and Massey (1969), 119–123; Games and Klare (1967), 321–329; and Winer (1962), 39–43.

12-10 Selecting a test for comparing two means of independent samples when their population variances are not known

Two assumptions that underlie the use of the t test of Section 12-12 for comparing the means of two samples are that their population distributions are normal and that their population variances are equal. In recent years there has been some interest in the importance of these assumptions and the effect on significance levels when they are violated.

A. Effect of Distribution Shapes

When you use Section 12-12, nonnormality of the population distributions can lead to rejection of the null hypothesis either too often or too seldom when H_0 is true. However, the amount of error may be reduced by selecting slightly larger samples (say N greater than 20). Samples drawn from populations with different shapes may lead to erroneous conclusions about significance, but this probably will not be an important factor unless a plot of the sample distributions shows them to be clearly different. With either small N's and nonnormal distributions, or distributions having different shapes, you should use the formula of Section 12-11.

B. Effect of Sample Sizes and Variances

The pooling formula of Section 12-12 assumes that the two samples are drawn from populations having the same standard deviation. If this is not true and the sample sizes are different, the tabulated values of t will not be appropriate, and Section 12-11 should be used. If either the sample sizes or the variances are about equal, then the formula of Section 12-12 is appropriate.

At the present time there is no consensus as to what the expression "about equal" means, although there is general agreement that the F test for equality of variances is too discriminating. If you wish a rough and arbitrary rule of thumb, decide that the N's are about equal or the variances are about equal if the smaller is more than two thirds the magnitude of the larger.

Reference See Hays (1963), 321–322.

12-11 Comparing the means of two independent samples when neither population variance is known

Relevant earlier section:
12-7 Two-sample z test

A. Formula and Computations

Here the design and hypotheses are identical to those of Section 12-7. The only difference in the two sections is that the population variances, σ_1^2 and σ_2^2, are not known in the present case, and cannot be assumed to be equal. Therefore, these values are estimated with the use of samples drawn from their respective populations, and a t test is used instead of the z test. The variable $(\overline{X}_1 - \overline{X}_2)$ is normally distributed when H_0 is true, with mean $(\mu_1 - \mu_2)$ and standard error $\sqrt{\sigma_1^2/N_1 + \sigma_2^2/N_2}$; but because σ_1^2 and σ_2^2 are not known, they must be approximated by $\hat{s}_1^2$ and $\hat{s}_2^2$. Therefore, formula 12-4 cannot be used. We use formula 12-10 instead and compare the calculated statistic to the t distribution to test hypotheses.

$$t = \frac{(\overline{X}_1 - \overline{X}_2) - (\mu_1 - \mu_2)_0}{\sqrt{\hat{s}_1^2/N_1 + \hat{s}_2^2/N_2}} \qquad (12\text{-}10)$$

In order to use this formula, both a mean and a standard deviation must be calculated for each sample. The resultant obtained value of t is then compared to the critical value(s) of t for the chosen level of significance and the hypotheses. If the observed t falls in the critical region, the null hypothesis is rejected; otherwise, it is not.

The approximate number of degrees of freedom for this design are given by the formula

$$df = \frac{(\hat{s}_1^2/N_1 + \hat{s}_2^2/N_2)^2}{(\hat{s}_1^2/N_1)^2/(N_1 + 1) + (\hat{s}_2^2/N_2)^2/(N_2 + 1)} - 2 \qquad (12\text{-}11)$$

(The task of calculating the degrees of freedom is smaller than might at first appear, because all the terms have to have been found already in order to compute the value of t.)

EXAMPLE   We wish to test the null hypothesis $H_0: \mu_1 = \mu_2$ at the .05 significance level, where

Sample	Size	$\bar{X}$	$\hat{s}_X$
1	5	85.4	10
2	25	76.2	20

The design is given by Figure 9-6. We select an α level of .05. We use the method of the present section, because $N_1 \neq N_2$ and $\hat{s}_1 \neq \hat{s}_2$. The calculations are

$$\frac{\hat{s}_1^2}{N_1} = \frac{100}{5} = 20 \qquad \frac{\hat{s}_2^2}{N_2} = \frac{400}{25} = 16$$

Then,

$$df = \frac{(20 + 16)^2}{20^2/6 + 16^2/26} - 2 = \frac{36^2}{400/6 + 256/26} - 2$$

$$= \frac{1296}{[26(400) + 6(256)]/6(26)} - 2 = \frac{156(1296)}{10,400 + 1536} - 2$$

$$= \frac{202,176}{11,936} - 2 \doteq 16.94 - 2$$

$$\doteq 15$$

The critical values are therefore $t_{.05}$(two tail, 15 df) $= \pm 2.131$.

$$\hat{s}_{\bar{X}_1 - \bar{X}_2} = \sqrt{\frac{100}{5} + \frac{400}{25}} = \sqrt{20 + 16} = \sqrt{36} = 6$$

$$t = \frac{(\overline{X}_1 - \overline{X}_2) - (\mu_1 - \mu_2)}{\hat{s}_{\overline{X}_1 - \overline{X}_2}} = \frac{85.4 - 76.2}{6} = \frac{9.2}{6} = 1.53$$

The observed t of 1.53 lies between the two critical values of t, so we do not reject H_0. There is no evidence for a difference between population means.

Perhaps you can save yourself some work in the future by noting that it really was not necessary to calculate the number of degrees of freedom for this problem. Examination of Table A-6 should convince you that the critical values of t could never be less than ± 1.96, no matter how many degrees of freedom you had. Since the observed t is between these values, it could not be significant.

B. Associated Correlational Method

An examination of the map for Chapter 16 will show you that the most reasonable correlation coefficient to use with these data is the point biserial, because the predictor (sample number) is dichotomous and the criterion (X) or dependent variable is continuous and interval-level. The point biserial is a special case of the Pearson coefficient, and the square of the correlation ($r_{p.bis.}^2$) tells the proportion of variance (error) in the dependent variable, X, reduced by knowledge of subjects' group memberships (as opposed to not knowing them).

References The basic reference for this section is Welsh (1947). Textbook discussions appear in Dixon and Massey (1969), 119; and Winer (1962), 36–39.

12-12 Comparing the means of two independent samples when the variance estimates can be pooled

A. Basic Formulas

The present section is an extension of Sections 12-7 and 12-11; discussions of design, assumptions, sampling, and hypotheses for the present case appear in the early paragraphs of these sections.

The form of the t test in Section 12-11 was given by formula 12-10:

$$t = \frac{(\overline{X}_1 - \overline{X}_2) - (\mu_1 - \mu_2)}{\sqrt{\hat{s}_1^2/N_1 + \hat{s}_2^2/N_2}}$$

When the two variances are equal, a common term can be removed from the radical in the standard-error term, which then becomes

$$\hat{s}_{\bar{X}_1 - \bar{X}_2} = \hat{s}_X \sqrt{\frac{1}{N_1} + \frac{1}{N_2}}$$

We would like to obtain an estimate of $\hat{s}_X$ which will make use of the information in both samples, because then the results will be more stable. However, if we merely use the formula for standard deviation,

$$\hat{s}_X = \sqrt{\frac{\sum (X - \bar{X})^2}{N - 1}}$$

we will have a problem in deciding what estimate to use for $\bar{X}$, for even under the null hypothesis we will expect the two means to be slightly different. And when the null hypothesis is false, taking all the deviations about the same mean would result in some scores having great deviations, simply because the mean that has been subtracted from them was not representative of the population from which the scores were drawn, and even less representative of the scores themselves. In this case, the standard deviations would be inflated, and we would have a too-small probability of rejecting a null hypothesis when it was false.

To avoid this problem, we can compare each score to the mean of the sample of which it is an element and then average the squared deviations. The formula for the pooled standard deviation thus is given by

$$\hat{s}_{X(\text{pooled})} = \sqrt{\frac{\sum (X_{i1} - \bar{X}_1)^2 + \sum (X_{i2} - \bar{X}_2)^2}{N_1 + N_2 - 2}} \qquad (12\text{-}12)$$

where subscripts on the scores indicate the samples they were drawn from. An alternative formula can be derived by noting that

$$\hat{s}_1^2 = \frac{\sum (X_{i1} - \bar{X}_1)^2}{N_1 - 1} \qquad \text{and} \qquad \hat{s}_2^2 = \frac{\sum_i (X_{i2} - \bar{X}_2)^2}{N_2 - 1}$$

are obtained by applying the basic formula for variance (formula 5-13) to each of the samples in turn. Then multiplying both sides by $N - 1$, we get

$$\sum (X_{i1} - \bar{X}_1)^2 = (N_1 - 1)\hat{s}_1^2 \qquad \text{and} \qquad \sum (X_{i2} - \bar{X}_2)^2 = (N_2 - 1)\hat{s}_2^2$$

Substituting these values in formula 12-12 gives

$$\hat{s}_{X(\text{pooled})} = \sqrt{\frac{(N_1 - 1)\hat{s}_1^2 + (N_2 - 1)\hat{s}_2^2}{N_1 + N_2 - 2}} \qquad (12\text{-}13)$$

Here the denominator of the standard deviation and the number of degrees of freedom are reduced by 2, because both means must be utilized in order to compute the two sets of deviations.

Either formula for the pooled standard deviation (12-12 or 12-13) can then be substituted in the formula for the standard error of the difference between means, and the latter substituted in the formula for t, giving

$$t = \frac{(\overline{X}_1 - \overline{X}_2) - (\mu_1 - \mu_2)_0}{\sqrt{\dfrac{(N_1 - 1)\hat{s}_1^2 + (N_2 - 1)\hat{s}_2^2}{N_1 + N_2 - 2}} \sqrt{\dfrac{1}{N_1} + \dfrac{1}{N_2}}} \qquad (12\text{-}14\text{a})$$

or

$$t = \frac{(\overline{X}_1 - \overline{X}_2) - (\mu_1 - \mu_2)_0}{\sqrt{\dfrac{\sum(X_{i1} - \overline{X}_1)^2 + \sum(X_{i2} - \overline{X}_2)^2}{N_1 + N_2 - 2}} \sqrt{\dfrac{1}{N_1} + \dfrac{1}{N_2}}} \qquad (12\text{-}14\text{b})$$

The degrees of freedom are $N_1 + N_2 - 2$, given by the denominator of the formula for the standard deviation.

Formula 12-14 is useful conceptually. The denominator emphasizes the distinction between the pooled standard deviation (see formula 12-13) and the term $\sqrt{1/N_1 + 1/N_2}$ used to obtain from $\hat{s}_{X(\text{pooled})}$ the standard error of the difference between means. For computational purposes, it is slightly easier to combine the two terms of the denominator; that way you only have to take one square root, as shown in formula 12-14c. Here, also, $(X_{i1} - \overline{X})$ has been replaced by x_1, and $(X_{i2} - \overline{X})$ by x_2.

$$t = \frac{(\overline{X}_1 - \overline{X}_2) - (\mu_1 - \mu_2)_0}{\sqrt{\left(\dfrac{\sum x_1^2 + \sum x_2^2}{N_1 + N_2 - 2}\right)\left(\dfrac{1}{N_1} + \dfrac{1}{N_2}\right)}} \qquad (12\text{-}14\text{c})$$

B. Three Simplifications

1. When $\hat{s}_1^2 = \hat{s}_2^2$,

$$\sqrt{\frac{(N_1 - 1)\hat{s}_1^2 + (N_2 - 1)\hat{s}_2^2}{N_1 + N_2 - 2}} = \sqrt{\frac{(N_1 - 1 + N_2 - 1)\hat{s}_1^2}{N_1 + N_2 - 2}} = \sqrt{\hat{s}_1^2} = \hat{s}_1$$

so formula 12-12 reduces to

$$t = \frac{(\overline{X}_1 - \overline{X}_2) - (\mu_1 - \mu_2)_0}{\hat{s}_1 \sqrt{1/N_1 + 1/N_2}} \qquad (12\text{-}15)$$

2. When $N_1 = N_2$,

$$\sqrt{\frac{(N_1 - 1)\hat{s}_1^2 + (N_2 - 1)\hat{s}_2^2}{N_1 + N_2 - 2}} = \sqrt{\frac{(N_1 - 1)(\hat{s}_1^2 + \hat{s}_2^2)}{2N_1 - 2}} = \sqrt{\frac{\hat{s}_1^2 + \hat{s}_2^2}{2}}$$

so formula 12-14 reduces to

$$t = \frac{(\overline{X}_1 - \overline{X}_2) - (\mu_1 - \mu_2)_0}{\sqrt{\frac{\hat{s}_1^2 + \hat{s}_2^2}{2}}\sqrt{\frac{2}{N_1}}} = \frac{(\overline{X}_1 - \overline{X}_2) - (\mu_1 - \mu_2)_0}{\sqrt{\frac{\hat{s}_1^2 + \hat{s}_2^2}{N_1}}} \qquad (12\text{-}16)$$

3. When both $\hat{s}_1^2 = \hat{s}_2^2$ and $N_1 = N_2$, formula (12-15) simplifies to

$$t = \frac{(\overline{X}_1 - \overline{X}_2) - (\mu_1 - \mu_2)_0}{\hat{s}_1 \sqrt{2/N_2}} \qquad (12\text{-}17)$$

In each case, the total number of degrees of freedom is equal to the total number of scores on which the pooled standard deviation is computed (usually $N_1 + N_2$), minus the number of means necessary to obtain the deviation scores (usually 2).

EXAMPLE A colleague has developed a test of "conscientiousness" for use in predicting college grades, which he says is fakeproof. As a partial check on his hypothesis, you define two populations of college applicants: a straight-take group (s), told to respond honestly, and a fake-take group (f), told to respond so as to fake their way into your college. Your hypotheses are $H_0: \mu_s = \mu_f$ and $H_m: \mu_s \neq \mu_f$. They are nondirectional, because you want to know whether the fakers do either better or worse (if they do worse, your colleague will so indicate in promotional material). You choose $\alpha = .05$ and decide to use 16 subjects in each of two samples (one sample from each population). Your design is that of Figure 9-3, and the critical values of t are $t_{.05}$ (30 df, two tail) $= \pm 2.042$. You draw two samples and find that the histograms have similar shapes. With $N_1 = N_2 = 16$, you are justified in using the analysis of the present section. The sample means are $\overline{X}_s = 36$, $\overline{X}_f = 54$ with variances $\hat{s}_s^2 = 320$, $\hat{s}_f^2 = 80$. Then from formula 12-16,

$$t = \frac{(54 - 36) - 0}{\sqrt{(320 + 80)/16}} = \frac{18}{\sqrt{400/16}} = \frac{18}{5} = 3.60$$

This observed value of t is greater than the positive critical value. Therefore, you reject the null hypothesis and conclude that $\mu_s < \mu_f$: students can improve their scores by faking.

EXAMPLE Suppose that you conduct an experiment in which you gather the following sets of scores from two independent groups of subjects: Group 1: 8, 8, 2, 4, 4, 4; and Group 2: 12, 10, 6, 6, 7, 7. $\alpha = .05$. Your null hypothesis is $\mu_1 = \mu_2$. You begin the analysis by constructing a table:

i	Group 1			Group 2		
	X	x	x^2	X	x	x^2
1	8	3	9	12	4	16
2	8	3	9	10	2	4
3	2	−3	9	6	−2	4
4	4	−1	1	6	−2	4
5	4	−1	1	7	−1	1
6	4	−1	1	7	−1	1
$\sum$	30	0	30	48	0	30

A graphing of the deviation scores for the two distributions does not indicate any great difference in their shapes (see Figure 12-12).

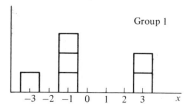

Group 1

Figure 12-12

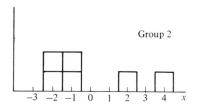

Group 2

Therefore, you apply formula 12-15 and obtain the following results to test the hypothesis of no difference in population means:

$$t = \frac{(5 - 8)}{\sqrt{6}\sqrt{\frac{1}{6} + \frac{1}{6}}} = \frac{-3}{\sqrt{6(\frac{1}{3})}} = \frac{-3}{\sqrt{2}} = \frac{-3\sqrt{2}}{2} \doteq -2.121$$

The critical values of t at the .05 level, for a nondirectional test with 10 df are ± 2.228. The observed value of 2.121 is between these two critical values, so the null hypothesis is not rejected: the two populations have equal means.

C. Degree of Relationship

As in Section 12-11, the degree of relationship between independent and dependent variables can be indicated using the point biserial correlation coefficient of Chapter 16. The square of the correlation $(r^2_{p.bis})$ tells the proportion of variance (error) in the dependent variable, X, reduced by knowing subjects' group memberships (as opposed to not knowing them).

References This test for comparing means is discussed in Dixon and Massey (1969), 116–119; Edwards (1973), 65–68; Games and Klare (1967), 327–338; and Winer (1962), 24–31.

Homework

1. An intelligence test is known to have a mean of 100 for the general population. You believe Flonk U. students to be, on the average, more intelligent than this.
   (a) What are your hypotheses?
   (b) Do you have a one-tail or a two-tail test?
   A random sample of 25 F.U. students has a mean of 106 on this test and an unbiased standard deviation of 15. You set α at .05.
   (c) What is your design (Section 9-3)?
   (d) Should you use a t or a z test?
   (e) What is/are your critical value(s)?
   (f) What is the calculated value of your test statistic?
   (g) What is your conclusion?
2. The mathematical section of the College Board Test has a mean of 500 and a standard deviation of 100 nationwide, for college applicants. In the past you have noticed that students coming to State U. from the southern half of the state are rather poorly prepared in math, and you predict that students from this region are significantly below the national average. You randomly select 16 applicants for admission from the southern region to test your hypotheses, setting α at .05.
   (a) What are your hypotheses?
   (b) Diagram your design as in Section 9-3.
   (c) Do you have a one-tail or a two-tail test?
   (d) What statistical formula should you use to test your hypotheses?
   (e) What is/are the critical value(s)?
   You collect your data and find that the mean for your subjects is 480.
   (f) What is the calculated value of your test statistic?
   (g) What is your conclusion?

For problems 3 through 5, select the most appropriate one of the formulas given in this chapter to test each hypothesis.

3. You wish to test the effect of testosterone injections on the personalities of 5-year-old males. You select 12 pairs of twins; one twin goes in the experimental group (treated), the other in the control group (no treatment). Compare groups using a personality test for masculine interests after 9 months of treatment.

4. You wish to test the effect of an anti-British message on attitudes toward the British; your hypothesis is that females will be more swayed by the message than males. You randomly and independently select 10 males and 10 females to fill your two samples, and test their attitudes toward the British one day after they have received the message.

5. A test of "attitude toward economic independence versus dependence on relatives" has an accepted mean of 15 and standard deviation of 3.82 for female college graduates. As part of a door-to-door survey, you administer the test to a large number of housewives. Later you remove the files for all college graduates in the sample and assign each to one of two groups: group A if the subject was a first-born or an only child, group B if she was a later-born child. You hypothesize that means for the two groups will differ.

6. We believe that the number of errors in running a maze will be a positive function of the proportion of a rat's cerebrum that has been destroyed. We have group A of 144 animals, each of which has had from 21 to 30 percent of its cerebral tissue destroyed and group B of 25 with 51 to 60 percent destroyed. We decide to use the design for comparing a sample to a population and to set α at .05; the mean number of errors for the first group is 41.4 and the standard deviation is 12.

(a) What are our hypotheses?

(b) What assumptions or approximations must we use in order to use the method of Section 12-4?

(c) Do we have a one-tail or a two-tail test?

(d) What is/are the critical value(s)?

The second group has a mean number of errors of 46.6.

(e) What is the calculated value of the test statistic, and what are our conclusions?

7. Consider problem 6 again, to determine how the method of Section 12-5 can be used to solve the problem.

(a) How do our assumptions, critical values, and calculated statistic change under these conditions?

(b) Is there a change in the conclusion? If so, in what way?

8. Suppose in problem 6 that we decide to treat the data as coming from two independent samples. We calculate the standard deviation of sample B to be 12 and find that the two distributions have similar shapes.

(a) What are our hypotheses?

(b) What is our design (Section 9-3)?

(c) What assumption can we relax by comparison with problem 7?
(d) Find the critical value(s).
(e) Calculate the statistic and test your hypotheses.

9. To find the effect of alcohol on typing skills, nine typists were selected at random from a typing pool; each was given a typing test before and after he or she had a martini. The scores given in the table represent the average number of mistakes made per letter

Subject	Before	After
1	1	6
2	1	6
3	3	4
4	3	4
5	2	7
6	2	7
7	3	4
8	1	2
9	2	5

Use a significance level of .01 to test whether the alcohol adversely affected typing skill.
(a) Diagram your design as in Section 9-3.
(b) State your formal hypotheses.
(c) What formula should you use?
(d) What is the critical value?
(e) What is the calculated value of the test statistic?
(f) What is your conclusion?

For problems 10 through 13 carry out a complete analysis, as required for problem 9.

10. An aggressiveness scale was given to a group of five girls and another group of 11 boys at a summer camp. The results are indicated below. Is there a sex difference in aggressiveness? Set α at .05.

Statistic	Girls	Boys
N	5	11
Mean	78	90
$\hat{s}$	5	11

11. Professor Anon has been hired by a government agency to try out a new brainwashing technique which is intended to make subjects considerably

more compliant. He first develops a compliance scale using a large sample ($N = 400$) of adult subjects, finding a mean of 40 and a standard deviation of 5. (A score of 100 indicates perfect compliance.) He decides to set α at .05. Because of the dangers, only one subject can be used; this subject is obtained by a random national sampling of the adult population of the United States. The treatment is administered, and the subject is tested, using the compliance scale. If the subject gets a compliance score of 48, would the treatment be effective when applied to the general population of the United States?

12. In an experiment on the effect of physical trauma on stimulus acceptance, a sample of nine normal subjects were administered one electroconvulsive shock each. Two days later they were given the Rosenfeld Gullibility Test. A control group of subjects who did not receive the shock also took the test. Results were as follows:

Treatment group	Control group
$\overline{X}_t = 20$	$\overline{X}_c = 24$
$N_t = 9$	$N_c = 9$
$\hat{s}_t^2 = 12$	$\hat{s}_c^2 = 13$

What assumptions must you make in order to use the formulas of Section 12-12? Complete the analysis and test at the .05 level.

13. An associate has dumped these data in your hands with the urgent request that you test them for significance:

Subject	Pretest	Post-test
1	10	11
2	9	12
3	8	10
4	7	13
5	6	13
6	8	12
7	7	11
8	10	13
9	8	14

Under the circumstances, what would be the safest hypotheses to test? Complete the analysis and test those hypotheses at the .01 level.

14. The winner of the Miss Flomertia Beauty Contest must have two attributes: talent, as demonstrated by her ability to yodel, and beauty, as indicated by large hip measurements. You believe that in the contest this

year, talent is being sacrificed to beauty. To test your idea, you draw a random sample of 4 contestants from the pool of semifinalists and compare their hip measurements to the hip measurements of a similar sample randomly drawn from last year's semifinalists. You hypothesize that this year's semifinalists will be more beautiful than last year's.

(a) State the null and motivated hypotheses.
(b) Is this a directional or nondirectional test? Why?
(c) What section of the text gives the formula to use?
(d) The data are as follows:

Last year		This year	
Contestant	Beauty	Contestant	Beauty
1	45	1	57
2	28	2	41
3	34	3	44
4	37	4	50

What is the critical value of the test statistic, at .05?

(e) What is the computed magnitude of your test statistic?
(f) Should you accept or reject the null hypothesis?
(g) Are this year's contestants more beautiful?
(h) Is talent being sacrificed to beauty?

Homework Answers

The answers for problems 1 through 8 are complete; those for problems 9 through 14 are partial.

1. (a) $H_0: \mu(\text{F.U.}) \leq 100$; $H_m: \mu(\text{F.U.}) > 100$. (b) One tail (c) Figure 9-1 (d) t (e) $+1.711$ (f) $+2$ (g) Reject H_0.
2. (a) $H_0: \mu_{\text{south}} \geq 500$; $H_m: \mu_{\text{south}} < 500$ (b) See Figure 9-1. (c) One tail (d) Formula 12-1. (e) -1.6449 (f) -0.80 (g) Do not reject H_0.
3. Formula 12-9
4. Formula 12-16
5. Formula 12-4
6. (a) If $X = $ number of errors, $H_0: \mu_0 \geq \mu_B$ and $H_m: \mu_0 < \mu_B$. (b) Approximate σ_0 by $\hat{s}_A$, μ_0 by $\overline{X}_A$; assume that $\sigma_A = \sigma_B$ and samples are drawn from the same initial population. (c) One tail (d) 1.6449 (e) 2.0; reject H_0.
7. (a) Let $\hat{s}_A$ estimate σ_0, with 143 df. Approximate μ_0 by $\overline{X}_A$, assume that $\sigma_A = \sigma_B$, and use critical value of 1.658. (b) No change
8. (a) $H_0: \mu_A \geq \mu_B$ and $H_m: \mu_A < \mu_B$ (b) Figure 9-3 (c) Do not approximate μ_A by $\overline{X}_A$. (d) $t_{.05} = 1.658$ (e) $t = 2$; reject H_0.
9. $t_{.01} = 2.896$ (e) 4.5 (f) Reject H_0.

10. (d) df $= 16$, $t_{.05} = \pm 2.120$ **(e)** $t_{\text{obs}} = 3$ **(f)** Reject H_0.

11. (d) Use either t or z : $t_{.05} \doteq z_{.05} = 1.645$ **(e)** 1.60 **(f)** Do not reject H_0.

12. (d) $t_{.05} = \pm 2.120$ **(e)** t(observed) $= -2.40$ **(f)** μ(treatment) $< \mu$(control)

13. (b) $H_0 : \mu_D = 0$ **(d)** $t_{.01} = \pm 3.355$ **(e)** t(observed) $= 6$ **(f)** There was an increase in something from pretest to post-test.

14. (a) $H_0 : \mu_T \leqq \mu_L$ **(b)** Directional: you won't complain if beauty is being sacrificed. **(c)** 12-12 **(d)** 1.943 **(e)** 2.4 **(f)** Reject **(g)** Yes **(h)** Who knows?; talent wasn't tested.

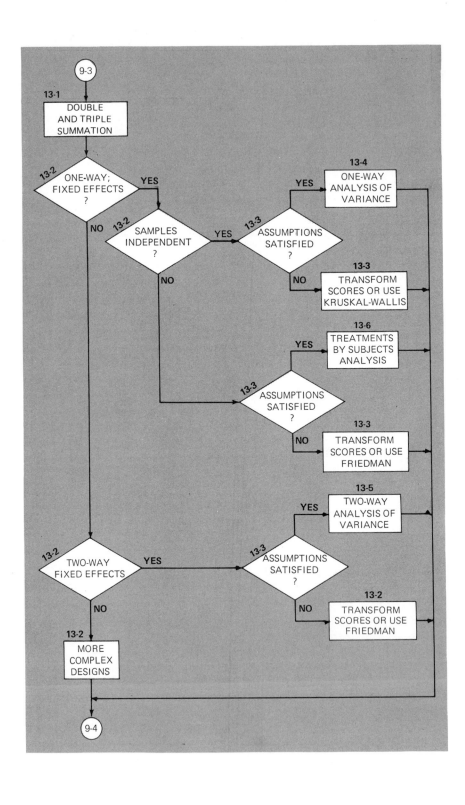

13
Comparing Means: Three or More Samples

In this chapter there are three sections that describe methods for analysis of data when there are more than two treatment or predictor categories. Section 13-4 indicates a way to analyze results when the treatments differ only in one respect (e.g., kind of reinforcement) and different subjects receive different treatments. Section 13-6 analyzes data in which subjects are matched, or in which the same subjects receive all treatments. In Section 13-5 a method is developed for dealing with results when there are two different independent variables (e.g., kind of reinforcement and sex) and one dependent variable.

Whichever section you use, the overall hypothesis testing procedure of Chapter 9 is to be followed.

The methods in this chapter are only an introduction to an extremely diverse area of statistics. Some of the ways in which more complicated designs can differ are suggested in Section 13-2. For details on these methods you will have to look in more advanced texts. Li (1964), Chapters 12, 14, 15, 18, and 19, uses an approach that is conceptually similar to the one followed here, but there are differences in notation. Other comprehensible sources include Edwards (1967), Chapters 7 and 8, and (1960), Chapters 9 through 17; and Hays (1963), Chapters 12 through 14. A more comprehensive treatment of analysis of variance designs can be found in Winer (1962).

13-1 Double and triple summation

When a data table has more than one row and more than one column, it may be called a *matrix* of scores. It is sometimes convenient to label each column of a matrix with a different letter and each row with a different subject number; in such cases, each score in the matrix can be uniquely identified by a combination of letter and subscript. This has been done in Table 13-1 (left). Example scores are A_1, B_5. F_{14}, etc.

A second method for labeling the scores in the data matrix is exemplified by Table 13-1 (right). Here, all scores in the matrix are given the same variable letter, X, and each score in the matrix is uniquely identified by a combination of two subscripts. In this example and throughout this text, the first subscript is used to indicate a score's row, the second its column. (Unfortunately, the order of subscripts used here is not universal, so in reading other works you will have

Table 13-1   *Two methods of labeling a data matrix*

	A	B	C	$\cdots$			1	2	3	$\cdots$	j	$\cdots$	J
1	A_1	B_1	C_1	$\cdots$		1	X_{11}	X_{12}	X_{13}	$\cdots$	X_{1j}	$\cdots$	X_{1J}
2	A_2	B_2	C_2	$\cdots$		2	X_{21}	X_{22}	X_{23}	$\cdots$	X_{2j}	$\cdots$	X_{2J}
3	A_3	B_3	C_3	$\cdots$		3	X_{31}	X_{32}	X_{33}	$\cdots$	X_{3j}	$\cdots$	X_{3J}
$\vdots$	$\vdots$	$\vdots$	$\vdots$			$\vdots$	$\vdots$	$\vdots$	$\vdots$		$\vdots$		
i	A_i	B_i	C_i	$\cdots$		i	X_{i1}	X_{i2}	X_{i3}	$\cdots$	X_{ij}	$\cdots$	X_{iJ}
$\vdots$	$\vdots$	$\vdots$	$\vdots$			$\vdots$	$\vdots$	$\vdots$	$\vdots$		$\vdots$		
N	A_N	B_N	C_N	$\cdots$		N	X_{N1}	X_{N2}	X_{N3}	$\cdots$	X_{Nj}	$\cdots$	X_{NJ}

to check to see which index comes first.) A particular but unspecified i will generally be a subject label, and j will represent some other mode of classifying scores. Thus, X_{ij} may represent the score, X, of the ith subject in the jth group; the score, X, of the ith person on the jth variable; or the score, X, of the ith person at time j; and so on.

A numerical example is the following:

		j			
		1	2	3	4
	1	5	2	6	9
	2	4	1	8	3
i	3	5	1	2	6
	4	3	9	8	7
	5	4	5	2	1

Here, $X_{14} = 9$, $X_{41} = 3$, $X_{23} = 8$, $X_{42} = 9$, etc. Again, you should note that the order of subscripts is important (e.g., $X_{14} \neq X_{41}$).

In a matrix of scores, you must be careful when indicating a summation to show which subscript is being summed over. The sum $\sum_{i=1}^{5} X_{ij}$ depends for its value on the value of j summed over; if $j = 2$, $\sum_{i} X_{ij} = 2 + 1 + 1 + 9 + 5 = 18$, but if $j = 4$, $\sum_{i=1}^{5} X_{ij} = 9 + 3 + 6 + 7 + 1 = 26$.

Similarly, the sum $\sum_{j=1}^{4} X_{ij}$ depends on what i is; if $i = 2$, $\sum_{j=1}^{4} X_{ij} = \sum_{j=1}^{4} X_{2j} = 4 + 1 + 8 + 3 = 16$, but if $i = 5$, then $\sum_{j=1}^{4} X_{ij} = 4 + 5 + 2 + 1 = 12$.

Finally, you can sum over both subscripts:

$$\sum_{j=3}^{4} \sum_{i=4}^{5} X_{ij} = \sum_{i=4}^{5} X_{i3} + \sum_{i=4}^{5} X_{i4}$$

$$= X_{43} + X_{53} + X_{44} + X_{54}$$

$$= 8 + 2 + 7 + 1 = 18$$

Triple-summation notation is used in this book in only one section, Section 13-5: Two-way analysis of variance. There we have two independent or classification variables and an outcome or dependent variable. We need to be able to distinguish among (1) the levels of the A variable, (2) the levels of the B variable, and (3) the subjects at each combination of a level of A with a level of B.

In this case, we generally construct a matrix of cells or boxes, each of which contains one or more scores, such as Table 13-2.

Table 13-2 *Data matrix with two variables of classification*

		Levels of $B = B_k$				
		B_1	B_2	B_3	B_4	B_5
Levels of A: A_j	A_1					
	A_2					
	A_3					
	A_4					

In this case, each score X in the matrix must have three subscripts: X_{ijk}. The j subscript tells what row of the matrix (level of A) the score appears in; the k tells which column (level of B) it falls in; thus, these two subscripts indicate a score's cell in the matrix. The i subscript distinguishes each score in a given cell from all others in the cell. Thus, X_{531} would appear in the third row and first column of a matrix, and it would be the fifth score in that cell.

Using triple summations really adds no new wrinkles to using double summation—you just have to be careful because there are more terms to confuse with one another.

13-2 Some distinctions among analysis-of-variance designs

The methods of this chapter are used to compare the means of two or more samples in order to infer whether the corresponding population means are the same or different. The expression "analysis of variance" may thus appear to be a misnomer, inasmuch as no hypotheses about population variances are tested as such. However, the expression is an accurate description of the manner in which hypotheses are tested rather than of the parameters of real interest. Hypotheses are tested by comparing the variances of means to the variances of raw scores to infer whether the means vary more than would be expected on the basis of random sampling. The exact methods for carrying out the analyses will be discussed in some detail in Sections 13-4 through 13-6.

The analysis-of-variance designs discussed in this chapter have the following characteristics: they are for fixed effects, fully crossed, one- and two-way analyses, and use either independent or related samples. If your problem fits into this set of categories, the present chapter can help you solve it; otherwise, you will have to go to more advanced texts. In any case, you should find out about these basic designs before tackling more advanced treatments of analysis of variance. Let us now consider in detail what the present designs are like and how they differ from other possible analyses.

A. One-Way, Two-Way, and Higher Analyses

This distinction among analyses simply refers to the number of independent, or classification, variables. In a one-way analysis you have one variable of classification and wish to know the effect of differences in this variable on scores on the dependent variable. Subjects might be drawn initially from the same population and used to test for differences among treatments. For example, to test for the relative effectiveness of four kinds of psychotherapy in reducing anxiety, we might randomly assign anxious subjects to four treatment groups and a control group that receives no therapy. We would have one predictor (kind of treatment), with five levels, and our dependent variable would be amount of tested anxiety after treatment.

In a two-way analysis, you have two variables of classification and wish to know their effect on some dependent variable. Thus, you might be interested in the amount of conformity that male and female subjects display under achievement-incentive conditions as opposed to affiliative-incentive conditions. The dependent variable would be amount of conformity and the variables of classification would be conditions, with two levels, and sex, with two levels.

Analyses that have three and more independent variables are not only possible but often found in the literature. They will not, however, be discussed in this text.

B. Fixed, Mixed, and Random Designs

Analyses can differ in the kinds of generalizations made from them. One-way analysis can be either fixed or random. In a *fixed-effects* analysis, the conclusions of the research apply only to the levels of the independent variable that are tested. In a *random-effects* analysis, levels of the independent variable are randomly selected and an inference is made, not only to populations of subjects, but also to the population of all levels of the independent variable. In a study of the effect of birth order on academic achievement, the levels of the independent variable might be "only child; first-born; second-born child; third or later child." This would be a fixed-effects analysis because no generalization would be needed to additional levels of the variable of classification. Suppose, on the other hand, that we want to test whether a new method of teaching third-grade mathematics varies in effectiveness from teacher to teacher. Our dependent variable might be students' math-test scores after receiving instruction by the new method, and our independent variable would be "teacher." To get an idea of teacher differences, we might obtain a random sample of 20 teachers from the population of third-grade teachers. There would be 20 levels to the independent variable, and they would be randomly chosen from a population of levels; therefore, this would be a random design. The results of the analysis would be generalized to the entire population of third-grade teachers.

Two-way and higher analyses are called *fixed* if all independent variables are fixed; that is, if no inference is to be made to any level of any independent variable that is not tested in the analysis. Two-way and higher analyses are called *random* if all variables of classification are random variables; that is, if the experiment constitutes a random sampling of levels on all the independent variables. Two-way and higher analyses are called *mixed* if one or more independent variables are fixed and one or more are random.

C. Independent Versus Related Samples

As indicated in Section 9-3C for the two-sample case, correlated data occur when there is a one-to-one matching of subjects in each sample with subjects in every other sample. When, as in the designs considered in this chapter, several samples are compared at once, correlation generally occurs through the use of the same subjects under all treatment conditions, and the design is often called a treatment-by-subjects analysis. Independent samples occur when there is no such matching of subjects across the samples (i.e., when the drawing of subjects for each sample is done independently of the drawing of subjects for every other).

Matching is done to reduce error and to give a more sensitive test of treatment effects. One illustration of such an increase in sensitivity was given previously, in Section 12-9, where a difference between two groups was shown to be nonsignificant if the groups were independent but significant if they were related. Similar examples could easily be developed for the methods of this chapter.

On the other hand, independent samples are used when the analysis compares different populations: the same subjects cannot be both male and female, or upper and middle and lower class, or first-born and later-born, etc. It is also necessary to use separate samples when the assumptions of a treatment-by-subjects analysis are untenable (see Section 13-3). For example, you cannot compare different methods of teaching the same course material and use the same subjects for all treatments, because some of the learning for one treatment would probably carry over to the next, and thus make later treatments look more effective than they really are. Although in these designs it might be possible to match samples and thus reduce random error without using the same subjects, many experimenters prefer, instead of matching samples on some additional variable, to control the additional variable using analysis of covariance (see part E).

D. Fully Crossed Versus Nested Designs

In two-way and higher analyses, when every level of each independent variable is paired with every level of each other independent variable, the design is said to be *fully crossed*. When some levels of variable B can only appear in conjunction with one level of variable A, then B is said to be *nested* under A. Suppose, for example, that we are trying out a new method of teaching freshman English in five high schools. In schools 1, 2, and 3 there are three freshman English teachers each, and in schools 4 and 5 there are four each. We want to test separately for effects due to high schools (variable A, with 5 levels) and to teachers (variable B, with 17 levels). Variables A and B cannot be fully crossed, because each teacher can only appear in conjunction with the high school at which he teaches. Teachers are said to be nested under high schools in this design.

In this text, only fully crossed designs will be considered.

E. Analysis of Covariance

Sometimes we may be worried that our experimental results will be confounded or obscured by an independent variable that does not really interest us. If the confounding variable is interval level, we may equate it statistically for our treatment groups and then proceed with the analysis. The method used to do this is called *analysis of covariance*, and the variable that is equated is called the *covariate*. For example, if we are trying to determine whether there are teacher differences in the use of a new teaching approach, the results may be difficult to interpret if the teachers have classes differing in intelligence. We might obtain a clearer test of teacher differences by treating student intelligence as a covariate and removing it statistically from affecting the outcome. For more information on this method, see more advanced texts.

References Fixed, mixed, and random models are discussed and contrasted in Dixon and Massey (1969), 150–156; and Hays (1963), 356–358, 413–418. Analysis of covariance is discussed in Dixon and Massey, Chapter 12; and Edwards (1960), Chapter 16.

13-3 Assumptions of the analysis of variance

The analysis of variance is predicated on several assumptions about the nature of the data to which it is applied. It is never possible to be certain that all the assumptions apply to your data, and you must analyze your data on the basis of reasonable judgments. When the assumptions of the analysis of variance are not met, it may still be possible to do the arithmetic prescribed, to complete the test, and to find an F ratio. The hitch comes in looking up a critical value in

Table A-8 of the Appendix to which to compare your calculated value. The tabulated value is based on those assumptions, and if your data don't meet them, the critical value tabulated may be either too great or too small for your data. As a result, your conclusion, based on a comparison of your obtained F to the tabulated value, may be wrong.

When the assumptions appear not to be met, you may be able to transform the data nonlinearly (see Section 6-6) so that they are met. However, when data are transformed, interpretation of results becomes more difficult. In many cases it is preferable to use a method of analysis which makes different assumptions, such as the Kruskal–Wallis (Section 14-6) or Friedman (Section 14-7) methods.

One assumption of the analysis of variance is that the dependent variable is interval-scaled; otherwise, the arithmetic manipulations performed on the data, and thus the results of the analysis, may be meaningless. If the data are ordinal but you are willing to assume that there is an underlying normal distribution on which they are based, you may wish to transform to z scores using the standard normal distribution (Section 7-3D). If you are not willing to make that assumption, you might be better advised to use the Kruskal–Wallis or Friedman test for your data.

A second assumption of the analysis of variance is that the dependent variable is normally distributed in each of the populations compared. As indicated in Chapter 10, we could carry out chi-square tests for the normality of the distributions; however, this is really not necessary. If the population distributions are nonnormal but all have about the same shape, the distribution of F will be relatively unaffected; alternatively, if the populations seem to have different shapes but the sample sizes are large, F will also be only slightly affected. It would therefore seem reasonable, each time several means are to be compared, to graph the sample scores for each sample. If they seem all to be roughly normal, go ahead; if they all seem to have about the same shape, go ahead; if they have different shapes, either (a) use another method of analysis, such as the Kruskal–Wallis or Friedman; (b) try to transform the data nonlinearly so that they will be similar in shape (see Section 6-6 for references); or (c) gather fairly large samples. The more the distributions depart from one another and from normality, the larger the samples should be.

A third assumption of the analysis of variance is that the populations from which the samples are drawn have equal variances. Statistical tests for equality of variances have been suggested in the past to determine whether the population variances are equal. However, statistical equality is more difficult to define and assess than statistical inequality (it involves type II error; see Section 9-4), and as the test has been used it has proven too sensitive. The F test has been disqualified by using this test in many applications where its use would have been very reasonable by other standards. Unequal variances have been shown to be a cause for concern only when sample sizes differ. If you anticipate this problem, you can simply draw samples with equal N's, which also has the advantage that it reduces the computations and generally increases the sensitivity of the test.

In the treatment by subject design, it must also be assumed that experimental errors are independent of one another and of treatments. When the same subjects are tested under several different treatment conditions, any of a number of extraneous variables may contribute to a systematic relationship between trial and performance and may be erroneously interpreted as showing a relationship between treatment and performance. Subjects may learn how to perform on the dependent variable as the experiment progresses, which may lead to less variability of performance for later trials. Their performance may also improve through learning, which may lead to systematically higher or lower scores for later treatments. Memory or fatigue may also operate and lead to a spuriously high correlation across trials. These effects tend to invalidate any correlated data analysis and must be handled through extrastatistical means, e.g., allow practice with the task of the dependent variable before the experiment begins, be careful about the spacing of trials, balance across subjects the order in which treatments are administered, and so on. If you cannot reduce such effects to negligible proportions, you would do best to avoid a correlated data design.

Finally, certain kinds of dependent variables generally require special transformations before the assumptions of the analysis of variance can be satisfied. Among these are proportions, scores that indicate amount of time to completion of a fixed task, and scores that indicate quantity of performance per unit time. Further discussions of these cases and transformations can be found in Sections 6-2 and 6-6.

13-4   One-way, fixed-effects analysis of variance

Relevant earlier section:
11-2   Comparing variances

A. Basic Theory
Analysis of variance is used when we wish to infer relationships among several population means on the basis of the relative magnitudes of means of samples drawn randomly from the populations. In the one-way case, the populations are categorized on a single independent variable. Hence, there is a single null hypothesis to the effect that different values of the independent variable have no effect on the values of the dependent variable:

$$H_0 : \mu_1 = \mu_2 = \mu_3 = \cdots = \mu_J \qquad H_m : \text{not } (\mu_1 = \mu_2 = \mu_3 = \cdots = \mu_J)$$

The motivated hypothesis states simply that there is some difference among the means of the several populations on the dependent variable (i.e., at least two of the means are not equal), which can be attributed to their different values on the independent variable.

For example, we might predict that four different-textured surfaces would appear to be at different distances when presented to subjects with other distance cues eliminated. Our independent variable would be texture and our dependent variable would be judged distance from subject to surface. We might have one sample of 10 subjects judge the distance to surface 1 under these conditions, a second sample judge the distance to surface 2, and so on, while all surfaces would actually be presented at the same distance. Our measure of perceived distance could be averaged for each sample, and our hypotheses could be written

$$H_0 : \mu_1 = \mu_2 = \mu_3 = \mu_4 \qquad H_m : \text{not } (\mu_1 = \mu_2 = \mu_3 = \mu_4)$$

where the μ's are means of judgments made by populations of subjects. The designs for this section are given in Section 9-3D1 and 2.

The basis for the test is the known relationship between the variance of raw scores in a population and the variance of the means of samples randomly drawn from the population, as indicated in Section 7-3C. For an infinite number of samples, all of size N,

$$\sigma_{\bar{X}}^2 = \frac{\sigma_X^2}{N}$$

Multiplying both sides of the equation by N, we get

$$N\sigma_{\bar{X}}^2 = \sigma_X^2$$

This formula says that we expect the variance of raw scores to be N times the variance of the means.

In most practical situations, we don't have population variances to work with; but we can substitute sample estimates in the above equation and obtain an approximate formula:

$$N\hat{s}_{\bar{X}}^2 \doteq \hat{s}_X^2$$

We could test this formula by defining a population distribution and randomly drawing a large number J of samples from it. If, for example, our samples were all of size $N = 10$, and if our population variance (σ_X^2) was 42, the sample variances $(\hat{s}_X^2)$ would each differ from 42 only randomly. If we then took the means from all the J samples and found their variance estimate $(\hat{s}_{\bar{X}}^2)$ by using the means themselves in the variance formula, this variance estimate would differ only randomly from $42/10 = 4.2$.

Suppose now that there are systematic nonrandom differences among the means—but not the variances—of the several samples. In this case, the formula for the standard error of the mean does not apply, $N\sigma_{\bar{X}}^2 \neq \sigma_X^2$, and similarly for the estimates: $N\hat{s}_{\bar{X}}^2 \neq \hat{s}_X^2$.

In analysis of variance, we compare several population means by comparing the means of samples drawn from the populations. If the populations all have the same mean, then the null hypothesis is true, the sample means differ only randomly, and $N\hat{s}_{\bar{X}}^2$ differs from $\hat{s}_{\bar{X}}^2$ only randomly. On the other hand, if the population means differ (but the population variances are the same), then the sample means will be too variable: there will be both random and systematic differences contributing to $\hat{s}_{\bar{X}}^2$, but only random differences contributing to $\hat{s}_X^2$. Hence, when the motivated hypothesis is true, $N\hat{s}_{\bar{X}}^2$ will be greater than $\hat{s}_X^2$. We might therefore rewrite the hypotheses in the following manner:

$$H_0 : N\sigma_{\bar{X}}^2 = \sigma_X^2 \qquad H_m : N\sigma_{\bar{X}}^2 > \sigma_X^2$$

where it is not possible for the left-hand term to be smaller than the right-hand term, although the sample estimate $N\hat{s}_{\bar{X}}^2$ might be smaller than the estimate $\hat{s}_X^2$ due to random sampling error. These hypotheses would then be equivalent to the null and motivated hypotheses given at the beginning of this section.

EXAMPLE　In the above example of judging distances to textured surfaces, we have four hypothetical populations of subjects and want to know if they have the same mean. We have good reason to assume that they have the same population variance. Our hypotheses are

$$H_0 : \mu_1 = \mu_2 = \mu_3 = \mu_4 \leftrightarrow H_0 : N\sigma_{\bar{X}}^2 = \sigma_X^2$$
$$H_m : \text{not } (\mu_1 = \mu_2 = \mu_3 = \mu_4) \leftrightarrow H_m : N\sigma_{\bar{X}}^2 > \sigma_X^2$$

Therefore, it follows that when H_0 is true, $N\hat{s}_{\bar{X}}^2 \doteq \hat{s}_X^2$, and when H_0 is false, $N\hat{s}_{\bar{X}}^2 > \hat{s}_X^2$. We test the null hypothesis by (1) first seeing whether N times the variance of the means is, in fact, greater than the variance of raw scores, and, if so, (2) testing to see whether the difference is greater than we could reasonably expect to happen by chance. We decide to set α at .05, then collect the data in Table 13-3 from samples drawn randomly from the four populations.

Table 13-3　*Example data for one-way analysis of variance*

Subject	Sample				Σ
	1	2	3	4	
1	5	1	3	1	
2	4	1	2	2	
3	5	2	2	3	
4	6	0	3	0	
ΣX	20	4	10	6	40
ΣX^2	102	6	26	14	148

There are four nominal-level categories to the independent variable, texture; we sometimes say that there are four *experimental conditions* or *treatment conditions.* Inasmuch as a different subject sample receives each experimental condition, we may also say there are four treatment groups. Within each treatment group we may obtain a variance of judgments by applying formula 5-13. For each group the formula is

$$\hat{s}_j^2 = \frac{\sum_i (X_{ij} - \overline{X}_j)^2}{N_j - 1} = \frac{\sum X_{ij}^2 - \left(\sum_i X_{ij}\right)^2 / N_j}{N_j - 1}$$

where X_{ij} is the score of subject i in group j, $\overline{X}_j$ is the mean of group j, and N_j is the number of subjects in group j. Then we fill in this formula for each of the groups in turn:

$$\hat{s}_1^2 = \frac{102 - (20)^2/4}{3} = \frac{102 - 400/4}{3} = \frac{102 - 100}{3} = \frac{2}{3}$$

$$\hat{s}_2^2 = \frac{6 - 4^2/4}{3} = \frac{6 - 16/4}{3} = \frac{6 - 4}{3} = \frac{2}{3}$$

$$\hat{s}_3^2 = \frac{26 - 10^2/4}{3} = \frac{26 - 100/4}{3} = \frac{26 - 25}{3} = \frac{1}{3}$$

$$\hat{s}_4^2 = \frac{14 - 6^2/4}{3} = \frac{14 - 36/4}{3} = \frac{14 - 9}{3} = \frac{5}{3}$$

There are some differences here, but we can reasonably assume that they could have occurred randomly. In this case, we can pool the variance estimates to get a single value which is based on more degrees of freedom than any of the four, and thus is a better estimate of the population variances, under the above assumption that all four populations have the same variance. If we extend the rationale of Section 12-12 for pooling the variance estimates, we get formula 13-1:

$$\hat{s}_{pooled}^2 = \frac{\sum x_1^2 + \sum x_2^2 + \sum x_3^2 + \sum x_4^2}{(N_1 - 1) + (N_2 - 1) + (N_3 - 1) + (N_4 - 1)}$$

$$= \frac{(N_1 - 1)\hat{s}_1^2 + (N_2 - 1)\hat{s}_2^2 + (N_3 - 1)\hat{s}_3^2 + (N_4 - 1)\hat{s}_4^2}{N_1 + N_2 + N_3 + N_4 - 4}$$

or, in general,

$$\hat{s}_{pooled}^2 = \frac{\sum_j (N_j - 1)\hat{s}_j^2}{\sum_j (N_j - 1)} \tag{13-1}$$

Substituting values from the present problem, we obtain

$$\hat{s}^2_{pooled} = \frac{2 + 2 + 1 + 5}{12} = \frac{10}{12} = \frac{5}{6} = .833$$

This calculated value $\hat{s}^2_{pooled}$ is the best estimate of σ^2_X that we can get from these data. Note that the pooled variance estimate is based on $N_j - 1$ degrees of freedom from each sample j, for a total of $\sum (N_j - 1)$ df. We can also get an estimate of $\sigma^2_{\bar{X}}$ from the means.

$$\bar{X}_1 = \frac{20}{4} = 5$$

$$\bar{X}_2 = \frac{4}{4} = 1$$

$$\bar{X}_3 = \frac{10}{4} = 2.5$$

$$\bar{X}_4 = \frac{6}{4} = 1.5$$

The variance of the means can be found by substituting $\bar{X}$'s in the general variance-estimate formula (5-13).

$$\hat{s}^2_{\bar{X}} = \frac{\sum_{j=1}^{J} (\bar{X}_{.j} - \bar{\bar{X}}_{..})^2}{J - 1} \tag{13-2}$$

where there are J samples having means $\bar{X}_{.j}$ and where $\bar{X}_{..}$ is the overall, grand mean. In each mean, a dot replaces the subscript averaged over: thus, $\bar{X}_{.j}$ is the mean, across subjects (i) for sample j. The grand mean is averaged over both subjects and groups:

$$\bar{X}_{..} = \frac{\sum_j \sum_i X_{ij}}{NJ} = \frac{40}{4(4)} = \frac{40}{16} = \frac{10}{4} = 2.5$$

Then, substituting the values for the four means and the grand mean into the variance formula gives

$$\hat{s}^2_{\bar{X}} = \frac{(5 - 2.5)^2 + (1 - 2.5)^2 + (2.5 - 2.5)^2 + (1.5 - 2.5)^2}{4 - 1}$$

$$= \frac{6.25 + 2.25 + 0 + 1}{3} = \frac{9.5}{3} = \frac{19}{6}$$

This estimate is based on $J - 1$ degrees of freedom (the denominator of formula 13-2), where, in this example, $J = 4$.

Remember that $N\hat{s}_{\bar{X}}^2$ is an unbiased estimate of the raw-score variance when the null hypothesis is true. Let us calculate its value in this case:

$$N\hat{s}_{\bar{X}}^2 = 4\left(\frac{19}{6}\right) = \frac{38}{3} = 12.667$$

From these results it is clear that the sample means are more variable than they should be for the standard-error formula to hold exactly: 12.667 is greater than .833. If the difference is significant, we can infer that the corresponding population means are unequal. If it is not, we must admit that the apparent difference may be due to random sampling. Now we can test to see if $N\hat{s}_{\bar{X}}^2$ is nonrandomly greater than $\hat{s}_X^2$.

When the null hypothesis is true, $N\hat{s}_{\bar{X}}^2$ and $\hat{s}_X^2$ are both unbiased estimates of the population parameter σ_X^2. Therefore, we can test the null hypothesis by comparing these two variance estimates, using the method of Section 11-2. The general formula is

$$F = \frac{\hat{s}_a^2}{\hat{s}_b^2}$$

with $(N_a - 1, N_b - 1 \text{ df})$, where $\hat{s}_a^2$ and $\hat{s}_b^2$ must be estimating the same parameter when the null hypothesis is true.

Here let $\hat{s}_a^2 = N\hat{s}_{\bar{X}}^2$ and $\hat{s}_b^2 = \hat{s}_X^2$ (under H_0, both estimate σ^2). Then

$$F = \frac{N\hat{s}_{\bar{X}}^2}{\hat{s}_X^2} \qquad \text{with } J - 1, \sum_{j=1}^{J} (N_j - 1) \text{ df} \qquad (13\text{-}3)$$

For this problem

$$F = \frac{38/3}{5/6} = \frac{38(6)}{3(5)} = \frac{76}{5} = 15.2$$

We obtain the critical value of F from the degrees of freedom:

$$F_{.05} \left(J - 1, \sum_j (N_j - 1)\right) = F_{.05}(3, 12) = 3.49$$

Notice that we do a one-tail test, because the motivated hypothesis is supported only when the numerator, $N\hat{s}_{\bar{X}}^2$, is larger than the denominator, $\hat{s}_X^2$.

The figure shows the distribution of F's we would expect if we were to replicate this experiment a large number of times, with the null hypothesis true. The critical value of 3.49 cuts off the upper .05 of this distribution. The value of F obtained from this experiment is in the rejection region, so

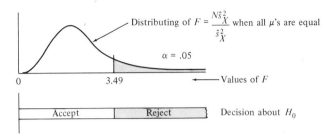

we conclude that it is so large that we don't think it could have happened by chance, with H_0 true. We therefore decide to reject H_0 and say that there are differences among the populations sampled.

B. Computational Format

In order to have computational procedures that will be generalizable to more complicated problems with a minimum of additional work, special formulas are used. Sums of squared deviations, called "sums of squares" (SS), are found, some of which form numerators of variance estimates used in testing the null hypothesis. The computational format results in the same F ratio as obtained in formula 13-3, as given in the right-hand column of Table 13-4.

Table 13-4   *Computational table for one-way analysis of variance*

Source of variance	Sum of squares	Degrees of freedom	Mean square	F
Between groups	SS_B	$df_B = J - 1$	$MS_B = SS_B/df_B$	$F = MS_B/MS_W$
Within groups	SS_W	$df_W = \sum_j (N_j - 1)$	$MS_W = SS_W/df_W$	
Total	SS_T	$df_T = (\sum_j N_j) - 1$		

The numerator of F, called the mean square between treatments or between groups (MS_B), is equal to $N\hat{s}_{\bar{X}}^2$ of formula 13-3; it is an unbiased estimate of raw-score variance, obtained from sample means, when H_0 is true. The denominator of F is the pooled within-groups variance estimate, $\hat{s}_X^2$, of formula 13-3.

Working backward in Table 13-4 we see that MS_B has two components, SS_B and df_B. By comparing formula 13-2 and the numerator of formula 13-3, we see that

$$MS_B = N\hat{s}_{\bar{X}}^2 = \frac{N \sum_{j=1}^{J} (\bar{X}_{.j} - \bar{\bar{X}}_{..})^2}{J - 1} \tag{13-4}$$

Formula 13-4 has $J - 1$ df, where J is the number of treatment means on which $\hat{s}_{\bar{X}}^2$ is based. $J - 1$ is called the between-groups degrees of freedom. The numerator of formula 13-4, $N \sum_{j=1}^{J} (\bar{X}._j - \bar{\bar{X}}..)^2$, is called the "sum of squared deviations between groups," or the "between-groups sum of squares (SS_B)." On page 327 this term was calculated from means; in this part of the section we will develop a formula for calculating it from totals, which is a little easier and involves less rounding error.

MS_W also has two components, SS_W and df_W. The former is called the "sum of squares within groups," or the "within sum of squares"; it is the numerator of formula 13-1. The within sum of squares may be written

$$SS_W = \sum_j \sum_i (X_{ij} - \bar{X}._j)^2$$

It is equal to the sum of the squared deviations of all scores about their respective group means. The number of degrees of freedom within groups is the sum of the numbers of degrees of freedom on which the pooled estimate is based.

It is convenient to define a total number of degrees of freedom, df_T, where

$$df_T = df_B + df_W$$

The total number of degrees of freedom is also equal to 1 less than the total number of subjects: $df_T = (\sum_j N_j) - 1$. A total sum of squares can also be defined as the sum of squared deviations of all scores X_{ij} about the grand mean $\bar{\bar{X}}..$.

$$SS_T = \sum_j \sum_i (X_{ij} - \bar{\bar{X}}..)^2$$

To ease the computational burden, a simplified formula is used, which will now be derived. Defining N_T as the total number of subjects and T_{++} as the sum of all scores, $\bar{\bar{X}}.. = \sum_j \sum_i X_{ij}/\sum_j N_j = T_{++}/N_T$. (In this notation, a '+' replaces the subscript summed over.) Then,

$$SS_T = \sum_j \sum_i \left(X_{ij} - \frac{T_{++}}{N_T}\right)^2$$

Expanding the squared term, we get

$$SS_T = \sum_j \sum_i \left(X_{ij}^2 - 2X_{ij}\frac{T_{++}}{N_T} + \frac{T_{++}^2}{N_T^2}\right)$$

Distributing the summation,

$$SS_T = \sum_j \sum_i X_{ij}^2 - 2\frac{T_{++}}{N_T}\sum_j \sum_i X_{ij} + \frac{N_T T_{++}^2}{N_T^2}$$

because T_{++} and N_T are constants. Then, simplifying,

$$SS_T = \sum_j \sum_i X_{ij}^2 - \frac{2T_{++}^2}{N_T} + \frac{T_{++}^2}{N_T}$$

and

$$SS_T = \sum_j \sum_i X_{ij}^2 - \frac{T_{++}^2}{N_T} \tag{13-5}$$

In this final formula you can find the terms of the definition. The second term is the grand mean with its numerator squared. The first term is the sum of squared raw scores, the terms whose deviations we are interested in.

This derivation shows how the total-sum-of-squares formula is obtained. It has the same general form as the derivations of all the other sums of squares for one-way and two-way analyses of variance. It is also identical to the derivation of the numerator of the variance, given in Section 5-11.

A computational formula for the between-groups sum of squares can also be obtained. We begin with $SS_B =$ sum of squared deviations of sample means about the grand mean. This definition is written below and then converted to total score notation.

$$SS_B = \sum_j \sum_i (\overline{X}_{\cdot j} - \overline{X}_{\cdot \cdot})^2 = \sum_j \sum_i \left(\frac{T_{+j}}{N_j} - \frac{T_{++}}{N_T}\right)^2$$

where T_{+j} is the sum of scores in group j and N_j is the number of subjects in group j. Through a derivation similar to that for the total sum of squares, the computational formula can be found to be

$$SS_B = \sum_j \frac{T_{+j}^2}{N_j} - \frac{T_{++}^2}{N_T} \tag{13-6}$$

You should be able to complete the derivation for yourself, if you wish, working by analogy with the derivation of formula 13-5.

The within-groups sum of squares SS_W was defined to be equal to the sum of squared deviations of each score about its own sample mean.

$$SS_W = \sum_j \sum_i (X_{ij} - \overline{X}_{\cdot j})^2 = \sum_j \sum_i \left(X_{ij} - \frac{T_{+j}}{N_j}\right)^2$$

Again, the computational formula can be derived by analogy with the derivation for the total sum of squares:

$$SS_W = \sum_j \left(\sum_i X_{ij}^2 - \frac{T_{+j}^2}{N_j} \right)$$

so

$$SS_W = \sum_j \sum_i X_{ij}^2 - \sum_j \frac{T_{+j}^2}{N_j} \qquad (13\text{-}7)$$

These are the only sums of squares to be calculated for the present design. A little algebra shows that $SS_T = SS_B + SS_W$. Substituting the computational formulas for SS_T, SS_B, and SS_W gives

$$SS_T \qquad = \qquad SS_B \qquad + \qquad SS_W$$

$$\left[\sum_j \sum_i X_{ij}^2 - \frac{T_{++}^2}{N_T} \right] = \left[\sum_j \frac{T_{+j}^2}{N_j} - \frac{T_{++}^2}{N_T} \right] + \left[\sum_j \sum_i X_{ij}^2 - \sum_j \frac{T_{+j}^2}{N_j} \right]$$

Clearly, the two $\sum_j (T_{+j}^2/N_j)$ terms on the right-hand side add to zero, and rearranging the remaining terms makes the two sides of the equation identical.

Because $SS_T = SS_B + SS_W$, it is not necessary to calculate all three sums of squares. The usual procedure is to find SS_T and SS_B, then subtract to find SS_W:

$$SS_W = SS_T - SS_B \qquad (13\text{-}8)$$

To solve a one-way analysis-of-variance problem, first calculate the sums of squares using formulas 13-5, 13-6, and 13-8. Then complete Table 13-4 by filling in values of J and N_j and doing the indicated arithmetic.

Let us now return to the example used earlier and solve it with the computational formulas just developed.

		Sample			
Subject	1	2	3	4	
1	5	1	3	1	
2	4	1	2	2	
3	5	2	2	3	
4	6	0	3	0	$\sum$
$T_{+j} = \sum X$	20	4	10	6	$40 = T_{++}$
$\sum X^2$	102	6	26	14	$148 = \sum\sum X^2$
N_j	4	4	4	4	$16 = N_T$

$$SS_T = \sum_j \sum_i X_{ij}^2 - \frac{T_{++}^2}{N_T}$$

$$= 148 - \frac{40^2}{16} = 148 - \frac{1600}{16}$$

$$= 148 - 100 = 48$$

$$SS_B = \sum \frac{T_{+j}^2}{N_j} - \frac{T_{++}^2}{N_T}$$

$$= \left(\frac{20^2}{4} + \frac{4^2}{4} + \frac{10^2}{4} + \frac{6^2}{4} \right) - \frac{40^2}{16}$$

$$= \frac{400 + 16 + 100 + 36}{4} - 100$$

$$= \frac{552}{4} - 100 = 138 - 100$$

$$= 38$$

$$SS_W = SS_T - SS_B$$

$$= 48 - 38 = 10$$

$$df_B = J - 1 = 4 - 1 = 3$$

$$df_W = \sum_j (N_j - 1) = 3 + 3 + 3 + 3 = 12$$

$$MS_B = \frac{SS_B}{df_B} = \frac{38}{3}$$

$$MS_W = \frac{SS_W}{df_W} = \frac{10}{12} = \frac{5}{6}$$

At this point, the test is completed with an F ratio of mean square between groups to the mean square within groups.

$$F = \frac{MS_B}{MS_W} \tag{13-9}$$

For this problem, the F ratio is

$$F = \frac{38/3}{5/6} = \frac{38 \cdot 6}{5 \cdot 3} = \frac{76}{5} = 15.2$$

Notice that the value of this F ratio is identical to the value of the F ratio calculated from the estimates of $N\hat{s}_{\bar{X}}^2$ and $\hat{s}_X^2$ in part A. This is because MS_B is identical to $N\hat{s}_{\bar{X}}^2$ and MS_W is identical to $\hat{s}_X^2$. The degrees of freedom are also the same (3 and 12), so the critical value is the same (3.49) and the conclusion is the same (the result is significant).

The analysis-of-variance table is as follows:

Source of variance	Sum of squares	Degrees of freedom	Mean square	F	$F_{.05}$	Significant?
Between groups	38	3	$\frac{38}{3}$	15.2	3.49	Yes
Within groups	10	12	$\frac{5}{6}$			
Total	48	15				

C. Review of One-Way Test

Let us now reconsider the design on which analysis of variance is usually based. We draw several samples from the same initial population and administer a different treatment to each sample (see Figure 13-1).

Figure 13-1 *Sampling design for one-way analysis of variance.*

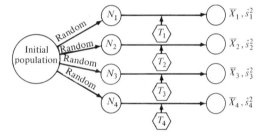

Our null hypothesis is that there are four post-treatment populations, one corresponding to each treatment sample, all with identical population means. In order to test this hypothesis, we imagine two distributions: a distribution of raw scores and a distribution of sample means. If the null hypothesis is true, the treatments have had no effect, and all the post-treatment samples could have been drawn from the same population. Therefore, under the null hypothesis, the following relationship holds between the variances of the two hypothetical distributions:

$$N\sigma_{\bar{X}}^2 = \sigma_X^2$$

On the other hand, if the null hypothesis is false, the means of the samples differ more than they would if drawn from the same populations. The effect of systematic treatment differences is to spread out the means. But we assume that the effect of a treatment is independent of a subject's initial score, so that only the means are affected, not the variances. Therefore, when there are systematic treatment effects,

$$N\sigma_{\bar{X}}^2 > \sigma_{X}^2$$

We use analysis of variance to provide estimates of the parameters that we are interested in:

$$MS_B = \text{est.}\,(N\sigma_{\bar{X}}^2) \qquad MS_W = \text{est.}\,(\sigma_{X}^2)$$

We test the hypothesis statistically by calculating an F ratio:

$$F = \frac{MS_B}{MS_W}$$

If this F is significant [greater than $F(df_B, df_W)$], we then reject the null hypothesis and conclude that the different treatments would produce different population means. We display our results in an analysis-of-variance table, which has the form of Table 13-4.

Let us now consider an example with unequal sample sizes to indicate the more general form of the one-way analysis.

EXAMPLE   In an experiment on the perception of apparent movement, subjects are placed in a darkened room. On the opposite wall are four lights located at the corners of a square. The size of the square can be varied as well as the order and pattern of presentation of the lights. Four different presentation patterns are used and the amount of reported apparent movement recorded. The data are given in the table. If a significance level of .05 is used, is there a significant effect of presentation pattern on the amount of reported movement?

$$SS_T = \sum_j \sum_i X_{ij}^2 - \frac{T_{++}^2}{N_T}$$

$$= 11{,}252 - \frac{(464)^2}{20} = 11{,}252 - \frac{215{,}296}{20}$$

$$= 11{,}252 - 10{,}764.8$$

$$= 487.2$$

$$SS_B = \sum_j \frac{T_{+j}^2}{N_j} - \frac{T_{++}^2}{N_T}$$

$$= \frac{(111)^2}{5} + \frac{(139)^2}{6} + \frac{(99)^2}{4} + \frac{(115)^2}{5} - 10{,}764.8$$

Subject	Treatment								
	1		2		3		4		
	X	X^2	X	X^2	X	X^2	X	X^2	
1	15	225	24	576	20	400	25	625	
2	20	400	22	484	22	484	18	324	
3	26	676	20	400	30	900	16	256	
4	26	676	21	441	27	729	32	1024	
5	24	576	34	1156	—	—	24	576	
6	—	—	18	324	—	—	—	—	Σ
$\sum X$	111	—	139	—	99	—	115	—	464
$\sum X^2$	—	2553	—	3381	—	2513	—	2805	11,252
N_k	5		6		4		5		20

$$= 2464.2 + 3220.2 + 2450.2 + 2645.0 - 10{,}764.8$$

$$= 14.8$$

$$\mathrm{df_B} = J - 1 = 4 - 1 = 3$$

$$\mathrm{df_w} = \sum_j (N_j - 1) = (5 - 1) + (6 - 1) + (4 - 1) + (5 - 1)$$

$$= 4 + 5 + 3 + 4 = 16$$

$$\mathrm{df_T} = N_T - 1$$

$$= 5 + 6 + 4 + 5 - 1 = 19$$

Source of variance	Sum of squares	Degrees of freedom	Mean square	F	$F_{.05}$	Significant?
Between groups	14.8	3	4.9	.167	3.24	No
Within groups	472.4	16	29.5			
Total	487.2	19				

The observed F is less than the corresponding critical value, so the results are nonsignificant. There is no evidence for a treatment effect.

EXAMPLE   In Chapter 7 it was claimed that either a t or an F distribution could be used to test whether two populations have the same mean. The assumptions of the general one-way analysis of variance of the present section are most similar to the assumptions on which the t test of Section 12-12 are based. Let us therefore reanalyze the last example in that section, which appears on page 307. The null hypothesis is that the two population means are equal, and the α level is .05. The data and preliminary calculations are as shown in the table.

	Group 1		Group 2		
	X	X^2	X	X^2	Σ
	8	64	12	144	
	8	64	10	100	
	2	4	6	36	
	4	16	6	36	
	4	16	7	49	
	4	16	7	49	
ΣX	30	—	48	—	78
ΣX^2	—	180	—	414	594

$$SS_T = 594 - \frac{(78)^2}{12}$$

$$= 594 - 507 = 87$$

$$SS_B = \frac{(30)^2}{6} + \frac{(48)^2}{6} - 507$$

$$= 150 + 384 - 507 = 27$$

$$SS_w = SS_T - SS_B = 87 - 27 = 60$$

Source of variance	Sum of squares	Degrees of freedom	Mean square	F
Between groups	27	1	27	4.5
Within groups	60	10	6	
Total	87	11		

Here the critical value of F at the .05 level for 1 and 10 df is 4.96, and the results are nonsignificant. There is no significant difference between the two population means.

In Chapter 7 it was stated that the value of t for b degrees of freedom was equal to the value of F for 1 and b degrees of freedom. The value of t found for these data in Section 12-12 was 2.121, which squares to 4.50, the calculated value of F in the present analysis. The critical value of t was ± 2.228, and 2.228 squares to 4.96, the critical value of F for this problem. In both analyses the observed sample statistic was less than the critical value, and the null hypothesis of no difference between population means was accepted. Thus, the results of the two analyses are identical.

D. Analysis-of-Variance Model

We will now construct a theoretical representation of a score on the dependent variable and use this representation to approach the analysis of variance from a slightly different point of view. This treatment may provide additional insights into the computations we have just concluded; but its primary purpose is to pave the way for more complex analysis-of-variance designs, especially that of the two-way analysis, to be covered in the next section.

When all scores are randomly drawn from the same population, we can consider each score as having two components: the population mean plus the difference between the score and the mean.

$$X_i = \mu + \epsilon_i \quad \text{or} \quad X_i - \mu = \epsilon_i$$

The epsilon (ϵ_i) in this formula represents error in prediction—the magnitude of error we would make if we predicted this subject's score knowing only that he came from the given population (i.e., knowing μ).

If we do not know the population mean, we can obtain an estimate of it by drawing a random sample from the population. Then we can obtain an estimate of the population mean from the sample mean and estimates of each error term from the deviation of the corresponding raw score about the sample mean.

$$\bar{X} = \text{est. } (\mu) \quad (X_i - \bar{X}) = \text{est. } (\epsilon_i)$$

By substitution, $\sigma_X^2 = \sigma_{\mu+\epsilon}^2$, but because μ is a constant, $\sigma_{\mu+\epsilon}^2 = \sigma_\epsilon^2$ (Section 6-4). Therefore, $\sigma_X^2 = \sigma_\epsilon^2$ (i.e., the error variance and the raw-score variance are identical).

It follows that the unbiased sample estimate of the raw-score variance is also an unbiased estimate of the error variance:

$$\hat{s}_X^2 = \text{est. } (\hat{\sigma}_X^2) = \text{est. } (\sigma_\epsilon^2)$$

Suppose now that we draw several samples and administer a different treatment to each, which is typically done in the one-way analysis of variance. Our model must become more complicated to reflect the possible systematic differences among means of treatment groups. We have a mean (μ_j) for each treatment population and a mean for the population of all treatments combined (μ). We indicate the difference between the mean of a given treatment group and the mean of the overall population by α_j.

$$\alpha_j = \mu_j - \mu$$

Now for any subject who has not yet been tested we would predict the mean of the treatment group to which he belongs, and we would have an error of prediction ϵ_{ij} equal to the difference between his raw score and the predicted score.

$$\epsilon_{ij} = X_{ij} - \mu_j$$
$$= X_{ij} - (\mu + \alpha_j)$$

Thus the raw score can be said to have three components:

$$X_{ij} = \mu + \alpha_j + \epsilon_{ij} \tag{13-10}$$

As in the case of the model in which all scores are randomly drawn from the same population, we can obtain sample estimates for the components of the present model:

$$\overline{X}._j = \text{est.} (\mu_j) = \text{est.} (\mu + \alpha_j)$$
$$\overline{\overline{X}}.. = \text{est.} (\mu)$$
$$\overline{X}._j - \overline{\overline{X}}.. = \text{est.} (\alpha_j)$$
$$X_{ij} - \overline{X}._j = \text{est.} (\epsilon_{ij})$$

Let us now obtain some of these estimates for the example of Table 13-1. The table is repeated below for convenience.

Subject within group	Group				
	1	2	3	4	
1	5	1	3	1	
2	4	1	2	2	
3	5	2	2	3	
4	6	0	3	0	Σ
$\sum X$	20	4	10	6	$\sum\sum X = 40$
N	4	4	4	4	$\sum N = 16$
$\overline{X}$	5	1	2.5	1.5	$\overline{\overline{X}} = 2.5$

Then

$$\text{est.} (\mu) = \overline{\overline{X}}.. = 2.5$$

$$\text{est.} (\alpha_1) = \overline{X}._1 - \overline{\overline{X}}.. = 5 - 2.5 = 2.5$$

$$\text{est.} (\alpha_2) = 1 - 2.5 = -1.5$$

$$\text{est. } (\alpha_3) = 2.5 = 2.5 = 0$$

$$\text{est. } (\alpha_4) = 1.5 - 2.5 = -1.0$$

Notice that $\sum_j \alpha_j = 0$, which is required by the fact that the α's are defined as deviation scores. The error terms can be estimated in a similar manner. Let us estimate the error terms for the scores in group 1.

$$\text{est. } (\epsilon_{11}) = X_{11} - \bar{X}_{\cdot 1} = 5 - 5 = 0$$

$$\text{est. } (\epsilon_{21}) = 4 - 5 = -1$$

$$\text{est. } (\epsilon_{31}) = 5 - 5 = 0$$

$$\text{est. } (\epsilon_{41}) = 6 - 5 = 1$$

Again, notice that $\sum_i \epsilon_{ij} = 0$, as is required by the definition of the ϵ's as deviation scores.

The "mean square within groups" line of the analysis of variance is a pooled, within-groups variance estimate. But according to this model, the differences within groups are due entirely to error. Therefore, the MS_w line of the analysis is an estimate of error variance:

$$MS_W = \text{est. } (\sigma_\epsilon^2) \tag{13-11}$$

When there are no treatment effects, the mean square between groups is also an estimate of the error variance, from the application of the formula for standard error of the mean. However, when there are treatment effects, the sample means are more variable; and the mean square between groups, which is based on the sample means, is larger. In more advanced texts, the mean square between groups is shown to be an estimate of a weighted sum of error and treatment variances:

$$MS_B = \text{est. } (\sigma_\epsilon^2 + N\sigma_\alpha^2) \tag{13-12}$$

where σ_α^2 is the variance of the population means of the several treatments.

When there are no treatment differences, the means of the treatment populations are identical ($\mu_1 = \mu_2 = \cdots = \mu_J = \mu$), and all the α's are zero. In this case, $\sigma_\alpha^2 = 0$, and

$$MS_B = \text{est. } (\sigma_\epsilon^2) \qquad \text{when } H_0 \text{ is true}$$

If we had the entire populations, we could determine whether there had been a treatment effect simply by comparing the population means. If there were any differences, there would have been treatment effects. Equivalently, we could compare the variance of raw scores within treatment groups to N times the variance of means; we would be comparing σ_ϵ^2 to $(\sigma_\epsilon^2 + N\sigma_\alpha^2)$.

To decide whether there have been treatment effects on the basis of sample data, we compare the mean square between groups to the mean square within, using the F ratio. In terms of the parameters estimated, the F ratio can be written

$$F = \frac{\mathrm{MS_B}}{\mathrm{MS_W}} = \frac{\text{est.}\,(\sigma_\epsilon^2 + N\sigma_\alpha^2)}{\text{est.}\,(\sigma_\epsilon^2)} \qquad (13\text{-}13)$$

Notice that the numerator of F estimates one term in addition to that estimated by the denominator. A significant F indicates that the numerator is larger than the denominator by a nonchance amount; by having only one additional theoretical component in the numerator, we are able to isolate the basis for a significant F when one occurs. In the present case, a significant F indicates that σ_α^2 is nonzero, and therefore that we probably would have found differences among the corresponding population means if we could have tested them. The principle of constructing F ratios so that there is one additional component in the numerator in addition to those in the denominator is a very general one that is particularly useful in complex analysis-of-variance designs.

E. Prediction and Effectiveness of Prediction

Should you want to predict a subject's status on the dependent variable from his or her status on the independent variable, your best estimate would be the mean of the sample receiving the same treatment, as indicated in Section 15-2. The coefficient to use to show the effectiveness of the experimental conditions in affecting scores on the dependent variable is the correlation ratio, eta, which is described in Section 17-1.

References This analysis can be found in Dixon and Massey (1969), 156–162; Edwards (1960), 117–125, and (1967), Chapter 7; Hays (1963), 356–385; and McNemar (1969), 288–306.

13-5 Two-way analysis of variance

> *Relevant earlier section:*
> **13-4** One-way analysis of variance

A. Model for Two-Way Analysis

In two way-analysis of variance, we have one dependent variable, X, and two independent variables, A and B. We wish to know whether the dependent variable X is affected by differences in the independent variables, either singly or in combination. In Table 13-5, variable A has two levels and variable B has three. Three scores appear in each cell of the table.

Table 13-5 *Data layout for two-way (2 × 3) analysis of variance*

	B_1	B_2	B_3
A_1	X_{111}	X_{112}	X_{113}
	X_{211}	X_{212}	X_{213}
	X_{311}	X_{312}	X_{313}
A_2	X_{121}	X_{122}	X_{123}
	X_{221}	X_{222}	X_{223}
	X_{321}	X_{322}	X_{323}

To account for all possible sources of differences in scores in this data table, we need a model of a test score that has five distinct components:

$$X_{ijk} = \mu + \alpha_j + \beta_k + \alpha\beta_{jk} + \epsilon_{ijk} \tag{13-14}$$

where i is the person subscript, j designates a level of variable A, and k indicates a level of variable B. In this model,

μ = overall, grand population mean
α_j = effect due to level j of variable A
β_k = effect due to level k of variable B
$\alpha\beta_{jk}$ = effect due to the specific combination of level j of variable A and level k of variable B in addition to the effects of α_j and β_k
ϵ_{ijk} = error, or difference between a particular subject's score and the population mean of all subjects receiving that treatment combination

None of these model components is directly calculable, because we do not have population data. However, we can obtain sample estimates of each of them. In order to do so, we obtain Table 13-6 from the data of Table 13-5. All the entries in Table 13-6 are means, with dots replacing the subscripts that have been averaged over.

Table 13-6 *Cell and marginal means in 2 × 3 analysis of variance*

	B_1	B_2	B_3	Mean
A_1	$\overline{X}_{\cdot 11}$	$\overline{X}_{\cdot 12}$	$\overline{X}_{\cdot 13}$	$\overline{X}_{\cdot 1 \cdot}$
A_2	$\overline{X}_{\cdot 21}$	$\overline{X}_{\cdot 22}$	$\overline{X}_{\cdot 23}$	$\overline{X}_{\cdot 2 \cdot}$
Mean	$\overline{X}_{\cdot \cdot 1}$	$\overline{X}_{\cdot \cdot 2}$	$\overline{X}_{\cdot \cdot 3}$	$\overline{X}_{\cdot \cdot \cdot}$

From the definitions of the model components, we can determine which components are estimated by the various means:

$\overline{X}\ldots$ is an estimate of the grand population mean, μ.

$\overline{X}_{.j.}$ is an estimate of the jth level of variable A in addition to the grand mean; hence, $\overline{X}_{.j.} = $ est. $(\mu + \alpha_j)$.

$\overline{X}_{..k}$ is an estimate of the effect of level k of variable B in addition to the grand mean; hence, $\overline{X}_{..k} = $ est. $(\mu + \beta_k)$.

$\overline{X}_{.jk}$ is affected by level j of variable A and level k of variable B and the effect specific to the particular treatment combination; hence

$$\overline{X}_{.jk} = \text{est.}\,(\mu + \alpha_j + \beta_k + \alpha\beta_{jk})$$

It is now possible to derive estimates of the various parameters of the model:

$$\overline{X}_{.j.} - \overline{\overline{X}}\ldots = \text{est.}\,(\mu + \alpha_j) - \text{est.}\,(\mu) = \text{est.}\,(\alpha_j)$$

$$\overline{X}_{..k} - \overline{\overline{X}}\ldots = \text{est.}\,(\mu + \beta_k) - \text{est.}\,(\mu) = \text{est.}\,(\beta_k)$$

$$\overline{X}_{.jk} - \overline{X}_{.j.} - \overline{X}_{..k} + \overline{\overline{X}}\ldots = \text{est.}\,(\mu + \alpha_j + \beta_k + \alpha\beta_{jk}) - \text{est.}\,(\mu + \alpha_j)$$

$$- \text{est.}\,(\mu + \beta_k) + \text{est.}\,(\mu)$$

$$= \text{est.}\,(\alpha\beta_{jk})$$

$$X_{ijk} - \overline{X}_{.jk} = \text{est.}\,(\mu + \alpha_j + \beta_k + \alpha\beta_{jk} + \epsilon_{ijk})$$

$$- \text{est.}\,(\mu + \alpha_j + \beta_k + \alpha\beta_{jk})$$

$$= \text{est.}\,(\epsilon_{ijk}).$$

Let us now consider a numerical example to illustrate estimation of the several parameters. Consider the following data table for an analysis of variance:

Raw scores

	B_1	B_2	B_3
A_1	19, 13, 16	13, 2, 6	17, 11, 11
A_2	37, 34, 34	28, 25, 22	31, 16, 22

Means

	B_1	B_2	B_3	Mean
A_1	16	7	13	12
A_2	36	25	23	28
Mean	26	16	18	20

Now the estimates can be obtained in accordance with the formulas given above:

$$\text{est. } (\mu) = \overline{\overline{X}}\ldots = 20$$

$$\text{est. } (\alpha_1) = (\overline{X}._1. - \overline{\overline{X}}\ldots) = 12 - 20 = -8$$

$$\text{est. } (\alpha_2) = (\overline{X}._2. - \overline{\overline{X}}\ldots) = 28 - 20 = 8$$

Est. $(\beta_1) = (\overline{X}.._1 - \overline{\overline{X}}\ldots) = 26 - 20 = 6$; similarly, est. $(\beta_2) = -4$ and est. $(\beta_3) = -2$.

Est. $(\alpha\beta_{11}) = (\overline{X}._{11} - \overline{X}._1. - \overline{X}.._1 + \overline{\overline{X}}\ldots) = 16 - 12 - 26 + 20 = -2$; similarly, est. $(\alpha\beta_{12}) = -1$; est. $(\alpha\beta_{13}) = 3$; est. $(\alpha\beta_{21}) = 2$; est. $(\alpha\beta_{22}) = 1$; and est. $(\alpha\beta_{23}) = -3$.

Est. $(\epsilon_{111}) = (X_{111} - \overline{X}._{11}) = 19 - 16 = 3$; est. $(\epsilon_{211}) = 13 - 16 = -3$; and so on for the other 16 subjects in the table.

Notice that, consistent with our definition of the model components as deviation scores, $\sum_j \text{est. } (\alpha_j) = 0$; $\sum_k \text{est. } (\beta_k) = 0$; $\sum_k \text{est. } (\alpha\beta_{jk}) = 0$ for each level of A, and $\sum_j \text{est. } (\alpha\beta_{jk}) = 0$ for each level of B; and $\sum_i \text{est. } (\epsilon_{ijk}) = 0$ within each cell of the table.

B. Hypotheses of the Two-Way Analysis of Variance

There are three distinct, systematic effects that we can test for: the A and B main effects and the interaction between A and B. The three corresponding null hypotheses are

$$H_A: \alpha_1 = \alpha_2 = \alpha_3 = \cdots = \alpha_J = 0$$

$$H_B: \beta_1 = \beta_2 = \beta_3 = \cdots = \beta_K = 0$$

$$H_{AB}: \alpha\beta_{11} = \alpha\beta_{12} = \cdots = \alpha\beta_{JK} = 0$$

Because each of the components in the hypotheses is a deviation score, all the α's can be equal only when they are equal to zero, and similarly for the β's and $\alpha\beta$'s. To say that all the α's are equal is to say that they have no variance; hence, the three hypotheses can be rewritten:

$$H'_A: \sigma_\alpha^2 = 0$$

$$H'_B: \sigma_\beta^2 = 0$$

$$H'_{AB}: \sigma_{\alpha\beta}^2 = 0$$

EXAMPLE   Consider the following 2×6 analysis of variance with three subjects per cell:

	B_1	B_2	B_3	B_4	B_5	B_6	Σ
A_1	16, 14, 15	13, 10, 10	14, 9, 10	10, 4, 7	13, 9, 11	14, 8, 11	198
A_2	11, 8, 8	12, 6, 9	8, 6, 7	11, 7, 9	13, 10, 10	12, 7, 8	162

From this table of raw scores we can derive a table whose entries are all means:

	B_1	B_2	B_3	B_4	B_5	B_6	Mean
A_1	15	11	11	7	11	11	11
A_2	9	9	7	9	11	9	9
Mean	12	10	9	8	11	10	10

As in the one-way analysis of variance, we can determine whether there is a B effect by finding the mean for each level of B, then using the formula for standard error of the mean:

$$\hat{s}_{\bar{X}}^2 \doteq \frac{\hat{s}_X^2}{N} \quad \text{or} \quad N\hat{s}_{\bar{X}}^2 \doteq \hat{s}_X^2$$

finally comparing the two sides of the equation using the F ratio:

$$F = \frac{N\hat{s}_{\bar{X}}^2}{\hat{s}_X^2}$$

If the null hypothesis is true, numerator and denominator will be approximately equal (except for random differences), and the most probable value of F is 1. If the null hypothesis is false, the means will be spread out, and the calculated value of F will be greater than 1.

As in the one-way analysis of variance, we can find the variance of the means and the within-cell variance (pooled), and then calculate an F ratio,

$$
\begin{aligned}
\hat{s}_{\bar{X}(B)}^2 &= \frac{\sum\limits_k (\bar{\bar{X}}{\cdot\cdot_k} - \bar{\bar{\bar{X}}}\ldots)^2}{K - 1} \\[2mm]
&= \frac{(12-10)^2 + (10-10)^2 + (9-10)^2 + (8-10)^2 + (11-10)^2 + (10-10)^2}{6-1} \\[2mm]
&= \frac{4 + 0 + 1 + 4 + 1 + 0}{5} = \frac{10}{5} = 2
\end{aligned}
$$

$$N\hat{s}_{\bar{X}(B)}^2 = 6(2) = 12$$

because each B mean is based on scores from six subjects. For the within-cells variance, 12 variance estimates are pooled. We assume that only the centrality parameter differs from cell to cell, so each cell's variance is an estimate of the variance of the undifferentiated population. We pool the 12 estimates to obtain an estimate that is more stable. The formula is formula 13-1.

$$\hat{s}^2_{pooled} = \frac{\sum x^2_{11} + \sum x^2_{21} + \sum x^2_{31} + \cdots + \sum x^2_{62}}{(N_{11} - 1) + (N_{21} - 1) + (N_{31} - 1) + \cdots + (N_{62} - 1)}$$

where

$$\sum x^2_{11} = (16 - 15)^2 + (14 - 15)^2 + (15 - 15)^2 = 1 + 1 + 0 = 2$$

$$\sum x^2_{21} = (13 - 11)^2 + (10 - 11)^2 + (10 - 11)^2 = 4 + 1 + 1 = 6$$

$$\sum x^2_{31} = (14 - 11)^2 + (9 - 11)^2 + (10 - 11)^2 = 9 + 4 + 1 = 14$$

and so on until

$$\sum x^2_{62} = (12 - 9)^2 + (7 - 9)^2 + (8 - 9)^2 = 9 + 4 + 1 = 14$$

These sums of squared deviations are then inserted into the pooled within-groups variance formula, giving

$$\hat{s}^2_{pooled} = \frac{2 + 6 + 14 + \cdots + 14}{2 + 2 + 2 + \cdots + 2} = \frac{120}{24} = 5$$

Finally, F is calculated to give

$$F = \frac{N\hat{s}^2_{\bar{X}(B)}}{\hat{s}^2_{pooled}} = \frac{12}{5} = 2.4$$

The numerator of this F ratio is based on $J - 1 = 6 - 1 = 5$ df and the denominator on $\sum_j \sum_k (N_{jk} - 1) = JK(N - 1) = 2 \cdot 6 \cdot 2 = 24$ independent values. The critical value of F at the .05 level is 2.62, so the test is nonsignificant. Differences in reinforcement do not produce differences in performance.

At this point you should be able, if you work by analogy with the analysis for the B main effect, to carry out a similar test of the A or interaction effect. Rather than continue in this manner, however, let us go on to the standard computational procedure for two-way analysis of variance, which includes tests for all three effects.

C. Computational Format for Two-Way Analysis of Variance

When all the null hypotheses are true, and the main effects and interaction terms are all zero, then all four variance estimates (A, B, interaction, and error) estimate the same parameter: the raw-score or error variance of the undifferentiated population. When the main effects and interaction terms are nonzero, only the error variance is an estimate of the original population variance.

Table 13-7 gives the general form of the two-way analysis of variance.

Table 13-7   *General form of the J × K analysis of variance*

Source of variance	Sum of squares	Degrees of freedom	Mean square	F
A	SS_A	$J - 1$	$MS_A = SS_A/df_A$	$F_A = MS_A/MS_W$
B	SS_B	$K - 1$	$MS_B = SS_B/df_B$	$F_B = MS_B/MS_W$
AB interaction	SS_{AB}	$(J - 1)(K - 1)$	$MS_{AB} = SS_{AB}/df_{AB}$	$F_{AB} = MS_{AB}/MS_W$
Cells	SS_{Cells}	$JK - 1$	—	—
Within	SS_W	$JK(N - 1)$	$MS_W = SS_W/df_W$	—
Total	SS_T	$NJK - 1$		

The sums of squares terms are the numerators of the corresponding variance estimates. They are formed as follows.

The total sum of squares is defined as the sum of squared deviations of all raw scores about the grand mean:

$$SS_T = \sum_k \sum_j \sum_i (X_{ijk} - \overline{\overline{X}}...)^2 = \sum_k \sum_j \sum_i \left(X_{ijk} - \frac{T_{+++}}{NJK}\right)^2$$

where, as in Section 13-4, a T represents a total or sum of scores, and a + replaces the subscript summed over. Thus, T_{+++} is the grand total, the sum of scores over all subjects in all cells of the table. By analogy with the derivation of SS_T for the one-way case, the computing formula is

$$SS_T = \sum_k \sum_j \sum_i X_{ijk}^2 - \frac{T_{+++}^2}{NJK} \tag{13-15}$$

The sum of squares for cells is defined as the sum of squared deviations of all cell means about the grand mean:

$$SS_{Cells} = \sum_k \sum_j \sum_i (\overline{X}._{jk} - \overline{\overline{X}}...)^2 = \sum_k \sum_j \sum_i \left(\frac{T_{+jk}}{N} - \frac{T_{+++}}{NJK}\right)^2$$

where a dot replaces the subscript i in the mean $\overline{X}._{jk}$ to indicate that subjects have been averaged over. Again, the computing formula is found using the rule of thumb given in Section 13-4. Starting with the expression that involves totals, remove the parentheses, apply the summation signs to the first term only, and square both numerators. The computing formula is thus

$$SS_{Cells} = \sum_k \sum_j \frac{T_{+jk}^2}{N} - \frac{T_{+++}^2}{NJK} \qquad (13\text{-}16)$$

The difference between SS_T and SS_{Cells} is that portion of differences among raw scores that cannot be accounted for by differences among the means. This is the variability within cells and the numerator of the error variance:

$$SS_{Within} = SS_T - SS_{Cells} \qquad (13\text{-}17)$$

For the main effects, the sums of squares are

$$SS_A = \sum_k \sum_j \sum_i (\bar{X}._{j}. - \bar{\bar{X}}...)^2 = \sum_k \sum_j \sum_i \left(\frac{T_{+j+}}{NK} - \frac{T_{+++}}{NJK} \right)^2$$

$$= \sum_j \frac{T_{+j+}^2}{NK} - \frac{T_{+++}^2}{NJK} \qquad (13\text{-}18)$$

$$SS_B = \sum_k \sum_j \sum_i (\bar{X}.._k - \bar{\bar{X}}...)^2 = \sum_k \sum_j \sum_i \left(\frac{T_{++k}}{NJ} - \frac{T_{+++}}{NJK} \right)^2$$

$$= \sum_k \frac{T_{++k}^2}{NJ} - \frac{T_{+++}^2}{NJK} \qquad (13\text{-}19)$$

The interaction variability is that which is left when the sums of squares for the main effects are subtracted from the sum of squares for cell means. It is those differences among cell means that cannot be accounted for by the separate effects of the two variables:

$$SS_{AB} = SS_{Cells} - SS_A - SS_B \qquad (13\text{-}20)$$

Tests of the three hypotheses require variance estimates, mean squares, which can be entered into F ratios. For each of the four sources of variance, the mean square is found by dividing the sum of squares by the number of degrees of freedom, or independent measurements, on which it is based. A total number of degrees of freedom can also be found which corresponds to the total sum of squares.

The total sum of squares was based on scores from NJK subjects, and one parameter (μ) was estimated from the data. Therefore, $df(T) = NJK - 1$. The sum of squares for cells was based on JK means, with deviations taken about $\bar{\bar{X}}$ instead of μ; therefore, $df(Cells) = JK - 1$. The sum of squares within cells was found by subtracting the sum of squares for cells from the total sum of squares, so df(Within) = df(T) − df(Cells) = $NJK − 1 − (JK − 1) =$ $JK(N − 1)$. Alternatively, the pooled, within-cells variance can be found from the variance estimates from all JK cells. There were $N − 1$ independent scores in each cell, so df(Within) = $JK(N − 1)$. Similarly, the number of degrees of freedom for factor A is 1 less than the number of levels of A, and the degrees of freedom for B is 1 less than the number of levels of B: df(A) = $J − 1$, and df(B) = $K − 1$. The number of degrees of freedom for the interaction term is given by df(AB) = df(Cells) − df(A) − df(B) = $JK − 1 − (J − 1) −$ $(K − 1) = (J − 1)(K − 1)$.

Because this is a fixed-effects model, we test for the main effects and interaction by forming F ratios in which the three mean squares are successively divided by the mean square for error, the within-cells mean square.

EXAMPLE   Consider the following table of data, which is the same as presented earlier in this section:

	B_1	B_2	B_3	B_4	B_5	B_6	Σ
A_1	16, 14, 15	13, 10, 10	14, 9, 10	10, 4, 7	13, 9, 11	14, 8, 11	198
A_2	11, 8, 8	12, 6, 9	8, 6, 7	11, 7, 9	13, 10, 10	12, 7, 8	162
Σ	72	60	54	48	66	60	360

We can obtain the total sum of squares directly from this table:

$$SS_T = \sum_k \sum_j \sum_i X_{ijk}^2 - \frac{T_{+++}^2}{NJK}$$

$$= 16^2 + 14^2 + 15^2 + 13^2 + 10^2 + \cdots + 7^2 + 8^2 - \frac{(360)^2}{36}$$

$$= 3876 - 3600$$

$$= 276$$

To simplify further calculations, let us work with the cell totals from the above table, as given below:

	B_1	B_2	B_3	B_4	B_5	B_6	Σ
A_1	45	33	33	21	33	33	198
A_2	27	27	21	27	33	27	162
Σ	72	60	54	48	66	60	360

From this let us first obtain the sum of squares for cells:

$$SS_{Cells} = \sum_k \sum_j \frac{T^2_{+jk}}{N} - \frac{T^2_{+++}}{NJK}$$

$$= \frac{45^2}{3} + \frac{33^2}{3} + \frac{33^2}{3} + \cdots + \frac{33^2}{3} + \frac{27^2}{3} - \frac{360^2}{36}$$

$$= 3756 - 3600$$

$$= 156$$

The Within or Error sum of squares can now be obtained by subtraction:

$$SS_{Within} = SS_T - SS_{Cells}$$

$$= 276 - 156$$

$$= 120$$

The effect of differences in values of A on the dependent variable is tested for using the A totals:

$$SS_A = \sum_j \frac{T^2_{+j+}}{NK} - \frac{T^2_{+++}}{NJK}$$

$$= \frac{198^2}{3 \cdot 6} + \frac{162^2}{3 \cdot 6} - \frac{360^2}{3 \cdot 2 \cdot 6}$$

$$= \frac{65,448}{18} - 3600 = 3636 - 3600$$

$$= 36$$

Similarly, differences in the column totals are used to test for differences in the effects of levels of variable B on the dependent variable:

$$SS_B = \sum_k \frac{T^2_{++k}}{NJ} - \frac{T^2_{+++}}{NJK}$$

$$= \frac{72^2}{6} + \frac{60^2}{6} + \frac{54^2}{6} + \frac{48^2}{6} + \frac{66^2}{6} + \frac{60^2}{6} - 3600$$

$$= 3660 - 3600$$

$$= 60$$

The interaction sum of squares is that portion of the variability of the means that is not due to either the A or the B main effect (acting alone):

$$SS_{AB} = SS_{Cells} - SS_A - SS_B$$
$$= 156 - 36 - 60$$
$$= 60$$

To complete the analysis, let us construct an analysis-of-variance table, like Table 13-7, using as degrees of freedom $df(T) = NJK - 1 = 36 - 1 = 35$; $df(Cells) = JK - 1 = 12 - 1 = 11$; $df(A) = J - 1 = 2 - 1 = 1$; $df(B) = K - 1 = 6 - 1 = 5$; $df(AB) = (J - 1)(K - 1) = 1 \cdot 5 = 5$; and $df(W) = JK(N - 1) = 2 \cdot 6 \cdot 2 = 24$.

From the analysis-of-variance table we can conclude that variable A does appear to affect the dependent variable, but that B does not appear to, and that A and B do not appear to interact to produce any additional effect over that of A operating alone.

Source of variance	Sum of squares	df	Mean square	F	$F_{.05}$	Significant?
A	36	1	36	7.2	4.26	Yes
B	60	5	12	2.4	2.62	No
AB	60	5	12	2.4	2.62	No
Within	120	24	5			
Total	276	35				

Notice that there is no line for cells. Because the sum of squares for cells can be partitioned completely into the sums of squares for main effects and interaction, there is no additional source of variance in cell means to be tested for. The cell and total sums of squares are merely computed to simplify the calculations for the within and interaction sums of squares.

D. Application of the Model to the Determination of F Ratios

As indicated in part A, each raw score in a two-way analysis of variance is considered to be the sum of five model components:

$$X_{ijk} = \mu + \alpha_j + \beta_k + \alpha\beta_{jk} + \epsilon_{ijk} \tag{13-21}$$

When we carry out the analysis, we actually are concerned with four different sources of variance. It is possible to determine the constitution of each of these variance estimates in terms of the parameter variances that they estimate.

First of all, the mean square within groups is actually a pooled, within-groups variance estimate that is assumed to be unaffected by differences in sample means. Therefore, the model components that apply only to sample means are not part of this variance: the mean square within groups only estimates error varriance.

$$\text{MS(W)} = \text{est.}\ (\sigma_\epsilon^2) \tag{13-22}$$

When there are no between-groups differences, the mean squares for the two main effects and interaction are all estimates of error variance, because even without treatment effects, the means still differ due to sampling fluctuations. The standard error of the mean formula applies, relating each of these mean squares to the mean square within groups. However, when there are treatment effects, these mean squares also are affected by main effects and interaction variances.

The α_j terms are deviation scores about the grand mean μ; it was indicated in Section 5-11 (population variances) that the sum of all deviations about a mean is always zero. The total for each level of variable B is found by adding over all levels of variable A, and the mean square for B is found from these totals. None of these totals can contain any α_j component, because the addition is across all levels of A. Therefore, there can be no σ_α^2 term in the B mean square. Analogously, there can be no σ_β^2 term in the A mean square. The $\alpha\beta_{jk}$ terms are constrained to sum to zero for every row and column of the table (otherwise, they could not be distinguished from the α and β terms); so interaction terms cannot appear in either main-effect mean square. The mean squares for A and B are thus:

$$\text{MS}(A) = \text{est.}\ (\sigma_\epsilon^2 + NK\sigma_\alpha^2) \tag{13-23}$$

$$\text{MS}(B) = \text{est.}\ (\sigma_\epsilon^2 + NJ\sigma_\beta^2) \tag{13-24}$$

The interaction mean square contains no main-effect components, because they are subtracted out in its calculation:

$$\text{MS}(AB) = \text{est.}\ (\sigma_\epsilon^2 + N\sigma_{\alpha\beta}^2) \tag{13-25}$$

F ratios are formed in accordance with Table 13-8. This table indicates that each of the first three lines of the analysis-of-variance table is used to test a different one of the null hypotheses. In each case, when the null hypothesis is true, the mean square is an estimate of error variance only. Thus the expected value of F under the null hypothesis is 1.0, because, as is required by the F statistic (Section 7-7), the numerator and denominator are estimates of the same parameter under this condition. Because the main effect and interaction mean squares each estimate only one parameter in addition to error, it is possible to make a separate judgment for each source of variance as to whether it affects the dependent variable.

Table 13-8 *Theoretical basis for two-way analysis of variance*

Source of variance	Expected mean square	Null hypothesis	F ratio	F-ratio estimates
A	$\sigma_\epsilon^2 + NK\sigma_\alpha^2$	$\sigma_\alpha^2 = 0$	MS_A/MS_W	$(\sigma_\epsilon^2 + NK\sigma_\alpha^2)/\sigma_\epsilon^2$
B	$\sigma_\epsilon^2 + NJ\sigma_\beta^2$	$\sigma_\beta^2 = 0$	MS_B/MS_W	$(\sigma_\epsilon^2 + NJ\sigma_\beta^2)/\sigma_\epsilon^2$
AB	$\sigma_\epsilon^2 + N\sigma_{\alpha\beta}^2$	$\sigma_{\alpha\beta}^2 = 0$	MS_{AB}/MS_W	$(\sigma_\epsilon^2 + N\sigma_{\alpha\beta}^2)/\sigma_\epsilon^2$
W	σ_ϵ^2			

E. Prediction and Variance Accounted for

For any combination of a level of A with a level of B, your best prediction of a subject's score on the dependent variable is the cell mean for the sample of subjects receiving that treatment combination, as indicated in Section 15-2. The squared correlation ratio (eta square, η^2) gives the proportion of the variance of the dependent variable that can be accounted for by knowledge of subjects' status on the independent variables, either separately or in combination. This coefficient is discussed in Section 17-1.

F. Two-Way Analysis with One Observation per Cell

This design is sometimes used in textbooks to introduce the analysis of variance, presumably because the computations are easier than for the design with several observations per cell. We work with the table of raw scores as we worked with the table of totals in part C of this section. However, the design presents special problems and should be avoided in practice, if possible. With only one observation per cell, it is impossible to obtain a within-cells mean square; hence, it is impossible to obtain a mean square that only estimates error variance. Instead, the main-effect lines are compared to the interaction line to test for significance, and no test for significant interaction is performed. This is fine if you know that there is no interaction between A and B in affecting scores on the dependent variable. If there *is* an interaction, then (1) you will never know, because you can't test for it; and (2) your tests for significance of the main-effect lines will be too conservative, because you will be comparing to both interaction and error, rather than just error, in the F ratio. The following example illustrates the computations; the formulas are the same as in part C except that $N = 1$ and the sum over the subscript i drops out.

EXAMPLE   Consider the following 2×6 matrix with one observation per cell:

	B_1	B_2	B_3	B_4	B_5	B_6	Σ
A_1	15	11	11	7	11	11	66
A_2	9	9	7	9	11	9	54
Σ	24	20	18	16	22	20	120

The testable hypotheses for this design are

$$H_A: \alpha_1 = \alpha_2; \qquad \text{that is, } \sigma_\alpha^2 = 0$$

$$H_B: \beta_1 = \beta_2 = \beta_3 = \beta_4 = \beta_5 = \beta_6; \qquad \text{that is, } \sigma_\beta^2 = 0$$

Let us set α at .05 and continue:

$$SS_A = \sum_j \frac{T_{j+}^2}{K} - \frac{T_{++}^2}{JK}$$

$$= \frac{66^2}{6} + \frac{54^2}{6} - \frac{120^2}{12}$$

$$= \frac{4356 + 2916}{6} - \frac{14{,}400}{12} = 1212 - 1200$$

$$= 12$$

$$SS_B = \sum_k \frac{T_{+k}^2}{J} - \frac{T_{++}^2}{JK}$$

$$= \frac{576}{2} + \frac{400}{2} + \frac{324}{2} + \frac{256}{2} + \frac{484}{2} + \frac{400}{2} - 1200$$

$$= 1220 - 1200$$

$$= 20$$

$$SS_{\text{Total}} = \sum_k \sum_j X_{jk}^2 - \frac{T_{++}^2}{JK}$$

$$= 15^2 + 11^2 + 11^2 + \cdots + 9^2 + 11^2 + 9^2 - 1200$$

$$= 1252 - 1200$$

$$= 52$$

Let us call that portion of the total variability which cannot be accounted for by differences in the main effects the residual (R) sum of squares.

$$SS_R = SS_T - SS_A - SS_B$$

$$= 52 - 12 - 20$$

$$= 20$$

The degrees of freedom are $df(A) = J - 1 = 1$; $df(B) = K - 1 = 5$; $df(T) = JK - 1 = 11$; $df(R) = df(T) - df(A) - df(B) = 5$. Then in accordance with Table 13-8 we obtain the following table:

Source of variance	Sum of squares	Degrees of freedom	Mean square	Expected mean square	F	$F_{.05}$
A	12	1	12	$\sigma_\epsilon^2 + 6\sigma_\alpha^2$	3	6.61
B	20	5	4	$\sigma_\epsilon^2 + 2\sigma_\beta^2$	1	5.05
Residual	20	5	4	$\sigma_\epsilon^2 + \sigma_{\alpha\beta}^2$		
Total	52	11				

Two problems should now be apparent with this design. First, relatively little information is obtained about the effects of the treatments, compared to a design having more than one observation per cell; and second, using the residual term in the denominators of F tests for the main effects is justifiable only when you know that there is no interaction effect. As you can see from the EMS column of the table, the residual line estimates both interaction and error. It is an appropriate error line for the A and B main effects only when $\sigma_{\alpha\beta}^2 = 0$. But it is difficult to think of a research situation in which you know enough to say that the interaction effect is zero and don't know enough to say whether the main effects are zero or not.

The raw scores in this problem were identical to the cell means for the example of part C. Yet where the A main effect is significant in part B, here it is nonsignificant. The calculated value of F is smaller and the critical value is larger.

References   The two-way analysis is discussed in Dixon and Massey (1969), 175–181; Hays (1963), 385–412; and McNemar (1969), 327–338. The single-observation-per-cell case is discussed in Dixon and Massey, 167–175.

13-6   Treatments-by-subjects analysis of variance

Relevant earlier section:
13-4   One-way analysis of variance

It is possible to compare the means of three or more populations using samples such that subjects in each sample are paired with subjects in every other sample. The most common instance of this design occurs when the same subject is tested under several different treatment conditions, and the results are used to make inferences about the effects of the treatments.

Appropriately used, the present design is generally more sensitive than the standard one-way analysis of variance for determining whether there are treatment effects. In the method of Section 13-4, samples differ not only in receiving different treatments, but also in having different subjects. In treatment-by-subjects design, differences in subjects from treatment to treatment are minimized.

Let us consider a generalized design in which N subjects are observed once under each of J treatment conditions. The data layout is given in Table 13-9.

Table 13-9 *Data table for treatment-by-subjects analysis of variance*

Subject (i)	Treatment (j) 1	2	3	$\cdots$	J	Sum	Mean
1	X_{11}	X_{12}	X_{13}	$\cdots$	X_{1J}	T_{1+}	$\bar{X}_{1\cdot}$
2	X_{21}	X_{22}	X_{23}	$\cdots$	X_{2J}	T_{2+}	$\bar{X}_{2\cdot}$
3	.	.	.	.	.	.	.
4	.	.	.	.	.	.	.
$\vdots$	.	.	.	.	.	.	.
N	X_{N1}	X_{N2}	X_{N3}	$\cdots$	X_{NJ}	T_{N+}	$\bar{X}_{N\cdot}$
Sum	T_{+1}	T_{+2}	T_{+3}	$\cdots$	T_{+J}	T_{++}	—
Mean	$\bar{X}_{\cdot1}$	$\bar{X}_{\cdot2}$	$\bar{X}_{\cdot3}$	$\cdots$	$\bar{X}_{\cdot J}$	—	$\bar{X}_{\cdot\cdot}$

We can specify a model to represent the theoretical differences among the scores. The raw score of person i under treatment condition j has several components:

$$X_{ij} = \mu + \pi_i + \tau_j + \pi\tau_{ij} + \epsilon_{ij} \tag{13-26}$$

where μ is the grand mean of the population of subjects over the J treatments; π_i is a component due to person i; τ_j is a component due to treatment j; $\pi\tau_{ij}$ is a systematic, specific effect of treatment j on person i; and ϵ_{ij} is the error in the score of person i under treatment j. This model will not be treated further here. It is possible to obtain unbiased estimates of some, but not all, of the components of the model. However, its purpose is primarily to provide a basis for the expected mean squares and F ratio of Table 13-9.

The total (T) sum of squares is defined as the sum of squared deviations of all observations in the data table about the grand mean:

$$SS_T = \sum_j \sum_i (X_{ij} - \bar{\bar{X}}_{\cdot\cdot})^2$$

$$= \sum_j \sum_i \left(X_{ij} - \frac{T_{++}}{NJ} \right)^2$$

$$= \sum_j \sum_i X_{ij}^2 - \frac{T_{++}^2}{NJ} \tag{13-27}$$

You may fill in this derivation if you wish, by analogy with the derivation of SS_T from Section 13-4B. The remaining sums of squares are derived in a similar manner.

Next, a sum of squares for differences between people (BP) may be defined by comparison of row means to the grand mean:

$$SS_{BP} = \sum_i (\bar{X}_i. - \bar{\bar{X}}..)^2$$

$$= \sum_i \left(\frac{T_{i+}}{J} - \frac{T_{++}}{NJ}\right)^2$$

$$= \sum_i \frac{T_{i+}^2}{J} - \frac{T_{++}^2}{NJ} \qquad (13\text{-}28)$$

A sum of squares within people (WP) can be defined as the variability of each subject's scores about that subject's average. These variabilities are then added over people:

$$SS_{WP} = \sum_i \left(\sum_j (X_{ij} - \bar{X}_i.)^2\right)$$

$$= \sum_i \left[\sum_j \left(X_{ij} - \frac{T_{i+}}{J}\right)^2\right]$$

$$= \sum_i \left[\sum_j X_{ij}^2 - \frac{T_{i+}^2}{J}\right]$$

$$= \sum_i \sum_j X_{ij}^2 - \sum_i \frac{T_{i+}^2}{J} \qquad (13\text{-}29)$$

With a little algebra you can demonstrate that $SS_{WP} = SS_T - SS_{BP}$. Thus all the differences among scores in the table may be partitioned into those that are due to differences among subjects and those that are not. The within-person sum of squares may be further partitioned into differences due to treatments and a remainder or residual.

The treatments sum of squares is the sum of squared differences of the treatment (tmt) means about the grand mean:

$$SS_{tmt} = \sum_j (\bar{X}._j - \bar{X}..)^2$$

$$= \sum_j \left(\frac{T_{+j}}{N} - \frac{T_{++}}{NJ}\right)^2$$

$$= \sum_j \frac{T_{+j}^2}{N} - \frac{T_{++}^2}{NJ} \qquad (13\text{-}30)$$

The residual (R) sum of squares is the remaining variation in the data:

$$SS_R = SS_{WP} - SS_{tmt} \qquad (13\text{-}31)$$

This residual sum of squares is actually an estimate of the combined effects of error and interaction of subjects and treatments (for further discussion of interaction, see Section 13-5).

The total sum of squares is based on NJ deviations about the grand mean. A degree of freedom is lost because deviations are taken about $\overline{X}..$ instead of the population mean μ, so the total number of degrees of freedom is $NJ - 1$. The sum of squares between persons is based on the deviations of N means about $\overline{X}..$, hence $df_{BP} = N - 1$. The sum of squares within persons is based on the deviations of J raw scores about a row mean for each of N subjects; hence $df_{WP} = N(J - 1)$. The treatment sum of squares is based on deviations of J column means about the grand mean, so $df_{tmt} = J - 1$. By subtraction, $df_R = N(J - 1) - (J - 1) = (N - 1)(J - 1)$.

The general form of the analysis-of-variance table is given by Table 13-10.

Table 13-10 *Analysis-of-variance table for treatments by subjects analysis*

Source of variance	Sum of squares	Degrees of freedom	Expected mean square	Mean square	F
Between people	SS_{BP}	$N - 1$	$\sigma_\epsilon^2 + J\sigma_\pi^2$	—	—
Within people	SS_{WP}	$N(J - 1)$	—	—	—
Treatment	SS_{tmt}	$J - 1$	$\sigma_\epsilon^2 + \sigma_{\pi\tau}^2 + N\sigma_\tau^2$	SS_{tmt}/df_{tmt}	MS_{tmt}/MS_R
Residual	SS_R	$(N - 1)(J - 1)$	$\sigma_\epsilon^2 + \sigma_{\pi\tau}^2$	SS_R/df_R	—
Total	SS_{Tot}	$NJ - 1$	—	—	—

The expected mean squares (EMS) column of this table indicates the parameters estimated by the between-people treatments, and residual mean squares. These formulas are derived from the model (formula 13-21), but the derivation will not be presented here. Although no EMS is presented for the WP line of the analysis, it clearly must be a compound of σ_ϵ^2, $\sigma_{\pi\tau}^2$, and σ_τ^2, since its sum of squares is partitioned into terms based entirely on these parameters. Similarly, the total mean square would estimate a composite of all the parameters estimated in the table.

There is thus no mean square that estimates σ_ϵ^2 alone and hence no error term for testing the significance of either MS_{BP} or MS_R. We are unable to obtain an estimate of σ_ϵ^2 alone because there is only one observation per cell (see Section 13-5E). The only line of the analysis for which there is an appropriate error term is the treatments line, because EMS(tmt) has one term more than EMS (residual), in accordance with Section 13-4D.

EXAMPLE Five subjects were each observed under four treatment conditions, with the following results. Determine whether there is a significant treatment effect.

Subject	Treatment 1	2	3	4	Σ
1	31	24	24	21	100
2	20	13	7	12	52
3	19	9	3	5	36
4	23	25	13	15	76
5	12	14	8	2	36
$\sum_i X$	105	85	55	55	300
$\sum_i X^2$	2395	1647	867	839	5748

Set α at .05; then,

$$SS_{total} = \sum_j \sum_i X_{ij}^2 - \frac{T_{++}^2}{NJ}$$

$$= 5748 - \frac{(300)^2}{20} = 5748 - 4500$$

$$= 1248$$

$$SS_{BP} = \sum_i \frac{T_{i+}^2}{J} - \frac{T_{++}^2}{NJ}$$

$$= \frac{100^2}{4} + \frac{52^2}{4} + \frac{36^2}{4} + \frac{76^2}{4} + \frac{36^2}{4} - 4500$$

$$= \frac{21{,}072}{4} - 4500 = 5268 - 4500$$

$$= 768$$

$$SS_{WP} = SS_{Total} - SS_{BP}$$

$$= 1248 - 768$$

$$= 480$$

$$SS_{tmt} = \sum_j \frac{T_{+j}^2}{N} - \frac{T_{++}^2}{NJ}$$

$$= \frac{105^2}{5} + \frac{85^2}{5} + \frac{55^2}{5} + \frac{55^2}{5} - 4500$$

$$= \frac{24,300}{5} - 4500 = 4860 - 4500$$

$$= 360$$

$$SS_R = SS_{WP} - SS_{tmt}$$

$$= 480 - 360 = 120$$

Then the analysis-of-variance table is as follows:

Source of variance	Sum of squares	Degrees of freedom	Mean square	F	$F_{.05}$	Significant ?
Between	768	4				
Within	480	15				
Treatment	360	3	120	12	3.49	Yes
Residual	120	12	10			
Total	1248	19				

Reference   This design can be found in Winer (1962), 105–124.

Homework

1. In the following examples, indicate the type of design using the terminology of Section 13-2. Indicate whether the analysis could be completed using the methods of the present chapter. Suggest what the null hypothesis or hypotheses might be.

   (a) Female subjects are divided into three groups on their perceived similarity to their respective mothers (self-report). The motivated hypothesis is that the three groups will differ in some way on degree of ego strength (ego-strength test assumed interval scale).

   (b) It is desired to determine the effects of sex and birth order (only, first-born, later-born) on tested anxiety.

   (c) A psychiatric team is sent to evaluate the care received by mental patients in state hospitals. Three hospitals are randomly selected within each of five New England states, and 10 subjects are randomly selected from the roster of patients at each hospital. A rating is then made of the care received by each patient. It is desired to determine whether there are differences among the states and among hospitals within the states. Assume the ratings are interval level.

2. Complete the analysis of the example of Section 13-5A, testing for main effects and interaction at α of .05.

For problems 3 through 9, each of the sets of data, perform a complete analysis of variance by (a) stating the hypotheses; (b) finding the critical value of F at an α level of .05; (c) computing the obtained value of F; and (d) stating your conclusion regarding the existence of a treatment effect.

3.

Group 1	Group 2	Group 3
11	8	11
9	8	7
15	11	11
13	12	7
12	6	9

4.

Group 1	Group 2	Group 3
2	10	6
4	5	5
5	6	3
5	6	6
4	3	

5.

Group 1	Group 2	Group 3
18	11	14
10	3	13
16	4	7
12	7	8
14	5	9
14	6	9

6.

	A_1	A_2	A_3
B_1	1	2	3
B_2	4	5	6
B_3	7	8	9

7.

	A_1	A_2	A_3	A_4
B_1	2	1	2	0
B_2	9	7	5	4
B_3	5	11	8	6

8.

	A_1	A_2
B_1	18	10
	14	4
	14	7
	14	7
B_2	9	8
	3	3
	3	3
	5	6

9.

	A_1	A_2	A_3
B_1	18	0	10
	6	8	2
	9	11	14
	12	2	8
	10	4	6
B_2	3	18	14
	3	20	9
	13	26	9
	3	14	9
	3	17	19

10. In a culinary experiment you chop up a grapefruit, drop it in a pot, and bring it to a boil over a well-regulated stove. You repeat this action several times over exactly the same heat as a test of the effects of different pot characteristics on the time it takes the grapefruit to boil. You do a fully crossed design, with three grapefruit per cell. The independent variables are pot size (1 quart, 2 quarts, and 3 quarts) and pot type (stainless steel, iron, porcelain, Teflon, and aluminum). In preliminary calculations you find the sum of squares for pot sizes to be 140 and the sum of squares for pot types to be 240. The sum of squares for all cells is 700, and the total sum of squares is 1300.

Complete the table and draw your conclusions about the research. Then state your conclusions in English.

Source	SS	df	MS	F	$F_{.05}$	Significant?
Size: S						
Type: T						
S × T						
Cells						
Within						
Total						

Homework Answers

1. **(a)** One-way, fixed effects, independent samples.
   (b) Two-way, fixed, fully crossed, two main effects and interaction.
   (c) Two-way nested, hospitals within states, states fixed, hospitals random.

2. $F_{.05}(1,\ 12\ \text{df}) = 4.75$, $F_A = 57.12$, F_A significant, $F_{.05}(2,12\ \text{df}) = 3.89$, $F_B = 8.33$, F_B significant, $F_{AB} = 2.08$, F_{AB} nonsignificant.

3. $F_{\text{obs}} = 3$, $F_{.05} = 3.89$, nonsignificant.

4. $F_{.05} = 3.98$, $F_{\text{obs}} = 1.44$, nonsignificant.

5. $F_{.05} = 3.68$, $F_{\text{obs}} = 12$, significant.

6. $F_A = \infty$, $F_B = \infty$, both significant because there is no error.

7. $F_A = 1.83$, $F_{.05}(3,\ 6\ \text{df}) = 4.76$, F_A nonsignificant, $F_B = 17.1$, $F_{.05}(2,\ 6\ \text{df}) = 5.14$, F_B significant.

8. $F_{.05}(1,\ 12\ \text{df}) = 4.75$, $F_A = 10.67$, $F_B = 24$, $F_{AB} = 10.67$, all significant.

9. $F_A = 2$ (nonsignificant), $F_B = 6$ (significant), $F_{AB} = 12.5$ (significant).

10. Size: $F_{\text{obs}} = 3.5$, $F_{.05} = 3.32$, significant; type: $F_{\text{obs}} = 3.0$, $F_{.05} = 2.69$, significant; interaction: $F_{\text{obs}} = 2.0$, $F_{.05} = 2.27$, nonsignificant. Pot size and pot type affect the time it takes a chopped grapefruit to boil, but they do not interact to affect boiling time.

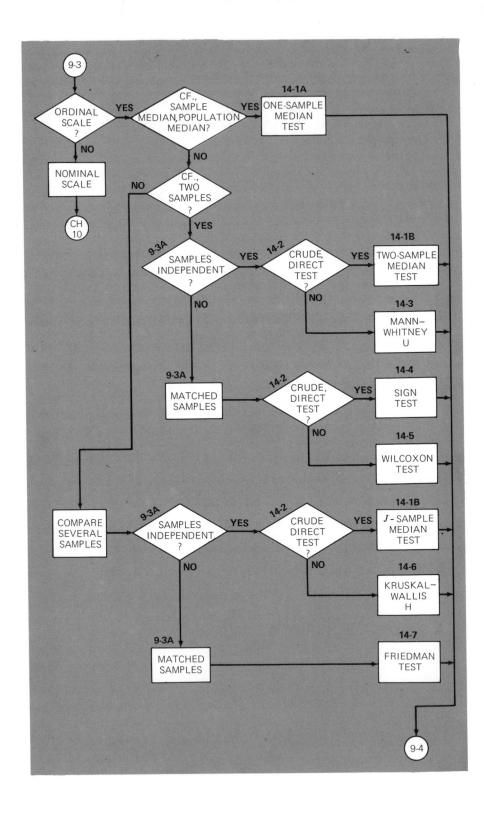

14 Hypothesis Tests with Ordinal Scales

THE map for this chapter is readily divided into conceptual regions based on the number of samples compared in testing a hypothesis. Because they are generally easy to apply, a wide range of nonparametric and distribution-free tests may be found in introductory texts. This chapter does not attempt to span that range of available techniques. Instead, it is limited to widely used methods consistent with the experimental designs of Section 9-3. For more comprehensive treatments, see Siegel (1956) or Walsh (1962).

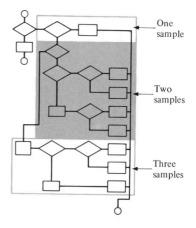

One sample

Two samples

Three samples

Because the distinction between ordinal and interval scales is not always easily made, you may wish also to consider methods of Chapter 12 for use with one or two samples, and Chapter 13 for comparison of three or more samples. Assumptions underlying t and F tests are given there and may be compared with assumptions underlying methods of the present chapter for relevance to your problem. Also, Section 5-7 compares the information available in means, medians, and modes, and may therefore be useful in selecting a test statistic.

14-1 Median test

Relevant earlier sections:
5-6 Median
8-7 Binomial distribution
10-9 Binomial test

A. Single Sample

This test can be used to determine whether a sample has been randomly drawn from a population with a specified median, Md_0. It is possible to do a nondirectional test (H_0: $Md = Md_0$) or a directional test (H_0: $Md \geq Md_0$ or H_0: $Md \leq Md_0$). The usual experimental designs are depicted in Figures 9-1 and 9-2.

To carry out the test, specify an a priori median, a significance level, and a null hypothesis. Draw your sample under the experimental conditions that you are interested in, and compare each observed score or rating to the specified median. Construct the following table:

	Frequency
Above Md_0	
Below Md_0	
$N =$	

Once the data are in this form, they can be analyzed using the binomial test of Section 10-9, with $p = q = .5$ under H_0.

EXAMPLE   The median family size in Friedman County (population: 18,000) is 3.80 (obtained by interpolation). In the town of Riverbankville, there are 14 families with four or more members and 22 with three or fewer. Are families in this town significantly smaller than families in the county as a whole? The frequency table is as follows:

	Frequency
Above the median	14
Below the median	22
$N = 36$	

For the solution, we must refer back to Sections 10-9 and 8-7. Let Md_e be the median of the hypothetical population from which the families in this town are actually drawn. Then the null hypothesis is H_0: $Md_e \geq Md_0$ and H_m: $Md_e < Md_0$. If we let π_0 be the population proportion of families of size greater than $Md_0 = 3.80$, then our hypotheses are H_0: $\pi_e \geq 0.50$ and H_m: $\pi_e < 0.50$. Under this null hypothesis, $N\pi_0 = N(1 - \pi_0) = 18$, which is greater than 10, so the normal approximation to the binomial distribution can be used. The area under the standard normal distribution which represents the proportion of samples with 14 or fewer families above the Md_0 is cut off by a z score corresponding to a frequency of 14.5. The computations for the normal approximation follow formulas 8-3 and 8-4.

$$\mu = N\pi_0 = 18$$

$$\sigma = \sqrt{N\pi_0(1 - \pi_0)} = \sqrt{36(.5)(.5)} = \sqrt{9} = 3$$

The graph of the problem is shown in the illustration. The z score that cuts off the proportion of samples with 14 or fewer families larger than the null median is

$$z = \frac{14.5 - 18}{3} = -\frac{3.5}{3} = -1.167$$

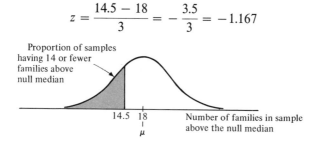

Proportion of samples having 14 or fewer families above null median

14.5 18   Number of families in sample
     μ above the null median

The critical value for a one-tail z test at $\alpha = .05$ is -1.64. The calculated value is greater than this, so there is no evidence that the families in this town are unusually small for the county.

B. Two or More Samples

In this case the median test is used to determine whether two or more samples have been drawn from populations with the same median. It is possible to do a nondirectional test with either the two-sample case ($H_0: Md_1 = Md_2$) or the several-sample case ($H_0: Md_1 = Md_2 = \cdots = Md_J$). A directional test, by contrast, is possible only with the two-sample case ($H_0: Md_1 \geq Md_2$ or $H_0: Md_1 \leq Md_2$). The usual experimental designs used with this test are diagrammed in Figures 9-3, 9-6, and 9-7.

The first step in carrying out the analysis is to combine scores from all samples into a single sample and find the grand median. Each score in each sample is then compared to the grand median and the results cast into a frequency table similar to that of part A of this section:

	Sample				
	1	2	3	$\cdots$	J
Above grand median	f_{1A}	f_{2A}	f_{3A}	$\cdots$	f_{JA}
Below grand median	f_{1B}	f_{2B}	f_{3B}	$\cdots$	f_{JB}
Σ	N_1	N_2	N_3	$\cdots$	N_J

Once the data are in this form, they can be analyzed using the methods of Chapter 10.

14-1 *Median test*

EXAMPLE A wine importer wants to determine whether Americans can distinguish among six red table wines he is thinking of carrying. To answer the question, he has 10 subjects sample all six wines and rate each on an 11-point scale. His null hypothesis is that all six wines will receive the same median rating; he selects an α level of .05. The data are as follows:

	Wine					
Subject	1	2	3	4	5	6
1	1	3	3	4	1	2
2	4	4	0	4	2	5
3	7	8	0	5	7	5
4	3	3	3	5	6	3
5	8	6	3	10	5	0
6	5	8	4	7	7	1
7	1	8	1	2	8	8
8	3	10	2	7	7	7
9	5	4	3	1	9	0
10	5	10	3	8	8	8

In this data table there are 30 ratings of 4 or less and 30 of 5 or more, so the median is 4.5. From this we construct a second table, giving the frequency of ratings above and below the overall median for each wine:

	Wine						
	1	2	3	4	5	6	Σ
Above grand median	5	6	0	6	8	5	30
Below grand median	5	4	10	4	2	5	30
Σ	10	10	10	10	10	10	60

For each cell of this table, the expected frequency is (10)(30)/60 = 5, estimating the expected frequencies from the marginals in accordance with Section 10-1. All E's are greater than or equal to 5, so the method of Section 10-3 is used on the given 2 × 6 table. Chi square is found to be

$$\chi^2 = \frac{(5-5)^2}{5} + \frac{(6-5)^2}{5} + \frac{(0-5)^2}{5} + \cdots + \frac{(5-5)^2}{5}$$

$$= \frac{72}{5} = 14.4$$

This observed chi square is greater than the critical value of $\chi^2_{.05}(5 \text{ df})$ = 11.0705, so the null hypothesis is rejected. The subjects could distinguish among the wines.

As indicated in the map for the chapter, a Kruskal–Wallis test could also be carried out with these data, and it would probably be more sensitive than the median test. Both the median and Kruskal–Wallis tests would be ignoring the fact that the data are correlated: the same subjects sample all six wines, so the six samples are not independent. Hence, a Friedman test could also be done with these data, or, if the assumptions seem satisfied, a treatment-by-subjects analysis of variance.

C. Analysis When Scores Are at the Median, and a Generalization

If the overall distribution of scores will not divide neatly at the median, then it is perfectly reasonable to divide it just above or just below the median and expect the results to indicate differences in centralities for two or more distributions. In fact, the general form of the analysis can be extended. There is no reason not to divide the overall distribution at the first or third quartile, if it seems reasonable to do so, or at any other interesting point. For comparing centrality measures, a split as near the overall median is optimal, however.

The analysis could also be generalized to, say, splitting the overall distribution at the median and first and third quartiles, giving a $J \times 4$ frequency table for the analysis.

D. Prediction and Correlation

Since the dependent variable is ordinal in this case, you can predict the median rank for each level of the independent variable, as indicated in Section 15-2. When there are two or more levels of the independent variable, a correlation can also be obtained, to show the degree of relationship between the independent variables. As indicated by the map for Chapter 17, if you think of the relationship between a nominal independent variable and an ordinal-level dependent variable, the coefficient of choice is theta (Section 17-4). Or you can treat the dependent variable as nominal and use lambda (Section 17-3).

References Additional textbook discussions of the median test can be found in Dixon and Massey (1969), 351–352; Guilford (1965), 257–258; Hays (1963), 620–625; Siegel (1956), 111–116, 179–184; and Walker and Lev (1953), 435–436. The extension of part C is discussed in Mattson (1965).

14-2 Choice of test

The median test (Section 14-1) and sign test (Section 14-4) make relatively few assumptions about the population distributions of raw scores, but they are not very powerful. Mean-rank tests such as the Mann–Whitney (Section 14-3), the Wilcoxon (Section 14-5), the Kruskal–Wallis (Section 14-6), and the

Friedman (Section 14-7) are more powerful when their assumptions hold than are the median and sign tests. However, they also make more assumptions—assumptions about the shapes of raw score or underlying interval-level distributions—and they are sensitive to assumption violations.

If the effect that you are seeking to establish appears to be great, or if samples are large, or if your score distributions don't satisfy the assumptions of the mean-rank tests, then the sign or median test should be used. If the effect you are looking for is probably small or if your samples are small, and if your distributions meet the assumptions of the appropriate mean-rank test, then it should be used.

You should make your choice by reading both sections and then choosing the one in which you have the greater confidence.

14-3   The Mann–Whitney U test

The Mann–Whitney test is used to compare two independent samples to determine whether subjects in one population have consistently higher ranks* than subjects in the second. For example, a spelldown ranks students in a class on their spelling ability. We might thus use a spelldown to determine whether boys or girls are better spellers. The design for this section follows Figure 9-3 or 9-6.

A. Hypotheses

The hypotheses tested by this method depend on the assumptions that you are willing to make about the data to which it is applied (Table 14-1). If you are

Table 14-1   *Testable hypotheses of the Mann–Whitney test as function of assumptions about population distributions*

	Comparison		
	(i)	(ii)	(iii)
No distribution assumptions	$\mu_{\mathscr{R}_1} = \mu_{\mathscr{R}_2}$†	$\mu_{\mathscr{R}_1} > \mu_{\mathscr{R}_2}$	$\mu_{\mathscr{R}_1} < \mu_{\mathscr{R}_2}$†
Population distributions of ordinal-level raw scores have same shapes	$Md_1 = Md_2$	$Md_1 > Md_2$	$Md_1 < Md_2$
Population distributions of interval-level raw scores have same shapes	$\mu_1 = \mu_2$	$\mu_1 > \mu_2$	$\mu_1 < \mu_2$

† We use $\mathscr{R}$ for ranks (just as we use X for raw scores). Then $\mu_{\mathscr{R}_1}$ is the mean of ranks for population 1. Do not confuse the use of $\mathscr{R}$ in the present context with the use of R as a general statistic in Chapters 9 and 12, or with the range R of Section 5-10.

* When subjects are ranked on an attribute, they are numbered consecutively from least to greatest or greatest to least on the attribute, starting with the number 1. If the data appear as ranks, they can be used as is; otherwise interval or ordinal level raw scores can be converted to ranks in the first step of the analysis.

willing to assume that the population distributions of ordinal raw scores have the same shapes,* then inferences can be made from the value of U calculated on your samples to the population medians; if you are willing to assume that the ranks used to calculate U are based on underlying interval level raw scores that have the same shaped distributions for the two samples, then you are justified in inferring to the population means of raw scores on the basis of the magnitude of U.

Nondirectional hypotheses are stated by pitting a comparison in column (i) against the other two columns. Thus, if you are willing to assume that the two populations have the same distribution of ranks, you could take your hypotheses to be $H_0: Md_1 = Md_2$ and $H_m: Md_1 \neq Md_2$ (i.e., $Md_1 > Md_2$ ⓥ $Md_1 < Md_2$). Or, you could frame your hypotheses in terms of mean ranks, since that would involve fewer assumptions. A directional or one-tailed null hypothesis is formed by combining (i) with either (ii) or (iii) for the assumptions you have chosen, and the motivated hypothesis is given by the remaining comparison. If you are willing to assume that the data at hand are based on an underlying interval scale on which the raw-score distributions for the two populations have the same shape, then your directional hypotheses might be given in terms of the means of those hypothesized raw-score distributions:

$$H_0: \mu_1 \geq \mu_2 \quad \text{and} \quad H_m: \mu_1 < \mu_2$$

B. Data and Calculations for Small Samples

Data for the comparison are obtained by combining the two samples and ranking subjects from a low of 1 to a high of $N_1 + N_2$. Thus, if there are 10 subjects in sample 1 and 7 in sample 2, ranks range from 1 to 17. When two or more subjects can't be distinguished, they are given the mean of the ranks they would have received if they could be ordered. For example, if the third and fourth lowest subjects are tied, the overall sequence of ranks would be 1, 2, 3.5, 3.5, 5, and so on.

The statistic used to test a one-tail hypothesis $H_0: \mu_{\mathcal{R}_2} \geq \mu_{\mathcal{R}_2}$ and H_m: $\mu_{\mathcal{R}_1} < \mu_{\mathcal{R}_2}$ is given by formula 14-1:

$$U(\overline{\mathcal{R}}_1 < \overline{\mathcal{R}}_2) = T_1 - \frac{N_1(N_1 + 1)}{2} \tag{14-1}$$

where $T_1 = \sum_1 \mathcal{R}_{i1}$ is the sum of ranks in sample 1. If $U(\overline{\mathcal{R}}_1 < \overline{\mathcal{R}}_2)$ is less than or equal to the critical value of U given in Table A-10 of the Appendix, reject the null hypothesis.

To test the two-tail hypothesis $H_0: \mu_{\mathcal{R}_1} = \mu_{\mathcal{R}_2}$, find both $U(\overline{\mathcal{R}}_1 < \overline{\mathcal{R}}_2)$ and $U(\overline{\mathcal{R}}_1 > \overline{\mathcal{R}}_2)$, and reject H_0 if either of these values is less than or equal to the critical value given in Table A-10. In carrying out this procedure, the easiest method is to use formula 14-2:

---

* That is, you treat the raw scores as though they are interval level. Then you assume that histograms or polygons of the underlying population distributions have the same shapes.

$$U(\overline{\mathscr{R}}_1 > \overline{\mathscr{R}}_2) = N_1 N_2 - U(\overline{\mathscr{R}}_1 < \overline{\mathscr{R}}_2) \tag{14-2}$$

so that only one of the two U's must be calculated from formula 14-1. Then, if the smaller of the two U's is less than or equal to the tabled value, the null hypothesis can be rejected.

EXAMPLE An experiment on the nature of hypnotism is run "blind." Experimental (hypnotized) and control subjects are used, and the experimenter is not told whether a given subject is hypnotized or not. Half of the subjects can be hypnotized easily; the other half cannot, but fake it. Order is randomized. As a check, the experimenter makes notes after running each subject, and at the end of the entire experiment, ranks all 12 subjects on his certainty that they actually were hypnotized. He gives rank 1 to the subject he is least certain was hypnotized, and rank 12 to the subject about whom he was most certain. His hypotheses are $H_0 : \mu_{\mathscr{R}_1} \geq \mu_{\mathscr{R}_2}$ and $H_m : \mu_{\mathscr{R}_1} < \mu_{\mathscr{R}_2}$, where subjects in group 1 are faking and subjects in group 2 are really hypnotized. His ratings are shown:

Subject	Group 1, faking	Group 2, hypnotized
1	1	2
2	10	7
3	4	12
4	5	9
5	11	3
6	6	8
Σ	37	41

When he applies formula 14-1 to these data, he gets

$$U(\overline{\mathscr{R}}_1 < \overline{\mathscr{R}}_2) = 37 - \frac{6(7)}{2}$$

$$= 16$$

The critical value of $U_{.05}$(one tail, $N_1 = 6$, $N_2 = 6$) from Table A-10 is 7. The observed U of 16 is greater than the critical value, so H_0 is not rejected.

If the experimenter had chosen to do a two-tail test, he would have found

$$U(\overline{\mathscr{R}}_1 > \overline{\mathscr{R}}_2) = 6(6) - 16$$

$$= 20$$

then compared 16 (the smaller of 16 and 20) to $U_{.05}$(two tail, $N_1 = 6$, $N_2 = 6$) from Table A-10. The critical value is 5, so again H_0 would not have been rejected, where in this case H_0 would have been: $\mu_{\mathscr{R}_1} = \mu_{\mathscr{R}_2}$.

C. Computation of U for Large Samples

If your samples are too large for Table A-10, you may use a normal approximation to find out whether your value of U is significant or not. The mean and standard deviation of the best-fitting normal distribution are given in formulas 14-3.

$$\mu_U = \frac{N_1 N_2}{2} \quad \text{and} \quad \sigma_U = \sqrt{\frac{N_1 N_2 (N_1 + N_2 + 1)}{12}} \tag{14-3}$$

Then the statistic

$$z_U = \frac{U - \mu_U}{\sigma_U}$$

can be compared to the table of the standard normal distribution (Table A-4) to test your hypotheses.

D. Prediction and Correlation

Like the median test, the Mann–Whitney is used to test for a significant relationship between two variables: a nominal (dichotomous) independent or predictor variable and an ordinal-level dependent or criterion variable. A significant U indicates that the two variables are not completely independent, in the population. If there is a significant relationship between the two variables, we can then ask, "What should we predict on the criterion for each level of the predictor?" We can obtain the prediction by finding the median rank on the dependent variable for each level of the independent variable as indicated in Section 15-2. To find out how strong the relationship is, we can use theta (Section 17-4), as suggested by the map for Chapter 17; or if we are willing to call the criterion an interval level, we can use eta (Section 17-1) or the point biserial (Section 16-3).

References The basic source is Mann and Whitney (1947). Textbook discussions can be found in Freeman (1965), 187–198; Hays (1963), 633–635; Siegel (1956), 116–127; and Walker and Lev (1953), 434–435.

14-4　The sign test

Relevant earlier sections:
8-7　Binomial distribution
10-9　Binomial test

This test is used to determine if scores in one of two matched sets of data are consistently greater than scores in the other. Data may be correlated through use of the same subjects under different conditions, or different samples matched subject by subject before the experiment is begun. Two common experimental designs are given in Figures 9-4 and 9-5.

This is called the "sign test" because it was originally suggested that you write a + each time a subject in sample 1 scored higher than the corresponding subject in sample 2, and a − when the reverse was true. Then the numbers of pluses and minuses were used to test the hypotheses.

A. Hypotheses

Statement of the comparisons possible with this test in terms of probabilities emphasizes the fact that the data are correlated. The comparisons are (i) $P\{X_{1i} > X_{2i}\} = .5$, (ii) $P\{X_{1i} > X_{2i}\} > .5$, and (iii) $P\{X_{1i} > X_{2i}\} < .5$. These comparisons are equivalent to the corresponding ones given in Section 14-3A, under the same restrictions. With no distribution assumptions, they are equivalent to hypotheses about mean ranks. With the assumption of similarly shaped underlying ordinal distributions, the hypotheses can be about medians. With an assumption that the ranks are based on similarly shaped, interval-level raw-score distributions, the hypotheses apply to raw-score means.

B. Data Collection and Analysis

For each pair of subjects (one from one sample, one from the other), determine which score is greater, and tally the pair in one cell of the following frequency table. In case of a tie, drop that pair from the analysis.

	Frequency
$X_{1i} > X_{2i}$	
$X_{1i} < X_{2i}$	
Σ	N pairs

State your hypotheses as indicated in part A, and test them using the binomial test of Section 10-9.

EXAMPLE In a study of reported feelings of male homosexuals, subjects are asked to compare their mothers and fathers on each of several variables and to indicate which parent had more of the characteristic in question. It was hypothesized that subjects would judge their mothers to more "active" and their fathers to be more "passive." The hypotheses, in terms of activity, were H_0: $P\{\text{Mother} > \text{Father}\} \leq .5$ and H_m: $P\{\text{Mother} > \text{Father}\} > .5$. Alpha was set at .05, and eight subjects were drawn. Responses were as follows:

More active	Frequency
Mother	7
Father	1
N	8

Note that in this case, an ordinal response was obtained from each subject, eliminating the need for a raw-data table and for the experimenter to do the ranking.

Under $H_0: N \cdot P\{\text{Mo} > \text{Fa}\} = 8(.5) = 4$ and $N \cdot P\{\text{Mo} < \text{Fa}\} = 8(.5) = 4$, both of which are less than 5 (Section 8-8B), so the method of Section 8-7 is used for the analysis. The probability that seven of eight mothers will be perceived as more active under the null hypothesis is

$$P\{7 \text{ Mo of } 8\} = {}_8C_7(.5)^7(.5)^1$$
$$= 8(\tfrac{1}{2})^8 = (\tfrac{1}{2})^5 = \tfrac{1}{32} = .031$$

The probability that 8 of 8 will be seen as more active is

$$P\{8 \text{ Mo of } 8\} = {}_8C_8(.5)^8(.5)^0$$
$$= \tfrac{1}{256} = .004$$

Therefore, the probability of a result as extreme or more extreme in the predicted direction under the null hypothesis is $.031 + .004 = .035$, which is less than .05. Therefore, the null hypothesis is rejected, and we conclude that for this population, their mothers were judged to be more active.

C. Prediction and Correlation

In the example of part B, the prediction can be made that these subjects see their mothers as more active; but because there is only one sample and one response per subject, a correlation (which shows the degree of relationship between two variables) cannot be obtained.

However, the sign test can also be used with data like those in the example of Section 14-5, where father and mother receive separate ratings. The test would be carried out on the sign of the difference scores (d_i), ignoring their magnitude. In that case, the procedures for prediction and correlation of Section 14-5D can be used with the data.

References Other discussions of this technique can be found in Dixon and Massey (1969), 335–340; Guilford (1965), 253–255; Hays (1963), 625–628; Siegel (1956), 68–75; and Walker and Lev (1953), 430–432.

14-5   Wilcoxon T

The Wilcoxon T statistic is used to determine whether scores in one sample are consistently higher than scores in another when data in the two samples are paired. Two experimental designs for which the test is appropriate are given in Figures 9-4 and 9-5.

A. Hypotheses

See Section 14-4A. The Wilcoxon differs from the sign test primarily in that it uses more information when it is available in the data. Thus, when both tests are appropriate, the Wilcoxon is the more sensitive to differences between samples.

B. Data Collection and Analysis

The basic data used in the calculation of T are difference scores which indicate the extent to which a subject's score in one sample exceeds the matched score from the other sample. These difference scores d_i may be obtained directly by the method of data collection, or they may be derived from pairs of raw scores, using a table like Table 14-2.

Table 14-2 *Data table for Wilcoxon test*

Subject pair	Sample 1	Sample 2	d_i	Rank d_i
1	X_{11}	X_{21}	$X_{11} - X_{21}$	
2	X_{12}	X_{22}	$X_{12} - X_{22}$	
$\vdots$	$\vdots$	$\vdots$	$\vdots$	
N	X_{1N}	X_{2N}	$X_{1N} - X_{2N}$	

In carrying out the analysis, the first step is to find a difference score for each pair of subjects: $d_i = X_{1i} - X_{2i}$, where the numerical subscripts refer to samples. Next, the d_i are ranked from smallest to largest in absolute magnitude (without regard to sign). If two or more d_i can't be distinguished in size, they are given the mean of the ranks they would have received if they could have been ordered.

For a one-tail test of the hypotheses $H_0: \mu_{\mathcal{R}_1} \geq \mu_{\mathcal{R}_2}$ and $H_m: \mu_{\mathcal{R}_1} < \mu_{\mathcal{R}_2}$, find the d_i with positive signs (for which $X_{1i} > X_{2i}$) and sum their ranks, as indicated in formula 14-4:

$$T(\overline{\mathcal{R}}_1 < \overline{\mathcal{R}}_2) = \sum \mathcal{R}(\text{positive } d_i) \qquad (14\text{-}4)$$

The motivated hypothesis is true if the sum $T(\overline{\mathcal{R}}_1 < \overline{\mathcal{R}}_2)$ is small. To determine if T(observed) is small enough, compare it to T(critical) as found in Table

A-11 of the Appendix. If the observed T is less than or equal to the tabled value, reject H_0; otherwise, accept it.

For a two-tail test of the hypotheses $H_0: \mu_{\mathscr{R}_1} = \mu_{\mathscr{R}_2}$ and $H_m: \mu_{\mathscr{R}_1} \neq \mu_{\mathscr{R}_2}$, compute both $T(\overline{\mathscr{R}}_1 < \overline{\mathscr{R}}_2)$ and $T(\overline{\mathscr{R}}_1 > \overline{\mathscr{R}}_2) = \sum \mathscr{R}(\text{negative } d_i)$. Let T_s be the smaller of these two values; compare it to the critical value of T in Table A-11 and reject H_0 if $T_s \leq T_\alpha$.

EXAMPLE   In a study of reported feelings of male homosexuals, 14 subjects are asked to rate their mothers and fathers on each of several 100-point scales. Ratings of "my mother" and "my father" are compared within each subject on the scale "active–passive" to determine whether, across subjects, one or the other parent is rated more active. The hypotheses are $H_0: \mu_{\mathscr{R}_1} = \mu_{\mathscr{R}_2}$ and $H_m: \mu_{\mathscr{R}_1} \neq \mu_{\mathscr{R}_2}$ and α is set at .05. The data are as shown in the table below. Since we are doing a two-tail test, we find $T_s = $ smaller (3.5, 101.5) = 3.5. The critical T at .05 level for a two-tail test with $N = 14$ is 21. The observed value of T_s is less than the critical value, so we reject the null hypothesis: $\mu_{\mathscr{R}_1} \neq \mu_{\mathscr{R}_2}$. By examination of the two sums of ranks we conclude that fathers are ranked less active by these subjects.

Subject	Activity rating of Father	Activity rating of Mother	$d_i = X_{Fa} - X_{Mo}$	Rank d_i	Positive d_i	Negative d_i
1	8	41	− 33	13		13
2	63	85	− 22	11		11
3	7	19	− 12	8.5		8.5
4	14	10	4	2.5	2.5	
5	36	48	− 12	8.5		8.5
6	64	70	− 6	5		5
7	8	37	− 29	12		12
8	26	34	− 8	6		6
9	19	18	1	1	1	
10	12	16	− 4	2.5		2.5
11	6	94	− 88	14		14
12	7	12	− 5	4		4
13	40	50	− 10	7		7
14	31	48	− 17	10		10
Σ					3.5	101.5

C. Analysis for Large Samples ($N > 100$)

If N is greater than, say 100, the distribution of T is closely approximated by a normal distribution with mean

$$\mu_T = \frac{N(N+1)}{4} \tag{14-5}$$

and standard deviation

$$\sigma_T = \sqrt{\frac{N(N+1)(2N+1)}{24}} \tag{14-6}$$

These values can be computed and used in the z formula,

$$z_T = \frac{T - \mu_T}{\sigma_T}$$

and compared to the critical values given in Table A-4.

EXAMPLE   Let us carry out the normal approximation for the example of part B of this section. Clearly, this application of the approximation is inappropriate because N is small, but will allow us to examine the method with a minimum of additional calculations. For this example,

$$\mu_T = \frac{14(15)}{4} = \frac{105}{2} = 52.5$$

$$\sigma_T = \sqrt{\frac{14(15)(29)}{24}} = \sqrt{\frac{1015}{4}} = \sqrt{253.75} \doteq 15.9$$

$$z_T = \frac{3.5 - 52.5}{15.9} = \frac{-49}{15.9} = -3.08$$

The critical value of z for a two-tail test at the .05 significance level is $z = \pm 1.96$. The observed value of -3.08 is less than -1.96, so H_0 would be rejected.

D. Prediction and Correlation

Following the suggestions of Section 15-2, our best guess of the rating to be received by a father for the example would be the median of the ratings of fathers in this sample: 14; and, similarly, the best guess for mothers would be 37, the median rating for mothers in this sample.

A correlation between "parent" (father or mother) and activity rating would tell us how different the two sets of ratings are. Given that the ratings are ordinal-level, our best coefficient would probably be theta (Section 17-4). However, if we are willing to assume that the ratings are interval level, we can use either the point biserial (Section 16-3) or eta (Section 17-1).

References   Other discussions of this method can be found in Guilford (1965), 255–256, 258–260; Hays (1963), 635–637; Siegel (1956), 75–83; Snedecor (1961), 115–120; and Walker and Lev (1953), 432–435.

14-6   The Kruskal–Wallis test

This statistic is used to compare three or more independent samples to determine whether they differ in average rank.* The experimental design related to this test is given in Figure 9-7.

A. Hypotheses

As in the case of the two-sample statistics (see Section 14-3A), the hypotheses you are able to test depend on the assumptions that you are willing to make about the score distributions in the populations you have sampled. Possible assumptions and the consequent testable hypotheses are listed in Table 14-3.

Table 14-3   *Testable hypotheses of the Kruskal–Wallis test as a function of assumptions about population distributions*

Assumption	Null hypothesis†	Motivated hypothesis†
No distribution assumptions	$\mu_{\mathscr{R}_1} = \mu_{\mathscr{R}_2} = \cdots = \mu_{\mathscr{R}_J}$	Not $(\mu_{\mathscr{R}_1} = \mu_{\mathscr{R}_2} = \cdots = \mu_{\mathscr{R}_J})$
Ordinal-level raw scores have same shape in population	$Md_1 = Md_2 = \cdots = Md_J$	Not $(Md_1 = Md_2 = \cdots = Md_J)$
Interval-level raw scores have same shape in population	$\mu_1 = \mu_2 = \cdots = \mu_J$	Not $(\mu_1 = \mu_2 = \cdots = \mu_J)$

† See footnote to Table 14-1.

B. Analysis of Data

The first step in the data analysis is to combine all subjects into a single sample and rank them together, from a low of 1 to a high of $N_1 + N_2 + \cdots + N_J$, where the N's are sample sizes of J samples. If there are ties in the ranking, all tied subjects receive the mean of the ranks they would have had if they had not been tied. Thus, if the four lowest subjects are tied, instead of receiving ranks 1, 2, 3, and 4, they all receive rank 2.5, and the ranking is 2.5, 2.5, 2.5, 2.5, 5, 6, etc.

Next, the ranks are cast in a subject-by-samples matrix such as the one in Table 14-4. For each sample, the sum of ranks $T_j = \sum_i \mathscr{R}_{ij}$ is found and then squared.

Then the test statistic is given by formula 14-7:

$$H = \frac{12}{N(N + 1)} \left(\sum_{j=1}^{J} \frac{T_j^2}{N_j} \right) - 3(N + 1) \tag{14-7}$$

where J is the number of samples, N_j is the number of subjects in sample j,

*See footnote on page 370.

Table 14-4   *General form of subject-by-samples matrix*

	Sample					
Subject		1	2	3	...	J
1		$\mathcal{R}_{11}$	$\mathcal{R}_{12}$	$\mathcal{R}_{13}$	...	$\mathcal{R}_{1J}$
2		$\mathcal{R}_{21}$	$\mathcal{R}_{22}$	$\mathcal{R}_{23}$	...	$\mathcal{R}_{2J}$
$\vdots$		$\vdots$	$\vdots$	$\vdots$	$\vdots$	$\vdots$
N_j		$\mathcal{R}_{N1}$	$\mathcal{R}_{N2}$	$\mathcal{R}_{N3}$	...	$\mathcal{R}_{NJ}$
T_j		T_1	T_2	T_3	...	T_J

T_j is the sum of ranks of sample j, and $N = \sum_{j=1}^{J} N_j$ is the total number of subjects.

If all sample sizes N_j are 5 or less, Table A-9 of the Appendix can be used to determine the probability of the calculated value of H under the null hypothesis. If the probability associated with H is less than α, reject H_0. If the sample sizes are too large for use of this table, then H is approximately distributed as chi square, with $J - 1$ degrees of freedom.

EXAMPLE   In an experimental junior high school, three different methods of teaching art are used, with eight students in each class. At the end of the school year, each student is given the same task — to draw a particular landscape. Two judges each rank order all 24 pictures with regard to creativity, reconciling any differences in rankings to produce a single rank order of all students. The null hypothesis is that the three methods of teaching lead to equal amounts of rated creativity in students; $H_0: \mu_{\mathcal{R}_1} = \mu_{\mathcal{R}_2} = \mu_{\mathcal{R}_3}$ and $H_m: \text{not}(\mu_{\mathcal{R}_1} = \mu_{\mathcal{R}_2} = \mu_{\mathcal{R}_3})$. An α level of .05 is selected. The data are as follows:

Student within method	Method of teaching		
	1	2	3
1	4	12	21
2	1	5	23
3	14	20	13
4	19	3	22
5	10	11	15
6	2	18	24
7	17	7	9
8	6	8	16
T_j	73	84	143
T_j^2	5329	7056	20449

Inserting these values in formula 14-7 gives

$$H = \frac{12}{24(25)} \frac{5329 + 7056 + 20{,}449}{8} - 3(25)$$

$$= \frac{32{,}834}{400} - 75$$

$$= 7.08$$

The critical value of chi square for 2 df is 5.99146 (Table A-7). The observed chi square is greater than the critical value, so the null hypothesis is rejected. We conclude that the three methods of teaching lead to a difference in the rated creativity of students.

C. Correction for Ties

If H is nonsignificant but close to the critical value, and if more than 10 percent of ranks are tied, use the correction given in formula 14-8:

$$C = 1 - \frac{\sum_i (t_i^3 - t_i)}{N^3 - N} \tag{14-8}$$

where t_i is the number of tied observations in a set of tied scores. Then the corrected value of H is

$$H_C = \frac{H}{C} \tag{14-9}$$

D. Prediction and Correlation

In accordance with Section 15-2, it seems most reasonable to predict the median rank on the dependent variable for each level of the independent variable. For the example of this section, we would predict rank 8 for method 1, 10 for method 2, and 18.5 for method 3 (see the first computational procedure for the median, Section 5-6). For this example, the mean ranks (9.13, 10.5, and 17.9) give the same information but are less variable than the medians.

In accordance with the map for Chapter 17, the most reasonable correlation coefficient is theta (Section 17-4). Eta (Section 17-1) could also be useful, but its assumption of an interval-level-dependent variable is almost certainly violated by the conversion to ranks.

References Basic sources are Kruskal (1952) and Kruskal and Wallis (1952). Textbook discussions include those of Dixon and Massey (1969), 344–345; Hays (1963), 637–639; Siegel (1956), 184–192; Walker and Lev (1953), 436–438; and Winer (1962), 622–623.

14-7   The Friedman test

The Friedman test is used to compare the effects of three or more treatments when the dependent variable is ordinal or the centrality measure is the median. It utilizes a subjects-by-treatments design in which a separate ranking of all treatments is obtained for each row of the table. Data in row i may consist of a ranking of a subject's performances or responses under J different treatment conditions, or a ranking of J matched subjects randomly assigned to J different treatment conditions, or a ranking of J stimuli by one subject.

A. Hypotheses

The null hypothesis is that $H_0: \mu_{\mathcal{R}_1} = \mu_{\mathcal{R}_2} = \cdots = \mu_{\mathcal{R}_J}$, that is, that the means of ranks are the same for all treatments (or stimuli). Under certain distribution assumptions, hypotheses about raw-score medians or means can also be tested (see Section 14-6A).

B. Data Analysis

The first step in the analysis is to cast the data into a subject-by-treatments table, with subjects varying from row to row and treatments varying from column to column, as indicated in Table 14-5. The ranking in each row is assumed to be separate and independent from the rankings in all other rows.

Table 14-5   *Data for Friedman analysis*

Subject	Treatment 1	Treatment 2	Treatment 3	$\cdots$	Treatment J	Σ
1	$\mathcal{R}_{11}$	$\mathcal{R}_{12}$	$\mathcal{R}_{13}$	$\cdots$	$\mathcal{R}_{1J}$	T_{1+}
2	$\mathcal{R}_{21}$	$\mathcal{R}_{22}$	$\mathcal{R}_{23}$	$\cdots$	$\mathcal{R}_{2J}$	T_{2+}
3	$\mathcal{R}_{31}$	$\mathcal{R}_{32}$	$\mathcal{R}_{33}$	$\cdots$	$\mathcal{R}_{3J}$	T_{3+}
$\vdots$	$\vdots$	$\vdots$	$\vdots$	$\cdots$	$\vdots$	$\vdots$
N	$\mathcal{R}_{N1}$	$\mathcal{R}_{N2}$	$\mathcal{R}_{N3}$	$\cdots$	$\mathcal{R}_{NJ}$	T_{N+}
Σ	T_{+1}	T_{+2}	T_{+3}		T_{+J}	T_{++}

In the present notation, a general rank in this table will be represented as $\mathcal{R}_{ij}$, where i is a person subscript and j is a treatment subscript. A sum will be represented by T, with a $+$ replacing the subscript summed over. Let us next define two sums of squares (SS):

$$SS_{\text{treatments}} = \sum_j \frac{T_{+j}^2}{N} - \frac{T_{++}^2}{NJ} \qquad (14\text{-}10)$$

$$SS_{\text{within subjects}} = \sum_j \sum_i \mathcal{R}_{ij}^2 - \sum_i \frac{T_{i+}^2}{J} \tag{14-11}$$

You may notice that these are identical in form to formulas 13-30 and 13-29. The hypothesis of equal treatment effects is then tested using the statistic

$$\chi_{\mathcal{R}}^2 = \frac{N(J-1)SS_{\text{tmt}}}{SS_{\text{ws}}} \tag{14-12}$$

and comparing the calculated value to the critical value obtained from the Appendix. If J is 3 or 4 and N's are small, the critical value is found in Table A-13. If either N or J is too large for Table A-13, $\chi_{\mathcal{R}}^2$ is distributed approximately as chi square with $J-1$ df, and the critical value is obtained from Table A-7. In case of ties, the tied cells are given the mean of the ranks they would have received if they were not tied.

When no ties are permitted in the ranking for any row, formula 14-12 reduces to

$$\chi_{\mathcal{R}}^2 = \frac{12}{NJ(J+1)} \sum_j T_{+j}^2 - 3N(J+1) \tag{14-13}$$

EXAMPLE   An experiment on the relationship between birth order and leadership takes as its basic unit a group of four freshman males, all four from different families, matched on age and intelligence. In each group, is an only child, one is first-born, one is second-born with an older brother, and one is second-born with an older sister. The group is given a difficult problem to solve, and judges rank the four subjects on relative amounts of leadership displayed. Six groups of this type are tested in the experiment. The null hypothesis is that the four kinds of subjects will receive equal mean ratings ($\mu_{\mathcal{R}_j}$) on leadership; α is set at .05. The data are as follows:

Group	Only	First	Second, brother	Second, sister	T_{i+}
			Birth order		
1	3	2	1	4	10
2	2	1	4	3	10
3	4	2	1	3	10
4	3	1	2	4	10
5	2	1	3	4	10
6	3	1	2	4	10
T_{+j}	17	8	13	22	60

There are no ties for any group, so we can use the simplified formula 14-13:

$$\chi_{\mathcal{R}}^2 = \frac{12}{6(4)(5)}(17^2 + 8^2 + 13^2 + 22^2) - 3(6)(5)$$

$$= \frac{1}{10}(1006) - 90$$

$$= 10.6$$

We could have used the formula 14-12 to solve this problem, although it does involve more work:

$$SS_{tmt} = \frac{17^2 + 8^2 + 13^2 + 22^2}{6} - \frac{60^2}{24} = \frac{1006}{6} - \frac{3600}{24}$$

$$= 167.667 - 150 = 17.667$$

$$SS_{ws} = (3^2 + 2^2 + 1^2 + 4^2 + 2^2 + \cdots + 4^2) - 6\left(\frac{100}{4}\right)$$

$$= 180 - 150 = 30$$

$$\chi_{\mathcal{R}}^2 = \frac{6(3)(17.667)}{30} = \frac{6(3)(53/3)}{30} = \frac{53}{3} = 10.6$$

The critical value $\chi_{.05}^2$(3 df) $= 7.81$ obtained from Table A-13 is less than the observed value, so the null hypothesis is rejected.

C. Prediction and Correlation

Probably the most reasonable thing to do in this case is to ignore the fact that the data are matched and follow the suggestions of Section 14-6: predict the median rank for each level of the independent variable, and use theta or eta to show degree of relationship between the two variables. Unlike the Kruskal–Wallis, the Friedman test does not require a complete ranking of all scores in the matrix, so the use of eta to show degree of relationship is somewhat more reasonable in this case.

References The basic sources are Friedman (1937, 1940). Textbook discussions include Hays (1963), 640–641; Siegel (1956), 166–172; Walker and Lev (1953), 438–440; and Winer (1962), 136–137.

Homework

1. In 1956 the median income of miners for Northern Maine Platinum, Inc., was $4200 per year. The union contends that salaries have decreased since

that time. In an attempt to test this motivated hypothesis, you randomly sample 16 miners, finding the distribution of salaries shown in the table:

Above $4200	4
$4200	0
Below $4200	12

(a) State your hypotheses formally.
(b) What figure in Section 9-3 best represents this design?
(c) What section of Chapter 14 should you use? What section of Chapter 10 is relevant to this problem? What section of Chapter 8 should be followed to carry out the calculations?
(d) Set α at .05, complete the test, and draw your conclusion.
(e) Suppose that you had used actual salaries of 50 miners in 1956 and today as your data (the same miners), but wanted to compare medians [see Huff (1954), Chapter 2]. What section of Chapters 10 to 14 would you have used?

2. In a study of the effect of birth order on leadership, male sibling pairs are sampled. Subjects are individually tested in a simulated work situation in which they must direct others to perform a group task. The performance of each subject is rated by observers and the results tabulated. Use these results to test the null hypothesis that older and younger brothers are equally good leaders.

Pair	Older brother	Younger brother
1	24	26
2	33	26
3	34	41
4	19	26
5	29	24
6	33	43
7	37	48
8	18	17
9	24	38
10	15	19

(a) What section of Chapter 14 should you use to solve this problem? If you could assume the ratings were interval-level, what section of Chapters 10 through 14 would you use?
(b) Continue with this problem in the light of this chapter, by stating your hypotheses formally.
(c) What figure of Section 9-3 best represents this design? How do you interpret the T's in this figure?

    (d) Let α be .05. What is the critical value of your test statistic?

    (e) What is the observed value of your statistic, and what is your conclusion?

3. In a classroom spelldown, the first student to miss a word gets rank 1, and so on until the last student, who gets rank 20. The results are as follows. Boys: 1, 2, 4, 7, 12, 14, 16, 17, 18, and 20. Girls: 3, 5, 6, 8, 9, 10, 11, 13, 15, and 19. Our null hypothesis is that boys and girls have equal ability on the average.

    (a) What figure from Section 9-3 best represents this design?

    (b) What section of Chapter 14 should you use to solve the problem? If you could assume that the rankings were interval level, what section of Chapters 10 through 14 would you use to complete the analysis?

    (c) Assume that the data are ordinal but that the true underlying distributions of spelling ability are similar in shape for boys and girls. In this case, state the null hypothesis that can be tested.

    (d) What is the critical value of your test statistic, if you let α be .05?

    (e) What is the observed value of your statistic, and what is your conclusion?

4. A high school principal has decided to use student ratings of faculty as an aid in making personnel decisions. Next year the mathematics department must be reduced by one member. To help decide which of the two newest teachers to retain, the principal polls all students who have taken courses from both, asking "From whom did you learn more?" The results are as follows: Ms. Stevenson, 12; Ms. Naturale, 2.

    (a) Which section of this chapter can be used to solve the problem?

    (b) State your hypotheses formally.

    (c) What section of Chapter 8 should be used to carry out the calculations?

    (d) Set α at .01, complete the test, and draw your conclusions.

5. As part of a study of consumer reaction to cigarette commercials and the Surgeon General's Report on smoking, subjects were asked to rank four cigarettes on the amount of cancer-producing materials they probably contained. Judging was done blindfolded, and subjects were required to puff on all four in the order ABCD for the first puff. After that, they could sample the cigarettes in any order they desired. The results were as follows:

	Brand			
i	A	B	C	D
1	4	2	1	3
2	3	2	1	4
3	4	2	1	3
4	1	2	4	3
5	1	2	3	4
6	2	1	3	4

You wish to determine if there are differences in average perceived amount of cancer-producing materials.

(a) What section of this chapter should you follow in solving this problem?

(b) State your hypotheses formally.

(c) What is the critical value of your test statistic, if α is set at .05?

(d) What is the observed value of your statistic, and what is your conclusion?

(e) Could you have used one of the sections of Chapter 13 to solve this problem? If yes, which one, and what additional assumptions would you have had to make? If no, why not?

6. The director of a dance studio has developed a rating system for evaluating students' performances in interpretive dancing, on a scale from 0 to 100. Each student is assigned to one of three teachers at the beginning of the year, on a random basis. At the end of the year, students are evaluated individually, using the director's system, by outside experts who do not know which teacher any student has had. The results of the evaluation are as follows:

A	B	C
19	62	20
21	69	36
18	91	22
93	92	35
37	48	45

A preliminary plot of scores shows that they depart severely from normality, and that the three groups differ in variability. You wish to know whether the three teachers are equally effective.

(a) Which section of this chapter should be followed in solving this problem?

(b) State your hypotheses formally.

(c) What figure from Section 9-3 best represents this design?

(d) Let α be .05; What is your critical value?

(e) What is the observed value of your chosen statistic, and what is your conclusion?

7. In a small high school all students take courses from the same four teachers. Five students selected at random are asked to rate the four teachers on teaching ability, where E = excellent, VG = very good, G = good, and P = poor. The results are as follows:

	Teacher			
Student	A	B	C	D
1	VG	P	E	G
2	E	P	VG	G
3	VG	G	E	P
4	G	P	VG	E
5	E	P	VG	G

You wish to test the hypothesis that all four teachers are considered to be equally good by their students.

(a) Which section of this chapter can be used to solve the problem?

(b) State your hypotheses formally.

(c) Let α be .05. What is the critical value of your test statistic?

(d) What is the observed value of your statistic, and what is your conclusion?

Homework Answers

1. (a) $H_0 : \pi \geq .5$; $H_m : \pi < .5$, where π is the proportion of miners earning more than \$4200

   (b) Figure 9-1

   (c) Section 14-1A; Section 10-9; Section 8-7

   (d) The probability of 4 or fewer above \$4200 is .0401 $< .05$, so reject H_0.

   (e) Section 14-5

2. (a) Section 14-5; Section 12-9

   (b) $H_0 : P\{X(\text{older}) > X(\text{younger})\} = .5$ and $H_m : P\{X(\text{older}) > X(\text{younger})\} \neq .5$

   (c) Figure 9-5; T's represent order of birth, not distinguishable treatments within the experiment itself.

   (d) $T_{.05} = 8$

   (e) $T_s = 11$; accept H_0.

3. (a) Figure 9-6

   (b) Section 14-3; Section 12-11

   (c) $H_0 : \mu_B = \mu_G$, $H_m : \mu_B \neq \mu_G$

   (d) 23 for N's of 10 and 10

   (e) 44; accept H_0.

4. (a) Section 14-4

   (b) $H_0 : P\{\text{Stevenson preferred to Naturale}\} = .50$; $H_m : P\{\text{Stevenson preferred to Naturale}\} \neq .50$

   (c) Section 8-8

   (d) The probability is about .0164 $> .01$ or .0082 $> .005$, so accept H_0 : neither is preferred.

5. (a) Section 14-7

(b) $H_0: \mu_{\mathcal{R}_1} = \mu_{\mathcal{R}_2} = \mu_{\mathcal{R}_3} = \mu_{\mathcal{R}_4}$; H_m: not $(\mu_{\mathcal{R}_1} = \mu_{\mathcal{R}_2} = \mu_{\mathcal{R}_3} = \mu_{\mathcal{R}_4})$

(c) 7.81473

(d) 5.60, do not reject H_0.

(e) See Section 13-6. No: it is unreasonable to assume that all subjects see all treatments as equally spaced.

6. (a) Section 14-6

(b) $H_0: \mu_{\mathcal{R}_1} = \mu_{\mathcal{R}_2} = \mu_{\mathcal{R}_3}$, H_m: not $(\mu_{\mathcal{R}_1} = \mu_{\mathcal{R}_2} = \mu_{\mathcal{R}_3})$

(c) Figure 9-7

(d) 5.99146

(e) 6.0; reject H_0.

7. (a) Section 14-7

(b) $H_0: \mu_{\mathcal{R}_1} = \mu_{\mathcal{R}_2} = \mu_{\mathcal{R}_3} = \mu_{\mathcal{R}_4}$, H_m: not $(\mu_{\mathcal{R}_1} = \mu_{\mathcal{R}_2} = \mu_{\mathcal{R}_3} = \mu_{\mathcal{R}_4})$

(c) 7.81473

(d) 9.24; reject H_0.

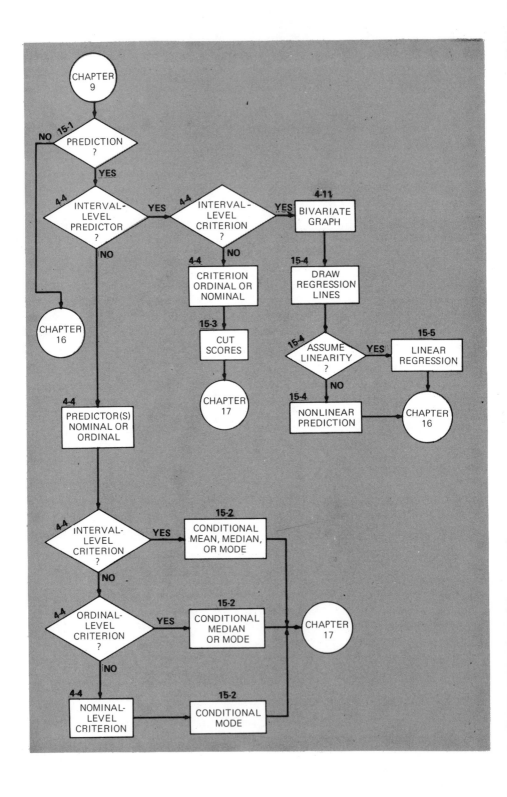

15 Prediction

THIS is the first of three chapters to be used in determining the form and magnitude of relationships among variables. In this chapter we examine the form of relationship between two variables: what value of the variable Y you should predict for a subject who has a particular score on another variable, X. In Chapters 16 and 17 we determine how accurate our predictions are.

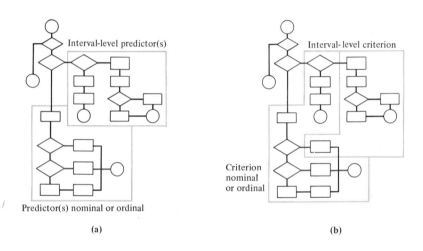

Interval-level predictor(s)

Predictor(s) nominal or ordinal

(a)

Interval-level criterion

Criterion nominal or ordinal

(b)

The map for this chapter can be divided into conceptual regions on the basis of the level of measurement of either the predictor(s) (see the first regional map) or the criterion (the variable predicted; see the second regional map).

15-1 The general problem of prediction

Prediction, as used in this text, will be defined as the use of all available information to guess a subject's status on a variable. The variable on which we are predicting scores will be called the *criterion*. Prediction as used here will not necessarily imply that we are predicting a future status on the criterion. It simply means that we are making a guess about an unknown value, which in some cases may be unknown because it occurs in the future.

This chapter will proceed on the assumption that you wish to make the most accurate predictions possible, although in many actual cases you may be willing to sacrifice accuracy for savings in time and expense. The accuracy with which you are able to predict criterion scores depends on the nature of your criterion and the amount of information that you have about it.

If you know nothing at all about the criterion, then you might as well guess randomly; but such is seldom the case. In Chapter 5, it was indicated that a measure of centrality is a good first characterization of a distribution. If you know something about the criterion distribution, and nothing about the relationships of the criterion to other variables, then some measure of centrality may be your best guess of the score of any individual on that criterion. For a choice of centrality measure, you should refer to Chapter 5.

If you are able to specify a relationship between the criterion and some other variable, that second variable is a *predictor*. In this chapter, we will consider methods of formulating relationships between two variables so that you can use knowledge of the nature of the variables, the relationship between them, and an individual's status on one to predict his status on the other. The methods are general in the sense that they allow a prediction of criterion status for each of the individuals on whom they are developed. If you construct a table for the prediction of Y scores from X scores on a sample of 100 subjects, the table gives optimal guesses of Y status for all 100 subjects in that sample. Also, if the subjects are randomly drawn from some population, then the prediction table can be used with any other subject randomly drawn from the same population.

There are three ways of evaluating the predictions suggested in this chapter. (1) You can determine how accurate the predicted scores are with respect to actual criterion status of subjects by finding the variability of observed criterion scores about the predicted values. (2) You can determine the extent to which knowledge of the predictor and the relationship between predictor and criterion increases your accuracy of prediction. (3) You can see whether the prediction formula is generalizable to the population from which your subject sample is drawn.

The first of these methods of evaluating predictions utilizes a measure of *variability of estimate*, just as a measure of variability indicates the representativeness of a centrality measure. Measures of variability of estimate will be discussed in conjunction with each of the prediction methods of this chapter. The second method of evaluation is the degree of relationship or correlation between two variables; it is elaborated in Chapters 16 and 17. The third relates to the significance of a relationship, which is discussed in conjunction with each of the correlation methods of Chapters 16 and 17.

15-2 Prediction table

If your predictor variable is nominal or ordinal, no general formula can be written relating the predictor and criterion variables. The best you can do is to make up a table that shows, for each observed score on the predictor, the score to expect on the criterion.

A. Nominal-Level Criterion

Relevant earlier sections:
5-5 Mode
5-9 Variation ratio

If the criterion is nominal, then for each level of the predictor, predict the conditional mode—the mode on the criterion for all subjects *at that level of the predictor*. Then your best indicator of the variability of estimate for each category of the predictor is the variation ratio for the criterion scores of subjects in that predictor category. Thus, for subjects in level *j* on the predictor *X*, the conditional variation ratio is given by

$$v_j = 1 - \frac{f_{Mo(Y_j)}}{N_j} \tag{15-1}$$

where $Mo(Y_j)$ is the modal *Y* score for those subjects who are in the *j*th level of *X*, and N_j is the number of subjects at level *j* of *X*.

An overall estimate of accuracy of predictions might be called the "variation ratio of estimate" and might be defined as the proportion of all criterion values that are at their respective conditional modes:

$$v_{YX} = 1 - \frac{\sum_j f_{Mo(Y_j)}}{N} \tag{15-2}$$

where *N* is the total number of subjects ($N = \sum_j N_j$). In this case, your best indicator of degree of relationship between predictor and criterion is lambda (Section 17-3) if the predictor is nominal and theta (Section 17-4) if it is ordinal.

EXAMPLE   At some time in the distant past of the Kingdom of Bortland, a series of genetic accidents occurred, as a result of which a substantial proportion of the population is now multiheaded. As part of a sociological survey you randomly sample adults from each of the five regions of the country, counting and reporting the heads of each person. The data are shown in the table.

| Region | Number of heads | | | | |
	One	Two	Three or more	None	Σ
North	360	40	0	0	400
South	190	20	40	50	300
Central	125	5	0	0	130
East	4	130	0	6	140
West	21	5	0	4	30
Σ	700	200	40	60	1000

Cell entries are frequencies. Both variables are nominal. Therefore, a table for predicting region from number of heads would predict modes. In the following table, calculation of conditional variation ratios is also shown:

Number of heads	Predicted region	Conditional variation ratio
One	North	$v = 1 - 360/700 = 340/700 \doteq 0.486$
Two	East	$v = 1 - 130/200 = 70/200 \doteq 0.350$
Three or more	South	$v = 1 - 40/40 = 0$
None	South	$v = 1 - 50/60 = \doteq 0.167$

Here it appears that predictions are not equally accurate for all genetic groups. They are perfectly accurate for the sample of three or more heads but accurate for only 65 percent of two-headed subjects. In this case, an overall measure of prediction accuracy is of doubtful value, but let us calculate it anyway:

$$v_{\text{region, no heads}} = 1 - \frac{360 + 130 + 40 + 50}{1000}$$

$$= 1 - \frac{580}{1000} = \frac{420}{1000}$$

$$= .42$$

B. Ordinal-Level Criterion

If the criterion is ordinal, then for each level of the predictor you can use either the criterion median or mode for subjects at that predictor level as your prediction (see Section 5-6 for a discussion of the choice between these two centrality measures). If you choose to use the conditional modes as your predicted values, refer to part A for further analyses. If you predict the conditional medians, then the measure of variability of estimate to use is an interquantile range for each predictor level, say the conditional semi-interquartile range (Q; see Section 5-10). For some level j of the predictor X,

$$Q_j = \tfrac{1}{2}(Q_{3j} - Q_{1j}) \tag{15-3}$$

where Q_{3j} is the third quartile on the criterion Y of all subjects at level j of X; and Q_{1j} is the first quartile on the criterion Y of all subjects at level j of X.

One possible measure of overall variability of estimate is the median of the semi-quartile ranges:

$$Q_{YX} = Md(Q_j) \tag{15-4}$$

Alternatively, we might transform all levels of the independent variable so that they have equal medians on the dependent variable, then find Q of the combined dependent variable scores. The best estimator of degree of relationship in this case is theta (Section 17-4) if the predictor is nominal or tau (Section 17-2) if it is ordinal.

EXAMPLE Suppose that our predictor is social class of subjects, which is divided into three ordinal levels: upper, middle, and lower. Our criterion is a 15-item test of attitude toward public support of parochial schools. Suppose that the criterion measure is a research instrument and that we are only willing to consider the criterion scores to be at ordinal level. The data are as follows:

	X = social class		
i	Upper	Middle	Lower
1	1	2	1
2	1	4	3
3	1	4	4
4	3	5	5
5	3	7	5
6	4	8	6
7	4	9	6
8	5	10	7
9	5	10	7
10	6	11	9
11	9	14	12
12	13	15	14

In this table, the 12 subjects in each level of the predictor have been listed so that their criterion (Y) scores are in increasing magnitude, for ease of computation. We can now construct a second table of results:

	Social class		
	Upper	Middle	Lower
Prediction	4	8.5	6
Q_1	2	4.5	4.5
Q_3	5.5	10.5	8
Q	1.75	3	1.75

These data indicate that the three groups of subjects differ on their median scores on the test. The middle-class subjects are most favorable to public support of parochial schools. The middle-class sample is also the most variable; it has a semi-interquartile range of 3. As an overall indicator of the variability of predicted scores, we can take the median of the semi-interquartile ranges, which is 1.75.

If we transform the attitude scores for upper- and lower-class respondents linearly so their medians are equal to 8.5 (the median for middle-class subjects), we get the left-hand table:

i	Upper $Y + 4.5$	Middle	Lower $Y + 2.5$	Score	Frequency
				2	1
1	5.5	2	3.5	3.5	1
2	5.5	4	5.5	4	2
3	5.5	4	6.5	5	1
4	7.5	5	7.5	5.5	4
5	7.5	7	7.5	6.5	1
6	8.5	8	8.5	7	1
7	8.5	9	8.5	7.5	4
8	9.5	10	9.5	8	1
9	9.5	10	9.5	8.5	4
10	10.5	11	11.5	9	1
11	13.5	14	14.5	9.5	4
12	17.5	15	16.5	10	2
				10.5	1
Md	8.5	8.5	8.5	11	1
				11.5	1
				13.5	1
				14	1
				14.5	1
				15	1
				16.5	1
				17.5	1

The combined frequency distribution can then be obtained, as shown in the right-hand portion of the table. For this distribution,

$$Q_Y = \frac{10.25 - 6.0}{2} = \frac{4.25}{2} = 2.125$$

C. Interval-Level Criterion

Relevant earlier sections:
5-7   Mean
5-11   Variance

If the criterion is interval, the preferred predicted value is usually the criterion mean for each level of the predictor, although it is sometimes better to use the conditional median or mode (for discussion and comparison of these three measures, see Section 5-7). If you predict medians, see part B for further analyses; if you predict modes, see part A. If you predict means, your measure of variability of estimate for each level j of the predictor X is the variance of Y scores for all subjects at that level of X, as indicated in formula 15-5:

$$\sigma_j^2 = \frac{\sum_i (Y_{ij} - \mu_{Y_j})^2}{N_j} \tag{15-5}$$

where

Y_{ij} = score on criterion Y of subject i who is at level j of the predictor X
μ_{Y_j} = mean of Y scores for all subjects at level j of X
N_j = number of subjects at level j of X

Notice that these values are population parameters. Use of parameters in these formulas is standard with populations and large samples. When small samples of subjects are used, replace the population values of μ_j and $\sigma_{Y_j}^2$ by $\overline{X}_j$ and $\hat{s}_j^2$, the unbiased variance estimate.

To find an overall measure of the error left after prediction, you can average conditional variances with the formula

$$\sigma_{YX}^2 = \frac{\sum_j N_j \sigma_j^2}{\sum_j N_j} = \frac{\sum_j \sum_i y_{ij}^2}{\sum_j N_j} = \frac{\sum_j \left[\sum_i Y_{ij}^2 - \frac{1}{N} \left(\sum_i Y_{ij} \right)^2 \right]}{\sum_j N_j} \tag{15-6}$$

The square root of this variability measure, σ_{YX}, is called the *standard error of estimate* for nonlinear prediction. If the predictor is nominal, the best estimate of relationship is eta (Section 17-1). If the predictor is ordinal, you can do one of two things; either treat it as a nominal scale and use eta, or assume an underlying interval scale and normal distribution, and linear regression between the underlying predictor and the criterion, then use the multiserial coefficient (Section 16-4).

When small samples are used, replace N_j by $N_j - 1$ and σ_j^2 by $\hat{s}_j^2$ in formula 15-6. This gives

$$\hat{s}_{yx}^2 = \frac{\sum_j (N_j - 1)\hat{s}_j^2}{\sum_j (N_j - 1)} = \frac{\sum_j \left[\sum_i Y_{ij}^2 - \frac{1}{N} \left(\sum_i Y_{ij} \right)^2 \right]}{\sum_j (N_j - 1)}$$

This is algebraically identical to formula 13-1 and will be referred to by that number.

EXAMPLE WITH LARGE SAMPLES   Sixth-grade boys in a small city are asked, "If you are drawing a poster for the school's spring carnival, and if you are going to be graded on the poster, would you rather work together with a boy or a girl?" Responses (X) were compared to scores on a measure of "Heterosexual interests" (HI). Results were as follows:

Response	N_j	σ_j^2	$N_j\sigma_j^2$	Prediction: μ_{HI_j}
Boy	100	5.00	500	22.0
Girl	180	4.00	720	20.16
Can't decide	120	5.20	624	19.26
Σ	400		1844	

Then,

$$\sigma^2_{HI \text{ from } X} = \frac{1844}{400} = 4.61$$

$$\sigma_{HI \text{ from } X} \doteq 2.15$$

EXAMPLE WITH SMALL SAMPLES   We are predicting status on an interval-level criterion Y from a nominal-level predictor X. The data are as follows:

X_j	Y_{ij}	N_j
1	9, 3, 1, 3	4
2	9, 4, 7, 5, 5	5
3	13, 12, 8, 9, 9, 9	6
4	7, 2, 2, 5	4
5	15, 7, 7, 12, 9	5

For level 1 of X, the predicted value (mean) is

$$\overline{Y}_1 = \frac{\sum_i Y_{i1}}{N_1} = \frac{9 + 3 + 1 + 3}{4} = \frac{16}{4} = 4$$

and the conditional variance is

$$\hat{s}_1^2 = \frac{N \sum Y^2 - (\sum Y)^2}{N(N - 1)} = \frac{4(81 + 9 + 1 + 9) - (16)^2}{12}$$

$$= \frac{400 - 256}{12} = \frac{144}{12}$$

$$= 12$$

The remaining conditional means and variances are as follows:

X_j	$\bar{Y}_j$	$\hat{s}^2_{Y_j}$	$\hat{s}_{Y_j}$
1	4	12	3.46
2	6	4	2
3	10	4	2
4	4	6	2.24
5	10	12	3.46

There are some differences in the $\hat{s}^2_{Y_j}$'s, but perhaps not enough to preclude pooling, depending on the use to which the predictions will be put. The variance of estimate, if desired, can be calculated from the conditional variances and sample sizes.

$$\hat{s}^2_{YX} = \frac{(N_1 - 1)\hat{s}^2_1 + (N_2 - 1)\hat{s}^2_2 + (N_3 - 1)\hat{s}^2_3 + (N_4 - 1)\hat{s}^2_4 + (N_5 - 1)\hat{s}^2_5}{N_1 + N_2 + N_3 + N_4 + N_5 - 5}$$

$$= \frac{36 + 16 + 20 + 18 + 48}{19}$$

$$= \frac{138}{19} \doteq 7.263$$

Then the standard error of estimate is the square root of this term:

$$\hat{s}_{YX} = \sqrt{7.263}$$

$$\doteq 2.695$$

The raw-score formula 15-6 looks like more work here, but that is because the calculations of $\bar{Y}_j$ and $\hat{s}^2_{Y_j}$ were not shown in the solution above. We construct the following table:

X_j	N_j	$\sum_i Y_{ij}$	$\sum_i Y^2_{ij}$	$\sum_i (Y_{ij})^2$	$\frac{1}{N_j} \sum_i (Y_{ij})^2$
1	4	16	100	256	64
2	5	30	196	900	180
3	6	60	620	3600	600
4	4	16	82	256	64
5	5	50	548	2500	500
$\sum$	24	172	1546		1408

Then we can use the raw-score formula (15-6) to calculate the standard error of estimate:

$$\hat{s}_{YX} = \sqrt{\frac{1546 - 1408}{19}} = \sqrt{\frac{138}{19}} = 2.695$$

as found above.

15-3   Predicting nominal-level criterion status from an interval-level predictor

In a wide variety of situations, your predictor may be interval level but your criterion dichotomous (two-valued), especially when a decision must be made on the basis of some input information. Examples occur in judging whether a student will complete college in five years or less on the basis of his incoming test score; in predicting political affiliation from income; or in obtaining an initial diagnosis of a client as normal or schizophrenic on the basis of a person-ality-test score. In each case, the objective is to find some predictor score X_c (called a cut score) so that you will be willing to say that any subject who scores higher than X_c should be considered in group I on the criterion, and any subject who scores lower than X_c should be considered in group II on the criterion.

Let us suppose that you wish to make as few errors of prediction as possible (for some applied problems, this may not be true, but instead, your value of X_c may depend on supply and demand or other considerations). Then the first step is to draw samples from the two criterion classes proportional in size to the relative sizes of their respective populations. For these two samples, plot their predictor scores as superimposed frequency polygons (see Section 4-11) as in Figure 15-1. X_c is the value along the predictor line at which the two polygons intersect because they have equal ordinates.

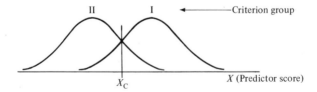

Figure 15-1   *Obtaining a cut score (X_c) which optimally separates two criterion groups.*

There are two kinds of errors possible here. Some members of group I may be classified in group II because their predictor scores fall below X_c; similarly, some members of group II may be incorrectly judged to be in group I, because their predictor scores fall above X_c. In general such errors cannot be avoided, but the choice of the intersection of the two polygons as cut score minimizes the total number of misclassifications. To see this, imagine moving the cut score a short distance to the left in Figure 15-1. A smaller portion of distribution I would be to the left of the cut score; hence, fewer members of group I would be misclassified. However, this decrease would be more than compensated for by the increase in misclassifications of members of group II (who would be to the right of the new cut score). Therefore, there would be a net increase in misclassifications. Similar results would follow after moving the cut score to the right, again yielding a net increase in misclassifications.

If you are willing to assume that both distributions have roughly the same shape and are symmetric, formula 15-7 may be used to find the cut score:

$$X_c = \overline{X}.. + \frac{.50 - p_{\mathrm{I}}}{p_{\mathrm{I}} p_{\mathrm{II}}} \cdot \frac{\hat{s}_x^2}{\overline{X}_{\mathrm{I}} - \overline{X}_{\mathrm{II}}} \tag{15-7}$$

where

$\overline{X}..$ = grand mean of all scores in the two distributions $\overline{X}_{\mathrm{I}} > \overline{X}_{\mathrm{II}}$
p_{I} = proportion of subjects in distribution I: $p_{\mathrm{I}} = N_{\mathrm{I}}/(N_{\mathrm{I}} + N_{\mathrm{II}})$
p_{II} = proportion of subjects in distribution II: $p_{\mathrm{II}} = N_{\mathrm{II}}/(N_{\mathrm{I}} + N_{\mathrm{II}})$
    $= 1 - p_{\mathrm{I}}$
$\hat{s}_x^2$ = variance of all scores in the two distributions taken about the grand mean $\overline{X}..$

When there are three or more levels to the criterion variable, a cut score must be obtained for each adjacent pair of criterion distributions (see Figure 15-2). If you wish to infer from the criterion categories to a hypothesized, continuous, underlying criterion variable, a different cut-score formula must be used [see Guilford (1965), 385].

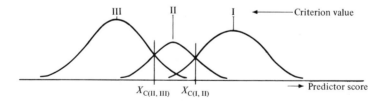

Figure 15-2   *Predicting a three-level nominal criterion from an interval-level predictor.*

EXAMPLE   The developer of the Psychic Status Scale, a 20-item self-report test, claims that it can clearly separate scores for normals, neurotics, and psychotics. In an attempt to calibrate the test as an initial screening device, you administer it to five normals, five neurotics, and four diagnosed psychotic patients. Results are as follows:

Subject	Diagnosis		
	Normal	Neurotic	Psychotic
1	6	6	9
2	2	6	7
3	6	9	13
4	2	10	11
5	4	4	—
$\sum X$	20	35	40
$\sum X^2$	96	269	420
N	5	5	4
$\overline{X}$	4	7	10

For the conclusion between neurotics and normals (the two lowest adjacent groups),

$$\overline{X} = \frac{20 + 35}{10} = 5.5$$

$.50 - p(\text{neurotic}) = 0$, so the second term of the formula for the cut score goes to zero, and

$$X_c = \overline{\overline{X}}.. = 5.5$$

For the cut score between the next two adjacent groups, neurotics and psychotics,

$$\overline{X} = \frac{35 + 40}{9} = \frac{75}{9} = 8.33$$

$$\hat{s}_x^2 = \frac{9(689) - (75)^2}{9(8)} = \frac{6201 - 5625}{72} = 8.0$$

$$X_c = 8.33 + \left(\frac{.50 - 4/9}{(5/9)(4/9)}\right)\left(\frac{8}{10 - 7}\right)$$

$$= 8.33 + .60 \doteq 8.93$$

Of course, these results will be unstable because of the small sample size, and it is really not possible to tell from a plot of the distributions whether the assumptions of the method are met. The results can be displayed in a prediction table:

Test score	Initial diagnosis
0–5.5	Normal
5.5–8.93	Neurotic
8.93–20	Psychotic

To evaluate the effectiveness of prediction, use phi (Section 16-5) if the criterion has only two levels, or lambda (Section 17-3) if there are more than two criterion values (as in the example above).

Reference   Another discussion of this problem can be found in Guilford (1965), 380–385.

15-4   Type of relationship assumed between interval-level variables

Relevant earlier section:
4-10   Two-dimensional plot

When we predict scores on one interval-level variable from scores on another interval-level variable, three options are available to us: linear regression, curvilinear regression, and nonlinear prediction. In linear regression, we find the equation of the straight line which best allows us to predict scores on the criterion, as indicated in Figure 15-3.

In curvilinear regression, we seek the equation of the best-fitting curved line for predicting criterion scores, as shown in Figure 15-4.

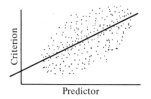

Figure 15-3   *Graph showing linear regression*

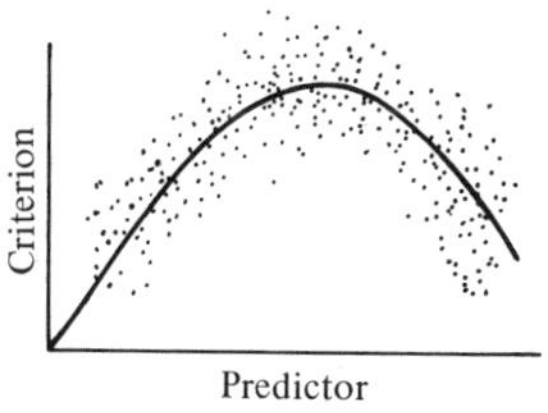

Figure 15-4   *Graph showing curvilinear regression.*

In nonlinear prediction, we predict the conditional mean of criterion scores for each level of the predictor. No equation is obtained, and the data are treated as though the predictor were nominal.

Any of several a priori considerations may enter into our decision about the kind of prediction we wish to make. These include:

1. For two interval-level variables, the best first guess of the form of relationship is generally a straight line, especially if both variables are roughly normally distributed.
2. Calculation of a straight-line relationship is generally easier to carry out than calculation of a curvilinear relationship.
3. On the other hand, there may be a priori reasons for expecting regression to be curvilinear, especially if it has been found to be curvilinear in previous research or there is a theoretical reason to expect curvilinearity.
4. You can't use the methods shown in this chapter with two or more predictors unless you can specify, in advance, the formula for combining predictors. In Figure 15-4, this means that you can see how well a particular line fits the data points, but you can't find the formula from the data. To find an equation that involves more than one predictor, you need to use multiple regression and correlation, which is discussed in Chapter 18.

Once the data have been gathered, a bivariate plot of scores can be useful in determining which kind of prediction will be most informative. There are methods for testing both for specific curvilinear relationships and for significant but unspecified nonlinearity (see Section 17-1).

EXAMPLE   Consider the data of Table 4-10, which are reproduced in Table 15-1 in a slightly different form. Let us plot the means for these data to determine whether regression should be considered linear. The test score X is the predictor and course grade Y is the criterion.

The conditional means are plotted in Figure 15-5 (the bivariate plot of raw scores appears in Figure 4-10). Lines have been drawn connecting criterion means of adjacent predictor values, and a straight line has been fitted visually to the points.

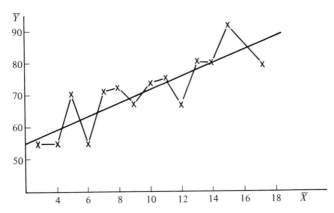

Figure 15-5   *Conditional means for the data of Table 15-1.*

It appears from this plot that there is a great deal of variability of criterion conditional means about the straight regression line, but no systematic curvilinear trends are obvious in these data. Therefore, linear regression appears preferrable to curvilinear regression. If we should decide to use nonlinear prediction, we would predict the conditional criterion mean $\overline{Y}_j$ for each predictor level X_j.

Table 15-1 *Data for predicting course grades from test scores*

Test score X_j	Course grade Y_{ij}	N_j	$\overline{Y}_j$
17	80	1	80
15	85, 95, 95	3	91.67
14	75, 85, 80	3	80
13	85, 70, 85	3	80
12	60, 70, 70	3	66.67
11	70, 90, 60, 80	4	75
10	70, 80, 70, 75	4	73.75
9	60, 70, 70, 55, 65, 70, 80	7	76.14
8	70, 75	2	72.5
7	60, 75, 70, 80	4	71.25
6	55	1	55
5	65, 65, 80	3	70
4	55	1	55
3	55	1	55

EXAMPLE   By contrast, consider Figure 15-6, in which the criterion, amplitude of galvanic skin response, is not just a linear function of number of reinforcement repetitions [Hull (1943), 103].

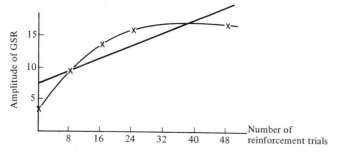

Figure 15-6 *Curvilinear relationship of GSR amplitude to number of reinforcement trials.* [From *Principles of Behavior* by Clark L. Hull. Copyright, 1943. Reproduced by permission of Appleton-Century-Crofts, Educational Division, Meredith Corporation. (From data published by Hovland.)]

In Figure 15-6 you can see that the data points differ consistently from linearity, starting with a steep slope but eventually approaching the horizontal. In this case, it would seem reasonable to look for a curvilinear-regression equation.

With the first of the examples above, you may feel compelled, on a priori considerations, to treat the data as curvilinear. This is prefectly legitimate, but it is also good to be aware that the magnitude of the systematic curvilinear trend in the data is apparently small. It is likewise legitimate to treat the data of the second example as though it were linear; however, to do so would be both to lose information in the data and to go against current knowledge of learning behavior, which leads us to expect curvilinear results from such data.

A regression equation is a formula for predicting some criterion variable, say Y, from one or more predictor variables, say X and Z.

The general form of a linear equation is given by

$$Y' = a + bX \qquad\qquad (15\text{-}8)$$

where a and b are numbers to be determined from the analysis. Properly obtained, these numbers give us the equation of a straight line which best fits our data. There are many possible formulas for curvilinear regression, including

$$Y' = a + bX + cX^2$$
$$Y' = a + bX + cX^2 + dX^3$$
$$Y' = a + b \log X$$

and so on, where in each case optimal values of the coefficients a, b, c, etc., are obtained from an analysis of our data. Our problem in regression analysis is first to choose a general form of equation and, second, to find optimal values for the necessary coefficients.

Once we have a regression equation in hand, then for any subject to whom it applies, we can use our knowledge of the subject's scores on the predictors to make an optimal prediction of his status on the criterion. If our regression equation is $Y' = 5 + 3X$ and some subject has a score of 6 on variable X, then his predicted score on Y is $Y' = 5 + 3(6) = 23$. If our equation is $Y' = X + X^2 - 16$ and a subject gets a score of 12 on X, then his predicted Y score is $Y' = 12 + 12^2 - 16 = 140$.

References Formulas for curvilinear regression are discussed in Ezekiel and Fox (1966). Chapter 6. McNemar (1969), 311–317, provides an analysis-of-variance test for curvilinearity. Before you look at these references, read Section 15-5.

15-5 Linear regression

A. Regression Line
We have perfect linear prediction from some variable X to a second variable Y, when all observed pairs of scores define points in a two-dimensional space

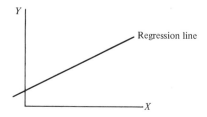

Figure 15-7 *Perfect linear regression of Y on X.*

that fall on a single straight line, as in Figure 15-7. Any such straight line that represents the relationship between Y and X can be defined by

$$Y' = a_{YX} + b_{YX}X \tag{15-9}$$

where Y' is the predicted value of Y for a given value of X, a_Y is the Y intercept (the value of Y at which the regression line crosses the Y axis), and b_{YX} is the slope of the regression line for predicting Y from X.

We can determine the slope of a given regression line by drawing a triangle under the line that has one side parallel to the X axis and the other parallel to the Y axis, as in Figure 15-8. The slope b_{YX} for predicting Y from X is defined by

$$b_{YX} = \frac{\text{rise}}{\text{run}} = \frac{\text{height}}{\text{length}} = \frac{\text{change in } Y}{\text{change in } X} = \text{change in } Y \text{ per unit change in } X \tag{15-10}$$

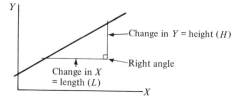

Figure 15-8 *Determining the slope of a regression line.*

Notice, from the formula for the regression line, that when $X = 0$, $Y' = a_{YX}$, the value of the Y intercept. This part of the equation takes into account the fact that Y' need not be zero when X is zero. An X score can be thought of as a change in X from zero. Therefore, a Y' score can be considered to be equal to the value of Y' when X is equal to zero plus the amount of change in Y' which corresponds to a given change in X from zero. In Figure 15-9, a value Y'_1 is shown which corresponds to a particular value X_1 of X. Because the opposite sides of rectangles are equal, it is clear from the figure that $Y'_1 = a_{YX} + H$ and that $L = X_1$. Line segments L and H form a triangle with the regression line and are parallel to the X and Y axes, respectively, so this triangle can be used to define the slope b_{YX} of the regression line: $b_{YX} = H/L$, and because $L = X_1$, $b_{YX} = H/X_1$.

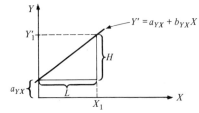

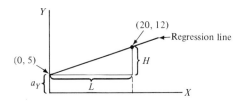

Figure 15-9 *Geometrical interpretation of a regression equation.*

From the regression formula, we know that $Y' = a_{YX} + b_{YX}X$; and substitution of the particular value X_1 into the general formula gives $Y'_1 = a_{YX}X_1$. If we replace b_{YX} by its equivalent, we have $Y'_1 = a_{YX} + (H/X_1)X_1 = a_{YX} + H$, as we already determined by inspection of Figure 15-9. Therefore, the formulas $Y' = a_{YX} + b_{YX}X$ and $b_{YX} = H/L$ adequately represent the relationships shown in the figure.

EXAMPLE   What is the formula for the line defined by the points $(X_1 = 0, Y_1 = 5)$ and $(X_2 = 20, Y_2 = 12)$?

ANSWER: The following figure shows the points and the line that passes through them:

The Y intercept, a_Y, is defined as the value of Y when $X = 0$, which in this case is known to be 5. The slope is given by the ratio of H to L, where in this case

$$H = Y_2 - Y_1 = (12 - 5) = 7$$

and

$$L = X_2 - X_1 = (20 - 0) = 20$$

Thus,

$$b_{YX} = \frac{H}{L} = \frac{7}{20} = .35 \quad \text{and} \quad Y' = a_{YX} + b_{YX}X = 5 + .35X$$

EXAMPLE   When neither point that defines the regression line is on the Y axis, we must do a little more algebra to obtain the regression formula. What is the regression equation for the line passing through the points (5, 12) and (9, 22)? Here we can begin by finding the slope:

$$H = (Y_2 - Y_1) = (22 - 12) = 10$$

and

$$L = (X_2 - X_1) = (9 - 5) = 4$$

so

$$b_{YX} = \frac{H}{L} = \frac{10}{4} = 2.5$$

This regression equation holds for all points along it and is known to pass through the two points specified. Therefore, we can substitute values for either point (X_1, Y_1) or (X_2, Y_2) into the regression formula to solve for a_y.

$$Y_1' = a_{YX} + b_{YX}X \qquad \text{and} \qquad 12 = a_{YX} + (2.5)(5) = a_Y + 12.5$$

so

$$a_{YX} = 12 - 12.5 = -.5$$

The final regression equation is therefore $Y' = -.5 + 2.5X$, or $Y' = 2.5X - .5$.

B. Regression Line When Prediction Is Not Perfect

When prediction is not perfect, it means that all points in the bivariate plot of X and Y do not fall on the same straight line. In this case, we proceed by finding the line that gives the most accurate predictions.

When all we know about a subject is that his score belongs in a particular distribution of scores on an interval-level criterion Y, then our best guess of his criterion status is the criterion mean $\overline{Y}$. When we have data on the relationship between some predictor X and the criterion Y, then for a subject who has predictor score X_j, our best guess of his criterion status is the conditional mean of all subjects who have that same predictor score: $\overline{Y}_j$ (see Section 15-2). This is illustrated in Figure 15-10.

Figure 15-10 *Bivariate graph showing distributions of Y scores for three levels of X, conditional Y means, and the regression line for predicting Y from X.*

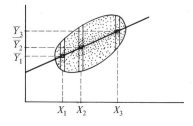

In Figure 15-10 each vertical ribbon or array represents a set of data points that differ in Y values but have the same value of X. The conditional means $\overline{Y}_j$ are indicated by small circles in the approximate centers of their respective arrays, and a straight line has been drawn to pass as close to all three conditional means as possible. This straight line represents the regression line for predicting Y from X.

The regression line for predicting X from Y is a best-fitting line to the means of the horizontal arrays, where each horizontal array corresponds to a different value of Y. Except in the case where prediction is perfect, the regression line for predicting Y from X will be different from the line for predicting X from Y. Figure 15-11 shows the regression line for predicting X from Y and also the line for predicting Y from X for the same set of data.

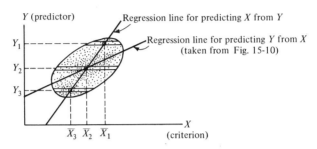

Figure 15-11 *Bivariate graph showing distributions of X values for each of three Y scores, conditional X means, and both regression lines.*

C. Computations

To determine the regression line for a set of raw data, you first find the slope, then the intercept. The slope for predicting Y from X is given by formula 15-11:

$$b_{YX} = \frac{\sum_i x_i y_i}{\sum_i x_i^2} = \frac{\sum_i (X_i - \overline{X})(Y_i - \overline{Y})}{\sum_i (X_i - \overline{X})^2} = \frac{N \sum XY - \sum X \sum Y}{N \sum X^2 - (\sum X)^2} \quad (15\text{-}11)$$

Once the slope has been determined, you can solve for the intercept. This is possible because the regression line always passes through the point defined by the means of the two variables. When b_{YX}, $\overline{X}$, and $\overline{Y}$ are substituted in the general regression formula, the equation has only one unknown, a_{YX}, which can then be solved for.

The regression line for predicting X from Y is found in a similar manner, using formula 15-12 for the slope:

$$b_{XY} = \frac{\sum_i x_i y_i}{\sum_i y_i^2} = \frac{\sum_i (X_i - \overline{X})(Y_i - \overline{Y})}{\sum_i (Y_i - \overline{Y})^2} = \frac{N \sum XY - \sum X \sum Y}{N \sum Y^2 - (\sum Y)^2} \quad (15\text{-}12)$$

EXAMPLE Find the regression line for predicting Y from X for the following data:

i	X	Y	X^2	Y^2	XY
1	7	2	49	4	14
2	9	4	81	16	36
3	5	5	25	25	25
4	6	5	36	25	30
5	3	8	9	64	24
Σ	30	24	200	134	129

SOLUTION   Usually only the raw scores X and Y will be given for each subject. In the table, values for X^2, Y^2, XY, and all the sums have been calculated and added to the table. Then, using formula 15-11,

$$b_{YX} = \frac{5(129) - (30)(24)}{5(200) - (30)^2} = \frac{645 - 720}{1000 - 900} = \frac{-75}{100} = -.75$$

The means are $\overline{X} = 30/5 = 6$ and $\overline{Y} = 24/5 = 4.8$. We know that the regression line passes through the point defined by the means, so

$$Y' = a_{YX} + b_{YX}X \qquad \text{becomes} \qquad \overline{Y} = a_{YX} + b_{YX}\overline{X}$$

Then, substituting for $b_{YX}\overline{X}$, and $\overline{Y}$ in the formula and solving for a_{YX}, we get

$$4.8 = a_{YX} + (-.75)(6) \qquad \text{and} \qquad a_{YX} = 4.8 + 4.5 = 9.3$$

The final regression equation is thus

$$Y' = 9.3 - .75X$$

EXAMPLE   To predict X from Y for these data, we first find the slope of the regression line:

$$b_{XY} = \frac{5(129) - (30)(24)}{5(134) - (24)^2} = \frac{-75}{94} \doteq -.80$$

Substituting b_{XY}, $\overline{X}$, and $\overline{Y}$ in the general regression formula for predicting X gives

$$\overline{X} = a_{XY} + b_{XY}\overline{Y}$$

$$6 = a_{XY} + (-.80)(4.8) = a_{XY} - 3.84$$

$$a_{XY} = 9.84$$

Therefore, the final regression equation is

$$X' = 9.84 - .80\ Y$$

D. Computations for Deviation Scores and Standard Scores

Relevant earlier section:
6-5   Standard scores

If we are predicting deviation scores on Y (y's) from deviation scores on X (x's), formula 15-9 becomes

$$y' = a_{YX} + b_{YX}x$$

As in the case of raw scores, the regression line must pass through the point defined by the two means. But $\bar{x} = 0$ and $\bar{y} = 0$ by definition of deviation scores (see Section 6-5A). Substituting in the equation and solving for the intercept, we get

$$0 = a_{yx} + 0 \qquad a_{yx} = 0$$

Thus, we can drop the intercept term from the equation for predicting deviation scores:

$$y' = b_{YX}x \tag{15-13}$$

where $b_{YX} = \sum x_i y_i / \sum x_i^2$, as defined in formula 15-11.

EXAMPLE   For the example of part C, the mean of X is 6 and the mean of Y is 4.8. Thus, the table of deviation scores and computations with them can be completed.

i	x	y	x^2	xy
1	+1	−2.8	1	−2.8
2	+3	−.8	9	−2.4
3	−1	+.2	1	−.2
4	0	+.2	0	.0
5	−3	+3.2	9	−9.6
$\sum$	0	.0	20	−15.0

We can then use the column sums from this table in formula 15-11 to obtain the slope of the regression line:

$$b_{YX} = \frac{-15}{20} = -.75$$

Inserting this value of the slope in the regression formula 15-13, we obtain

$$y' = -.75x$$

Note in this problem that the slope for the prediction of deviation scores is the same as that for predicting raw scores.

If we are predicting standard scores on Y from standard scores on X, the intercept a_{YX} is again zero, because the mean of any complete set of standard scores is zero (Section 6-5). The regression equation is then given by

$$z'_Y = \beta_{YX} z_X \tag{15-14}$$

where the usual practice is to use β instead of b to represent the slope in the z-score equation. Beta is defined in the same way as any other slope, only substituting z scores for deviation scores in the formula (see formula 15-15, with formula 15-11 for raw-score or deviation-score slopes). So

$$\beta_{YX} = \frac{\sum_i z_{X_i} z_{Y_i}}{\sum_i z^2_{X_i}} \tag{15-15}$$

is the *standard regression coefficient*, or the slope of the regression line for predicting standard scores on one variable from standard scores on another. It can easily be shown that the denominator of the standard regression coefficient $\sum_i z^2_{X_i}$ is equal to N.

$$\sum_i z^2_{X_i} = \sum_i \left(\frac{x_i}{\sigma_X}\right)^2 \qquad \text{(by formula 6-1)}$$

$$= \sum_i \frac{x^2_i}{\sigma^2_X}$$

$$= \frac{\sum_i x^2_i}{\sigma^2_X} \qquad \text{(by rule III, Section 5-1, because } \sigma^2 \text{ is a constant)}$$

But $\sigma_X^2 = \sum_i x_i^2/N$, from formula 5-10. Then, substituting, we obtain

$$\sum_i z_{X_i}^2 = \frac{\sum_i x_i^2}{\sum_i x_i^2/N} = \frac{1}{1/N} = N$$

Substituting this result in formula 5-15, we obtain

$$\beta_{YX} = \frac{\sum_i z_X z_Y}{N}$$

But this is the same expression that will be defined as the Pearson product-moment correlation coefficient (r_{XY}) between X and Y in formula 16-1. It is an indicator of the degree of relationship between the two variables X and Y. Therefore, we can rewrite formula 15-14 as

$$z_Y' = r_{XY} z_X \tag{15-16}$$

where r_{XY} is the Pearson product-moment correlation between X and Y, as discussed in Section 16-2.

E. Standard Error of Estimate

The measure of variability or error of prediction is the standard error of estimate. It is the standard deviation of criterion scores about the regression line. It is shown in Figure 15-12. Here the oval in the center represents an entire set of points, such as in Figure 15-10. Only one point has been drawn and labeled (X_1, Y_1) to show that it might be the point for the two scores of subject 1. On the Y axis is drawn a polygon to represent the distribution of Y scores, when subjects' scores on X are ignored. A dashed line runs from (X_1, Y_1) to the Y axis to show where subject 1's Y score might fall in this distribution. Through the graph runs the regression line, found using the methods of Part C of this section. On the right is a second distribution, of $Y_i - Y_i'$, the difference between each raw score on Y for a subject and the value Y' that we would predict

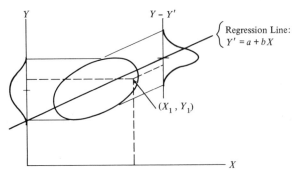

Figure 15-12

for that subject using the subject's X score and the regression equation. Subject 1's point is slightly below the regression line on the bivariate graph. A dashed line runs parallel to the regression line and up to the distribution of $Y - Y'$ to show how subject 1 fits into this distribution.

If we must predict scores for subjects and the only basis we have for the prediction is the overall distribution on Y, then our best prediction for each subject is the mean of Y, $\overline{Y}$. Our indicator of error of prediction in this case is the standard deviation of Y scores, s_Y. If we also know subjects' X scores and the regression equation, our best prediction for each subject is Y' and our indicator of the error of prediction is the standard error of estimate, s_{YX}. This is represented on the graph as the spread of $Y - Y'$ scores, the standard deviation of scores about the regression line. When we are predicting to Y, the formula for the standard error of estimate is

$$s_{YX} = \sqrt{\frac{\sum (Y - Y')^2}{N}} \tag{15-17}$$

where Y' is taken from the regression equation. The standard error of estimate may also be interpreted as a pooled, within-groups variability estimate, where each group is a set of criterion scores for some level of the predictor, and again, where all deviations are taken about the regression line.

The computational formulas of part C of this section ensure that the sum of squared deviations about the regression line is less than the sum of squared deviations about any other straight line that you could draw.

An unbiased estimate of the population standard error of estimate can be obtained using

$$\hat{s}_{YX} = s_{YX} \sqrt{\frac{N}{N - 2}} \tag{15-18}$$

where s_{YX} is obtained in accordance with formula 15-17.

It is possible to derive a computing formula for the standard error of estimate using terms already defined, but the formula is unnecessarily messy. The simplest formula expresses the standard error in terms of the standard deviation of the criterion and the Pearson correlation coefficient (see Section 16-2):

$$s_{YX} = s_Y \sqrt{1 - r_{XY}^2} \quad \text{and} \quad s_{XY} = s_X \sqrt{1 - r_{XY}^2}$$

for the biased standard errors, with the unbiased values found subsequently using formula 15-18. Because the correlation coefficient has not yet been defined, further discussion of this topic will be deferred to a later section.

References Linear regression is discussed in Blommers and Lindquist (1960), 407–422, 435–442; Dixon and Massey (1969), 193–197; Edwards (1973), Chapter 11; and Games and Klare (1967), 378–393.

Homework

1. What is the equation of the regression line in the illustration?

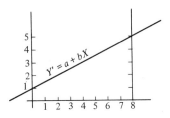

2. The regression line for predicting Y from X passes through the point $(X = 8, Y = 5)$. If $\bar{X} = 18$ and $\bar{Y} = 8$, what is the formula for predicting Y from X?

For each of the following data sets, assume that all the data are interval-level. Then (a) find the regression equation for predicting Y from X; and (b) using the equation, predict the score on Y when a subject's score on X is 15.

3.

i	X	Y
1	14	19
2	6	11
3	12	7
4	8	18
5	10	5

4.

i	X	Y
1	17	12
2	31	2
3	25	2
4	23	2
5	14	7

5.

i	X	Y
1	12	26
2	1	14
3	10	12
4	3	18
5	4	25

6.

i	X	Y
1	14	30
2	3	24
3	12	18
4	5	16
5	6	22

7. Consider again the example of Section 10-1. Suppose that you wish to predict whether subjects are field-independent or field-dependent on the basis of their grade-point average.

	Grade					
	F	D	C	B	A	Σ
FI	10	20	30	18	12	90
FD	18	36	28	12	6	100
Σ	28	56	58	30	18	190

(a) Find the conditional mode and conditional variation ratio for each grade.

(b) What is the variation ratio of estimate?

For problems 8 through 10, decide whether it is more reasonable to use linear or curvilinear prediction of Y from X. In each case, assume that both X and Y are interval level.

8.

i	X	Y
1	56	41
2	52	43
3	60	50
4	55	46
5	65	51
6	62	46
7	55	49
8	57	48
9	66	53
10	62	54
11	57	54
12	59	45
13	62	54
14	64	53
15	56	45
16	53	46
17	52	45
18	59	49
19	58	51
20	62	51
21	62	55
22	53	49
23	58	48

9.

i	X	Y
1	25	32
2	15	23
3	30	34
4	15	29
5	11	21
6	20	32
7	17	30
8	27	31
9	14	32
10	31	35
11	23	30
12	27	34
13	14	27
14	12	23
15	17	32
16	14	30
17	23	32
18	13	29
19	12	29
20	25	33
21	13	26
22	23	34
23	20	33

10.

i	X	Y
1	101	312
2	116	288
3	119	235
4	108	330
5	114	345
6	108	308
7	96	355
8	91	381
9	96	312
10	99	337
11	90	400
12	122	220
13	86	388
14	102	350
15	110	280
16	110	250
17	106	293
18	102	251
19	93	310
20	117	250
21	90	349
22	111	273
23	106	266

11. A small private college had decided to begin selecting among students at the time of admission, using an aptitude test that is reported to give interval-level scores. A new freshman class is tested on admission and all students are accepted. At the end of 5 years, 8 of the students have graduated and 12 have not. Their admission scores are checked and found to be

Graduating	47, 37, 39, 40, 43, 45, 45, 46
Not graduating	33, 35, 36, 36, 37, 38, 38, 39, 39, 40, 43, 44

Based on these data, what cut score used in the future will optimally predict whether a student will graduate?

Homework Answers

1. $Y' = 1 + \frac{1}{2}X$

2. $Y' = 2.6 + .3X$

3. $Y' = 9.5 + .25X$; 13.25

4. $Y' = 16 - .5X$; 23.50

5. $Y' = 17 + \frac{1}{3}X$; 22

6. $Y' = 12 + X$; 27

7. (a) $Mo(A) = FI$; $Mo(B) = FI$; $Mo(C) = FI$; $Mo(D) = FD$, $Mo(E) = FD$;
   $v(A) = .33$; $v(B) = .40$; $v(C) \doteq .48$; $v(D) \doteq .36$; $v(E) \doteq .36$

   (b) $v_{YX} \doteq .40$

8. Linear   9. Curvilinear   10. Linear   11. 1.56

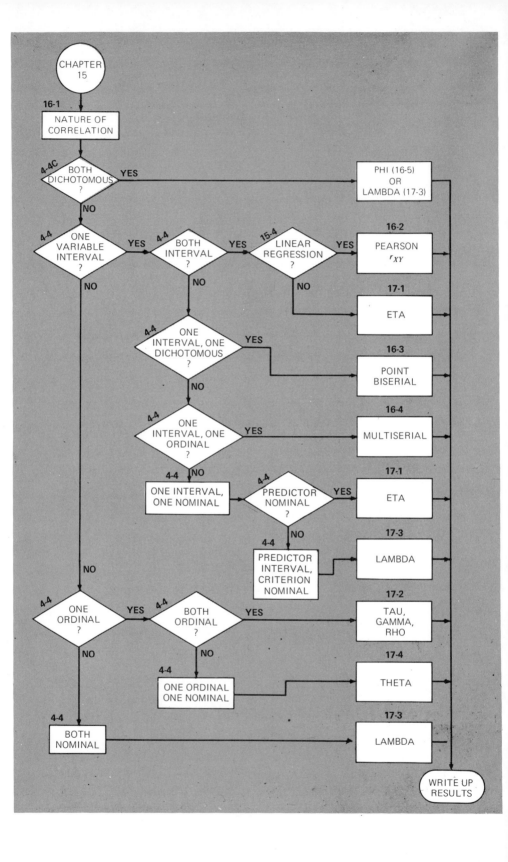

16 Correlation: Pearson and Related Formulas

METHODS for correlating variables are distinguished from one another primarily by two kinds of considerations: the number of variables correlated and their levels of measurement. Whenever you have two variables to correlate, the methods of Chapters 16 and 17 apply. When you want to deal with more than two variables at a time, you can consider them pairwise, using the methods of these chapters; you may interrelate them using the methods of Chapter 18; or you may choose to use advanced statistical methods such as cluster analysis, factor analysis, and multiple regression. These more advanced techniques generally assume dichotomous or interval-level variables and are used when the Pearson, point-biserial, and phi coefficients are used to obtain all the pairwise correlations. References are provided at the end of Section 18-1.

The second basis for selecting among coefficients is level of measurement of the variables related. Like number of variables, this criterion is often more advisory than essential. It is always possible, although wasteful, to ignore information in your data and treat interval data as ordinal and ordinal or interval data as nominal. It is sometimes possible to treat ordinal data as though it were interval-level, and the results may very well be meaningful.

Different coefficients applied to the same data often give different values for the degree of relationship. When you compare the degree of relationship in one set of data to that in another it is therefore usually advisable to use the same coefficient for both, even though for one set of data the coefficient may not be optimal.

Chapter 16 contains the Pearson correlation coefficient and three others—multiserial, point biserial and phi—which are derived from it. The remaining two-variable coefficients appear in Chapter 17.

After each description of a coefficient will be a significance test to decide whether it is reasonable to say that there is some relationship among the variables in the entire population from which you have sampled. These parts of sections assume that you are already familiar with Chapter 9 and one or more sections in Chapters 10 through 14. If you are not already familiar with these earlier portions of the book, don't try to read them now: wait until the need arises and the section(s) are referred to by the present chapter.

16-1 General problem of correlation

A correlation coefficient is a statistical formula that allows you to express the degree of relationship between two variables, usually represented as X

and Y. It allows you to answer the question, "To what extent am I getting similar or equivalent information about subjects from these two variables?"

For ordinal- or interval-level variables, degree of similarity in information has to do with the extent to which high scores on X tend to be associated with high scores on Y, and, at the same time, low scores on X tend to be associated with low scores on Y. With tau or gamma, a correlation is high to the extent that the two variables rank all subjects in the same way, from highest to lowest. With the Pearson coefficient, the correlation is large to the extent that the bivariate plot of data points approaches a straight line.

Another way of putting the definition is that a correlation coefficient indicates the extent to which two variables classify subjects in the same or equivalent ways. If subjects get the same ranks on two variables, they are being classified in the same way; and if one set of scores is a linear transformation of the other, subjects are being classified in similar ways (although the actual numbers may differ).

If two variables give similar or equivalent information about subjects, then knowing subjects' scores on X, you have a better idea of their status on Y than you would if you didn't know their X scores or the relationship between X and Y. Using the methods of Chapter 15, you can determine *what* to predict for subjects on Y from knowledge of their scores on X; and using the methods of Chapters 16 through 18, you can find out *how well* you can predict: how much difference the prediction makes. If the correlation is large, then making the prediction helps you determine subjects' Y scores; if the correlation is small, it doesn't make much difference.

Different coefficients have different limits. Some, such as eta and lambda, range from 0.0 when the two variables are independent of each other, to 1.0 when there is a perfect relationship between them. Others, like gamma and the Pearson coefficient, range from -1.0 for a perfect negative relationship to $+1.0$ for a perfect positive relationship, with zero again indicating independence of the two variables.

To compute the magnitude of a correlation, you ordinarily need a data table that has a different row for each subject, and two columns, one for each variable. Each subject must have a score (or other index of status) on *both* variables. You choose your coefficient primarily on the level of measurement of the two variables, but also according to the kind of information you want the coefficient to reflect and the other coefficients to which it will be compared. An eta of .50 and a Pearson correlation of .50 do not usually reflect the same degree or kind of relationship, even when they are calculated on the same set of data.

Most of the correlation coefficients can be interpreted as showing the proportion of error in predicting the dependent or criterion variable that can be reduced by knowing subjects' status on the predictor and the relationship between the two variables:

$$\text{correlation} = \frac{\text{error}(Y) - \text{error}(Y \text{ given } X)}{\text{error}(Y)}$$

If knowing the relationship between X and Y doesn't help you predict status on Y, then error(Y given X) is just as great as error(Y), the numerator of the coefficient goes to zero, and the correlation is zero also. If prediction knowing Y is perfect, then there is no error(Y given X), the numerator and denominator of the fraction are equal, and the correlation is 1.0. If you can reduce some of the error in predicting subjects' status on Y from knowing their status on X, then the correlation falls between 0 and 1, its magnitude depending on the amount of error you have gotten rid of.

EXAMPLE Suppose that you want to determine the degree of relationship between subjects' handedness and footedness (the foot they prefer to kick a ball with). Both variables are dichotomous, so you could use either phi or lambda. The two coefficients are not equivalent, and the one you chose would depend on whether you wanted to compare it to other lambdas or to other phis. Suppose that you chose to use phi and the correlation turned out to be .90. This would mean that (1) right-handed people tend to be right-footed, and left-handed people tend to be left-footed; so (2) you can predict a person's footedness quite well, although not perfectly, from his or her handedness; and (3) the number of mistakes you will make in predicting subjects' footedness will be substantially reduced if you are told their handedness before you make your prediction.

In this discussion, nothing has been said about one variable affecting scores on the other, or causing subjects to have a particular status on the other. All you can say from finding a nonzero correlation is that the two variables are related, to a specified degree; you don't know what leads to that correlation. In the example, handedness and footedness are not the same thing, even though the correlation is quite high between them; nor can we say that either is the source or cause of the other. If you can specify a reason for the relationship, you do so based on information other than that provided by the correlation itself.

16-2 Pearson product-moment correlation coefficient

Relevant earlier sections:
4-10 Two-dimensional plot
6-5 *z* transformation

A. General Description of the Pearson Coefficient

The Pearson coefficient is easily the most widely used indicator of degree of relationship between two variables that is found in social science. The coefficient can be calculated whenever a sample of subjects receives a numerical score on each of two variables to be correlated. However, if the variables are not interval-scaled, it is not clear what effect the arithmetic operations used have on the scores, and the correlation becomes difficult to interpret. When used in conjunction with linear regression, the Pearson coefficient indicates the degree of accuracy in predicting from one variable to the other. When nonlinear regression is called for, this statistic does not adequately indicate the accuracy of prediction, because it is dependent on the assumption of linearity.

The coefficient may be defined as follows:

$$r_{XY} = \frac{\sum_i z_{X_i} z_{Y_i}}{N} \tag{16-1}$$

where N is the number of subjects, z_{X_i} is the standard score of subject i on variable X, and z_{Y_i} is the standard score of subject i on variable Y. In Section 6-5 it was indicated that z scores are defined in terms of population means and standard deviations. This might lead us to seek an unbiased estimate to use with sample data. However, sample means and biased standard deviations can be used directly in place of the corresponding parameters without greatly biasing the estimate of the correlation coefficient; this is the procedure followed by social scientists when working with sample data.

The first thing to note in this formula is that a correlation is invariant under a linear transformation of either variable. That is, the raw scores on variable X can be altered by any linear transformation (addition, subtraction, multiplication, division, or any combination of these), and the z scores will not be affected. Because the correlation coefficient is determined only by the z scores, raw-score transformations will not affect its magnitude. Because the slopes and intercepts of the regression lines are affected by the means and variances of the variables related, this means that the correlation coefficient gives at least some information that is different from the information given by the regression lines. In fact, for a correlation of $+1.00$ (a perfect positive correlation), the slope may have any positive value at all between (but not including) 0 and $+\infty$.

A second notable property of the Pearson correlation is that it is symmetric: $r_{XY} = r_{YX}$. This can be seen from formula 16-1, because the order of terms in a product is irrelevant: $5 \times 2 = 2 \times 5$, and $z_{X_i} z_{Y_i} = z_{Y_i} z_{X_i}$. The coefficient r_{YX} would differ from r_{XY} only because it had a different order of terms in its numerator, so the two coefficients must have the same magnitude for any set of data.

The five graphs in Figure 16-1 indicate various possibilities for the correlation coefficient and show a correspondence between some possible shapes of the bivariate plots and values of r_{XY}.

A correlation of $+1$ indicates that perfect prediction can be made from X to Y and Y to X. Such a correlation is usually found only when X and Y are

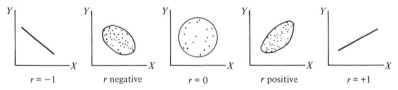

Figure 16-1 *Bivariate plots and magnitudes of r_{XY}.*

different scales for measuring the same identical variable, as height in inches and height in centimeters or temperature in degrees Fahrenheit and temperature in degrees centigrade. At the opposite extreme, a correlation of -1 indicates that a perfect negative prediction can be made between X and Y. Such a relationship might be expected to obtain between subjects' heights in inches and shortness measured in inches down from the ceiling.

It can easily be shown that the limits of r_{XY} are $+1$ and -1. Two variables have a maximum positive relationship when every subject has the exact same z score on both variables. In this case, $z_{Y_i} = z_{X_i}$, for all i, so that

$$(r_{XY})_{max} = \frac{\sum z_X z_Y}{N} = \frac{\sum z_X^2}{N}$$

But $z_{X_i} = x/s_X$, and s_X is a constant; therefore,

$$(r_{XY})_{max} = \frac{\sum x^2}{N s_X^2}$$

However, $\sum x^2/N$ is the formula for the biased variance estimate, s_X^2. Therefore, we have $(r_{XY})_{max} = s^2/s^2 = 1$, and when there is a perfect positive relationship between X and Y, $r_{XY} = 1$. You also can show easily that a perfect negative relationship gives $r_{XY} = -1$, by replacing z_Y by $-z_X$ and carrying through the same derivation.

B. Computation of the Correlation Coefficient

Let us now illustrate the use of the definitional formula for the Pearson correlation coefficient with some raw-score data. We have a sample of 10 subjects for whom scores are obtained on both X and Y. The data table is given as Table 16-1.

Table 16-1 *Hypothetical data to illustrate computation of the correlation coefficient*

i	X	Y	x	y	x^2	y^2	z_X	z_Y	$z_X z_Y$
1	2	23	-6	11	36	121	-1.50	1.83	-2.75
2	10	9	2	-3	4	9	.50	$-.50$	$-.25$
3	5	12	-3	0	9	0	$-.75$	.00	.00
4	15	9	7	-3	49	9	1.75	$-.50$	$-.875$
5	14	3	6	-9	36	81	1.50	-1.50	-2.25
6	10	12	2	0	4	0	.50	.00	.00
7	6	21	-2	9	4	81	$-.50$	1.50	$-.75$
8	5	15	-3	3	9	9	$-.75$	.50	$-.375$
9	8	5	0	-7	0	49	.00	-1.17	.00
10	5	11	-3	-1	9	1	$-.75$	$-.17$	1.25
$\sum$	80	120	0	0	160	360			-7.125

We need to obtain z_X and z_Y for each subject in order to get their product for each subject and then the sum of their products. To find the z scores, we must first find the standard deviations for both variables and the deviation score of each subject on each variable. To find the deviation scores we will need the means of both variables. The means are given by

$$\bar{X} = \frac{\sum X}{N} = \frac{80}{10} = 8 \quad \text{and} \quad \bar{Y} = \frac{\sum Y}{N} = \frac{120}{10} = 12$$

Then the deviation scores can be found from $x_i = X_i - \bar{X}$ and $y_i = Y_i - \bar{Y}$. The deviation scores are shown in the columns labeled x and y in Table 16-1. The deviation-score formulas will be used to calculate the biased standard deviations:

$$s_X = \sqrt{\frac{\sum x^2}{N}} = \sqrt{\frac{160}{10}} = \sqrt{16} = 4$$

and

$$s_Y = \sqrt{\frac{\sum y^2}{N}} = \sqrt{\frac{360}{10}} = \sqrt{36} = 6$$

The standard score is found for each subject on each variable by dividing his deviation score on the variable by the standard deviation of the variable. These are listed in the columns labeled z_X and z_Y of Table 16-1. The product $z_X z_Y$ is then found for each subject and these products summed over subjects. Finally, entering the sum of products into the correlation formula gives the value of the correlation:

$$r_{XY} = \frac{\sum z_X z_Y}{N} = \frac{-7.125}{10} \doteq -.71$$

Clearly this has been a tedious procedure, but fortunately it is unnecessarily so. Alternative formulas can be derived from the definition, formulas that permit greater ease of calculation when the z scores are not immediately given. Let us now derive some of these formulas so that later it will be possible to select an easiest formula for whatever form the data may be in. Beginning with

$$r_{XY} = \frac{\sum z_X z_Y}{N} = \frac{1}{N} \sum z_X z_Y$$

we note that $z_X = x/s_X$ and $z_Y = y/s_Y$. Substituting, we obtain

$$r_{XY} = \frac{1}{N} \sum \left(\frac{x}{s_X}\right)\left(\frac{y}{s_Y}\right)$$

but s_X and s_Y are constants, so

$$r_{XY} = \frac{\sum xy}{N s_X s_Y} \tag{16-2}$$

The expression $\sum xy/N$ is called the *covariance* of X and Y, sometimes written $\text{cov}(xy)$. Using this expression, formula 16-2 may be rewritten

$$r_{XY} = \frac{\text{cov}(xy)}{s_X s_Y} = \frac{\text{cov}(xy)}{\sqrt{V_X V_Y}} \tag{16-2a}$$

where $V_X = s_X^2$ is the variance of X and $V_Y = s_Y^2$ is the variance of Y.

Next, we recall that $s_X = \sqrt{\sum x^2/N}$ and $s_Y = \sqrt{\sum y^2/N}$, so formula 16-2 can be reduced to

$$r_{XY} = \frac{\sum xy}{N\sqrt{\sum x^2/N}\sqrt{\sum y^2/N}} = \frac{\sum xy}{N\sqrt{\sum x^2 \sum y^2/N^2}}$$

Simplifying gives

$$r_{XY} = \frac{\sum xy}{\sqrt{\sum x^2 \sum y^2}} \tag{16-3}$$

This formula is stated entirely in deviation scores. To obtain a raw-score formula, we recall (Section 5-12) that

$$\sum x^2 = \sum X^2 - \frac{(\sum X)^2}{N}$$

Similar derivations could be used to show that

$$\sum y^2 = \sum Y^2 - \frac{(\sum Y)^2}{N} \quad \text{and} \quad \sum xy = \sum XY - \frac{(\sum X)(\sum Y)}{N}$$

Substituting in the deviation score formula 16-3, we obtain

$$r_{XY} = \frac{\sum XY - (\sum X)(\sum Y)/N}{\sqrt{[\sum X^2 - (\sum X)^2/N][\sum Y^2 - (\sum Y)^2/N]}}$$

This formula can be further simplified by multiplying numerator and denominator by N:

$$r_{XY} = \frac{N \sum XY - (\sum X)(\sum Y)}{N \sqrt{[\sum X^2 - (\sum X)^2/N][\sum Y^2 - (\sum Y)^2/N]}}$$

$$= \frac{N \sum XY - (\sum X)(\sum Y)}{\sqrt{N[\sum X^2 - (\sum X)^2/N]N[\sum Y^2 - (\sum Y)^2/N]}}$$

and, finally,

$$r_{XY} = \frac{N \sum XY - (\sum X)(\sum Y)}{\sqrt{[N \sum X^2 - (\sum X)^2][N \sum Y^2 - (\sum Y)^2]}} \tag{16-4}$$

Any of these formulas might have been used to obtain the value of the correlation coefficient. For example, we might have used an xy column instead of the columns for z_X, z_Y, and $z_X z_Y$ in Table 16-1.

i	xy
1	-66
2	-6
3	0
4	-21
5	-54
6	0
7	-18
8	-9
9	0
10	3
$\sum$	-171

Then using formula 16-2 we would have

$$r_{XY} = \frac{\sum xy}{N \sigma_X \sigma_Y} = \frac{-171}{10(4)(6)} = \frac{-171}{240} = \frac{-57}{80} \doteq -.71$$

If we used formula 16-3, we would not have needed to calculate σ_X or σ_Y:

$$r_{XY} = \frac{\sum xy}{\sqrt{\sum x^2 \sum y^2}} = \frac{-171}{\sqrt{160(360)}} = \frac{-171}{\sqrt{(16)(10)(36)(10)}} = \frac{-171}{4(10)(6)} \doteq -.71$$

Alternatively, we might have used raw scores exclusively:

i	X	Y	X^2	Y^2	XY
1	2	23	4	529	46
2	10	9	100	81	90
3	5	12	25	144	60
4	15	9	225	81	135
5	14	3	196	9	42
6	10	12	100	144	120
7	6	21	36	441	126
8	5	15	25	225	75
9	8	5	64	25	40
10	5	11	25	121	55
Σ	80	120	800	1800	789

Then, using the raw-score formula,

$$r_{XY} = \frac{N \sum XY - (\sum X)(\sum Y)}{\sqrt{[N \sum X^2 - (\sum X)^2][N \sum Y^2 - (\sum Y)^2]}}$$

$$= \frac{10(789) - (80)(120)}{\sqrt{[10(800) - (80)^2][10(1800) - (120)^2]}}$$

$$= \frac{7890 - 9600}{\sqrt{[8000 - 6400][18000 - 14400]}}$$

$$= \frac{-1710}{\sqrt{[1600][3600]}} = \frac{-1710}{40(60)}$$

$$= \frac{-171}{240} \doteq -.71$$

The reason that the various formulas give the same value for the correlation is that deviation scores and standard (z) scores are linear transformations on raw scores, and the correlation coefficient, as indicated above, is independent of changes in scale.

To emphasize this point further, the scores have been represented in a bivariate plot in Figure 16-2, where all three scales are given in the margins. The points therefore can serve interchangeably to represent relationships among raw scores, deviation scores, and z scores.

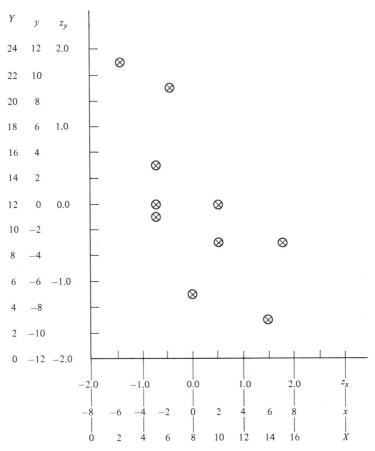

Figure 16-2 *Bivariate graph of data for correlation example.*

C. Variance Accounted for and the Standard Error of Estimate

If we are predicting scores of subjects in a sample and we have no information about specific subjects except that they are members of the sample or the
population from which it was drawn, then the best guess of each subject's score
is the sample mean $\overline{Y}$, and an indicator of the amount of error in the guesses
is the biased estimate of the population variance,

$$s_Y^2 = \frac{\sum (Y - \overline{Y})^2}{N} = \frac{\sum y^2}{N} \tag{16-5}$$

This is depicted graphically as the spread of scores on the vertical (Y) axis
at the left of Figure 16-3.

If, on the other hand, we know the relationship between Y scores and scores
on a predictor X, and we know, for each subject, his X score, then our best
guess is his predicted score, under the assumption of linearity of regression,

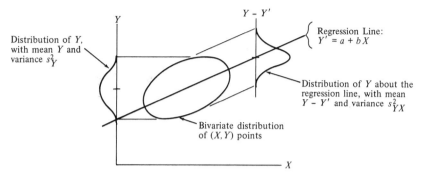

Figure 16-3

$$Y' = a_Y + b_{YX}X$$

$$= \frac{\sum xy}{\sum x^2}(x) + \bar{X} \tag{16-6}$$

An indicator of the amount of error in these guesses is given by the squared standard error of estimate, taken about the regression line. The standard error of estimate is a measure of variability around predicted Y values, so we obtain the formula by substituting Y' for $\bar{Y}$ in formula 16-5.

$$s_{YX}^2 = \frac{\sum (Y - Y')^2}{N} \tag{16-7}$$

One way that we might indicate the effectiveness of using regression is to determine whether use of the regression line reduces error of prediction, or variance. The amount of reduction in error due to using the regression line can be indicated by the amount of reduction in variance, from s_Y^2 to s_{YX}^2: we subtract the variance about the regression line from the variance about the mean of Y. If we then express this difference as a proportion of the total variance of Y, we can write another formula for the (squared) correlation coefficient.

$$r_{XY}^2 = \frac{s_Y^2 - s_{YX}^2}{s_Y^2} \tag{16-8}$$

This can be shown to be equivalent to the other Pearson formulas by substituting equivalent expressions in the right-hand side of formula 16-8 and deriving any one of the formulas of part B. Let us substitute formula 16-5 for s_Y^2; then because we know that the Pearson formula is invariant under linear transformation, we can use a deviation-score formula for s_{YX}^2:

$$s_{YX}^2 = \frac{\sum (y - y')^2}{N} = \frac{\sum (y - b_{yx}x)^2}{N}$$

(This came from using deviation scores in formula 16-7, then replacing y' by $b_{yx}x$, according to formula 15-13.)

Now, substituting in formula 16-8, we get

$$r_{XY}^2 = \frac{\sum y^2/N - \sum (y - b_{yx}x)^2/N}{\sum y^2/N}$$

$$= \frac{\sum y^2 - \sum (y - b_{yx}x)^2}{\sum y^2}$$

when we multiply numerator and denominator by N. Note that, although it is customary to use biased variance estimates in these formulas, we could just as well have used unbiased estimates: the $(N - 1)$ would have divided out just as neatly as the N did. Now, completing the square and simplifying, we get

$$r_{XY}^2 = \frac{\sum y^2 - \sum y^2 + 2 \sum yb_{yx}x - \sum b_{yx}^2 x^2}{\sum y^2}$$

$$= \frac{2b_{yx} \sum xy - b_{yx}^2 \sum x^2}{\sum y^2}$$

Finally, substituting the deviation-score formula for b_{yx} and simplifying again, we find that

$$r_{XY}^2 = \frac{2(\sum xy/\sum x^2)(\sum xy) - [(\sum xy)^2/(\sum x^2)^2] \sum x^2}{\sum y^2}$$

$$= \frac{2(\sum xy)^2/\sum x^2 - (\sum xy)^2/\sum x^2}{\sum y^2}$$

$$= \frac{(\sum xy)^2}{\sum x^2 \sum y^2}$$

which is the square of formula 16-3; and from formula 16-3 all the other Pearson formulas can be derived. Thus, the squared Pearson formula, as defined in parts A and B, tells the proportion of Y variance that can be reduced, or accounted for, by knowledge of X and the relation between X and Y.

Just as the correlation is symmetric ($r_{XY} = r_{YX}$), so is the squared correlation. Thus, r squared also gives the proportion of the variance of X that can be accounted for by Y:

$$r_{XY}^2 = r_{YX}^2 = \frac{s_X^2 - s_{XY}^2}{s_X^2}$$

Let us now return to formula 16-8 to obtain a simplified formula for the standard error of estimate s_{YX}. First we multiply both sides of formula 16-8 by s_Y^2 and get

$$r_{XY}^2 s_Y^2 = s_Y^2 - s_{YX}^2$$

Then we add $s_{YX}^2 - r_{XY}^2 s_Y^2$ to both sides:

$$s_{YX}^2 = s_Y^2 - r_{XY}^2 s_Y^2$$
$$= s_Y^2 (1 - r_{XY}^2)$$

Finally, we obtain the formula for the standard error by taking the square root of both sides of this equation:

$$s_{YX} = s_Y \sqrt{1 - r_{XY}^2} \tag{16-9}$$

Formula 16-9 is also presented in Section 15-5E and related to the regression equation.

D. Unbiased Estimates of a Population Correlation

Biases appear in the Pearson formulas at two levels. First, you may have been surprised that the formulas of parts B and C used biased variance estimates. This is standard notation, although some texts use population parameters μ and σ instead of sample estimates. It is really unimportant to most of the formulas, however, because, as shown in part C, the N or $N - 1$ divides out in the derivation of the computing formulas.

Regardless of whether you use N or $N - 1$ in the formula, however, the formula as a whole is biased as an estimate of the population correlation. A corrected or unbiased formula can be developed,* but the error in the biased formula is small, even for small samples ($25 \leq N \leq 50$). On the other hand, the amount of random error in correlations based on N's less than 50 is fairly great, so the correction of a small systematic error seems pointless.

E. Test for Significance

> *Relevant earlier sections:*
> **7-4** t distribution
> **Chapter 9** Hypothesis testing

When it seems reasonable to assume that for the subject population the joint distribution of X and Y is bivariate normal and homoscedastic,† then it is

---

* See Peters and Van Voorhis (1940), 153.

† The distribution of X and Y is bivariate-normal if (a) the overall distribution of X is normal, (b) the overall distribution of Y is normal, (c) Y scores are normally distributed for each level of X (see Figure 15-10), and (d) X scores are normally distributed for each level of Y (see Figure 15-11). The distribution is homoscedastic if Y scores have the same variance for each level of X (Figure 15-10) and if X scores have the same variance for each level of Y (Figure 15-11).

possible to test the hypothesis of a zero population correlation using either a two-tail test ($H_0: \rho_{XY} = 0$ and $H_m: \rho_{XY} \neq 0$) or a one-tail test ($H_0: \rho_{XY} \geq 0$ and $H_m: \rho_{XY} < 0$; or $H_0: \rho_{XY} \leq 0$ and $H_m: \rho_{XY} > 0$).

The formula used to test for a nonzero population correlation is

$$t = \frac{r_{XY}\sqrt{N-2}}{\sqrt{1 - r_{XY}^2}} \quad \text{with } N-2 \text{ df} \tag{16-10}$$

EXAMPLE   Let us test the hypothesis $H_0: \rho_{XY} = 0$ at the .05 level, using the data of Table 16-1. The correlation calculated from these data was $-.71$, and N was 10; so

$$t = \frac{(-.71)\sqrt{10-2}}{\sqrt{1 - (-.71)^2}} = \frac{(-.71)(2.828)}{\sqrt{.4959}} = \frac{-2.008}{\sqrt{.4959}}$$

$$= -2.85$$

The critical values of $t_{.05}(8 \text{ df})$ are ± 2.306. The observed t of -2.85 lies outside these values, so the null hypothesis is rejected: the population correlation is negative.

F. Correction for Grouped Data

As indicated in Section 4-5, it is sometimes convenient to group data into categories that have more than one scale unit each. When one or both of the variables that you are correlating have been grouped, the correlation obtained from the grouped data will be smaller in absolute magnitude than the correlation obtained from the raw-score data would be.

When data have been grouped, it may be necessary to calculate the correlation on either midpoints of intervals or means of intervals. It is possible then to estimate the magnitude of the correlation that you would have obtained if you had correlated the raw scores. Formula 16-11 may generally be used for this purpose, when scores on X have been grouped.

$$_c r_{XY} = \frac{r_{\bar{X}Y}}{r_{X\bar{X}}} \tag{16-11}$$

where $_c r_{XY}$ is the inferred correlation between the raw scores on Y and the assumed raw scores on X, $r_{\bar{X}Y}$ is the obtained correlation between raw scores on Y and means (or midpoints) on X, and $r_{X\bar{X}}$ is the assumed correlation between the underlying raw scores on X and the means (or midpoints) actually used in calculating $r_{\bar{X}Y}$. If both X and Y have been grouped, the correction formula is

$$_c r_{XY} = \frac{r_{\bar{X}\bar{Y}}}{r_{X\bar{X}} r_{Y\bar{Y}}} \tag{16-12}$$

Case 1: Equal intervals You may be willing to assume that the intervals have equal widths throughout the range of the grouped variable, and that the underlying distribution of the grouped variable is either normal or rectangular.

In either of these cases, values of $r_{X\bar{X}}$ and $r_{Y\bar{Y}}$ are given in Table 16-2.

Table 16-2 *Correction terms to be used in the correction of correlations for grouped data†*

Number of categories of grouped variable	Normal distribution		Rectangular distribution; correlation of raw scores and either means or midpoints
	Correlation of raw scores and means	Correlation of raw scores and midpoints	
2	.798	.816	.866
3	.859	.859	.943
4	.915	.916	.968
5	.943	.943	.980
6	.959	.960	.986
7	.970	.970	.990
8	.976	.977	.992
9	.981	.982	.993
10	.985	.985	.995
11	.987	.988	.996
12	.989	.990	.996
13	.991	.991	.997
14	.992	.992	.997
15	.993	.994	.998

† Peters and Van Voorhis (1940), 398.

EXAMPLE   Suppose you read in a research report that variables X and Y were assumed normally distributed, and that Y scores ranged from 80 to 198. The experimenter trichotomized Y into categories 80 to 119, 120 to 159, and 160 to 199, and the Pearson correlation was then found to be .43. The correlation that the experimenter probably would have found if he had correlated raw scores on both X and Y can be found using formula 16-11 and obtaining $r_{Y\bar{Y}}$ from Table 16-2. It is not clear from the information given whether the experimenter used category means or midpoints to calculate $r_{X\bar{Y}}$, but in either case, $r_{Y\bar{Y}}$ given by Table 16-2 is .859. Therefore, whichever method was used, our estimate of r_{XY} is

$$_c r_{XY} = \frac{.430}{.859} \doteq .50$$

Case 2: Unequal intervals If the grouped variable has not been grouped into intervals of equal width, Table 16-2 cannot be used. However, the corrections of formulas 16-11 and 16-12 can still be carried out. If calculations are based on category means and you want to infer to raw scores, use

$$r_{X\bar{X}} = \frac{s_{\bar{X}}}{s_X} \quad \text{and/or} \quad r_{Y\bar{Y}} = \frac{s_{\bar{Y}}}{s_Y} \tag{16-13}$$

Notice that these formulas require you to have an estimate of raw-score variance. If you can obtain an estimate of s_X and/or s_Y, the calculation of r_{XY} from grouped data (as opposed to raw scores) generally becomes harder to justify (it is less precise, and the calculations are generally no easier). One instance when use of grouped data is fairly plausible is when the data collected are in the form of ranks, and conversion to z scores has been made with the table of the standard normal distribution. In this case, the standard deviation of the raw scores on the grouped variable is taken as 1, simplifying the correction (see Section 16-4).

If the calculations are based on the midpoints of intervals and you want to infer to the raw-score distributions, use

$$r_{X\bar{X}} = \frac{\sqrt{s_{\bar{X}}^2 - 1/12}}{s_{\bar{X}}^2} \quad \text{and} \quad r_{Y\bar{Y}} = \frac{\sqrt{s_{\bar{Y}}^2 - 1/12}}{s_{\bar{Y}}^2} \tag{16-14}$$

in formulas 16-11 and 16-12.

A general formula for the correlation of raw scores and means with two categories and an assumed underlying normal distribution is given by formula 16-18.

Examples of the use of these corrections are given in Sections 16-3C and 16-4.

References   The Pearson correlation coefficient is discussed in Blommers and Lindquist (1960), 388–406, 423–435; Edwards (1973), Chapter 11; Freeman (1965), 89–106; Games and Klare (1967), 340–368, 395–405; and Guilford (1965), Chapter 6. The Fisher Z' transformation for testing hypotheses about the Pearson coefficient can be found in Edwards (1960), Chapter 6, and (1973), Chapter 12.

16-3   Point-biserial and biserial correlation coefficients

Relevant earlier section:
16-2   Pearson correlation

When we wish to find the correlation between two variables, one of which is dichotomous, we may use the Pearson coefficient or we may use a second formula that is a simplification of the Pearson formula for this restricted case. The *point-biserial* formula assumes that you are really interested in using the second variable as a dichotomy (regardless of whether it is a true dichotomy, like sex, or a dichotomized variable, like subjects' popularity). On the other hand, you may have dichotomized one variable out of convenience or have been forced to work with it in dichotomous form although you are really interested in the correlation between the two continuous variables. When you wish to infer the correlation between two continuous-interval variables on the basis of data that have been dichotomized for one of the variables, then the point-biserial coefficient is not appropriate; the *biserial* correlation is the one you should use. The biserial coefficient will be discussed as a special case of the point-biserial coefficient in part C of this section. In cases where the point biserial is an appropriate coefficient, it is also algebraically and numerically equivalent to the correlation ratio, eta (Section 17-1).

A. Point Biserial

Let us assume that X is the dichotomous variable and Y is the continuous one. We know that we can carry out a linear transformation on either variable without affecting the magnitude of the coefficient, so let us transform the X variable so that one of its dichotomous categories has a score of 0 and the other a score of 1. Then using formulas for the mean and standard deviation of a dichotomous variable in the Pearson formula we can derive a slightly simplified formula:

$$r_{pb} = \frac{(\overline{Y}_1 - \overline{Y}_0)\sqrt{N_0 N_1}}{s_Y(N_0 + N_1)} \tag{16-15}$$

where

$$\begin{aligned}
N_0 &= \text{number of subjects for whom } X = 0 \\
N_1 &= \text{number of subjects for whom } X = 1 \\
\overline{Y}_0 &= \text{mean of } Y \text{ scores for subjects whose } X \text{ score is 0} \\
\overline{Y}_1 &= \text{mean of } Y \text{ scores for subjects whose } X \text{ score is 1} \\
s_Y &= \text{biased sample standard deviation of } Y
\end{aligned}$$

The derivation can be found in a number of intermediate statistics texts, and other formulas are available as well [see especially, Walker and Lev (1963), 262–267]. Here, however, let us simply use an example to show the method of computing r_{pb} and to illustrate its equivalence to the Pearson coefficient.

EXAMPLE   First let us calculate the Pearson correlation coefficient for the following sets of scores, X and Y:

i	X	Y	X^2	Y^2	XY
1	1	7	1	49	7
2	1	9	1	81	9
3	0	5	0	25	0
4	1	6	1	36	6
5	0	3	0	9	0
$\sum$	3	30	3	200	22

$$r_{XY} = \frac{N\sum XY - (\sum X)(\sum Y)}{\sqrt{[N\sum X^2 - (\sum X)^2][N\sum Y^2 - (\sum Y)^2]}}$$

$$= \frac{5(22) - 3(30)}{\sqrt{[5(200) - (30)^2][5(3) - (3)^2]}} = \frac{110 - 90}{\sqrt{100(6)}}$$

$$= \frac{20}{10\sqrt{6}} = \frac{2}{\sqrt{6}} = \frac{2\sqrt{6}}{6} = \frac{\sqrt{6}}{3} \doteq .816$$

To compute the point-biserial correlation, we need a slightly different data table. The X^2 and XY columns are no longer needed, and in their place we have a column of Y scores for subjects whose X score is 0 and a column of Y scores for subjects whose X score is 1. Y given $X = 0$ and Y given $X = 1$.

i	X	Y	Y^2	Y when $X = 0$	Y when $X = 1$
1	1	7	49	—	7
2	1	9	81	—	9
3	0	5	25	5	—
4	1	6	36	—	6
5	0	3	9	3	—
$\sum$	3	30	200	8	22

$$s_Y = \sqrt{\frac{\sum Y^2 - (\sum Y)^2/N}{N}} = \sqrt{\frac{N\sum Y^2 - (\sum Y)^2}{N^2}}$$

$$= \sqrt{\frac{5(200) - (30)^2}{25}} = \sqrt{\frac{1000 - 900}{25}} = \sqrt{\frac{100}{25}} = \sqrt{4} = 2$$

$$\bar{Y}_0 = \frac{\sum(Y \text{ given } X = 0)}{N_0} = \frac{8}{2} = 4$$

$$\overline{Y}_1 = \frac{\sum (Y \text{ given } X = 1)}{N_1} = \frac{22}{3} = 7.33$$

$$r_{pb} = \frac{(\overline{Y}_1 - \overline{Y}_0)\sqrt{N_0 N_1}}{s_Y(N_0 + N_1)} = \frac{(22/3 - 4)\sqrt{3(2)}}{2(3 + 2)}$$

$$= \frac{(22 - 12)\sqrt{6}}{3 \cdot 2 \cdot 5} = \frac{10\sqrt{6}}{3 \cdot 10} = \frac{\sqrt{6}}{3} \doteq .816$$

which is identical to the value obtained from the Pearson formula applied to the same data.

B. Significance of a Point-Biserial Coefficient

> *Relevant earlier sections:*
> **7-4** t distribution
> **Chapter 9**   Hypothesis testing
> **16-2E**   Significance of Pearson coefficient

Let us call X the dichotomous variable. When the variance of Y for $X = 0$ is the same as the variance of Y for $X = 1$ in the population, and when the two distributions of Y are both normally distributed, then it is possible to test the hypothesis of a zero population correlation. The formula to use is

$$t = \frac{r_{pb}\sqrt{N - 2}}{\sqrt{1 - r_{pb}^2}} \qquad \text{with } N - 2 \text{ df} \qquad (16\text{-}16)$$

This formula can be used to test either a nondirectional null hypothesis ($\rho_{pb} = 0$) or a directional null hypothesis ($\rho_{pb} \geq 0$ or $\rho_{pb} \leq 0$). For the example above,

$$t = \frac{.816\sqrt{3}}{\sqrt{1 - \frac{6}{9}}} = \frac{.816\sqrt{3}}{\sqrt{\frac{1}{3}}} = \frac{.816\sqrt{3}}{\sqrt{3/3}} = 3(.816) = 2.45$$

With $N - 2 = 5 - 2 = 3$ df, $t_{.05} = \pm 3.182$, for a two-tail test. The observed t is less than $+3.182$, so the observed correlation is not significantly different from zero.

C. Correction of the Point Biserial for Grouping: The Biserial Coefficient
As has already been indicated in this section, the point-biserial coefficient is equivalent to the Pearson coefficient applied to two variables, one of which is dichotomous. If we assume that there is a continuous variable underlying the dichotomous one, we can use the correction for grouping (Section 16-2F) to make an inference about the correlation between our continuous variable and the continuous variable hypothesized to underlie the dichotomy.

16-3 *Point-biserial and biserial correlation coefficients*

The underlying distribution of the dichotomized variable may be assumed to be either rectangular or normal and the dichotomization may have produced categories that are equal or unequal in size. In any case, when X is the dichotomized variable, the correction formula is

$$_c r_{XY} = \frac{r_{pb}}{r_{X\bar{X}}} \tag{16-17}$$

and the formulas for $r_{X\bar{X}}$ are as given in the earlier section.

One case that is of general interest occurs when the dichotomized variable is assumed to have an underlying normal distribution. Then the correlation of raw scores and group means for the dichotomized variable is given by

$$r_{X\bar{X}} = \frac{y(N_0 + N_1)}{\sqrt{N_0 N_1}} \tag{16-18}$$

where N_0 and N_1 are defined as above, and y is the ordinate of the standard normal distribution at the point that separates it into two proportions, $N_0/(N_0 + N_1)$ and $N_1/(N_0 + N_1)$. Substitution of formulas 16-15 and 16-18 into formula 16-17 gives

$$_c r_{XY} = r_{bis} = \frac{(\bar{Y}_0 - \bar{Y}_1)N_0 N_1}{y(N_0 + N_1)^2 s_Y} \tag{16-19}$$

which is known as the biserial correlation coefficient.

EXAMPLE   The biserial correlation coefficient can be calculated for the example of this section. In these data, $N_0/(N_0 + N_1) = 2/5 = .40$ and $N_1/(N_0 + N_1) = 3/5 = .60$. From Table A-4, the ordinate of the standard normal distribution at the point where the larger area is 0.60, is $y = .3863$. Using formula 16-19, we find the biserial correlation to be

$$r_{bis} = \frac{(22/3 - 4)(3)(2)}{(.3863)(5)^2(2)} = \frac{3(22 - 12)/3}{25(.3863)}$$

$$= \frac{10}{25(.3863)} = \frac{1}{1.9315} = 1.03$$

We can also correct the point biserial value for grouping to get r_{bis}. We assume that the underlying distribution of X is normal and that 1 and 0 are category means. Applying formula 16-18, we get $r_{X\bar{X}} = .795$. We already know that $r_{pb} = .816$ for these data, so, from formula 16-11,

$$r_{bis} = {}_c r_{pb} = \frac{.816}{.795} \doteq 1.03$$

This example also illustrates the fact that the biserial coefficient can be greater than 1. Our first tendency in such a case is to look for some error, an error in calculations, an assumption not met. As in the present case, however, there may be no obvious error. In such a case, it is safest to say that if the Pearson coefficient were calculated on the underlying distribution of raw scores, it would probably be between .816 (the value of the point biserial) and 1.00, which is its maximum possible value.

References The point-biserial coefficient is discussed in Guilford (1965), 322–325; McNemar (1969), 218–219; Peatman (1963), 143–146; and Walker and Lev (1953), 262–267. The biserial can be found in Guilford, 317–322; McNemar, 215–218; Peatman, 126–128; and Walker and Lev, 267–271.

16-4   The multiserial correlation coefficient

> *Relevant earlier sections:*
> **7-3**   Normal distribution
> **16-2**   Pearson correlation coefficient

The multiserial correlation coefficient allows us to obtain an indication of the degree of relationship between an ordinal and an interval variable when we are willing to make the assumption that the ordinal variable is based on an underlying interval-scaled, normally distributed variable. If we are not willing to make this assumption, then it is better to handle the data in some other fashion, say by treating the ordinal variable as interval and using the Pearson coefficient (Section 16-2), or by degrading the interval-level variable to ordinal and finding the relationship between two ordinal variables (Section 17-2).

A. Calculation of the Multiserial Correlation Coefficient
The general procedure followed is to convert the ordinal variable to an interval scale and then to use the Pearson coefficient to indicate degree of relationship between the two interval-level variables. Let us call the variable that is interval-scaled X and the ordinal variable W.

We begin by assuming that the ranks on W are based on an underlying interval scale, and that the underlying set of scores are normally distributed. We know that for any normally distributed variable it is possible to convert from percentages of subjects to z scores and back, using the table of the standard normal distribution (Table A-4; for the method, see Section 7-3). Each subject in a given category of W is assumed to be at the mean of that category on the underlying interval scale. The mean of the underlying scale for that category is given by

$$\bar{z}_j = \frac{y_{Lj} - y_{Uj}}{p_j}$$

where j is the category label, y_{Uj} is the ordinate of the standard normal distribution corresponding to the upper bound of the category; y_{Lj} is the ordinate of the standard normal distribution corresponding to the lower bound of the category; and p_j is the proportion of subjects in the category $p_j = N_j/N$, where N_j is the number of subjects in the category and N is the total number of subjects.

Let us double subscript each score, using j to represent the level of the ordinal variable W and i as a subject subscript within levels of W. Using this notation in the deviation-score formula for the Pearson coefficient (formula 16-3) gives

$$r_{wx} = \frac{\sum_j \sum_i w_{ij} x_{ij}}{\sqrt{(\sum_j \sum_i x_{ij}^2)(\sum_j \sum_i w_{ij}^2)}} \tag{16-20}$$

If we assume that each subject in category j of W is at the mean $\bar{z}_j$ of the underlying variable for that category, then we can replace w_{ij} by $\bar{z}_j$ for each subject. Thus,

$$\sum_j \sum_i w_{ij}^2 = \sum_j \sum_i \bar{z}_{ij}^2 = \sum_j N_j \bar{z}_j^2 \tag{16-21}$$

and

$$\sum_j \sum_i w_{ij} x_{ij} = \sum_j \sum_i \bar{z}_j x_{ij} = \sum_j \left(\bar{z}_j \sum_i x_{ij} \right) \tag{16-22}$$

Inserting formulas 16-21 and 16-22 in formula 16-20 gives the formula for the multiserial (*ms*) correlation:

$$r_{ms} = \frac{\sum_j \left(\bar{z}_j \sum_i x_{ij} \right)}{\sqrt{(\sum_j \sum_i x_{ij}^2)(\sum_j N_j \bar{z}_j^2)}} \tag{16-23}$$

This is the formula to use when generalization is to be made to the same set of ordinal categories that has the same definitions as those the calculations are based on.

It is also possible to develop a correction formula that allows you to estimate the correlation between the interval-level variable (X) and the second interval-level variable hypothesized to underlie the ordinal variable (W) on which your calculations are made. The correction formula is formula 16-11, into which formula 16-13 has been inserted. The hypothesized z scores underlying the ordinal variable W have $\mu_z = 0$ and $\sigma_z = 1$ by definition of z scores. Thus, formula 16-13 becomes

$$r_{z\bar{z}} = \frac{\sigma_{\bar{z}}}{\sigma_z} = \frac{\sigma_{\bar{z}}}{1.0} = \sigma_{\bar{z}}$$

When this is inserted in formula 16-11, we get the corrected multiserial coefficient of

$$_c r_{ms} = \frac{r_{ms}}{\sigma_{\bar{z}}} \qquad (16\text{-}24)$$

where $\sigma_{\bar{z}}$ is the standard deviation of the $\bar{z}$ scores used in the calculation of the multiserial coefficient. In practice, the sample standard deviation is used, calculated as in

$$s_{\bar{z}} = \sqrt{\frac{\sum_j N_j \bar{z}_j^2}{N}} \qquad (16\text{-}25)$$

Substituting formulas 16-23 and 16-25 in formula 16-24 gives a formula for the multiserial coefficient which has the correction built in:

$$_c r_{ms} = \frac{\sum_j \left(\bar{z}_j \sum_i x_{ij} \right)}{\sum_j N_j \bar{z}_j^2 \sqrt{\sum_j \sum_i x_{ij}^2 / N}} \qquad (16\text{-}26)$$

EXAMPLE   Fifteen students in a sociology class are graded on a term paper and a final examination. The examination grades are numerical, but only letter grades are given on the term papers. The instructor wishes to know the degree of relationship between the two sets of marks, under the assumption that the underlying ability continuum is interval level and normally distributed. The marks are listed in the first two columns of the table.

Term paper (W)	Exam (X)	i	$x_{ij} = X_{ij} - \bar{X}..$	x^2
A	19	1	4	16
A	18	2	3	9
A	22	3	7	49
B	19	1	4	16
B	20	2	5	25
B	18	3	3	9
B	18	4	3	9
C	16	1	1	1
C	15	2	0	0
C	12	3	-3	9
C	13	4	-2	4
C	16	5	1	1
D	6	1	-9	81
D	8	2	-7	49
F	5	1	-10	100
$\sum$	225			378
Mean ($\bar{X}..$)	15			

In this table, students are grouped according to their marks on the term paper (the ordinal variable W), with j varying from A to F (here we are letting j take on letter values rather than numbers). The subscript i distinguishes among the students having the same term-paper mark. Now let us obtain a second table.

j	$\sum_i x_{ij}$	N_j	p_j	y_L	y_U	$\bar{z}$	$\bar{z}^2$	$\bar{z}_j \sum_i x_{ij}$	$N_j \bar{z}_j^2$
A	14	3	.200	.2800	.0000	1.40	1.96	19.60	5.88
B	15	4	.267	.3974	.2800	.44	.194	6.60	.78
C	−3	5	.333	.2800	.3974	−.35	.122	1.05	.61
D	−16	2	.133	.1268	.2800	−1.15	1.32	18.40	2.64
F	−10	1	.067	.0000	.1268	−1.89	3.57	18.90	3.57
$\sum$		15						64.55	13.48

Next are example calculations for obtaining the values in the top row of the second table from the first table:

$$\sum_i x_{ij} = \sum_i x_{iA} = x_{1A} + x_{2A} + x_{3A} = 4 + 3 + 7 = 14$$

$$p_j = p_A = \frac{N_A}{N} = \frac{3}{15} = .200$$

y_L is the ordinate of the standard normal distribution for the z score below which .80 of the distribution falls and y_U is the ordinate of the standard normal distribution for the z score below which 1.00 of the distribution falls, so

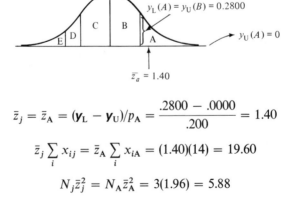

$$\bar{z}_j = \bar{z}_A = (y_L - y_U)/p_A = \frac{.2800 - .0000}{.200} = 1.40$$

$$\bar{z}_j \sum_i x_{ij} = \bar{z}_A \sum_i x_{iA} = (1.40)(14) = 19.60$$

$$N_j \bar{z}_j^2 = N_A \bar{z}_A^2 = 3(1.96) = 5.88$$

From the values in the two tables, the value of the multiserial coefficient can be calculated:

$$r_{ms} = \frac{\sum_j \left(\bar{z}_j \sum_i x_{ij}\right)}{\sqrt{\left(\sum_j \sum_i x_{ij}^2\right)\left(\sum_j N_j \bar{z}_j^2\right)}} = \frac{64.55}{\sqrt{(378)(13.48)}}$$

$$= \frac{64.55}{\sqrt{5095}} = \frac{64.55}{71.38} \doteq .90$$

Finally, we may use formulas 16-25 and 16-24 to estimate the correlation that we would have obtained if we had had a measure of the hypothesized interval-level, normally distributed variable underlying the term-paper marks:

$$s_z = \sqrt{\frac{\sum_j N_j \bar{z}_j^2}{N}} = \sqrt{\frac{13.48}{15}} \doteq .948$$

$$_c r_{ms} = \frac{r_{ms}}{s_z} = \frac{.90}{.948} \doteq .95$$

B. Significance Test for the Multiserial Correlation Coefficient

The multiserial coefficient is a form of the Pearson coefficient, and the exact same formula is used to test its significance as was presented in Section 16-2E:

$$t = \frac{r_{ms}\sqrt{N-2}}{\sqrt{1-r_{ms}^2}} \quad \text{with } N-2 \text{ df} \tag{16-27}$$

As before, this statistic can be used either to test a nondirectional hypothesis ($H_0: \rho_{ms} = 0$) or a directional null hypothesis ($H_0: \rho_{ms} \geq 0$ or $H_0: \rho_{ms} \leq 0$).

EXAMPLE   For the computational example of this section, $r_{ms} = .90$ and $N = 15$. Let us test the motivated hypothesis that the two variables (exam score and term-paper mark) are positively correlated. The null hypothesis is $H_0: \rho_{ms} \leq 0$. Let us choose an α level of .05, giving a critical t of 1.771. The calculated value of t is

$$t = \frac{.90\sqrt{13}}{\sqrt{1-.81}} = \frac{3.24}{.44} = 7.44$$

which is significant at the .05 level. Thus, we say that the two variables are positively correlated in the population.

References   The multiserial coefficient was developed by Jaspen (1946). Another discussion of it can be found in Freeman (1965), Chapter 12.

16-5 Phi correlation coefficient

Relevant earlier sections:
7-6 Chi square
16-2 Pearson correlation coefficient

A. Use of the Phi Coefficient

The phi coefficient is used to indicate degree of relationship between two dichotomous (two-valued) variables. It is probably most often used to correlate pairs of test items, where items are scored right–wrong, pro–anti, etc.; but it might also be used to correlate sex and handedness, or party preference (Republican–Democrat) with attitude toward, say, government control of the oil industry (for–against). Phi can be used to correlate continuous variables that have been forced into a dichotomy; when this is done, it is common practice to divide both variables at their respective medians. However, the resulting phi coefficient does not accurately reflect the correlation that would have been obtained by application of the Pearson coefficient to the original variables, and the loss of information in dichotomizing generally makes this procedure undesirable.

When phi is appropriate, lambda (Section 17-3) can also be used. Lambda has the advantage of being used also with larger $(J \times K)$ frequency tables. Phi's advantages have to do with its being better known and its being a special case of the Pearson coefficient.

B. Computation of Phi

The phi coefficient actually amounts to a simplification of the Pearson correlation when applied to the case where both variables are dichotomous. Rather than give the derivation, however, the formula will be presented, and a comparison of the two coefficients will be made in the subsequent example.

Table 16-3 *Data for computing phi*

Variable 1	Variable 2		
	I	II	Σ
I	a	b	$a + b$
II	c	d	$c + d$
Σ	$a + c$	$b + d$	N

The data arrangement is shown in Table 16-3. Here $a, b, c,$ and d are frequencies of subjects giving the four response combinations, where $a + b + c + d = N$, the total number of subjects. The correlation coefficient is then

$$r_\phi = \frac{ad - bc}{\sqrt{(a + b)(c + d)(a + c)(b + d)}} \qquad (16\text{-}28)$$

EXAMPLE A hundred subjects take an examination, and we wish to find the correlation between two of the items, here labeled X and Y. The data are:

	Y		
X	Right	Wrong	Σ
Right	30	30	60
Wrong	10	30	40
Σ	40	60	100

Let us begin by applying the Pearson formula to these data. In the following table a sum of each of the needed variables and combinations is given for each row, to avoid making a table 100 lines long. Rights are scored 1, wrongs scored 0.

Subjects	X	Y	f	$\sum X$	X^2	$\sum X^2$	$\sum Y$	Y^2	$\sum Y^2$	XY	$\sum XY$
1–30	1	1	30	30	1	30	30	1	30	1	30
31–60	1	0	30	30	1	30	0	0	0	0	0
61–70	0	1	10	0	0	0	10	1	10	0	0
71–100	0	0	30	0	0	0	0	0	0	0	0
Σ			100	60		60	40		40		30

Using the Pearson formula, we obtain

$$r_{XY} = \frac{N \sum XY - (\sum X)(\sum Y)}{\sqrt{[N \sum X^2 - (\sum X)^2][N \sum Y^2 - (\sum Y)^2]}}$$

$$= \frac{100(30) - (60)(40)}{\sqrt{[100(60) - (60)^2][100(40) - (40)^2]}}$$

$$= \frac{3000 - 2400}{\sqrt{[6000 - 3600][4000 - 1600]}} = \frac{600}{\sqrt{(2400)(2400)}}$$

$$= \frac{600}{2400} = .25$$

Applying formula 16-28 to the same data, we obtain

$$r_\phi = \frac{(30)(30) - (30)(10)}{\sqrt{(60)(40)(40)(60)}} = \frac{900 - 300}{(60)(40)} = \frac{600}{2400}$$

$$= .25$$

We have avoided the need for the second table, and the resulting computations are easier for phi than for the Pearson coefficient.

C. Maximum and Minimum Values of Phi

Like the Pearson coefficient from which it is derived, phi can range between $+1$ and -1; but the limits can be achieved only under unusual conditions. Phi can equal $+1$ only when the total $a + b$ equals the total $a + c$, and -1 only when the total $a + b$ equals the total $b + d$. In the above example a phi of $+1$ cannot be achieved, because with the given totals, the largest number of subjects who can be right for both items is 40, the total right for item Y. Then if we fill in the cells to give the highest possible correlation for the given marginal totals, we get the following table:

		Y		
X	Right	Wrong	Σ
Right	40	20	60
Wrong	0	40	40
Σ	40	60	100

The magnitude of phi calculated from this table is as follows:

$$r_{\phi(\text{max})} = \frac{40 \cdot 40}{\sqrt{40 \cdot 60 \cdot 60 \cdot 40}} = \frac{1600}{2400} = .67$$

One example of this situation occurs with achievement-test items, where Y is harder than X, because fewer people get Y right. The difference in item difficulties is a result of the fact that at least some people get X right and Y wrong. But the correlation can be $+1$ only if everyone who gets X right gets Y right, and everyone who gets X wrong gets Y wrong. Therefore, differences in item difficulty make a perfect correlation impossible.

It should be noted that phi could achieve a minimum value of -1 for this example if $b = 60$ and $c = 40$.

D. Significance of Phi

We can test the null hypothesis $H_0 : \rho_\phi = 0$, that is, that the population phi correlation is equal to zero, using the statistics for a 2×2 matrix, as indicated in the decision map for Chapter 10. In practice, however, it seldom pays to calculate phi when N is so small that the Fisher test is appropriate, because in this case r is very unstable from sample to sample drawn from the same population. For large N, the appropriate formula is

$$\chi^2 = \frac{N\left(|ad - bc| - \dfrac{N}{2}\right)^2}{(a + b)(c + d)(a + c)(b + d)} \qquad df = 1 \qquad (16\text{-}29)$$

If $|ad - bc| < (N/2)$, set chi square equal to zero. Except for Yates' correction, required because of the chi-square approximation, this would reduce to $\chi^2 = Nr_\phi^2$. As it is, some of the calculations made in obtaining r may be used to obtain χ^2.

In the example of part B,

$$|ad - bc| = 600 \qquad \text{and} \qquad \sqrt{(a + b)(c + d)(a + c)(b + d)} = 2400$$

$N = 100$, so

$$\chi^2 = \frac{100(600 - 50)^2}{(2400)^2} = \frac{100(302,500)}{5,760,000}$$

$$= \frac{3025}{576} = 5.25$$

The critical value of chi square for 1 df at an α of .05 is 3.841. The observed chi square is greater than the tabulated value, so we reject the null hypothesis. We conclude that the population correlation is positive.

References Phi is discussed in Guilford (1965), 333–339; Hays (1963), 604–605; and McNemar (1969), 225–227. The tetrachoric coefficient can be found in Guilford, 326–332; and McNemar, 221–225.

Homework

1. The following bivariate plots indicate possible relationships between variables X and Y. Only the regression lines are presented. What is the correct order of magnitude of the correlations, from greatest to least?

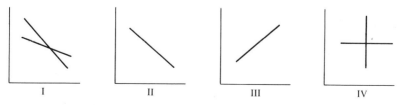

I   II   III   IV

Arbitrarily assume that r of .25 is greater than r of $-.25$

2. For each of the following, state what correlation coefficient you would use to indicate accuracy of prediction and why.

(a) Subjects: college undergraduates.
   Predictor: anxiety as measured by a standard, 80-item multiple-choice test.
   Criterion: score on a 40-item test of complex analogies. (*Note:* A *moderate* amount of anxiety is known to be optimal for performance on this test.)

(b) Subjects: fourth-grade children.
   Predictor: letter grade in spelling.
   Criterion: class standing in a spelldown.

(c) Subjects: male heads of households.
   Predictor: race (white, black, oriental, and American Indian).
   Criterion: income.

(d) Subjects: four-year-olds.
   Predictor: sex.
   Criterion: preferred parent.

(e) Subjects: college undergraduate math majors.
   Predictor: scores on math portion of College Board Test.
   Criterion: grade average in math courses at end of two years.

(f) Subjects: adults.
   Predictor: nationality.
   Criterion: beverage preferred with main meal of day.

(g) Subjects: undergraduates playing a game in a psychology laboratory.
   Predictor: treatment condition : competitive or cooperative.
   Criterion: amount of money won in the game after a large number of trials.

For problems 3 through 6, find the correlation between X and Y and determine whether it is significantly different from zero (two-tail test, $\alpha = .05$). Use the methods of Chapter 16 only. These samples are too small to check for bivariate normality and homogeneity, so assume that they apply.

3.

i	X	Y
1	7	-5
2	9	-5
3	5	0
4	6	-5
5	3	-5

4.

i	x	y
1	6	2
2	6	1
3	-6	-1
4	-4	-1
5	-2	-1

5.

i	z_x	z_y
1	1.42	1.68
2	.07	.00
3	-2.50	-.62
4	-.15	-.48
5	.46	-.76
6	.70	.18

6.

i	X	Y
1–20	1	1
21–30	1	2
31–40	2	1
41–60	2	2

7. The same subjects are administered three tests, A, B, and C. If the correlation $r_{AB} = .42$, and $r_{AC} = .25$, what is r_{BC}?

8. The following information is known about the relationship between two variables, X and Y: $\bar{Y} = 2$, $s_X = 10$, $s_Y = 2$, and the regression line for predicting Y is given by $Y' = .16X + .40$. Find the regression line for predicting X from Y and the correlation between the variables.

9. For a sample of 62 subjects, what is the smallest Pearson correlation that is significantly different from zero at an α of .05?

10. Problem 2 asked you to state the order of magnitude of several Pearson correlation coefficients when only the regression lines were given. Show that this can be done mathematically by computing a correlation when $b_{XY} = 3$ and $b_{YX} = .27$.

11. In a survey of undergraduate psychology majors at two universities, students were asked whether they thought the greatest advances in psychology in the next five years would come from research done "in the laboratory" or "in the field." In university A, 10 said "lab" and 40 said "field"; in university B, 40 said "lab" and none said "field." What is the degree of relationship between university and student judgment on this issue?

12. For the following data, assume that the assumptions of the Pearson coefficient are met:

Then,
(a) Find r_{XY}.
(b) Find the two regression equations.
(c) Find the two standard errors of estimate.
(d) Find the percent of variance in each variable accounted for by the other.

i	X	Y
1	10	5
2	11	3
3	17	1
4	11	3
5	11	3

Homework Answers

1. III, IV, I, II
2. Two things are important: the reasonableness of your assumptions about the data and consistency between your assumptions and the coefficient

that you choose. I chose: (a) eta, (b) gamma, (c) eta, (d) phi, (e) Pearson, (f) lambda, (g) point biserial.

3. $r_{pb} = -.25$; $t_{.05}(3 \text{ df}) = \pm 3.182$; $t_r = -.45$, which is nonsignificant (NS)

4. $r_{XY} \doteq .94$; $t_{.05}(3 \text{ df}) = \pm 3.182$; $t_r = 4.7^*$

5. $r_{XY} \doteq .63$; $t_{.05}(4 \text{ df}) = \pm 2.776$; $t_r = 1.62(\text{NS})$

6. $r_\phi = .33$; $\chi^2_{.05}(1 \text{ df}) = 3.84146$; $\chi^2(\text{obs}) = 5.4^*$

7. There is not sufficient information to solve this problem.

8. $r_{XY} = .80$; $X' = 4Y + 2$

9. $r_{XY} = \pm .25$

10. .90

11. $r_\phi = .80$ (the sign is arbitrary)

12. (a) $r_{XY} = -.875$

(b) $X' = 17.25 - 1.75Y$; $Y' = 8.26 - .438X$

(c) $s_{YX} \doteq .612$; $s_{XY} \doteq 1.22$

(d) $r^2 \doteq .77$

---

* Significant.

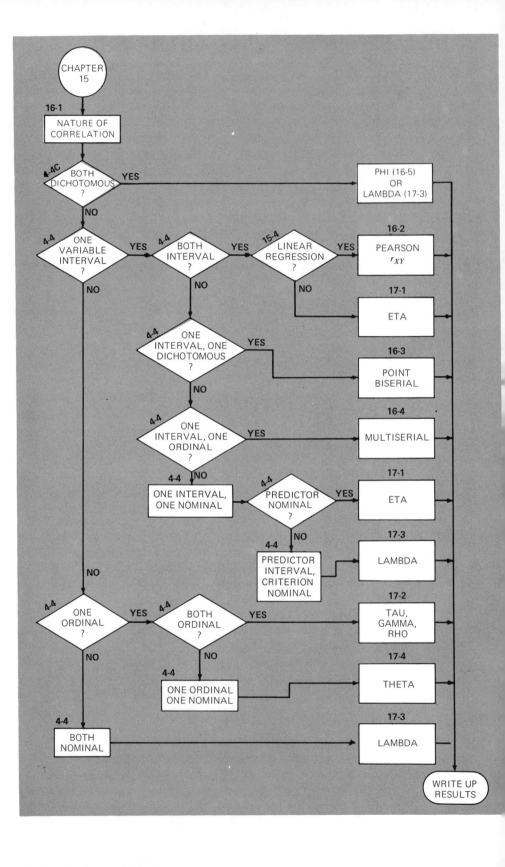

17

Other Two-Variable
Correlation Indices

THIS chapter describes the correlation ratio and methods for correlating pairs
of nominal- and ordinal-level variables. Not here are the Pearson coefficient
and Pearson-derived coefficients, which appear in Chapter 16; and methods
for correlating three or more variables, which appear in Chapter 18.

A general introduction to correlation methods appears in the introduction
to Chapter 16 and in Section 16-1. You should read these before continuing.

17-1   Eta, the correlation ratio

Relevant earlier sections:
16   Introduction
16-1   General problem of correlation

A. Uses of the Correlation Ratio

Eta (η) is a coefficient that can be used to indicate the degree of effectiveness
in predicting an interval-scaled dependent or criterion variable from a predictor
or independent variable that may be interval, ordinal, or nominal. If the predictor is interval-scaled, the regression line for predicting the dependent variable
may be straight or curved or even discontinuous. Thus, eta may be used to
indicate degree of effectiveness in predicting authoritarianism test scores
(interval) from religion (nominal), grade-point average (interval) from undergraduate major (nominal), scores on the Scholastic Aptitude Test (interval)
from grades in fourth-year English (ordinal?), achievement-test performance
from anxiety-test scores (both interval but with possible curvilinear regression),
or the relationship between independent and dependent variables in any experiment where a *t* test or an analysis of variance might be used to test for equality
of means.

The major distinction between eta and the Pearson coefficient is that the
Pearson tells how effective it is to predict Y scores using values of X and the
regression line between X and Y; eta tells how effective it is to predict, using the
conditional mean of Y values for each of the levels of X, as in Section 15-2C.
When all the conditional means fall on a straight line, eta and the Pearson will
be identical (assuming that X is interval level); and to the extent that the conditional means fall away from the regression line, the Pearson will be smaller
than eta.

When there are only two levels of X, the Pearson coefficient is the same as the point biserial. In this case, the two conditional means must fall on a straight line, because from geometry we know that a straight line can be drawn through any two points, and on a graph the two means are represented as points, one point for each of the levels of X. Therefore, in this case, eta is equal to both the Pearson and to the point biserial.

B. Definition of the Correlation Ratio

Eta is the positive square root of eta square. Eta square is an indicator of the extent to which error of prediction of individual scores on the dependent or criterion variable can be reduced through knowledge of subjects' scores on the predictor variable, and the relationship between the two variables.

Suppose that we have on hand a population of individuals for whom we have scores on two variables, X and Y. Suppose that we now draw an individual at random from this population. If we do not look at either his X or his Y score, our best predictor of his status on Y is the mean μ_Y of that distribution; and our indicator of error of prediction is σ_Y^2, the variance of Y scores.

If, on the other hand, we know his X score, our best predictor of his status on Y is the mean of the distribution of Y scores for all subjects in the population who have the same X score that he has. In this case, our best indicator of the error of prediction is $\sigma_{YX_j}^2$, the variance of Y scores for all subjects who have the same predictor score X_j as the subject for whom the prediction is being made. To obtain the overall value of error of prediction from X to Y, we pool the variances $\sigma_{YX_j}^2$ to get σ_{YX}^2.

From Section 16-1, we define our correlation coefficient as the proportion of error of prediction that is eliminated by knowledge of X scores and the relationship between X and Y. Then using σ_{YX}^2 and σ_Y^2 as our indicators of error of prediction with and without knowledge of the relationship between X and Y, respectively, we can define eta square as the proportion of the total variance of Y that can be eliminated by, or accounted for by, knowledge of subjects' status on X and the mean $\overline{Y}_j$ of Y scores for each level X_j of X:

$$\eta^2 = \frac{\sigma_Y^2 - \sigma_{YX}^2}{\sigma_Y^2} \tag{17-1}$$

The correlation ratio, eta, is the positive square root of this formula.

C. Sample Estimates of a Population Correlation Ratio

A sample estimate of a population eta square can be obtained by replacing the population variances in formula 17-1 by unbiased sample variance estimates:

$$\hat{\eta}^2 = \frac{\hat{s}_Y^2 - \hat{s}_{YX}^2}{\hat{s}_Y^2} \tag{17-2}$$

and, as before, the estimated correlation ratio is the positive square root of this expression.

There are several methods for carrying out the calculations, depending on the form of the data you have to work with.

1. Raw-Score Method: In this case, you find values of $\hat{s}_Y^2$ and $\hat{s}_{YX}^2$ and enter them directly into equation 17-2. First find $\hat{s}_Y^2$, using the formula

$$\hat{s}_Y^2 = \frac{\sum_j \sum_i (Y_{ij} - \overline{Y}..)^2}{N_T - 1}$$

where Y_{ij} is the score on Y of some subject i who gets a score X_j on variable X, $\overline{Y}..$ is the grand mean of Y scores, regardless of X scores, and N_T is the total number of subjects.

Next find $\hat{s}_{YX_j}^2$, the variance of Y scores for some level of j of X, and repeat for each other level of X, using the usual formula for an unbiased variance estimate:

$$\hat{s}_{YX_j}^2 = \frac{\sum_i (Y_{ij} - \overline{Y}._j)^2}{N_j - 1}$$

where N_j is the number of subjects getting the score X_j. The values of the several $\hat{s}_{YX_j}^2$ are then pooled using the formula

$$\hat{s}_{YX}^2 = \frac{\sum_j (N_j - 1)\hat{s}_{YX_j}^2}{N_T - J}$$

which has the same general form as formulas 12-13 and 13-1. The values of $\hat{s}_Y^2$ and $\hat{s}_{YX}^2$ are then inserted in formula 17-2 to find the value of eta square, and the square root is taken to give the estimated correlation ratio.

EXAMPLE   Consider the data table, which has been arranged in three subtables to aid in calculations.

i	X_1	Y	Y^2	i	X_2	Y	Y^2	i	X_3	Y	Y^2
1	7	11	121	1	12	8	64	1	31	11	121
2	7	9	81	2	12	8	64	2	31	7	49
3	7	15	225	3	12	11	121	3	31	11	121
4	7	13	169	4	12	12	144	4	31	7	49
5	7	12	144	5	12	6	36	5	31	9	81
$\sum$		60	740	$\sum$		45	429	$\sum$		45	421

The calculations for these data are

$$\hat{s}_1^2 = \frac{N_1 \sum Y_{i1}^2 - (\sum Y_{i1})^2}{N_1(N_1 - 1)} = \frac{5(740) - 3600}{20} = \frac{3700 - 3600}{20} = 5$$

$$\hat{s}_2^2 = \frac{5(429) - (45)^2}{20} = \frac{2145 - 2025}{20} = \frac{120}{20} = 6$$

$$\hat{s}_3^2 = \frac{5(421) - 2025}{20} = \frac{80}{20} = 4$$

$$\hat{s}_{YX}^2 = \frac{4(5) + 4(6) + 4(4)}{4 + 4 + 4} = \frac{60}{12} = 5$$

$$\hat{s}_Y^2 = \frac{15(1590) - (150)^2}{15(14)} = \frac{23,850 - 22,500}{210} = \frac{1350}{210}$$

$$\doteq 6.43$$

$$\eta^2 = \frac{6.43 - 5}{6.43} = \frac{1.43}{6.43} \doteq .222$$

$$\eta = \sqrt{.222} \doteq .47$$

2. One-Way Analysis-of-Variance Method: Suppose that we had been presented with these data, not as a correlation problem, but as a one-way analysis of variance (Section 13-4). We would have set it up and solved it first to obtain a value of F and to determine whether differences in levels of X had a significant effect on Y scores. The analysis-of-variance table has the following form:

Source of variance	Sum of squares	Degrees of freedom	Mean square	F
Between	SS_B	$J - 1$	$SS_B/(J - 1)$	MS_B/MS_W
Within	SS_W	$J(N_j - 1)$	$SS_W/J(N_j - 1)$	
Total	SS_T	$JN_j - 1$		

The within mean square is the pooled variance within levels of the predictor and hence is equal to $\hat{s}_{YX}^2$. However, the estimate of total variance $\hat{s}_Y^2$ of the dependent variable is not usually calculated in an analysis-of-variance table. To find it, simply construct an additional entry for "total mean square":

$$MS_T = \frac{SS_T}{df_T} \qquad (17\text{-}3)$$

A review of Section 13-4 will show that the total mean square so defined is an unbiased estimate of the variance of all scores on the dependent variable Y about the grand mean $\overline{\overline{Y}}$. Thus, $\hat{s}_Y^2$ and $\hat{s}_{YX}^2$ in the formula for eta square can be replaced by MS_{Total} and MS_W, respectively, giving

$$\hat{\eta}^2 = \frac{MS_T - MS_W}{MS_T} \qquad (17\text{-}4)$$

EXAMPLE   The analysis of Section 13-4 applied to the data used for illustration of the first method for calculating the unbiased estimate of eta square gives these results:

$$SS_T = \sum_j \sum_i Y_{ij}^2 - \frac{\left(\sum_j \sum_i Y_{ij}\right)^2}{N} = 1590 - \frac{(150)^2}{15}$$

$$= 90$$

$$SS_B = \sum_j \frac{\left(\sum_i Y\right)^2}{N_j} - \frac{\left(\sum_j \sum_i Y_{ij}\right)^2}{N} = \frac{60^2}{5} + \frac{45^2}{5} + \frac{45^2}{5} - 1500$$

$$= 30$$

$$SS_W = SS_T - SS_B = 90 - 30$$

$$= 60$$

The analysis-of-variance table is thus as follows:

Source	Sum of squares	Degrees of freedom	Mean square	F	$F_{.05}$	Conclusion
Between	30	2	15	3	3.89	Not significant
Within	60	12	5			
Total	90	14	6.43			

The finding of a nonsignificant effect of treatment means that there is no evidence for any relationship at all between the treatment variable and the criterion in the population. It therefore seems pointless to estimate the degree of relationship in the population. However, let us complete the analysis in order to illustrate the calculations:

$$\hat{\eta}^2 = \frac{MS_T - MS_W}{MS_T} = \frac{90/14 - 60/12}{90/14} = \frac{6.43 - 5}{6.43}$$

$$= .222$$

and

$$\hat{\eta} = \sqrt{.222} \doteq .47$$

Using the definitions of mean squares as given in this section and Section 13-4, you can easily derive the following computational formula for eta square from formula 17-4:

$$\hat{\eta}^2 = \frac{SS_{betw.} - df_{betw.} MS_W}{SS_T} \qquad (17\text{-}5)$$

Formula 17-5 has two advantages over formula 17-4: you don't have to compute MS_T; and it generalizes easily to two-way analyses. Applying formula 17-5 to the example above gives

$$\hat{\eta}^2 = \frac{30 - 2(5)}{90} = \frac{20}{90} = .222$$

which is the same value found using formula 17-4.

Eta square can also be calculated from a formula using the final F ratio:

$$\hat{\eta}^2 = \frac{(F - 1)df_{betw.}}{F\,df_{betw.} + df_W} \qquad (17\text{-}6)$$

and this formula can be further modified by substituting $J - 1$ for $df_{betw.}$ and $N_T - J$ for df_W. You can verify for yourself that applying formula 17-6 to the example of this section also gives an $\hat{\eta}^2$ of .222. The special advantage of formula 17-6 is that it can be used to evaluate the effectiveness of treatments in published reports, where mean squares and sums of squares are generally not reported.

3. Eta Square for Two-Way Analyses of Variance: A two-way, fixed-effects analysis can be regarded as an extension of the one-way analysis in which the "between" or "cells" line is split into three lines, A, B, and AB (interaction). As indicated in Section 13-5.

$$SS_{cells} = SS_A + SS_B + SS_{AB}$$

and

$$df_{cells} = df_A + df_B + df_{AB}$$

It turns out that we can find the magnitude of eta square for each line of the two-way analysis by replacing $SS_{betw.}$ and $df_{betw.}$ in formula 17-5 by the sum of squares and degrees of freedom for the line we are interested in. For example,

$$\hat{\eta}_A^2 = \frac{SS_A - df_A MS_W}{SS_T} \tag{17-7}$$

and similarly for the B main effect and the interaction. Then it also turns out that

$$\hat{\eta}_A^2 + \hat{\eta}_B^2 + \hat{\eta}_{AB}^2 = \hat{\eta}_{cells}^2 \tag{17-8}$$

where $\hat{\eta}_{cells}^2$ is the two-way equivalent of $\hat{\eta}_{betw.}^2$.

EXAMPLE   In the example of Section 13-5C, the following summary table was presented:

Source	SS	df	MS	F	$F_{.05}$	Significant?
A	36	1	36	7.2	4.26	Yes
B	60	5	12	2.4	2.62	No
AB	60	5	12	2.4	2.62	No
Cells	156	11				
Within	120	24	5			
Total	276	35				

Applying formula 17-7 to the A line of the analysis gives

$$\hat{\eta}_A^2 = \frac{36 - 1(5)}{276} = \frac{31}{276} = .112$$

Thus, we can account for 11 percent of the variance of the dependent variable by knowledge of subjects' status on the A variable. Now let us go on to apply formula 17-7 to the B and interaction lines of the analysis.

$$\hat{\eta}_B^2 = \frac{60 - 5(5)}{276} = \frac{35}{276} = .127$$

and, since the interaction line has the same SS and df as the B main-effect line for this problem, $\hat{\eta}_{AB}^2 = .127$ also.

Here you can see a good example of the fact that you don't get the same information from a significance test and from a correlation. In the example, the two nonsignificant lines of the analysis each account for more of the variance of the dependent variable than the one line that is significant! We cannot tell the importance of an independent variable in accounting for differences on the dependent variable from the significance test alone; and we cannot tell whether we can generalize a relationship to the population just from the size of the correlation between independent and dependent variables. Because you get different, complementary information from correlations and significance tests, most authorities who consider the issue recommend that, wherever possible, you do both.

For this analysis, we could also obtain an estimate of eta square for the cells line, using a modification of formula 17-7:

$$\hat{\eta}^2_{\text{cells}} = \frac{156 - 11(5)}{276} = .366$$

Then equation 17-8 is verified numerically by noting that $.112 + 2(.127) = .366$. The variance of the dependent variable that can be accounted for by differences in cell means can be partitioned into three independent components: row effects, column effects, and interaction effects.

4. Eta Square for the Two-Sample t Test: When the means of two independent samples have been compared using the methods of Section 12-12, a modified version of formula 17-6 can be used to find the effectiveness of group membership in accounting for differences in the dependent variable. In the two-sample case, it can be shown that $t^2 = F$, as indicated in Section 7-7; and a further example of the relationship was given by the reanalysis of the example of Section 12-12 in Section 13-4. We can therefore replace F by t^2, df_B by 1, and df_W by $N_1 + N_2 - 2$ in formula 17-6, giving formula 17-9:

$$\hat{\eta}^2 = \frac{t^2 - 1}{t^2 + N_1 + N_2 - 2} \tag{17-9}$$

EXAMPLE For the example of page 307, $N_1 = N_2 = 6$ and $t = 2.121$. Then $t^2 = 4.50$, and inserting these values in formula 17-8 we get

$$\hat{\eta}^2 = \frac{4.50 - 1}{4.50 + 6 + 6 - 2} = \frac{3.50}{14.50} = .2414$$

Thus, differences between groups account for 24 percent of the variance of the dependent variable.

D. Some Comments About the Correlation Ratio

First, it is important to realize that the sign of eta is always positive regardless of the shape or direction of the regression line for predicting either variable from the other. In the case of the Pearson coefficient, the slope of the regression line has a meaningful relation to the value of the coefficient; here it does not. Consider Figure 17-1(a) and (b), in both of which X has five levels and Y is interval-scaled. Any decision about the sign of eta for these would probably not be generalizable to other graphs, because it would be arbitrary.

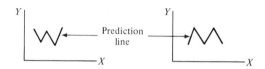

Figure 17-1   *Two possible bivariate graphs for X and Y.*

Second, it is important to note that, when two variables are related and both are interval-scaled, the eta for predicting one variable will not in general be the same as the eta for predicting the other: $\eta_{YX} \neq \eta_{XY}$. This principle can be illustrated in Figure 17-2. In this graph two smooth curves have been drawn: one that connects the Y means for different values of X, and one that connects the X means for different values of Y. An examination of the lines and plot of points indicates that the Y means are better predictors of Y values than the X means are of X scores. Therefore, the eta for predicting Y scores will be greater than the eta for predicting X scores.

Figure 17-2   *Bivariate graph with regression lines drawn in.*

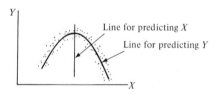

η_{YX} is an appropriate indicator of ability to predict subjects' status on Y from their status on X when predictions are made in the following way: whatever a subject's score on X, find all the other subjects who have that same X-score. Compute the mean Y-score for the selected subsample of subjects, and predict that value on Y for the given subject.

The variance $\hat{s}_Y^2$ in formula 17-2 was shown to be the variability of all the Y-scores about $\overline{Y}..$, the grand mean. This can be considered in two parts: the variabilities of the raw Y-scores about the predicted (conditional mean) values, s_{YX}^2, and the variability of the conditional means about the grand mean.

When all the raw Y-values are at the conditional means, $\hat{s}_{YX}^2 = 0$, and $\hat{\eta}^2 = \hat{s}_Y^2/\hat{s}_Y^2 = 1.0$. Another way of saying it is that when prediction is perfectly accurate, eta is 1.

At the other extreme, when all the conditional means $\overline{Y}._j$ are the same, the component of $\hat{s}_Y^2$ due to variability of conditional means is zero, and $\hat{s}_Y^2 = \hat{s}_{YX}^2$. Substituting $\hat{s}_Y^2$ for $\hat{s}_{YX}^2$ in formula 17-2, we get

463

$$\hat{\eta}^2 = \frac{\hat{s}_Y^2 - \hat{s}_Y^2}{\hat{s}_Y^2} = \frac{0}{\hat{s}_Y^2} = 0.$$

Thus, when we make the same Y-prediction for all subjects, regardless of their X-scores, we have gained nothing from the prediction, and eta is 0.

Recall that in using linear regression and the Pearson coefficient, predictions are made using a straight line that best fits the points in the bivariate graph. Now, if it should happen that all the conditional Y-means fall on a straight line, then by predicting, using the conditional means, we would actually be making linear predictions. In this case $\eta_{YX} = r_{YX}$. On the other hand, the difference $(\eta_{YX} - r_{YX})$ is an index of the extent to which the graph points depart systematically from a straight line.

The value of the eta that you obtain is in part a function of the fineness of the predictor scale in relation to the criterion scale and the number of subjects. If there is only one X score, eta must be zero; if there are only two, eta is equal to the point-biserial correlation (Section 16-3). Only if there are three or more X scores is it possible to show curvilinearity. At the other extreme, if there is a different X score for each different Y score observed in the sample, eta will be 1.0, even though a plot of Y as a function of X might be absurdly complicated and the predictions were so unstable as to be worthless when applied to another sample. As a minimum you should have two criterion scores for each level of the predictor, and preferably more. You cannot obtain a variance estimate unless there is the possibility of variance. The more scores each estimate is based on, the more stable it will be.

E. Correction of Eta for Grouping

As indicated in part D, the magnitude of eta is affected by the number of levels of the predictor. For fixed sample size, the fewer the predictor categories, the more accurately the conditional criterion means are estimated, but this may also increase the conditional variabilities, especially for small samples. Because eta does not make use of any regression line or other device for relating the predicted criterion value of one level of the predictor to the criterion scores for any other level of the predictor, grouping predictor levels is often necessary in order to have stable predictions. Grouping is often necessary for computing eta, therefore, when it is not necessary for calculating the Pearson coefficient.

A reasonably accurate correction may be made by assuming linear regression within predictor levels and applying formula 16-11 to eta, which gives

$$_c\hat{\eta}_{YX} = \frac{\hat{\eta}_{YX}}{r_{X\bar{X}}} \tag{17-10}$$

for the small-sample estimate. The magnitude of $r_{X\bar{X}}$ is estimated using the methods of Section 16-2F.

If both predictor and criterion have been grouped before calculation of eta, a double correction can be performed:

$$_c\hat{\eta}_{YX} = \frac{\eta_{YX}}{r_{X\bar{X}}r_{Y\bar{Y}}} \qquad (17\text{-}11)$$

Of course, it is inappropriate to correct the predictor when it is nominal, as is the case in analysis of variance and some other applications.

EXAMPLE For the first example of part C, let us suppose that the X variable represents a forced-choice ranking of subjects into three equal-sized categories, and we wish to know what the correlation would have been if we had been able to achieve a complete rank order on X. Thus, X is assumed to have a rectangular distribution; $\hat{\eta} = .47$, so the correction for grouping on X is

$$_c\hat{\eta}_{YX} = \frac{\hat{\eta}_{YX}}{r_{X\bar{X}}} = \frac{.47}{.943} \doteq .50$$

where .943 is taken from the last column of Table 16-2, page 435. The correction is small both because the uncorrected eta is large (.47) and because the correction is not in general very great.

F. Significance Test

Relevant earlier section:
7-7 *F* distribution

You can use a sample eta to determine whether any relationship exists between two variables in the population from which your subjects were drawn. The hypotheses are

$$H_0: \eta^2 = 0 \text{ in the population} \qquad H_m: \eta^2 > 0 \text{ in the population}$$

(A two-tail test is not possible, inasmuch as a population eta cannot be negative.)

If you have estimated eta square from the data of an analysis of variance, then η is significant whenever the overall F from the analysis of variance is significant:

$$F = \frac{\text{MS}_B}{\text{MS}_\text{W}} \qquad (17\text{-}12)$$

If you have estimated eta square from a small sample but do not have an analysis of variance on which to base the significance test, use the formula

$$F = \frac{\hat{\eta}^2(N - J) + (J - 1)}{(1 - \hat{\eta}^2)(J - 1)} \quad \text{with } J - 1 \text{ and } N - J \text{ df} \quad (17\text{-}13)$$

EXAMPLE For the example of part C, $\hat{\eta}^2 = .222$, $N = 15$, and $J = 3$. From formula 17-12,

$$F = \frac{.222(15 - 3) + (3 - 1)}{(1 - .222)(3 - 1)} = \frac{2.667 + 2}{1.556} = \frac{4.667}{1.556}$$

$$= 3.00$$

which is identical to the value of F obtained from the analysis of variance of these data in method 2 of part D.

G. Test for Curvilinearity

Relevant earlier sections:
15-4 Assume linearity?
16-2 Pearson correlation

When both variables are interval-level, a test for curvilinearity of regression can be made by comparison of eta with the Pearson coefficient, which assumes linearity. If the two coefficients take on about the same values, there is no evidence for curvilinearity, but if eta is much greater than the Pearson coefficient, there is evidence that the population regression curve is not linear.

When the large-sample formula 17-1 has been used to define eta, the test for curvilinearity is given by

$$F = \frac{(\eta^2 - r^2)/(J - 2)}{(1 - \eta^2)/(N - J)} \quad \text{with } J - 2 \text{ and } N - J \text{ df} \quad (17\text{-}14)$$

where N is the total number of subjects in the sample, J is the number of categories of the predictor, and r is the Pearson correlation for the same set of data. When the small-sample estimate (formula 17-2) of a population eta has been calculated, the test for curvilinearity is

$$F = \frac{(N - J)\hat{\eta}^2 - (N - 1)r^2 + (J - 1)}{(1 - \eta^2)(J - 2)} \quad \text{with } J - 2 \text{ and } N - J \text{ df} \quad (17\text{-}15)$$

By comparing the magnitude of η, which does not depend on the assumption of linear regression, with the magnitude of the Pearson coefficient, which does, we can determine whether departures from that assumption are systematic or not.

References Eta is discussed in Freeman (1965), 120–130; Guilford (1965), 308–317; and McNemar (1969), 231–232, 306–308. The method in part D for obtaining a sample estimate of η^2 was developed by Kelley (1935) and recommended by Peters and Van Voorhis (1940). Hays (1963, Chapters 10 and 12) developed an alternative estimate, which he called est. ω^2. Glass and Hakstian (1969) compared and criticized the two estimates.

17-2   Correlating two ordinal scales

> *Relevant earlier sections:*
> **16**   Introduction
> **16-1**   General problem of correlation

In this section, three different coefficients will be considered for indicating the degree of relationship between two sets of ranks. Two of these coefficients, tau (τ) and gamma (γ), are discussed in some detail; the third, here labeled rho (r_s), will be mentioned briefly in part E.

A. Tau and Gamma When There Are No Tied Scores

Two ordinal variables may be said to be in perfect agreement if for every object i rated by the two scales, every other object rated higher on variable X is also rated higher on variable Y. There is perfect disagreement between X and Y if, for every object i, every other object rated higher on X is rated lower on Y, or equivalently, if every object rated lower on X is rated higher on Y.

Let us call A_i the number of agreements of the two variables for object i; it is equal to the number of other objects rated higher on both X and Y. The total number of agreements of the two scales for a set of objects will then be given by $A = \sum_i A_i$. If D_i is the number of disagreements of the two variables for object i, D_i is equal to the number of other objects rated higher than i by X and lower than i by Y. The total number of disagreements for a set of objects is equal to $D = \sum_i D_i$.

When there are no tied scores on either variable, we may express the degree of agreement between the variables as the proportion of agreements in the two sets of rankings minus the proportion of disagreements. In this case (no ties), the total number of comparisons between pairs of scores is given by

$$A + D = {}_N C_2 = \frac{N(N-1)}{2} \tag{17-16}$$

The proportion of comparisons for which there are agreements minus the proportion of comparisons for which there are disagreements is called either tau or gamma.

$$\tau = \gamma = \frac{A}{A + D} - \frac{D}{A + D} \tag{17-17}$$

$$= \frac{A - D}{N(N - 1)/2} \tag{17-18}$$

The numerator of tau or gamma is customarily labeled S; it is this statistic that we will use in significance testing (parts C and D of this section). When there are no ties, a simplified form of S is possible:

$$S = A - D = 2A - \frac{N(N - 1)}{2} \tag{17-19}$$

It is possible to obtain a simplified formula for tau by substituting formula 17-19 into formula 17-18, but it is generally preferable to carry out explicit calculation of S.

The usual procedure for obtaining tau or gamma when there are no ties is first to find S using formula 17-19, then insert this value in formula 17-18. When this is done, only two values are needed: the number of objects, N, and the number of agreements, A. An easy way to find A is to construct a table such as Table 17-1, in which values of one variable are listed horizontally and values of the other are listed vertically. Here, the X and Y values may be either (ordinal) raw scores or ranks (from 1 to N). A single tally is made for each object to indicate its rank on both variables.

Table 17-1 *Data layout for calculating tau and gamma*

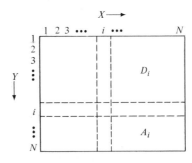

In Table 17-1, the row and column for object i have been indicated, as have areas for A_i and D_i. The values of A_i and D_i are given by the numbers of tallies in the indicated sections of the table. In terms of the marginal values, A_i is the number of other objects rated higher than i on both variables, and D_i is the number rated higher on X and lower on Y.

468

We could also label A_i' the number ranked lower than i on both variables and D_i' the number rated lower on X and higher on Y. If so, we would find that $\sum_i A_i = \sum_i A_i'$ and $\sum_i D_i = \sum_i D_i'$.

EXAMPLE In an amateur-night contest, two judges rank each of the contestants on "overall musical talent." The winner (subject E) is the one receiving the highest sum of ranks. Later, we decide to determine the extent to which the two judges agreed. The data are as shown:

		Contestant							
		A	B	C	D	E	F	G	H
Rank given by judge 1		5	8	2	6	7	4	1	3
Rank given by judge 2		3	4	1	7	8	5	2	6

For purposes of analysis, we construct a table similar to that of Table 17-1:

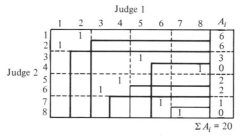

For each tally in this table, the agreements appear to the right and below the tally. The appropriate area for A_i is marked off by solid lines and the number of tallies in the area is given in the right-hand column. For example, subject G was rated 2 by judge 1 and 1 by judge 2; in the table, there are six tallies below and to the right of the tally for contestant G, and $A_g = 6$ is given in the right-hand column. $A = 20$, and

$$S = 2(20) - \frac{8(7)}{2} = 40 - 28 = 12$$

$$\tau = \gamma = \frac{12}{28} = \frac{6}{14} \doteq .428$$

B. Correlation When There Are Tied Ranks

When scores are tied on one or both variables, any of three approaches can be used:

1. If one of the rankings contains no ties and is serving as an objective measure for evaluation of the second set of rankings (as in testing the accuracy of a judge Y against the known status of the objects ranked X), the uncorrected

measure should be used, as in part A, because ties will then be evaluated properly as poor discrimination on the part of the second ranking procedure.
2. If a tau measure is desired, then a correction can be applied to the denominator of tau to allow for the fact that, with ties, the maximum value of the uncorrected tau is less than 1. A tau measure would be wanted here primarily if your anticipated audience is already familiar with tau. In this case it is necessary to calculate A and D separately, and to correct the denominator of formula 17-15. The formula of tau corrected for ties is

$$_c\tau = \frac{S}{\sqrt{\left[\frac{N(N-1)}{2} - C_1\right]\left[\frac{N(N-1)}{2} - C_2\right]}} \qquad (17\text{-}20)$$

where

$$C_1 = \tfrac{1}{2} \sum_j N_j(N_j - 1)$$

N_j = number of subjects tied at level X_j of variable X

$$C_2 = \tfrac{1}{2} \sum_k N_k(N_k - 1)$$

N_k = number of subjects at level Y_k of variable Y

Notice that when there are no ties on either X or Y, C_1 and C_2 are zero, and formula 17-20 reduces to formula 17-18.

EXAMPLE In an attempt to understand a couple's marital difficulties, a marriage counselor asks both husband and wife to rate 20 stimulus words on a scale from passive (1) to active (7). The results are as follows, where each tally represents a stimulus word:

Wife (Y)	Husband (X)							$\sum A_i$	$\sum D_i$	N_k	$N_k(N_k - 1)$
	1	2	3	4	5	6	7				
1	|							18	0	1	0
2		||					|	28	0	3	6
3			|		|			14	2	2	2
4		|		||		|		22	8	4	12
5	|		|		|||			16	20	5	20
6				|	|	|		3	9	3	6
7					|		|	0	8	2	2
								$\sum 101$	47		48
N_j	2	3	2	4	5	2	2				
$N_j(N_j - 1)$	2	6	2	12	20	2	2	46			

In this case, $\sum A_i$ and $\sum D_i$ are summed across all stimuli in a row of the table. For row 1, there is only one stimulus word, at $(X = 1, Y = 1)$, so $\sum A_i = A_i$ for that stimulus, which is equal to 18. For the second row, there are three stimuli, so $A_i = 2(14) + 0$; that is, 14 for each of the stimuli at $(X = 2, Y = 2)$, plus 0 for the stimulus at $(X = 7, Y = 2)$, because there are no tallies to the right and below the tally for this stimulus. D_i values are calculated by the number of tallies above and to the right of each stimulus, that is, other stimuli rated higher by the husband and lower by the wife. Then,

$$S = A - D = 101 - 47 = 54$$

$$A + D = 101 + 47 = 148$$

$$C_1 = \frac{46}{2} = 23$$

$$C_2 = \frac{48}{2} = 24$$

$$\frac{N(N - 1)}{2} = \frac{20(19)}{2} = 190$$

so the corrected value of tau is found to be

$$_c\tau = \frac{54}{\sqrt{(190 - 23)(190 - 24)}} = \frac{54}{\sqrt{(167)(166)}}$$

$$\doteq 54/167 \doteq .324$$

3. A third alternative is to compute gamma, which has the advantage of being more easily calculated and interpreted than tau under these conditions. Whereas the denominator of tau is based on the total number of pairs of rankings, the denominator of gamma is based only on the untied pairs of rankings. The formula for gamma is

$$\gamma = \frac{S}{A + D} \tag{17-21}$$

The distinction between formulas 17-15 and 17-18 lies only in the fact that, when there are ties $(A + D) < {}_NC_2$, so that, in the present case, formula 17-21 does not reduce to 17-16.

EXAMPLE For the example on which the corrected tau was computed, $S = 54$ and $A + D = 148$. Therefore,

$$\gamma = \frac{54}{148} = .365$$

C. Significance Test for Tau or Gamma When There Are No Ties

Relevant earlier section
Chapter 9   Hypothesis testing

For a given sample size N, the information as to whether two variables are related resides in the numerators S of both tau and gamma. Therefore, to compare either statistic against a null hypothesis of zero relationship, it is necessary only to test S for zero magnitude.

In Table 17-2, the three possible relationships of parameters tau and gamma to zero are listed, together with the corresponding relationship of S to zero. Here, capital tau and gamma are used to represent the population correlations because the lowercase Greek letters have already been used for the sample values. For a two-tail test, the null hypothesis is (ii) that the population correlation (T or Γ) is zero; the motivated hypothesis is (i or iii) that the correlation is nonzero. For a directional test, the motivated hypothesis is either that the correlation is greater than zero (i), or that it is less than zero (iii).

Table 17-2   *Relationships and motivated hypotheses about tau and gamma*

	Coefficient		
Relationship	Tau	Gamma	Test statistic
i	$T > 0$	$\Gamma > 0$	$S > 0$
ii	$T = 0$	$\Gamma = 0$	$S = 0$
iii	$T < 0$	$\Gamma < 0$	$S < 0$

Case 1: $N \leq 10$ When the number of objects ranked is small and there are no ties, the numerator S may be tested for significance by comparison with Table A-14. The tabulated values are positive if $H_m : T > 0$ or $H_m : \Gamma > 0$, negative if $H_m : T < 0$ or $H_m : \Gamma < 0$, and either positive or negative if $H_m : T \neq 0$ or $H_m : \Gamma \neq 0$. Tau or gamma is significant at the specified α level if the observed value of S is greater than the tabulated value.

EXAMPLE   Suppose in the example of part A that we wish to test $H_0 : T \leq 0$, $H_m : T > 0$ at an α of .05. For the example, $S = 12$. From Table A-14 we find that $S_{.05}$(one tail, $N = 8$) is 16. The observed S is less than the critical value, so H_0 is not rejected. There is insufficient relationship in the sample to indicate a correlation in the population.

Case 2: $N > 10$ When the number of subjects ranked is large and there are no ties, S is approximately normally distributed under H_0, and hypotheses can be tested using the statistic

$$z = \frac{S^*}{\sigma_S} \qquad (17\text{-}22)$$

where $S^* = S - 1$ is corrected for continuity, and

$$\sigma_S = \sqrt{\frac{N(N-1)(2N+5)}{18}} \qquad (17\text{-}23)$$

This value of z can then be compared to the table of the standard normal distribution to test either directional or nondirectional hypotheses.

EXAMPLE Two judges each rank 12 test protocols on amount of expressed need for affiliation. You wish to know whether, over a population of such protocols, their rankings would probably be correlated. Your hypotheses are $H_0 : T \leq 0$ and $H_m : T > 0$. You set α at .05 and collect these data:

	Protocol											
	1	2	3	4	5	6	7	8	9	10	11	12
Judge X	7	2	10	4	1	9	12	8	5	11	3	6
Judge Y	9	1	11	3	5	8	7	6	10	12	4	2

Application of the computational method of part A gives $S = 30$, $\tau = 30/66 = .455$. Then $S^* = 30 - 1 = 29$, and

$$\sigma_S = \sqrt{\frac{12(11)(29)}{18}} = \sqrt{212.67} = 14.6$$

$$z = \frac{29}{14.6} \doteq 1.99$$

$z_{.05}$(one tail) $= 1.64$. The observed value of 1.99 is greater than the critical value, so the null hypothesis is rejected. The judges' rankings would probably correlate over a population of such protocols.

D. Significance Test for Tau or Gamma When There Are Tied Scores in One or Both Variables

Relevant earlier section:
Chapter 9 Hypothesis testing

Case 1: $N \leq 10$, all $N_j \leq 3$, only one variable has ties In this case, S is calculated as indicated in part B and the result is compared to Table

A-15 of the Appendix. In general, the instructions for this case are the same as for case 1 of part C of this section. In Table A-15, p_2 is the number of times that two things receive the same rank, and p_3 is the number of times that three things receive the same rank.

Case 2: $N \leq 10$, some $N_j > 3$ or ties in both rankings A significance test is not feasible for this case; the most reasonable thing to do is to collect more data, or find some way to break the ties.

Case 3: $N > 10$: In this case, S is approximately normally distributed under H_0, and hypotheses can be tested using the statistic

$$z = \frac{S}{\sigma_S} \tag{17-24}$$

where

$$
\begin{aligned}
\sigma_S^2 = \frac{1}{18} \Big\{ & N(N-1)(2N+5) - \sum_j N_j(N_j-1)(2N_j+5) \\
& - \sum_k N_k(N_k-1)(2N_k+5) \\
& + \frac{\sum_j N_j(N_j-1)(N_j-2) \sum_k N_k(N_k-1)(N_k-2)}{9N(N-1)(N-2)} \\
& + \frac{\sum_j N_j(N_j-1) \sum_k N_k(N_k-1)}{2N(N-1)} \Big\}
\end{aligned}
\tag{17-25}
$$

N_j is the number of objects at level j of variable X, N_k is the number of objects at level k of variable Y, N is the total number of objects rated, and σ_S is the positive square root of σ_S^2.

EXAMPLE Suppose for the example of part B that we were testing the two-tail hypotheses $H_0: \Gamma = 0$ and $H_m: \Gamma \neq 0$, at $\alpha = .05$. We would need the following calculations in addition to the ones already completed.

X_j	$N_j(N_j - 1)$	$N_j(N_j - 1)(2N_j + 5)$	$N_j(N_j - 1)(N_j - 2)$
1	2	18	0
2	6	66	6
3	2	18	0
4	12	156	24
5	20	300	60
6	2	18	0
7	2	18	0
Σ	46	594	90

Y_k	$N_k(N_k - 1)$	$N_k(N_k - 1)(2N_k + 5)$	$N_k(N_k - 1)(N_k - 2)$
1	0	0	0
2	6	66	6
3	2	18	0
4	12	156	24
5	20	300	60
6	6	66	6
7	2	18	0
Σ	48	624	96

$$\sigma_S^2 = \frac{1}{18}\left\{ 20(19)(45) - 594 - 624 + \frac{90(96)}{9(20)(19)(18)} + \frac{46(48)}{2(20)(19)} \right\}$$

$$= \frac{1}{18}(17,100 - 1218 + 0.14 + 2.90)$$

$$= \frac{15,885.04}{18} = 882.5$$

$$\sigma_S = \sqrt{882.5} \doteq 29.7$$

Then $z = 54/29.7 = 1.82$. The critical values of the standard normal distribution, taken from Table A-4, are $z_{.05}$(two tail) $= \pm1.96$. The computed values of z lies between these limits, so the null hypothesis is not rejected.

E. A Third Coefficient: Spearman's Rho

The Spearman coefficient (here called r_s) may be derived by assuming that two complete sets of ranks constitute interval scales and applying the Pearson coefficient to them. Rho would seem to be preferred to tau or gamma only where your anticipated audience is much more familiar with it than with the other coefficients.

If the two sets of ranks correlated are $\mathscr{R}_X$ and $\mathscr{R}_Y$, then the difference in ranks on the two variables for some subject i is given by

$$d_i = \mathscr{R}_{X_i} - \mathscr{R}_{Y_i}$$

From the fact that the values being correlated are complete rank orders, $\sum \mathscr{R}_{X_i}, \sum \mathscr{R}_{Y_i}, \sum \mathscr{R}_{X_i}^2$, and $\sum \mathscr{R}_{Y_i}^2$ can be specified a priori. When this is done, formula 16-4 can be shown to reduce to

$$r_s = 1 - \frac{6 \sum_i d_i^2}{N^3 - N} \tag{17-26}$$

where N is the total number of subjects. [For the derivation, see Hays (1963), 644–645, or Siegel (1956), 203–204.]

When $N \geq 10$, r_s is distributed approximately as t and may be tested for significance using formula 17-27:

$$t = \frac{r_s \sqrt{N - 2}}{\sqrt{1 - r_s^2}} \quad \text{with } N - 2 \text{ df} \tag{17-27}$$

An unbiased estimate of the population correlation ρ_s is given by

$$\hat{r}_s = \frac{(N + 1)r_s}{N - 2} - \frac{3\tau}{N - 2} \tag{17-28}$$

where τ is computed according to part A or B of this section.

For treatment of ties, or for hypothesis testing with small samples, see Siegel (1956), 207–213.

EXAMPLE   For the example of part A of this section, we have

i	$\mathcal{R}_X$	$\mathcal{R}_Y$	d	d^2
1	5	3	2	4
2	8	4	4	16
3	2	1	1	1
4	6	7	−1	1
5	7	8	−1	1
6	4	5	−1	1
7	1	2	−1	1
8	3	6	−3	9
Σ				34

$$r_s = 1 - \frac{6(34)}{512 - 8} = 1 - \frac{204}{504} = .595$$

If we wished to test the one-tail hypothesis $H_0 : \rho_s = 0$ at $\alpha = .05$ or $.01$, we could compare this value to Table P of Siegel [(1956), 284]. The critical value given there for $N = 8$, $\alpha = .05$, is 643; so the observed correlation, being less than the critical value, is not significant.

An unbiased estimate of ρ_s is given by

$$\hat{r}_s = \frac{9(.595)}{7} - \frac{3(.428)}{6} \doteq .551$$

where $\tau = .428$ was obtained in part A.

References Tau is discussed in Hays (1963), 647–655; and Siegel (1956), 213–223. Gamma can be found in Freeman (1965), 79–87; and Hays, 655–656. Spearman's rho is discussed in most texts, for example, Guilford (1965), 305–308; Hays, 643–647; McNemar (1969), 232–234; and Siegel, 202–213.

17-3   Methods for correlating two nominal variables

Relevant earlier sections:
5-5   Mode
5-9   Variation ratio
15-2   Prediction table
16   Introduction
16-1   General problem of correlation

With nominal variables two different kinds of coefficients are in use: symmetric coefficients, which indicate degree of relationship between variables, and predictive coefficients, which indicate the extent to which knowledge of a predictor variable reduces error in estimating criterion status. The predictive coefficient to be discussed here is lambda (parts A through C). Symmetric coefficients include symmetric lambda, the contingency coefficient, and phi prime, all of which are discussed in part D. For the 2 × 2 case, the phi coefficient (Section 16-5) can also be used.

A. A Predictive Coefficient: Lambda

Lambda is a coefficient which indicates the degree to which a criterion variable Y can be predicted from some predictor X. The raw data for Y may be nominal, ordinal, or interval, but for purposes of the analysis the criterion data are treated as nominal. The predictor may be nominal or ordinal (see Section 15-2), or it may be interval (see Section 15-3).

If all we know about a subject is that he is a member of some group, and we want to predict his status on some nominal variable Y when we know the distribution of scores on Y for the entire group, then our best guess of the subject's score is the mode of the group on Y. An estimate of the amount of error in predicting the mode is given by the variation ratio v_Y, which is equal to the proportion of nonmodal cases in the group.

If, for our group of subjects, we have a table indicating the joint distribution of scores on both Y and some predictor X, and if we want to predict a subject's status on Y from his score on X, then our best guess is the conditional mode: the modal Y value for that subgroup of subjects who have the same X score as the subject for whom we are predicting. An indicator of the amount of error of prediction is given by the variation ratio of estimate v_{YX}, which is the proportion of subjects in the group who are not at their respective modes.

Lambda is then defined as the proportion of error of prediction reduced by knowledge of the relationship between X and Y and subjects' status on

X (so that predictions can be made on the basis of conditional modes rather than the overall criterion mode).

$$\lambda_{YX} = \frac{v_Y - v_{YX}}{v_Y} \tag{17-29}$$

Thus, lambda for predicting Y from X is equal to the variation ratio of Y minus the variation ratio of estimate; the difference is then divided by the Y variation ratio. Since the variation ratio is an indicator of variability or error of prediction when predictions are made using the mode, formula 17-29 is another specific instance of the general correlation formula presented in Section 16-1. Lambda ranges from zero, when the predictor doesn't help and $v_{YX} = v_Y$, to 1.0, when prediction is perfect and $v_{YX} = 0$ (all scores are at the conditional modes).

Let us adopt the following notation: f'_j is the frequency of subjects at the conditional mode of Y for those subjects at level j of the predictor, f' is the frequency of subjects at the overall mode of Y (when no distinction is made as to X scores), and N is the total number of subjects. Then, from formula 5-7, the overall variation ratio is given as

$$v_Y = 1 - \frac{f'}{N} = \frac{N - f'}{N} \tag{17-30}$$

and from formula 15-2, the variation ratio of estimate is

$$v_{YX} = 1 - \frac{\sum\limits_j f'_j}{N} = \frac{N - \sum\limits_j f'_j}{N} \tag{17-31}$$

Inserting formulas 17-30 and 17-31 in formula 17-29 gives

$$\lambda_{YX} = \frac{(N - f') - (N - \sum\limits_j f'_j)}{N - f'} \tag{17-32}$$

Now, the term $(N - f')$ is the number of errors of prediction made in a given sample when X is not known, and prediction is based solely on the mode of Y. The term $(N - \sum\limits_j f'_j)$ is the total number of errors made in predicting Y from X, using the conditional modes. Thus, if we are going from a situation in which we do not have knowledge of the predictor X, to the case in which we do, we can rewrite the formula for lambda as

$$\lambda_{YX} = \frac{\text{number of errors reduced}}{\text{original number of errors}} \tag{17-33}$$

Alternatively, we can complete the subtraction in the numerator of formula 17-33, giving

$$\lambda_{YX} = \frac{\sum_j f'_j - f'}{N - f'} \tag{17-34}$$

as a computational formula based only on the frequencies of subjects at the overall mode, the frequencies at the various conditional modes, and the total number of subjects. This formula is least satisfying conceptually but easiest computationally.

B. Sample Estimate of Lambda

Sample estimates of population lambdas are found by using sample values in the formulas of part A. The obtained value is indicated to be a sample estimate of lambda by labeling it with an uppercase L instead of the corresponding Greek letter. Thus, for a sample the coefficient corresponding to formula 17-34 is

$$L_{YX} = \frac{\sum_j f'_j - f'}{N - f'} = \text{est. } (\lambda_{YX}) \tag{17-35}$$

where all terms are defined with respect to the sample.

EXAMPLE Let us return to the example of Section 15-2A, in which the problem was to predict the region of Bortland that subjects come from, given their number of heads. The data table is as follows:

Region	One	Two	Three or more	None	$\sum$
			Number of heads		
North	360	40	0	0	400
South	190	20	40	50	300
Central	125	5	0	0	130
East	4	130	0	6	140
West	21	5	0	4	30
$\sum$	700	200	40	60	1000

From this table, let us make a second table of predictions. A separate prediction is made for each level of the predictor (number of heads).

Number of heads	Predicted region (conditional mode)	Frequency (f_j) at conditional mode	Number of errors
One	North	360	340
Two	East	130	70
Three or more	South	40	0
None	South	50	10
Σ		580	420

The modal region is North, $f' = 400$, and the number of errors in using this mode to predict for all the subjects in the sample is $1000 - 400 = 600$. Then using sample values in formula 17-32 or 17-33,

$$L_{YX} = \frac{600 - 420}{600} = \frac{180}{600} = .30$$

Alternatively, we could use formula 17-35 to obtain

$$L_{YX} = \frac{580 - 400}{1000 - 400} = \frac{180}{600} = .30$$

Thus, in using number of heads to predict region of origin, we reduce our number of errors of prediction by 30 percent over not using the predictor.

Lambda is not a symmetric coefficient; that is, it is not in general the case that $\lambda_{YX} = \lambda_{XY}$. Let us determine for the above example the estimated value of λ for prediction in the other direction, to emphasize the point. We again make up a table of predictions, this time for predicting number of heads from region:

Region	Predicted number of heads (conditional mode)	Frequency (f_j) at conditional mode
North	One	360
South	One	190
Central	One	125
East	Two	130
West	One	21
Σ		826

The modal number of heads is 1, $f' = 700$. Then using formula 17-35,

$$L_{YX} = \frac{826 - 700}{1000 - 700} = \frac{126}{300} = .42$$

Thus, using region to predict number of heads, we reduce errors by 42 percent over trying to predict subjects' number of heads without knowledge of the predictor.

C. Significance of Lambda

> *Relevant earlier sections:*
> **Chapter 9** Hypothesis testing
> **7-3** Normal distribution

It is possible to use the sample estimate of lambda, L_{YX}, to test hypotheses about the value of the parameter, provided that the sample size is large ($N \geq 50$).
If the null hypothesis is $H_0: \lambda = 0$, accept H_0 if $L = 0$, reject it if $L \neq 0$. If the null hypothesis is $H_0: \lambda = 1$, reject it if $L \neq 1$, accept it if $L = 1$.
For values of lambda between zero and 1, the null hypothesis $H_0: \lambda_{YX} = \lambda_0$ can be tested using the formula

$$z = \frac{L_{YX} - \lambda_0}{\sqrt{\dfrac{(N - \sum_j f'_j)(\sum_j f'_j + f' - 2\sum_a f'_j)}{(N - f')^3}}} \qquad (17\text{-}36)$$

where $\sum_a f'_j$ is found by looking down the row or column in which f' appears, and adding all the values of f'_j that appear in that row or column.
All other values are as previously defined (part A). This statistic is normally distributed, with mean zero and standard deviation of 1. Hence, significance can be tested by comparison with Table A-4 of the Appendix.

EXAMPLE   Consider the example of part B of this section for predicting region of origin (Y) from number of heads (X). Suppose that, before collecting the data used in this problem we had predicted $\lambda_{YX} = \lambda_{XY} = .50$. We then collected our data and calculated $L_{YX} = .30$. Let us test our prediction at the .05 level of significance. For a two-tail test, $z_{.05}$ is found from Table A-4 to be ± 1.96. Next, let us calculate $\sum_a f'_j : f' = 400$, the sum of frequencies in the North row. The only f'_j in this row is for the column One, so $\sum_a f'_j = 360$. Inserting the values found in part B into formula 17-36 gives

$$z = \frac{.30 - .50}{\sqrt{(1000 - 580)(580 + 400 - 2 \cdot 360)/(1000 - 400)^3}}$$

$$= \frac{-.20}{\sqrt{420(260)/(600)^3}}$$

$$\doteq -8.90$$

The observed value of -8.90 is less than the lower critical value of -1.96. Therefore, the prediction (null hypothesis) is rejected: lambda is less than .50 in the population.

D. Symmetric Coefficients

1. Symmetric lambda   If we want an index of the degree of relationship between two variables which is not specific to either direction of prediction, we may use the following modified version of lambda (see formula 17-33):

$$\lambda_s = \frac{\text{number of errors reduced in both variables}}{\text{original number of errors in both variables}} \qquad (17\text{-}37)$$

and the sample estimate is defined in the same way.

EXAMPLE for the example of part A, the original number of errors in predicting from number of heads to region of origin was 600, and in predicting from region to number of heads, 300. The reduction in errors for predicting from number of heads to region of origin was 180, and for predicting from region to number of heads, 126. Therefore, lambda for the degree of relationship between the two variables is estimated by

$$L_s = \frac{180 + 126}{600 + 300} = \frac{306}{900} = .34$$

It is possible to test L_s for significance; however, the test is messy and will not be presented here [see Goodman and Kruskal (1963), 321].

2. The contingency coefficient   This is the best-known coefficient (C) for indicating the degree of symmetric relationship between two (or more) nominal variables. The formula is

$$C = \sqrt{\frac{\chi^2}{\chi^2 + N}} \qquad (17\text{-}38)$$

where χ^2 is computed according to formula 10-2 and N is the total number of observations.

This index has two disadvantages. Its theoretical limits are zero for no relationship and 1 for perfect association, but the actual limits are a function of
sample size and the theoretical limits can't be attained by any real data. In
addition, values of C are not comparable across studies when the number of
categories of the variables in the studies differ. Because χ^2 applies to any number
of variables, C may also be used to indicate the overall degree of relationship
among several variables (see Section 18-3). C is significant whenever chi square
is significant.

3. Phi prime   In order to overcome difficulties of the contingency coefficient,
Cramer (1946) developed the statistic ϕ', which can be used when formula
10-2 is appropriate for relating two variables. For any two-way table in which
formula 10-2 applies, the maximum possible value of χ^2 can be shown to be
$N(B - 1)$, where B is the number of rows or the number of columns in the
table, whichever is smaller, and N is the total number of observations. Phi
prime is then defined as the square root of the ratio of the observed chi square
to the maximum possible value of chi square:

$$\phi' = \sqrt{\frac{\chi^2}{N(B - 1)}} \tag{17-39}$$

This statistic can achieve a value of $+1$ for any number of rows and columns,
but it is difficult to interpret. Like C, ϕ' is significant whenever χ^2 is significant.

EXAMPLE   Let us consider the second example of Section 10-3. Here
we were comparing the grade distributions of field-independent (FI) and
field-dependent (FD) subjects. The frequency table was as follows:

Perceptual Style (P)	Grade (G)					Σ
	F	D	C	B	A	
FI	10	20	30	18	12	90
FD	18	36	28	12	6	100
Σ	28	56	58	30	18	190

The computed value of χ^2 was 9.63.

The contingency coefficient relating these two variables is

$$C_{P,G} = \sqrt{\frac{9.63}{9.63 + 190}} = \sqrt{\frac{9.63}{199.63}} \doteq .22$$

For the same data the phi-prime coefficient is found to be

$$\phi' = \sqrt{\frac{9.63}{190(1)}} = .22$$

The estimates of directional lambdas are

$$L_{GP} = \frac{(30 + 36) - 58}{190 - 58} = \frac{8}{132} \doteq .06$$

and

$$L_{PG} = \frac{(18 + 36 + 30 + 18 + 12) - 100}{190 - 100} = \frac{14}{90} \doteq .16$$

Finally, the estimate of symmetric lambda is found to be

$$L_s = \frac{8 + 14}{132 + 90} = \frac{22}{222} \doteq .10$$

References   The contingency coefficient is discussed by Goodman and Kruskal (1954); McNemar (1969), 227–231; and Siegel (1956), 196–202. Cramer (1949) and Goodman and Kruskal (1954) discuss ϕ', and λ can be found in Freeman (1965), 71–78; Goodman and Kruskal; and Hays (1963), 606–610.

17-4   Theta

The coefficient theta (θ), called the coefficient of differentiation by its originator [Freeman, (1965), Chapter 10], is used to indicate degree of relationship between a nominal and an ordinal variable. The magnitude of theta is based on the extent to which subjects at one level of the nominal variable tend to receive higher ratings on the ordinal variable than subjects at another level of the nominal variable do.

A. Computation
First, arbitrarily number the levels of the nominal variable. Then consider pairs of levels (j and k) such that $j < k$. Suppose that the nominal variable is sex, and levels 1 and 2 are male and female. (This restriction prevents us from comparing males with males, or both males with females and females with males.) Then theta is computed by formula 17-40:

$$\theta = \frac{\sum\limits_{j}\sum\limits_{k>j} |A_{jk} - B_{jk}|}{\sum\limits_{j}\sum\limits_{k>j} N_j N_k} \tag{17-40}$$

where A_{jk} is the total number of times a subject in level j of the nominal variable is above a subject in level k on the ordinal variable, B_{jk} is the number of times a subject in level j of the nominal variable is below some subject in level k on the ordinal variable. N_j is the number of subjects at level j, and N_k is the number of subjects at level k.

Perhaps the use and interpretation of this coefficient are best shown with a couple of short examples. Suppose that we have a spelldown in a class that contains two girls and three boys. We want to know the degree of relationship between sex (b or g) and spelling ability (rank 1 through 5, with 1 the worst speller). The first variable is nominal, the second ordinal. We only have two levels of the nominal variable, so we let j = boys and k = girls. Then $N_j = 3$ and $N_k = 2$. We focus on the boys; A_{jk} is the number of times a boy does better than a girl, and B_{jk} is the number of times a boy does worse. Formula 17-40 can be reduced to

$$\theta = \frac{|A - B|}{6}$$

We would have a zero correlation between the two variables if boys did equally well with girls overall. One example of this outcome would be the ranking $b_1 g_1 b_2 g_2 b_3$, in which the first boy (b_1) misses first, then girl g_1 drops out, and so on until boy b_3 wins. Here b_1 is above no girl and below two girls, b_2 is above one and below one, and b_3 is above two and below none. Therefore, $A = 0 + 1 + 2$, and $B = 2 + 1 + 0$, and ·

$$\theta = \frac{|3 - 3|}{6} = \frac{0}{6} = 0$$

You can verify that theta would also be zero for the ranking $g_1 b_1 b_2 b_3 g_2$, where, again, boys and girls rank equally well overall.

We would have a perfect relationship between the two variables only if all boys ranked higher than all girls, or all girls ranked higher than all boys. If the order was $b_1 b_2 b_3 g_1 g_2$, $A = 0$, $B = 6$, and $\theta = 1.0$.

Theta is always positive because, by definition, a nominal scale has no direction. Therefore, there could be no conceptual difference between positive and negative values of theta.

If we take spelldown rank as an indication of spelling ability, then in the first of these two examples, we could say that there was no consistent difference between boys and girls on spelling ability, in the second example we could say that 100 percent of the differences favored the girls. In the rank order $b_1 g_1 g_2 b_2 b_3$, $A = 4$, $B = 2$, and $\theta = .33$. We note that

$$\frac{|A - B|}{N_b N_g} = \left| \frac{A}{N_b N_g} - \frac{B}{N_b N_g} \right|$$

This may be interpreted as the proportion of comparison among individuals that favors boys minus the proportion that favors girls.

Let us now go on to some more complicated examples.

EXAMPLE Eight male and eight female students receive the following grades in a sociology class. Males: A, D, A, F, C, C, F, A. Females: A, A, D, A, B, C, B, A. The nominal variable is sex, the ordinal variable is grade. We can make a frequency distribution for the two variables as follows:

Sex	F	D	C	B	A	N
1. Male	2	1	2	0	3	8
2. Female	0	1	1	2	4	8

Each of the three males who received an A was higher on grade than four females; each of the two males who received a C was higher on the ordinal variable than the one female who got a D. Therefore,

$$A_{12} = 3(2 + 1 + 1) + 2(1) = 14$$

Similar considerations give us

$$B_{12} = 2(8) + 1(7) + 2(6) = 35$$

There is only one pair of levels of the nominal variable. Let us construct a computation table for that pair:

j	k	A_{jk}	B_{jk}	$|A_{jk} - B_{jk}|$	$N_j N_k$
1	2	14	35	21	64

Because there is only one row to this table, no summation is necessary.

$$\theta = \frac{21}{64} \doteq .33$$

Referring back to the frequencies, we see that females tend to receive higher grades than males. Therefore, our conclusion is that female students received higher grades than males 33 percent more often than they received lower grades.

As indicated above, the more times subjects at one level of the nominal variable receive higher ratings than subjects at the other, the greater the difference between A_{jk} and B_{jk}. If all the females had received A's and all the males B's and lower, $\theta = 1$, because then the males and females would have been perfectly differentiable.

EXAMPLE A child-guidance clinic specializes in patients with the following symptoms: depression, stealing, truancy, and lying. On intake, each patient is rated on prognosis in therapy, from 5 for good to 1 for poor. 65 patients accepted for therapy in 1969 had primary symptoms and prognoses as indicated in the table of frequencies.

	Prognosis					
Primary symptom	1	2	3	4	5	N
1. Truancy (T)	2	1	1	3	7	14
2. Lying (L)	5	6	4	2	2	19
3. Stealing (S)	3	2	8	5	2	20
4. Depression (D)	6	2	3	0	1	12
Σ						65

Because all seven truants getting rating 5 are each above all the liars rated 1, 2, 3, and 4, in that comparison alone there are $7(5 + 6 + 4 + 2) = 119$ cases of a subject in level 1 being above a subject in level 2. The value of A_{12} is then

$$A_{12} = 7(2 + 4 + 6 + 5) + 3(4 + 6 + 5) + 1(6 + 5) + 1(5)$$
$$= 119 + 45 + 11 + 5$$
$$= 180$$

Similarly,

$$B_{12} = 2(6 + 4 + 2 + 2) + 1(4 + 2 + 2) + 1(2 + 2) + 3(2)$$
$$= 46$$

These and the remaining values are then entered in the computation table, where values of the nominal variable have been abbreviated:

j	k	A_{jk}	B_{jk}	$\|A - B\|$	$N_j N_k$
1. T	2. L	180	46	134	266
	3. S	173	62	111	280
	4. D	124	20	104	168
2. L	3. S	100	207	107	380
	4. D	112	60	52	228
3. S	4. D	153	39	114	240
Σ				622	1562

Now theta can be computed easily:

$$\theta = \frac{622}{1562} \doteq .398$$

We conclude that 40 percent of the comparisons among patients with different symptoms show consistent differences in rated prognosis.

B. Significance of Theta

If there are two levels of the nominal variable, as in the first example of part A, the Mann–Whitney U test can be used to test for significance. If there are more than two nominal levels, as in the second example, use the Kruskal–Wallis test.

All three statistics, Freeman's θ, the Mann–Whitney U, and the Kruskal–Wallis H, are used to compare a nominal-level variable to an ordinal-level variable. Theta indicates the degree of relationship between the two variables in the sample. U and H are used to determine whether differences observed in the samples can be generalized to the populations from which they were drawn.

Homework

1. Consider the following data table:

Subject	X	Y
1	26	87
2	8	89
3	2	86
4	29	98
5	28	124

(a) Indicate why a Pearson correlation coefficient might not be a best index of degree of relationship for these data.

(b) Calculate values for ρ, τ, and γ.

(c) Test τ and γ for a positive population correlation, at $\alpha = .05$.

2. For the following population of nine subjects, what is the maximum proportion of the variance of Y accounted for by knowledge of X if Y is assumed to be interval level but linear regression is not assumed?

i	X	Y
1	5	8
2	2	3
3	2	4
4	7	7
5	4	8
6	7	3
7	7	3
8	2	5
9	4	4

3. The following bivariate frequency distribution indicates a hypothetical relationship between diagnosis of disturbed children and diagnosis of same-sexed parents. Select an appropriate coefficient to show the degree to which children's diagnosis can be predicted from knowledge of parent's diagnosis.

Diagnosis of child	Diagnosis of parent				Σ
	Neurotic	Schizophrenic	Paranoid	Hysterical	
Neurotic	48	3	6	3	60
Schizophrenic	0	15	3	2	20
Paranoid	0	1	8	1	10
Hysterical	8	5	2	25	40
Normal	44	6	1	19	70
Σ	100	30	20	50	200

Compute the value of the coefficient that you have selected.

4. For the following sample data, estimate the population squared correlation ratio for predicting Y from X. (*Hint:* The analysis-of-variance approach involves the least arithmetic.)

i	X	Y
1	1	1
2	3	5
3	3	4
4	1	1
5	4	1
6	2	3
7	4	2
8	2	2
9	1	0
10	1	2
11	4	3
12	4	0
13	3	5
14	2	2
15	2	3
16	3	6

5. Two judges have rated five different student oil paintings on the amount of creativity displayed by the artists. The data are given in the following table. Find the correlation between the judges as an indication of the reliability of the ratings.

	Painting				
	a	b	c	d	e
Judge 1	5	2	1	3	4
Judge 2	4	1	2	3	5

6. In a study of the relationship between number of siblings and need for achievement, 56 subjects are used. There are six levels of classification for "number of siblings" and 30 for "need achievement." Eta is .50, while the Pearson correlation is equal to .30. Can we say, at an α of .05, that the prediction of need achievement from number of siblings departs significantly from linearity?

7. Consider again the data table for problem 1, Chapter 10.

	A	B	Σ
1	33	7	40
2	9	11	20
3	33	7	40
Σ	75	25	100

Find the degree of relationship between the two variables, using each of the symmetric coefficients.

8. You are interested in the degree of relationship between intelligence and creativity in a highly select group of art students. Their scores on the tests are as follows:

Student	Intelligence	Creativity
1	126	35
2	145	57
3	132	37
4	125	45
5	135	40
6	128	65

Choose an appropriate index of relationship between the two variables. Calculate its magnitude and justify your choice.

9. Find the degree of relationship between method of teaching and rated creativity in the example of Section 14-6B.

Homework Answers

1. $r_s = .80$; $\tau = \gamma = .60$
2. $\eta^2 = .425$
3. $L_{child/parent} = .20$
4. $\hat{\eta}^2 \doteq .74$
5. $\tau = .60$; $r_s = .80$
6. $F = 2.67 > F_{.05}(4, 50) = 2.61$; reject H_0.
7. $C = .33$, $\phi' \doteq .35$, $L_s \doteq .07$
8. $r_s = \tau = .20$
9. $\theta \doteq .52$

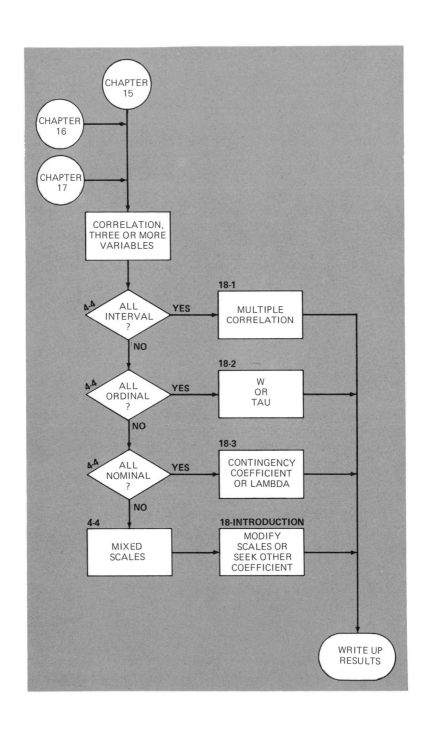

18

Correlating Three or More Variables

E ACH of the correlation coefficients discussed in this chapter indicates degree of relationship among three or more variables when all are of the same level of measurement.

Each index of relationship may be classified as either symmetric or predictive. The symmetric indices, such as the coefficient of concordance (Section 18-2) and the contingency coefficient (Section 18-3) indicate the overall, or average, relationship among the variables compared. The predictive indices, such as multiple correlation, multiple tau, or multiple lambda, give the extent to which one variable can be predicted from the others.

Each of the indices discussed in this chapter assumes that all the measures correlated are based on the same kind of scale. All are nominal (lambda or the contingency coefficient), ordinal (coefficient of concordance or tau), or interval (multiple correlation). However, the coefficients are more flexible than you might assume. Remember that any interval-scale data may be treated as nominal or ordinal, and ordinal-scale data may be treated as nominal. For example, you may lose information when you treat an interval-level variable as ordinal, but you can then predict an ordinal variable from one ordinal and one interval variable, using multiple tau. Under more limited circumstances, you may decide to treat an ordinal variable as interval, in order to use multiple regression and correlation. And don't forget that a dichotomous variable such as sex can be treated as nominal, ordinal, or interval, and used with any of the coefficients in this chapter.

If you have nominal-level predictors and an interval-level criterion, you may with to look into multiple eta (Section 17-1 part C3, or Kendall, 1947). A more flexible approach to the treatment of mixed nominal and interval data can be found in Cohen (1968) and in Kerlinger and Pedhazur (1973), part II.

18-1 Multiple correlation and regression

Relevant earlier sections:
15-5 Linear regression
16-2 Pearson correlation coefficient

A. Multiple Regression
The problem of this section is analogous in many respects to the problems of linear regression and the Pearson correlation coefficient described in earlier

sections, except that here we are interested in making the best prediction of a criterion variable from two or more predictors.

In order to handle this more complicated situation easily, we will modify the notation slightly. The criterion variable will be called Y and the predictors $X_1, X_2, X_3, \cdots, X_J$. When the predictor variables are used as subscripts, only the numbers will be retained, for example, r_{12} is the Pearson correlation between predictors X_1 and X_2, and r_{Y4} is the Pearson correlation between the criterion Y and predictor X_4.

With the present notation, formula 15-14 for predicting standard scores on the criterion Y from a single predictor X_1 becomes

$$z'_Y = \beta_{Y1} z_1$$

where, as was shown in Section 15-5D, the standard regression coefficient β_{Y1} is actually the correlation r_{Y1} between the two variables.

In order to predict raw scores on the criterion variable Y from raw scores on the predictor variable X_1, we must use a form of formula 15-9:

$$Y' = a_Y + b_{Y1} X_1$$

where the regression coefficient or slope is

$$b_{Y1} = \frac{\sum yx_1}{\sum x_1^2} = \beta_{Y1} \left(\frac{\sigma_Y}{\sigma_1} \right) = r_{Y1} \left(\frac{\sigma_Y}{\sigma_1} \right)$$

and the constant or intercept is

$$a_{Y1} = \mu_Y - b_{Y1} \mu_1$$

Simple linear regression equations can be interpreted graphically as in Figure 18-1, which is similar to the figures in Section 15-5. The pair of scores on X and Y for each subject can be represented as a point in a two-dimensional, XY space. A collection of such points is represented in Figure 18-1 by an oval,

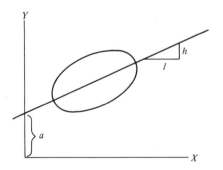

Figure 18-1

or ellipse, drawn around them. The regression equation is represented by a line in the two-dimensional space, a line that best fits all the points in the sense that the sum-of-squared vertical distances from the points to the line are less than to any other straight line that could be drawn in the space. In this equation, a_Y is the Y intercept: the point where the line crosses the Y axis, and where $X = 0$; b_{YX} is the slope of the line: the height of any triangle whose hypotenuse is the line and whose sides are parallel to the X and Y axes, divided by its length.

The three-variable case proceeds in a directly analogous manner. The formula for predicting standard scores on the criterion variable Y from standard scores on the two predictor variables X_1 and X_2 is given as follows:

$$z_Y' = \beta_{Y1\cdot2}z_1 + \beta_{Y2\cdot1}z_2 \tag{18-1}$$

This formula indicates that Y can be considered to have two parts. It is partly predictable from X_1 and partly from X_2, and each of the two predictors potentially gives some independent information about the criterion.

Analogously to the two-variable case, the regression line for predicting Y from X_1 and X_2 must pass through the point defined by the means of the three variables. But by definition of standard scores (Section 6-5) the mean of each distribution of z scores is zero. Thus the regression line must pass through the point $(z_Y = 0, z_1 = 0, z_2 = 0)$. The terms $\beta_{Y1\cdot2}$ and $\beta_{Y2\cdot1}$ are the standard regression coefficients, or "beta weights." They give the slopes of the regression line as well as the weights that z scores for the predictors are to be multiplied by to obtain the best prediction of z scores on the criterion. The term $\beta_{Y1\cdot2}$ is the beta weight for predicting variable Y from variable X_1 with variable X_2 held constant, and $\beta_{Y2\cdot1}$ is the beta weight for predicting z_Y from z_2 with z_1 held constant. Graphically, $\beta_{Y1\cdot2}$ is the slope of the regression line for predicting z_Y in the plane of z_Y and z_1; $\beta_{Y2\cdot1}$ is the slope of the regression line for predicting z_Y in the plane of z_Y and z_2 (independent of z_1).

The computing formulas for $\beta_{Y1\cdot2}$ and $\beta_{Y2\cdot1}$ are derived through the use of methods that are beyond the scope of this text [see Baggaley (1964), Chapter 4; Kerlinger and Pedhazur (1974), or McNemar (1969), 188–189]. The formulas are as follows:

$$\beta_{Y1\cdot2} = \frac{r_{Y1} - r_{Y2}r_{12}}{1 - r_{12}^2}$$

$$\beta_{Y2\cdot1} = \frac{r_{Y2} - r_{Y1}r_{12}}{1 - r_{12}^2} \tag{18-2}$$

In predicting raw scores on the criterion variable from raw scores on the predictor variables, we can again work by analogy with the two-variable case. The regression equation is given by

$$Y' = a_{Y\cdot12} + b_{Y1\cdot2}X_1 + b_{Y2\cdot1}X_2 \tag{18-3}$$

where, again analogously to the two-variable case, the regression coefficients (slopes) are

$$b_{Y1\cdot2} = \beta_{Y1\cdot2}\left(\frac{\sigma_Y}{\sigma_1}\right)$$

$$b_{Y2\cdot1} = \beta_{Y2\cdot1}\left(\frac{\sigma_Y}{\sigma_2}\right)$$

(18-4)

and the intercept is found by substituting means into the general equation:

$$a_{Y\cdot12} = \mu_Y - b_{Y1\cdot2}\mu_1 - b_{Y2\cdot1}\mu_2$$

(18-5)

Multiple linear regression with two predictors can be represented graphically as in Figure 18-2, although it is difficult to represent a three-dimensional space on a two-dimensional page. The data points have three coordinates each and in general tend to form an ellipsoidal or football-like shape. This is a solid extension of the oval or elliptical shape for the single predictor case. The regression equation, instead of being a line, now is a plane intersecting the Y axis at $a_{Y\cdot12}$. This plane has two slopes: $b_{Y1\cdot2}$, the slope in the X_1Y plane (when $X_2 = 0$), and $b_{Y2\cdot1}$, the slope in the X_2Y plane (when $X_1 = 0$). Although the development of multiple regression is in terms of parameters, application uses sample values: sample means and biased variance estimates.

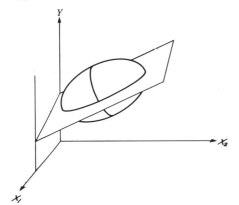

Figure 18-2

EXAMPLE Suppose that we wish to find the formula for predicting variable Y from X_1 and X_2 when $\bar{Y} = 10$, $\bar{X}_1 = 40$, $\bar{X}_2 = 20$, $s_Y = 4$, $s_1 = 8$, $s_2 = 2$, and the intercorrelations are $r_{Y1} = .30$, $r_{Y2} = .50$, and $r_{12} = .40$. We begin by finding the beta weights:

$$\beta_{Y1\cdot2} = \frac{r_{Y1} - r_{12}r_{Y2}}{1 - r_{12}^2} = \frac{.3 - (.4)(.5)}{1 - .16} = \frac{.10}{.84} = .119$$

$$\beta_{Y2\cdot1} = \frac{r_{Y2} - r_{Y1}r_{12}}{1 - r_{12}^2} = \frac{.5 - (.3)(.4)}{1 - .16} = \frac{.38}{.84} = .4523$$

Next the regression coefficients (slopes) are found:

$$b_{Y1\cdot 2} = \beta_{Y1\cdot 2}(s_Y/s_1) = (.119)\left(\frac{4}{8}\right) = .059$$

$$b_{Y2\cdot 1} = \beta_{Y2\cdot 1}(s_Y/s_2) = (.4523)\left(\frac{4}{2}\right) = .9046$$

These values can now be inserted in the regression equation:

$$Y' = a_{Y\cdot 12} + .059X_1 + .9046X_2$$

Substitution of the means of the three variables into this latter equation permits us to solve for the intercept a_Y:

$$
\begin{aligned}
a_{Y\cdot 12} &= \overline{Y} - .059\overline{X}_1 - .9046\overline{X}_2 \\
&= 10 - (.059)(40) - (.9046)(20) \\
&= 10 - 2.360 - 18.092 \\
&= -10.452
\end{aligned}
$$

Now the completed regression equation can be written

$$Y' = .059X_1 + .9046X_2 - 10.452$$

From this equation, we can make a "best" prediction of any subject's score on variable Y if we are given his scores on X_1 and X_2. Thus, for a subject who obtains a score of 50 on X_1 and 26 on X_2, we would predict his score on Y to be

$$
\begin{aligned}
Y' &= (.059)(50) + (.9046)(26) - 10.452 \\
&= 2.950 + 23.5196 - 10.452 \\
&= 16.0176 \qquad \text{or about 16}
\end{aligned}
$$

The case of predicting a criterion variable from a single predictor can be considered a special case of multiple regression: the formulas of Section 15-5 can be derived from the formulas of the present section if we assume that $r_{Y2} = 0$ and $r_{12} = 0$.

Be careful here, though, because the regression line in the $X_1 Y$ plane when there are two predictors is *not* the same as the regression line in the $X_1 Y$ plane when there is only one predictor. You can see this numerically in the example: when X_1 is the only predictor,

$$b_{Y1} = r_{Y1}(s_Y/s_1) = .30(4/8) = .15$$

$$a_{Y1} = \overline{Y} - b_{Y1}\,\overline{X}_1 = 10 - .15(40) = 10 - 6 = 4$$

so the prediction equation is

$$Y' = 4 + .15X_1$$

When there are two predictors, X_1 and X_2 in this example, and we set $X_2 = 0$, the regression line in the X_1Y plane is

$$Y' = .059X_1 - 10.452$$

This is the two-predictor equation when $X_2 = 0$.

Graphically the difference between the two regression equations comes from their representing different projections of the three dimensional ellipsoid into the X_1Y plane, as shown in Figure 18-3. These two projections are like two shadows of the set of points, and they have different positions and shapes, as though caused by light sources from different positions.

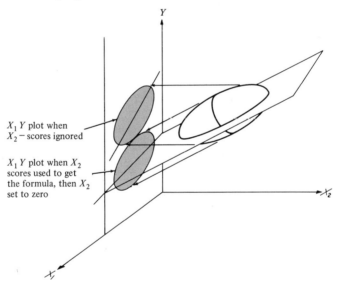

X_1Y plot when X_2 – scores ignored

X_1Y plot when X_2 scores used to get the formula, then X_2 set to zero

Figure 18-3

When simple linear regression is used to predict Y from X_1, ignoring X_2, this is like projecting the ellipsoid of points vertically onto the X_1Y plane, giving the upper shadow in Figure 18-3. When a multiple regression equation is obtained from X_1, X_2, and Y, and then X_2 is set to zero. This is like projecting the points down the ellipsoid's regression plane, giving the lower shadow in Figure 18-3.

The formulas developed so far can in themselves be considered to be special cases of a more general regression problem. If there are J predictors ($J \geq 1$), the general standard regression equation can be written as follows:

$$z'_Y = \beta_{Y1\cdot23\ldots J}z_1 + \beta_{Y2\cdot134\ldots J}z_2 + \cdots + \beta_{YJ\cdot12\ldots J-1}z_J \quad (18\text{-}6)$$

where $\beta_{Y1\cdot234\ldots J}$ is the standard regression weight (beta weight) for predicting z_Y from z_1, holding predictors z_2 through z_J constant, and similarly for the other beta weights. The regression formula for predicting raw scores is then

$$Y' = b_{Y1\cdot234\ldots J}X_1 + \cdots + b_{YJ\cdot123\ldots J-1}X_J + a_{Y\cdot123\ldots J} \quad (18\text{-}7)$$

where the first regression weight (slope) is given by

$$b_{Y1\cdot234\ldots J} = \beta_{Y1\cdot234\ldots J}\left(\frac{\sigma_Y}{\sigma_1}\right) \quad (18\text{-}8)$$

and so on. Methods for calculating the beta weights for three or more predictors will not be given here. The student who must calculate them by hand can refer to Baggaley (1964), Chapter 5, McNemar (1969), 199–203, or Walker and Lev (1953), 324–336. However, computer programs are generally available for doing this sort of drudgery and should be used if at all possible.

B. Multiple Correlation

Just as the Pearson coefficient r_{XY} indicates how well a criterion variable Y can be predicted, assuming linear regression, from a predictor X, the multiple-correlation coefficient $R_{Y\cdot123\ldots J}$ indicates how well the criterion variable Y can be predicted from an optimal combination of the predictors $X_1, X_2, \cdots, X_J$. Just as the Pearson coefficient indicates how well the regression equation $Y' = a_Y + b_{YX}X$ of Section 15-5 works, the multiple-correlation coefficient indicates how well the multiple regression of part A works.

The formula for multiple R is a function of the number of predictors. When there are two predictors, X_1 and X_2, the formula is

$$R_{Y\cdot12} = \sqrt{\frac{r_{Y1}^2 + r_{Y2}^2 - 2r_{Y1}r_{Y2}r_{12}}{1 - r_{12}^2}} \quad (18\text{-}9)$$

where r_{Y1}, r_{Y2}, and r_{12} are the pairwise correlations. Only three of the coefficients of correlation discussed previously are suitable to be used in this formula: the Pearson coefficient (Section 16-2) and two derivatives of it, the point-biserial coefficient (Section 16-3) and the phi coefficient (Section 16-5). Multiple R is always positive or zero ($0 \leq R \leq 1$).

The formula may be operated upon algebraically to provide an alternative form:

$$R_{Y \cdot 12} = \sqrt{\frac{(r_{Y1}^2 - r_{Y1}r_{Y2}r_{12}) + (r_{Y2}^2 - r_{Y1}r_{Y2}r_{12})}{1 - r_{12}^2}}$$

$$= \sqrt{\frac{r_{Y1}(r_{Y1} - r_{Y2}r_{12})}{1 - r_{12}^2} + \frac{r_{Y2}(r_{Y2} - r_{Y1}r_{12})}{1 - r_{12}^2}}$$

But from part A of this section, we know that

$$\beta_{Y1 \cdot 2} = \frac{r_{Y1} - r_{Y2}r_{12}}{1 - r_{12}^2} \quad \text{and} \quad \beta_{Y2 \cdot 1} = \frac{r_{Y2} - r_{Y1}r_{12}}{1 - r_{12}^2}$$

so that

$$R_{Y \cdot 12} = \sqrt{r_{Y1}\beta_{Y1 \cdot 2} + r_{Y2}\beta_{Y2 \cdot 1}}$$

a formula that has the advantage of being generalizable to any number of predictors:

$$R_{Y \cdot 12 \cdots J} = \sqrt{r_{Y1}\beta_{Y1 \cdot 23 \cdots J} + \cdots + r_{YJ}\beta_{YJ \cdot 123 \cdots J-1}} \qquad (18\text{-}10)$$

As indicated in part A of this section, however, when there are more than two predictors, it usually doesn't pay to try to calculate the beta weights by hand. For this reason, no discussion is given here for their calculation. For more than two predictors, your best strategy is to use a computer to do the calculations for you.

EXAMPLE   Suppose that we have three variables, Y, X_1, and X_2, with $r_{Y1} = .30$, $r_{Y2} = .50$, and $r_{12} = .40$. Then the multiple correlation for predicting Y from X_1 and X_2 is given by

$$R_{Y \cdot 12} = \sqrt{\frac{.09 + .25 - 2(.30)(.40)(.50)}{1 - .16}}$$

$$= \sqrt{\frac{.22}{.84}} = \sqrt{.26}$$

$$\doteq .51$$

C. Coefficient of Multiple Determination and Standard Error of Estimate

Like the Pearson coefficient, the squared multiple correlation coefficient also has the interpretation

$$R^2_{Y \cdot 123 \, \cdots \, J} = \frac{\sigma^2_Y - \sigma^2_{Y \cdot 123 \, \cdots \, J}}{\sigma^2_Y} \tag{18-11}$$

where σ^2_Y is the variance of all Y scores about μ_Y, and $\sigma^2_{Y \cdot 123 \, \cdots \, J}$ is the variance of Y scores about the best-fitting regression line, given by formula 18-7. The squared multiple correlation coefficient is also called the *coefficient of determination*. It has the same interpretation as r^2_{XY}, that proportion of criterion variance accounted for by knowledge of the predictors and the regression formula for predicting Y from them.

It is possible to obtain a simple formula for the standard error of estimate by a derivation similar to that of formula 16-9. The resultant formula is

$$\sigma_{Y \cdot 123 \, \cdots \, J} = \sigma_Y \sqrt{1 - R^2_{Y \cdot 123 \, \cdots \, J}} \tag{18-12}$$

EXAMPLE In the example used in parts A and B, s_Y was given as 4, and $R^2_{Y \cdot 12}$ was found to be .26. Thus, we can say that we can account for 26 percent of the variance of Y using the formula $Y' = .059X_1 + .9046X_2 - 10.452$. We can find the standard error of estimate to be

$$s_{Y \cdot 12} = 4\sqrt{1 - .26} = 4\sqrt{.74} = 4(.86)$$

$$= 3.44$$

D. Test for Nonzero Population Multiple Correlation

> *Relevant earlier sections:*
> **7-7** F distribution
> **Chapter 9** Hypothesis testing

If you are willing to assume that, throughout the range of predictions, the observed criterion values Y are normally distributed about the predicted values Y' (from equation 18-7), with a constant variance for all Y', it is possible to test for a zero-population multiple correlation. The hypotheses are

$$H_0 : R = 0 \text{ in the population} \quad \text{and} \quad H_m : R > 0 \text{ in the population}$$

(Recall that a negative R is not defined.) The formula used to test this pair of hypotheses is

$$F = \frac{R^2/J}{(1 - R^2)/(N - J - 1)} \qquad \text{with } J, N - J - 1 \text{ df* } \qquad (18\text{-}13)$$

where J is the number of predictors.

EXAMPLE   Let us test, at the .05 significance level, the hypothesis of zero population correlation, for the example of part B of this section. Suppose that the total number of subjects, N, was 103. There were two predictors, and the obtained R^2 was .26.

$$F_{.05}(2, 100 \text{ df}) = 3.07 \qquad \text{and} \qquad F = \frac{.26/2}{.74/100} = \frac{.13}{.0074} \doteq 17.57$$

which is significant. Thus, there is evidence for a positive population correlation.

It is interesting to note that when $J = 1$, multiple R reduces to the Pearson correlation between two variables. Then the significance test for zero correlation becomes

$$F = \frac{R^2/1}{(1 - R^2)/(N - 2)} = \frac{r^2}{(1 - r^2)/(N - 2)} \qquad \text{with 1 and } N - 2 \text{ df}$$

But from Section 7-7D we know that t^2 with b df is equal to F with 1 and b df. Multiplying numerator and denominator by $N - 2$ and taking the square root of both sides gives

$$t = \sqrt{F} = \frac{r\sqrt{N - 2}}{\sqrt{1 - r^2}} \qquad \text{with } N - 2 \text{ df}$$

which is identical to formula 16-10.

E. Test for Significant Increment in R From Additional Predictors

Suppose that we have some set of predictor variables A, containing predictors $X_1, X_2, \ldots, X_a$, and that we obtain the multiple correlation for predicting Y from these (a) predictors. Suppose that we are considering adding to our regression equation several new predictors, and we want to know whether R will increase. Let us call the new set of predictors B, having (b) predictors X_{a+1}, $X_{a+2}, \ldots, X_{a+b}$. Thus, we want to know whether the multiple correlation that

---

* A common alternative notation uses X_1 as the criterion and $X_2, X_3, \cdots, X_K$ as the predictors. Apparent discrepancies in formulas between that notation and the one here adopted may be resolved by observing that the number of predictors, J in the present notation, is given by $K - 1$ in the other system, and by adjusting one or the other set of formulas accordingly.

uses both sets A and B is significantly greater than the multiple correlation using only set A of the predictors. Let us use the abbreviations

$$R^2_{Y \cdot A} = R^2_{Y \cdot 123 \cdots a} \qquad \text{and} \qquad R^2_{Y \cdot A,B} = R^2_{Y \cdot 123 \cdots a(a+1) \cdots (a+b)}$$

Then the hypotheses are

$$H_0 : R_{Y \cdot A} = R_{Y \cdot A,B} \qquad \text{and} \qquad H_m : R_{Y \cdot A} < R_{Y \cdot A,B}$$

and the test is

$$F = \frac{(R^2_{Y \cdot A,B} - R^2_{Y \cdot A})/b}{(1 - R^2_{Y \cdot A,B})/(N - a - b - 1)} \qquad \text{with } b \text{ and } N - a - b - 1 \text{ df} \qquad (18\text{-}14)$$

EXAMPLE Suppose, in the example of part B, that we have 103 subjects, and we first ask whether there is a significant correlation between predictor X_1 and the criterion, $t_{.05}$(two tail, 101 df) $= \pm 2.00$, and

$$t = \frac{.30\sqrt{101}}{\sqrt{1 - .09}} = \frac{.30(10.05)}{\sqrt{.91}} = \frac{3.01}{.95} = 3.17$$

which is significant. Now let us ask whether predictor X_2 increases our ability to predict the criterion, also at the .05 level. $F_{.05}(1,100 \text{ df}) = 4.00$, and

$$F = \frac{(.26 - .09)/1}{(1 - .26)/100} = \frac{.17}{.0074} = 23.0$$

which is also significant. Thus, X_2 adds predictability over X_1.

A problem to be kept in mind when carrying out this type of analysis is that the order of adding variables to the multiple-regression equation affects which ones will be determined to be significantly related to the criterion. In the example above we might have concluded that both X_1 and X_2 should be included in the multiple-regression equation on the basis of the analysis. But this conclusion has not been entirely determined by the data; when we decided to first test for the significance of the correlation of X_1 with the criterion, we attributed to X_1 whatever predictability X_1 and X_2 have in common (and they do correlate .40). Suppose, by contrast, that we had added X_1 and X_2 in reverse order: what would the results have been? It is clear that if a .30 correlation of X_1 with the criterion is significant, a .50 correlation of X_2 with the criterion would also have been significant. But what about the increment in predictability of X_1 over X_2?

$$F = \frac{(.26 - .25)/1}{(1 - .26)/100} = \frac{.01}{.0074} = 1.35$$

which is not significant. Thus, X_1 does not add predictability over X_2, and we would not have two predictors in the regression equation.

Because the order in which predictors are tested affects your conclusions as to which ones and how many are to be retained in the regression equation, it is a good idea to have an explicit strategy for order of testing them.

One such strategy would be to take the order of testing variables on the basis of their relative importance for whatever theory you are working with; these decisions would be made a priori. Having decided on order of testing, you begin with the first predictor, find its correlation with the criterion, and test for significance. If the correlation is nonsignificant, you try the other predictors in order until you find one that is significantly correlated with the criterion or you run out of variables. Once you find a significant correlation with the criterion, you start adding variables and calculating the multiple correlation, testing the multiple R for each new variable, and retaining only those that raise R significantly.

Another strategy is to add, at each step in building a multiple-regression equation, the predictor that produces the greatest increment in R. Computer programs are widely available which will carry out this procedure for you automatically. If what you want is efficient prediction, then this is an optimal strategy. It is clearly mutually exclusive with the previous strategy of deciding on an order for testing a priori. It is also a poor method for theory construction, because the most effective predictor will not necessarily be the most theoretically meaningful one.

F. Use of Multiple Regression in Curvilinear Regression and Correlation

As was indicated in Sections 15-4 and 17-1, a kind of nonlinear prediction and correlation can be carried out by predicting conditional criterion means for each level of the predictor and evaluating the degree of relationship using the correlation ratio, η. A major drawback of that analysis is that it generates no equation for predicting criterion values; a separate and distinct prediction is made for each level of the predictor.

The first step in generating an equation for Y as a curvilinear function of some predictor, say W, is to select the components of the regression line. For example, Y might vary as a function of any two or more of W, W^2, W^3, $\sqrt{W}$, $\sqrt[3]{W}$, e^{2W+3}, $\log_{10} W$, $1/(4W - 2)$, $10^{W - .27}$, $W^{-(1/4)}$, etc. The choice of terms could either be made on theoretical grounds or on the basis of an examination of the plot of conditional criterion means.

Once the components of the equation have been selected, multiple regression can be used to determine their optimal weights in an equation for predicting the criterion. Suppose, for example, that you have decided to find Y

as a function of W, W^2, and W^3. There is one criterion and three predictors; let $X_1 = W$, $X_2 = W^2$, and $X_3 = W^3$, and use the method of multiple correlation, with criterion Y and predictors X_1, X_2, and X_3.

$$Y' = a + b_{YW \cdot W^2 W^3} W + b_{YW^2 \cdot WW^3} W^2 + b_{YW^3 \cdot WW^2} W^3$$

You can either treat the data in a single regression and correlation problem, as in parts A and B of this section, or use the stepwise method of part E, which has the advantage of giving you a statistical criterion for the number of predictors. For further discussion of this and related uses of multiple regression and correlation, see Cohen (1968), or Kerlinger and Pedhazur (1973), Chapter 9.

G. Correction for Shrinkage

The multiple correlation based on a sample of subjects is a best fit to the data of that sample. However, as an estimate of the population multiple correlation, it will tend to be too large, especially if the sample is small ($N \leq 100$).

One way of obtaining a more accurate estimate of the population multiple correlation is to apply the obtained multiple-regression equation to a new sample of subjects and calculate, for the second sample, the standard error of estimate about the regression line obtained from the first sample. This value, squared, can then be used in equation 18-11 to give a more reasonable estimate of the population R.

This method, called *cross validation*, may not be possible if you have a limited number of subjects. Alternatively, an estimate can be found from a single sample, using the correction formula

$$_cR_{Y \cdot 12 \cdots J} = \sqrt{1 - \frac{(1 - R^2_{Y \cdot 12 \cdots J})(N - 1)}{N - J - 1}} \tag{18-15}$$

The corrected standard error of estimate can either be found from the corrected multiple R,

$$_c\sigma_{Y \cdot 123 \cdots J} = \sigma_Y \sqrt{1 - {_cR^2_{Y \cdot 123 \cdots J}}} \tag{18-16}$$

or from the uncorrected multiple R, using the formula

$$_c\sigma_{Y \cdot 123 \cdots J} = \sigma_Y \sqrt{(1 - R^2_{Y \cdot 123 \cdots J})\left(\frac{N - 1}{N - J - 1}\right)} \tag{18-17}$$

EXAMPLE   For the example carried through this section, $R^2_{Y \cdot 12} = .26$, $R_{Y \cdot 12} = .51$, $s_Y = 4$, $N = 103$, and $s_{Y \cdot 12} = 3.44$. From this information, corrected values of the multiple correlation and the standard error of estimate may be obtained:

$$_cR^2_{Y\cdot 12} = 1 - (1 - .26)\left(\frac{103 - 1}{103 - 3}\right)$$

$$= 1 - .74\left(\frac{102}{100}\right)$$

$$= 1 - .755$$

$$= .245$$

$$_cR_{Y\cdot 12} = \sqrt{.245} \doteq .49$$

This corrected value of the multiple correlation is slightly smaller than the uncorrected value of .51, indicating that there was some slight over-estimate from using sample values. For the standard error of estimate,

$$_c\sigma_{Y\cdot 12} = 4\sqrt{1 - .245}$$

$$= 4\sqrt{.755} = 4(.87)$$

$$= 3.48$$

This corrected value of the standard error of estimate is somewhat larger than the uncorrected value of 3.44, showing that predictions are not quite as accurate as the uncorrected values indicated.

The difference between uncorrected and corrected values lies in the ratio $(N - 1)/(N - J - 1)$. Therefore, the correction will be greater for smaller samples and more predictors than used in the present problem.

References   There are several commonly used techniques for dealing with three or more variables at a time. In multiple regression you optimally combine several predictor variables to predict a single criterion. This is discussed in Guilford (1965), Chapter 16; Hays (1963), 566–577; Kerlinger and Pedhazur (1973); and McNemar (1969), Chapter 11. In factor analysis, you start with a large number of variables and try to derive from the analysis a reduced number of variables (called *factors*) that contain most of the information in the full set. It is discussed in Gorsuch (1974) and in Mulaik (1972). Cluster analysis allows you to group variables or subjects on degree of similarity to one another; see Anderberg (1973) or Hartigan (1975). Other multiple variable techniques can be found in Cooley and Lohnes (1962), Overall and Klett (1972), and Tatsuoka (1971).

18-2   Relationship among three or more ordinal variables

Relevant earlier section:
17-2   Tau and gamma

When you wish to know the degree of relationship among three or more variables, all of which you wish to consider to be ordinal-level, either of two kinds of coefficient can be used. When you wish to know how well you can predict ranks on a criterion variable from ranks on two predictors, the appropriate coefficient is multiple tau, an extension of the tau coefficient of Section 17-2. When, by contrast, you wish to know the overall degree of consistency of several rankings in a nonpredictive sense, the indicator to use is the coefficient of concordance, which is related to the rho coefficient of Section 17-2.

A. Coefficient of Concordance, W

The coefficient of concordance is used to indicate the extent to which several rank orders agree. It might, for example, be used to determine the extent to which several judges are responding similarly to a set of stimuli or to the behavior of a set of subjects.

Table 18-1   *Data layout for coefficient of concordance*

	Stimuli						
Judge	1	2	3	4	5	$\cdots$	N
1							
2							
3							
$\vdots$							
J							
Σ	T_1	T_2	T_3	T_4	T_5	$\cdots$	T_N

In Table 18-1, each judge must rank all the stimuli from 1 to N (or ranks must be derived from whatever scores you do obtain). Then a sum of ranks is obtained for each stimulus by adding ranks down its column. If there is no consistent agreement among the judges about the order of the stimuli, then the totals will differ only randomly, and the variance of the totals will be near zero. If there is perfect agreement among the judges, the totals will range from J, for the stimulus receiving 1's from all judges, to NJ, for the stimulus receiving N's from all judges, and the variance of the totals will be at a maximum. Thus, the variance of the totals can be used as an indicator of the degree of relationship among the judges' ratings.

The biased variance estimate for the set of totals (s_T^2) is given (Section 5-12) by the formula

$$s_T^2 = \frac{\sum_i T_i^2 - (\sum_i T_i)^2/N}{N}$$

and the maximum possible value of this variance is equal to

$$s_{max}^2 = \frac{J^2(N^2 - 1)}{12}$$

The coefficient of concordance is simply the ratio of the observed variance of T's to their maximum possible variance:

$$W = \frac{\sum_i T_i^2 - (\sum_i T_i)^2/N}{NJ^2(N^2 - 1)/12} \tag{18-18}$$

The coefficient of concordance is related to the rho (r_s) coefficient of Section 17-2E in the following manner. If you were to correlate the rankings for each pair of judges and then find the average correlation for all pairs of judges, it would be related to W via the formula

$$\text{average } r_s = \frac{JW - 1}{J - 1} \tag{18-19}$$

EXAMPLE Patients at a psychiatric clinic are seen on intake by a team of one psychiatrist, one psychologist, and a social worker for diagnosis and recommendation as to type of therapy. At the end of every month, the staff members of a team rank all patients seen that month on prognosis in therapy. The results for one team for one month are as follows:

					Patient					
	1	2	3	4	5	6	7	8	9	Σ
Psychiatrist	8	4	3	9	1	5	7	2	6	
Psychologist	7	6	3	8	2	4	9	1	5	
Social worker	9	1	5	8	2	7	6	3	4	
Total (T_i)	24	11	11	25	5	16	22	6	15	135
T_i^2	576	121	121	625	25	256	484	36	225	2469

$$W = \frac{2469 - (135)^2/9}{9(3^2)(9^2 - 1)/12} = \frac{2469 - 2025}{9(9)(80)/12}$$

$$= \frac{444}{540} \doteq .822$$

Thus, there appears to be substantial agreement among the three staff members about the prognosis of the given patients. The average rho between pairs of staff members is

$$\text{average } r_s = \frac{3(.822) - 1}{3 - 1} = \frac{1.466}{2} = .733$$

B. Correction of W For Ties

When one or more judges give tied ranks, it is not possible for formula 18-18 to achieve unity. However, a correction can be applied which reduces the maximum variance of the T_i's, and makes possible a perfect value of W. First, find a correction term for each judge who has given tied ranks:

$$C_j = \sum (t^3 - t) \qquad (18\text{-}20)$$

where t is the number of objects tied at a given rank by judge j, and summation is over the groups of ties in j's ranking. Then the corrected W is given by the formula

$$_cW = \frac{\sum_i T_i^2 - (\sum_i T_i^2)/N}{N J^2 (N^2 - 1)/12 - N \sum_j C_j/12} \qquad (18\text{-}21)$$

EXAMPLE   In a small group of seven subjects, suppose that each subject rates every other on a 5-point scale, from 1, for very submissive, to 5, for very aggressive. Each must rate himself as 3 on the scale. The results are as follows:

Subject as judge	Subject as stimulus						
	1	2	3	4	5	6	7
1	3	5	5	1	3	3	3
2	4	3	4	1	4	3	4
3	1	4	3	2	4	1	3
4	4	4	5	3	4	3	3
5	2	2	3	2	3	4	4
6	5	4	4	4	4	3	3
7	3	5	4	1	1	2	3

These data are most reasonably considered ordinal, but they do not constitute a rank order for each row. We therefore put them in the desired form, and give tied scores the average rank.

Subject as judge	Subject as stimulus							t_1 t_2 t_3	C_j
	1	2	3	4	5	6	7		
1	3.5	6.5	6.5	1.0	3.5	3.5	3.5	4 2 —	66
2	5.5	2.5	5.5	1.0	5.5	2.5	5.5	2 4 —	66
3	1.5	6.5	4.5	3.0	6.5	1.5	4.5	2 2 2	18
4	5.0	5.0	7.0	2.0	5.0	2.0	2.0	3 3 —	48
5	2.0	2.0	4.5	2.0	4.5	6.5	6.5	3 2 2	36
6	7.0	4.5	4.5	4.5	4.5	1.5	1.5	2 4 —	66
7	4.5	7.0	6.0	1.5	1.5	3.0	4.5	2 2 —	12
T_i	29.0	34.0	38.5	15.0	31.0	20.5	28.0	$\sum_j C_j = 312$	
T_i^2	841.0	1156.0	1482.25	225.0	961.0	420.25	784.0		

The ranks of judge 1 are obtained first by putting the ratings in numerical order: 1, 3, 3, 3, 3, 5, 5. The rating of 1 becomes rank 1; the four 3's are given rank 3.5 (the mean of ranks 2, 3, 4, and 5), and the two 5's are given rank 6.5. Because there are four tied at 3.5, $t_1 = 4$; and because there are two tied at 6.5, $t_2 = 2$. Then $C_1 = (4^3 - 4) + (2^3 - 2) = 66$.

Add across the two bottom rows of the table to get $\sum_i T_i = 196$ and $\sum_i T_i^2 = 5869.50$. Then the uncorrected W is

$$W = \frac{5869.50 - (196)^2/7}{7(49)(49 - 1)/12} = \frac{381.50}{1372} = .278$$

The correction term is

$$N \sum_j \frac{C_j}{12} = \frac{7(312)}{12} = 182$$

and the value of W corrected for ties is

$$_cW = \frac{381.50}{1372 - 182} = \frac{381.50}{1190} \doteq .32$$

C. Testing the Significance of W

> *Relevant earlier sections:*
> **7-6** Chi-square distribution
> **Chapter 9** Hypothesis testing

It is possible to test the null hypothesis $H_0: W = 0$ (that there is no agreement among the judges) against the motivated hypothesis $H_m: W > 0$ (that there is some agreement). These two alternatives exhaust the possibilities, because it is not possible for W to be negative.

Case 1: J between 3 and 20, and N between 3 and 7   Because the denominator of W is a function only of J and N, significance can be tested using only the numerator: $S = \sum T_i^2 - (\sum_i T_i)^2/N$. This value is compared directly to the critical values given in Table A-16 of the Appendix. If the observed S is greater than the critical value at the chosen level of significance, H_0 is rejected.

EXAMPLE   For the example of part B, $N = 7$, $J = 7$, and $S = 381.50$. The critical value of S at $\alpha = .05$ is obtained by interpolation: $S_{.05} = (335.2 + 453.1)/2 = 394.2$. The observed S of 381.50 is less than the critical value, so the null hypothesis is not rejected.

Case 2: N greater than 7   This case may be tested using the uncorrected value of W in the formula

$$\chi^2 = J(N - 1) W \tag{18-22}$$

which is approximately distributed as χ^2 with $N - 1$ df. The observed value of chi square can then be compared to the critical value given in Table A-7 of the Appendix.

EXAMPLE   For the example of part A, $N = 9$, and the present method can be applied. Let us set α at $.05$; $W = .822$, so

$$\chi^2 = 3(8)(.822) = 19.728$$

From Table A-7, $\chi^2_{.05} = 15.5073$. The observed χ^2 is greater than the tabulated value, so H_0 is rejected.

D. Multiple Tau

There has been very little theoretical or applied work in the prediction of an ordinal criterion from two or more ordinal predictors. However, it is possible to specify an equation for multiple tau when there are two predictors:

$$\tau_{Y\cdot12} = \sqrt{\frac{\tau_{Y1}^2 + \tau_{Y2}^2 - 2\tau_{Y1}\tau_{Y2}\tau_{12}}{1 - \tau_{12}^2}} \qquad (18\text{-}23)$$

where τ_{Y1}, τ_{Y2}, and τ_{12} are calculated in accordance with Section 17-2. Variables X_1 and X_2 are predictors and variable Y is the criterion.

Suppose that you know the correlation τ_{12} between predictors and imagine that criterion ranks are random. Your best guess of the multiple tau for predicting criterion ranks is given by formula 18-24 (Moran, 1951):

$$\text{est. } T^2_{(\text{random})\cdot12} = \frac{2[2(2N + 5) - 6\tau_{12}^2 - 4(N + 1)\tau_{12}r_{s(12)}]}{9N(N - 1)(1 - r_{s(12)}^2)} \qquad (18\text{-}24)$$

where $r_{s(12)}$ is the rho correlation (Section 17-2E) between variables X_1 and X_2.

This value can serve as a base line for evaluating the magnitude of any $\tau_{Y\cdot12}$ obtained from real data. Formula 18-24 can also be used in testing for significance of an obtained multiple tau. To do this, define the term e as indicated by

$$e = \frac{2}{\text{est. } T^2_{1\cdot23}} + 1 \qquad (18\text{-}25)$$

Then the one-tail hypotheses $H_0 : T^2_{Y\cdot12} = T^2_{(\text{random})\cdot12}$ and $H_m : T^2_{Y\cdot12} > T^2_{(\text{random})\cdot12}$ may be tested using the statistic

$$F = \frac{\tau_{Y\cdot12}^2/2}{(1 - \tau_{Y\cdot12}^2)/(e - 3)} \qquad \text{at 2 and } e - 3 \text{ df} \qquad (18\text{-}26)$$

This F test is an approximate solution obtained theoretically. There has not yet been sufficient work done with this test to make it clear how large N must be for the approximation to be reasonably accurate.

EXAMPLE   Truancy cases are often returned to the custody of the parents by a juvenile court, with supervision by a court social worker. Twenty-one such cases are ranked by two social workers on the amount of expected truancy in the next year. At the end of the year they are ranked on amount of actual truancy and the following correlations are obtained: (Y = amount of truancy, S_1 is rating by social worker 1 and S_2 is rating by worker 2), $\tau_{Y1} = .6$, $\tau_{Y2} = .4$, $\tau_{12} = .2$, and $r_s(S_1, S_2) = .3$. Then

$$\tau_{Y\cdot12} = \sqrt{\frac{.36 + .16 - 2(.6)(.4)(.2)}{1 - .04}} = \sqrt{\frac{.424}{.96}}$$

$$= \sqrt{.442} = .66$$

$$\text{est. } T^2_{\text{random} \cdot 12} = \frac{2[2(42 + 5) - 6(.04) - 4(22)(.2)(.3)]}{9(21)(20)(.91)}$$

$$= \frac{2(94 - .24 - 5.28)}{3439.8}$$

$$= \frac{177}{3024} = .051$$

From this, e is found to be

$$e = \frac{2}{.053} + 1 = 39.22 + 1 = 40.22$$

and

$$F = \frac{.442/2}{(1 - .442)/37.22} = \frac{.221}{.558/37.22} = \frac{.221}{.015} \doteq 14.74$$

The critical value of F for this problem is $F_{.05}(2, 40 \text{ df}) = 3.23$. The observed value of 14.74 is greater than the critical value, so we conclude that the results are significant: the magnitude of multiple tau is greater than chance in the population.

References   The coefficient of concordance is discussed in Hays (1963), 656–658; McNemar (1969), 435–437; and Siegel (1956), 229–238. Multiple tau is developed by Moran (1951).

18-3   Relationship among three or more nominal variables

> *Relevant earlier sections:*
> **15-2**   Prediction table
> **17-3**   Lambda

A. Predictive Association
Prediction of a nominal-level variable from two or more predictors is carried out with only a slight generalization from the single predictor case presented in Sections 15-2 and 17-3. As before, the conditional mode is predicted, here for each combination of levels of the predictors. If there are J levels of predictor X_1 and K levels of predictor X_2, then the prediction table has JK rows.

The formula for computing multiple lambda is identical to that of formula 17-34. Here it is modified slightly for the case of two predictors only to emphasize the number of conditional modes to be found:

$$\lambda_{Y\cdot 12} = \frac{\sum_j \sum_k f'_{jk} - f'}{N - f'} \tag{18-27}$$

for prediction of Y from X_1 and X_2, with f'_{jk} the frequency of subjects at the conditional mode for the combination of level j of X_1 and level k of X_2; and f' the frequency of subjects at the overall mode of Y (when no distinction is made as to levels of the predictors). Again, the sample estimate L of λ is identical to it in form except that sample, not population, frequencies are used:

$$L_{Y\cdot 12} = \frac{\sum_j \sum_k f'_{jk} - f'}{N - f'} \tag{18-28}$$

EXAMPLE   Consider the hypothetical data in the table, in which sex (S) and birth order (B) are used to predict students' undergraduate majors (M). Criterion abbreviations are physical science (PS), social science (SS), humanities (H), fine arts (FA), and engineering (E).

Sex	Birth Order	Undergraduate major					
		PS	SS	H	FA	E	Σ
Male	Only	50	14	4	2	30	100
	First	20	12	6	4	28	70
	Later	26	3	7	19	25	80
Female	Only	20	56	3	7	14	100
	First	3	40	53	2	2	100
	Later	1	25	7	16	1	50
Σ		120	150	80	50	100	500

In this example, there are two levels of the first predictor (sex) and three levels of the second (birth order). Thus, there are $2 \times 3 = 6$ predictor levels, and 5 levels of the criterion. The following prediction table also gives the frequencies of subjects at each predictor level:

Sex	Birth order	Predicted major	f'_{jk}
Male	Only	PS	50
	First	E	28
	Later	PS	26
Female	Only	SS	56
	First	H	53
	Later	SS	25
Σ			238

If we did not have the predictor information but only the overall criterion distribution, our best prediction would be SS for every subject, and $f' = 150$. $N = 500$, so

$$L_{\text{M·BS}} = \frac{238 - 150}{500 - 150} = \frac{88}{350} = .251$$

Therefore, knowledge of sex and birth order reduce error of prediction by .252.

B. Significance Test for Multiple Lambda

Relevant earlier sections:
7-3   Normal distribution
Chapter 9   Hypothesis testing

Multiple lambda has the same form as the two-variable lambda. Hence, it is tested for significance using the method of Section 17-3C.

$$z = \frac{L_{\text{M·BS}} - \lambda_0}{\sqrt{(N - \sum_j \sum_k f'_{jk})(\sum_j \sum_k f'_{jk} + f' - 2 \sum_a f'_{jk})/(N - f')^3}} \qquad (18\text{-}29)$$

EXAMPLE   In the problem of part A of this section, $L_{\text{M·BS}} = .251$. Suppose that we had been testing the null hypothesis $H_0: \lambda_0 \leq .10$, at $\alpha = .05$. Then $f' = 150$, which appears in the SS column. Two values of f'_{jk} appear in this column: f' (female, only child) $= 56$, and f' (female, later-born) $= 25$. $\sum_a f'_{jk} = 56 + 25 = 81$, and $\sum_j \sum_k f'_{jk} = 238$ has already been found. Then,

$$z = \frac{.251 - .100}{\sqrt{(500 - 238)[238 + 150 - 2(81)]/(500 - 150)^3}}$$

$$= \frac{.151}{\sqrt{(162)(226)/(350)^3}} = \frac{.151}{\sqrt{36,612/42,875,000}} \doteq 5.17$$

From Table A-5 we find the critical value of the standard normal distribution to be 1.6449. The observed z score is greater than this value, so the null hypothesis is rejected.

C. Contingency Coefficient

The contingency coefficient for relating three or more variables is defined by equation 17-38, as follows.

$$C = \sqrt{\frac{\chi^2}{\chi^2 + N}}$$

where χ^2 is computed according to equation 10-2 and N is the total number of observations. As before, C is significant when χ^2 is significant.

EXAMPLE   Consider the following example, in which the attitudes of subjects to a popular musical group are compared. The data are as follows:

Age of males	Attitude		Age of females	Attitude	
	Like	Dislike		Like	Dislike
Over 30	10	40	Over 30	35	15
Under 30	20	30	Under 30	25	25

Chi square is computed and found to be 8.93. The contingency coefficient is then

$$C = \sqrt{\frac{8.93}{8.93 + 200}} = \sqrt{\frac{8.93}{208.93}} = \sqrt{.0427} \doteq .21$$

Homework

1. If $r_{Y1} = .60$, $r_{Y2} = .30$, $r_{12} = .50$, $s_Y = 8$, $s_1 = 12$, $s_2 = 7$, $\bar{Y} = 6$, $\bar{X}_1 = 7$, $\bar{X}_2 = 4$, and $N = 35$,
   (a) What is the multiple correlation for predicting Y from X_1 and X_2 within this sample?
   (b) What is the formula for predicting Y from X_1 and X_2?
   (c) Is there a nonzero population multiple correlation? Test at an α of .01.
   (d) What is your best estimate of the population multiple correlation?
   (e) What is your best estimate of the population standard error of estimate?

2. If $r_{Y1} = .80$, $r_{Y2} = .80$, $r_{12} = .60$, and $N = 123$, and if we wish to determine whether we can predict Y more accurately using both X_1 and X_2 than we can using X_1 alone,
   (a) What are our null and motivated hypotheses?
   (b) Show that the appropriate test statistic is

$$F = \frac{R^2_{Y\cdot12} - r^2_{Y1}}{(1 - R^2_{Y\cdot12})/(N-3)} \quad \text{with 1 and } N-3 \text{ df}$$

   (c) Calculate $R_{Y\cdot12}$.
   (d) Test the hypothesis at the .05 level and draw your conclusion.

3. In the example of Section 14-7 (pp. 383–384), freshman males were assigned to experimental groups so that each group contained one only child, one first-born, one second-born with an older brother, and one second-born with an older sister. The same judges ranked subjects in each of six groups on leadership ability. We might ask to what extent the kinds of subjects receive the same rankings in the several groups. The data are as follows:

		Birth order		
Group	Only	First	Second, brother	Second, sister
1	3	2	1	4
2	2	1	4	3
3	4	2	1	3
4	3	1	2	4
5	2	1	3	4
6	3	1	2	4

   Find the extent to which the rank orders are the same in the several rows.

4. Five racing drivers have competed in the same races three times this year, finishing in the following orders relative to one another:

	Driver				
Occasion	A	B	C	D	E
1	3	5	4	1	2
2	3	5	2	1	4
3	1	5	4	2	3

(a) What is the magnitude of the squared correlation which indicates your ability to predict ranks on the third contest from ranks on the first two?

(b) Is this correlation significant?

5. In Section 18-3C the contingency coefficient was used to indicate overall relationship among age, sex, and attitude toward a popular musical group. If, instead, you were predicting the subjects' attitudes from their age and sex,

(a) What coefficient would you use?

(b) What would its magnitude be?

6. Given that the assumptions of the Pearson correlation coefficient are satisfied for each pair of variables in the following table,

(a) Find those pairwise correlations.

(b) Find the multiple correlation for predicting X from Y and Z.

i	X	Y	Z
1	9	2	16
2	11	4	19
3	7	5	13
4	7	4	19
5	6	10	13

Homework Answers

1. (a) .60 (b) $Y' = 3.2 + 0.4X_1$ (c) Yes (d) $_cR_{Y\cdot 12} = .57$ (e) 6.56

2. (a) $H_0: R_{Y\cdot 12} = r_{Y1}$, $H_m: R_{Y\cdot 12} > r_{Y1}$ (c) $R^2 = .80$ (d) $F = 96$; reject H_0.

3. $W = .59$, average $r_s = .61$

4. (a) $\tau^2_{3\cdot 12} = .39$ (b) $F = 0.635 < 1.0$, not significant

5. (a) Lambda or multiple R with r_ϕ for pairwise comparisons

(b) $L_{\text{attitude}\cdot\text{age, sex}} = .22$, $R_{\text{attitude}\cdot\text{age, sex}} = .30$

6. (a) $r_{XY} = -.625$, $r_{XZ} = .625$, $r_{YZ} = -.583$ (b) $R_{X\cdot YZ} \doteq .70$

Appendix

General Tables

Critical values

Table A-1 *The Greek alphabet*

A	α	alpha
B	β	beta
Γ	γ	gamma
Δ	δ	delta
E	ϵ	epsilon
Z	ζ	zeta
H	η	eta
Θ	θ	theta
I	ι	iota
K	κ	kappa
Λ	λ	lambda
M	μ	mu
N	ν	nu
Ξ	ξ	xi
O	o	omicron
Π	π	pi
P	ρ	rho
Σ	σ	sigma
T	τ	tau
Υ	υ	upsilon
Φ	ϕ	phi
X	χ	chi
Ψ	ψ	psi
Ω	ω	omega

Table A-2 *Squares, square roots, and reciprocals of numbers 1 to 1000**

N	N^2	$\sqrt{N}$	$\sqrt{10N}$	$1/N$	$1/\sqrt{N}$	$1/\sqrt{10N}$
1	1	1.0000	3.1623	1.000000	1.0000	.31623
2	4	1.4142	4.4721	.500000	.7071	.22361
3	9	1.7321	5.4772	.333333	.5774	.18257
4	16	2.0000	6.3246	.250000	.5000	.15811
5	25	2.2361	7.0711	.200000	.4472	.14142
6	36	2.4495	7.7460	.166667	.4082	.12910
7	49	2.6458	8.3666	.142857	.3780	.11952
8	64	2.8284	8.9443	.125000	.3536	.11180
9	81	3.0000	9.4868	.111111	.3333	.10541
10	1 00	3.1623	10.0000	.100000	.3162	.10000
11	1 21	3.3166	10.4881	.090909	.3015	.09535
12	1 44	3.4641	10.9545	.083333	.2887	.09129
13	1 69	3.6056	11.4018	.076923	.2774	.08771
14	1 96	3.7417	11.8322	.071429	.2673	.08452
15	2 25	3.8730	12.2474	.066667	.2582	.08165
16	2 56	4.0000	12.6491	.062500	.2500	.07906
17	2 89	4.1231	13.0384	.058824	.2425	.07670
18	3 24	4.2426	13.4164	.055556	.2357	.07454
19	3 61	4.3589	13.7840	.052632	.2294	.07255
20	4 00	4.4721	14.1421	.050000	.2236	.07071
21	4 41	4.5826	14.4914	.047619	.2182	.06901
22	4 84	4.6904	14.8324	.045455	.2132	.06742
23	5 29	4.7958	15.1658	.043478	.2085	.06594
24	5 76	4.8990	15.4919	.041667	.2041	.06455
25	6 25	5.0000	15.8114	.040000	.2000	.06325
26	6 76	5.0990	16.1245	.038462	.1961	.06202
27	7 29	5.1962	16.4317	.037037	.1925	.06086
28	7 84	5.2915	16.7332	.035714	.1890	.05976
29	8 41	5.3852	17.0294	.034483	.1857	.05872
30	9 00	5.4772	17.3205	.033333	.1826	.05774
31	9 61	5.5678	17.6068	.032258	.1796	.05680
32	10 24	5.6569	17.8885	.031250	.1768	.05590
33	10 89	5.7446	18.1659	.030303	.1741	.05505
34	11 56	5.8310	18.4391	.029412	.1715	.05423
35	12 25	5.9161	18.7083	.028571	.1690	.05345
36	12 96	6.0000	18.9737	.027778	.1667	.05270
37	13 69	6.0828	19.2354	.027027	.1644	.05199
38	14 44	6.1644	19.4936	.026316	.1622	.05130
39	15 21	6.2450	19.7484	.025641	.1601	.05064
40	16 00	6.3246	20.0000	.025000	.1581	.05000
41	16 81	6.4031	20.2485	.024390	.1562	.04939
42	17 64	6.4807	20.4939	.023810	.1543	.04880
43	18 49	6.5574	20.7364	.023256	.1525	.04822
44	19 36	6.6332	20.9762	.022727	.1508	.04767
45	20 25	6.7082	21.2132	.022222	.1491	.04714
46	21 16	6.7823	21.4476	.021739	.1474	.04663
47	22 09	6.8557	21.6795	.021277	.1459	.04613
48	23 04	6.9282	21.9089	.020833	.1443	.04564
49	24 01	7.0000	22.1359	.020408	.1429	.04518
50	25 00	7.0711	22.3607	.020000	.1414	.04472

* Reprinted from *Fundamental Statistics in Psychology and Education*, by J. P. Guilford. Copyright © 1965 by McGraw-Hill, Inc. Used with permission of McGraw-Hill Book Company.

N	N^2	$\sqrt{N}$	$\sqrt{10N}$	$1/N$	$1/\sqrt{N}$	$1/\sqrt{10N}$
51	26 01	7.1414	22.5832	.019608	.1400	.04428
52	27 04	7.2111	22.8035	.019231	.1387	.04385
53	28 09	7.2801	23.0217	.018868	.1374	.04344
54	29 16	7.3485	23.2379	.018519	.1361	.04303
55	30 25	7.4162	23.4521	.018182	.1348	.04264
56	31 36	7.4833	23.6643	.017857	.1336	.04226
57	32 49	7.5498	23.8747	.017544	.1325	.04189
58	33 64	7.6158	24.0832	.017241	.1313	.04152
59	34 81	7.6811	24.2899	.016949	.1302	.04117
60	36 00	7.7460	24.4949	.016667	.1291	.04082
61	37 21	7.8102	24.6982	.016393	.1280	.04049
62	38 44	7.8740	24.8998	.016129	.1270	.04016
63	39 69	7.9373	25.0998	.015873	.1260	.03984
64	40 96	8.0000	25.2982	.015625	.1250	.03953
65	42 25	8.0623	25.4951	.015385	.1240	.03922
66	43 56	8.1240	25.6905	.015152	.1231	.03892
67	44 89	8.1854	25.8844	.014925	.1222	.03863
68	46 24	8.2462	26.0768	.014706	.1213	.03835
69	47 61	8.3066	26.2679	.014493	.1204	.03807
70	49 00	8.3666	26.4575	.014286	.1195	.03780
71	50 41	8.4261	26.6458	.014085	.1187	.03753
72	51 84	8.4853	26.8328	.013889	.1179	.03727
73	53 29	8.5440	27.0185	.013699	.1170	.03701
74	54 76	8.6023	27.2029	.013514	.1162	.03676
75	56 25	8.6603	27.3861	.013333	.1155	.03651
76	57 76	8.7178	27.5681	.013158	.1147	.03627
77	59 29	8.7750	27.7489	.012987	.1140	.03604
78	60 84	8.8318	27.9285	.012821	.1132	.03581
79	62 41	8.8882	28.1069	.012658	.1125	.03558
80	64 00	8.9443	28.2843	.012500	.1118	.03536
81	65 61	9.0000	28.4605	.012346	.1111	.03514
82	67 24	9.0554	28.6356	.012195	.1104	.03492
83	68 89	9.1104	28.8097	.012048	.1098	.03471
84	70 56	9.1652	28.9828	.011905	.1091	.03450
85	72 25	9.2195	29.1548	.011765	.1085	.03430
86	73 96	9.2736	29.3258	.011628	.1078	.03410
87	75 69	9.3274	29.4958	.011494	.1072	.03390
88	77 44	9.3808	29.6648	.011364	.1066	.03371
89	79 21	9.4340	29.8329	.011236	.1060	.03352
90	81 00	9.4868	30.0000	.011111	.1054	.03333
91	82 81	9.5394	30.1662	.010989	.1048	.03315
92	84 64	9.5917	30.3315	.010870	.1043	.03297
93	86 49	9.6437	30.4959	.010753	.1037	.03279
94	88 36	9.6954	30.6594	.010638	.1031	.03262
95	90 25	9.7468	30.8221	.010526	.1026	.03244
96	92 16	9.7980	30.9839	.010417	.1021	.03227
97	94 09	9.8489	31.1448	.010309	.1015	.03211
98	96 04	9.8995	31.3050	.010204	.1010	.03194
99	98 01	9.9499	31.4643	.010101	.1005	.03178
100	1 00 00	10.0000	31.6228	.010000	.1000	.03162

N	N²	√N	√10N	1/N	1/√N	1/√10N
101	1 02 01	10.0499	31.7805	.009901	.0995	.03147
102	1 04 04	10.0995	31.9374	.009804	.0990	.03131
103	1 06 09	10.1489	32.0936	.009709	.0985	.03116
104	1 08 16	10.1980	32.2490	.009615	.0981	.03101
105	1 10 25	10.2470	32.4037	.009524	.0976	.03086
106	1 12 36	10.2956	32.5576	.009434	.0971	.03071
107	1 14 49	10.3441	32.7109	.009346	.0967	.03057
108	1 16 64	10.3923	32.8634	.009259	.0962	.03043
109	1 18 81	10.4403	33.0151	.009174	.0958	.03029
110	1 21 00	10.4881	33.1662	.009091	.0953	.03015
111	1 23 21	10.5357	33.3167	.009009	.0949	.03002
112	1 25 44	10.5830	33.4664	.008929	.0945	.02988
113	1 27 69	10.6301	33.6155	.008850	.0941	.02975
114	1 29 96	10.6771	33.7639	.008772	.0937	.02962
115	1 32 25	10.7238	33.9116	.008696	.0933	.02949
116	1 34 56	10.7703	34.0588	.008621	.0928	.02936
117	1 36 89	10.8167	34.2053	.008547	.0925	.02924
118	1 39 24	10.8628	34.3511	.008475	.0921	.02911
119	1 41 61	10.9087	34.4964	.008403	.0917	.02899
120	1 44 00	10.9545	34.6410	.008333	.0913	.02887
121	1 46 41	11.0000	34.7851	.008264	.0909	.02875
122	1 48 84	11.0454	34.9285	.008197	.0905	.02863
123	1 51 29	11.0905	35.0714	.008130	.0902	.02851
124	1 53 76	11.1355	35.2136	.008065	.0898	.02840
125	1 56 25	11.1803	35.3553	.008000	.0894	.02828
126	1 58 76	11.2250	35.4965	.007937	.0891	.02817
127	1 61 29	11.2694	35.6371	.007874	.0887	.02806
128	1 63 84	11.3137	35.7771	.007813	.0884	.02795
129	1 66 41	11.3578	35.9166	.007752	.0880	.02784
130	1 69 00	11.4018	36.0555	.007692	.0877	.02774
131	1 71 61	11.4455	36.1939	.007634	.0874	.02763
132	1 74 24	11.4891	36.3318	.007576	.0870	.02752
133	1 76 89	11.5326	36.4692	.007519	.0867	.02742
134	1 79 56	11.5758	36.6060	.007463	.0864	.02732
135	1 82 25	11.6190	36.7423	.007407	.0861	.02722
136	1 84 69	11.6619	36.8782	.007353	.0857	.02712
137	1 87 69	11.7047	37.0135	.007299	.0854	.02702
138	1 90 44	11.7473	37.1484	.007246	.0851	.02692
139	1 93 21	11.7898	37.2827	.007194	.0848	.02682
140	1 96 00	11.8322	37.4166	.007143	.0845	.02673
141	1 98 81	11.8743	37.5500	.007092	.0842	.02663
142	2 01 64	11.9164	37.6829	.007042	.0839	.02654
143	2 04 49	11.9583	37.8153	.006993	.0836	.02644
144	2 07 36	12.0000	37.9473	.006944	.0833	.02635
145	2 10 25	12.0416	38.0789	.006897	.0830	.02626
146	2 13 16	12.0830	38.2099	.006849	.0828	.02617
147	2 16 09	12.1244	38.3406	.006803	.0825	.02608
148	2 19 04	12.1655	38.4708	.006757	.0822	.02599
149	2 22 01	12.2066	38.6005	.006711	.0819	.02591
150	2 25 00	12.2474	38.7298	.006667	.0816	.02582

Table A-2 *Squares, square roots, and reciprocals (continued)*

N	N^2	$\sqrt{N}$	$\sqrt{10N}$	$1/N$	$1/\sqrt{N}$	$1/\sqrt{10N}$
151	2 28 01	12.2882	38.8587	.006623	.0814	.02573
152	2 31 04	12.3288	38.9872	.006579	.0811	.02565
153	2 34 09	12.3693	39.1152	.006536	.0808	.02557
154	2 37 16	12.4097	39.2428	.006494	.0806	.02548
155	2 40 25	12.4499	39.3700	.006452	.0803	.02540
156	2 43 36	12.4900	39.4968	.006410	.0801	.02532
157	2 46 49	12.5300	39.6232	.006369	.0798	.02524
158	2 49 64	12.5698	39.7492	.006329	.0796	.02516
159	2 52 81	12.6095	39.8748	.006289	.0793	.02508
160	2 56 00	12.6491	40.0000	.006250	.0791	.02500
161	2 59 21	12.6886	40.1248	.006211	.0788	.02492
162	2 62 44	12.7279	40.2492	.006173	.0786	.02485
163	2 65 69	12.7671	40.3733	.006135	.0783	.02477
164	2 68 96	12.8062	40.4969	.006098	.0781	.02469
165	2 72 25	12.8452	40.6202	.006061	.0778	.02462
166	2 75 56	12.8841	40.7431	.006024	.0776	.02454
167	2 78 89	12.9228	40.8656	.005988	.0774	.02447
168	2 82 24	12.9615	40.9878	.005952	.0772	.02440
169	2 85 61	13.0000	41.1096	.005917	.0769	.02433
170	2 89 00	13.0384	41.2311	.005882	.0767	.02425
171	2 92 41	13.0767	41.3521	.005848	.0765	.02418
172	2 95 84	13.1149	41.4729	.005814	.0762	.02411
173	2 99 29	13.1529	41.5933	.005780	.0760	.02404
174	3 02 76	13.1909	41.7133	.005747	.0758	.02397
175	3 06 25	13.2288	41.8330	.005714	.0756	.02390
176	3 09 76	13.2665	41.9524	.005682	.0754	.02384
177	3 13 29	13.3041	42.0714	.005650	.0752	.02377
178	3 16 84	13.3417	42.1900	.005618	.0750	.02370
179	3 20 41	13.3791	42.3084	.005587	.0747	.02364
180	3 24 00	13.4164	42.4264	.005556	.0745	.02357
181	3 27 61	13.4536	42.5441	.005525	.0743	.02351
182	3 31 24	13.4907	42.6615	.005495	.0741	.02344
183	3 34 89	13.5277	42.7785	.005464	.0739	.02338
184	3 38 56	13.5647	42.8952	.005435	.0737	.02331
185	3 42 25	13.6015	43.0116	.005405	.0735	.02325
186	3 45 96	13.6382	43.1277	.005376	.0733	.02319
187	3 49 69	13.6748	43.2435	.005348	.0731	.02312
188	3 53 44	13.7113	43.3590	.005319	.0729	.02306
189	3 57 21	13.7477	43.4741	.005291	.0727	.02300
190	3 61 00	13.7840	43.5890	.005263	.0725	.02294
191	3 64 81	13.8203	43.7035	.005236	.0724	.02288
192	3 68 64	13.8564	43.8178	.005208	.0722	.02282
193	3 72 49	13.8924	43.9318	.005181	.0720	.02276
194	3 76 36	13.9284	44.0454	.005155	.0718	.02270
195	3 80 25	13.9642	44.1588	.005128	.0716	.02265
196	3 84 16	14.0000	44.2719	.005102	.0714	.02259
197	3 88 09	14.0357	44.3847	.005076	.0712	.02253
198	3 92 04	14.0712	44.4972	.005051	.0711	.02247
199	3 96 01	14.1067	44.6094	.005025	.0709	.02242
200	4 00 00	14.1421	44.7214	.005000	.0707	.02236

Table A-2 *Squares, square roots, and reciprocals (continued)*

N	N²	√N	√10N	1/N	1/√N	1/√10N
201	4 04 01	14.1774	44.8330	.004975	.0705	.02230
202	4 08 04	14.2127	44.9444	.004950	.0704	.02225
203	4 12 09	14.2478	45.0555	.004926	.0702	.02219
204	4 16 16	14.2829	45.1664	.004902	.0700	.02214
205	4 20 25	14.3178	45.2769	.004878	.0698	.02209
206	4 24 36	14.3527	45.3872	.004854	.0697	.02203
207	4 28 49	14.3875	45.4973	.004831	.0695	.02198
208	4 32 64	14.4222	45.6070	.004801	.0693	.02193
209	4 36 81	14.4568	45.7165	.004785	.0692	.02187
210	4 41 00	14.4914	45.8258	.004762	.0690	.02182
211	4 45 21	14.5258	45.9347	.004739	.0688	.02177
212	4 49 44	14.5602	46.0435	.004717	.0687	.02172
213	4 53 69	14.5945	46.1519	.004695	.0685	.02167
214	4 57 96	14.6287	46.2601	.004673	.0684	.02162
215	4 62 25	14.6629	46.3681	.004651	.0682	.02157
216	4 66 56	14.6969	46.4758	.004630	.0680	.02152
217	4 70 89	14.7309	46.5833	.004608	.0679	.02147
218	4 75 24	14.7648	46.6905	.004587	.0677	.02142
219	4 79 61	14.7986	46.7974	.004566	.0676	.02137
220	4 84 00	14.8324	46.9042	.004545	.0674	.02132
221	4 88 41	14.8661	47.0106	.004525	.0673	.02127
222	4 92 84	14.8997	47.1169	.004505	.0671	.02122
223	4 97 29	14.9332	47.2229	.004484	.0670	.02118
224	5 01 76	14.9666	47.3286	.004464	.0668	.02113
225	5 06 25	15.0000	47.4342	.004444	.0667	.02108
226	5 10 76	15.0333	47.5395	.004425	.0665	.02104
227	5 15 29	15.0665	47.6445	.004405	.0664	.02099
228	5 19 84	15.0997	47.7493	.004386	.0662	.02094
229	5 24 41	15.1327	47.8539	.004367	.0661	.02090
230	5 29 00	15.1658	47.9583	.004348	.0659	.02085
231	5 33 61	15.1987	48.0625	.004329	.0658	.02081
232	5 38 24	15.2315	48.1664	.004310	.0657	.02076
233	5 42 89	15.2643	48.2701	.004292	.0655	.02072
234	5 47 56	15.2971	48.3735	.004274	.0654	.02067
235	5 52 25	15.3297	48.4768	.004255	.0652	.02063
236	5 56 96	15.3623	48.5798	.004237	.0651	.02058
237	5 61 69	15.3948	48.6826	.004219	.0650	.02054
238	5 66 44	15.4272	48.7852	.004202	.0648	.02050
239	5 71 21	15.4596	48.8876	.004184	.0647	.02046
240	5 76 00	15.4919	48.9898	.004167	.0645	.02041
241	5 80 81	15.5242	49.0918	.004149	.0644	.02037
242	5 85 64	15.5563	49.1935	.004132	.0643	.02033
243	5 90 49	15.5885	49.2950	.004115	.0642	.02029
244	5 95 36	15.6205	49.3964	.004098	.0640	.02024
245	6 00 25	15.6525	49.4975	.004082	.0639	.02020
246	6 05 16	15.6844	49.5984	.004065	.0638	.02016
247	6 10 09	15.7162	49.6991	.004049	.0636	.02012
248	6 15 04	15.7480	49.7996	.004032	.0635	.02008
249	6 20 01	15.7797	49.8999	.004016	.0634	.02004
250	6 25 00	15.8114	50.0000	.004000	.0632	.02000

Table A-2 *Squares, square roots, and reciprocals (continued)*

N	N²	$\sqrt{N}$	$\sqrt{10N}$	$1/N$	$1/\sqrt{N}$	$1/\sqrt{10N}$
251	6 30 01	15.8430	50.0999	.003984	.0631	.01996
252	6 35 04	15.8745	50.1996	.003968	.0630	.01992
253	6 40 09	15.9060	50.2991	.003953	.0629	.01988
254	6 45 16	15.9374	50.3984	.003937	.0627	.01984
255	6 50 25	15.9687	50.4975	.003922	.0626	.01980
256	6 55 36	16.0000	50.5964	.003906	.0625	.01976
257	6 60 49	16.0312	50.6952	.003891	.0624	.01973
258	6 65 64	16.0624	50.7937	.003876	.0623	.01969
259	6 70 81	16.0935	50.8920	.003861	.0621	.01965
260	6 76 00	16.1245	50.9902	.003846	.0620	.01961
261	6 81 21	16.1555	51.0882	.003831	.0619	.01957
262	6 86 44	16.1864	51.1859	.003817	.0618	.01954
263	6 91 69	16.2173	51.2835	.003802	.0617	.01950
264	6 96 96	16.2481	51.3809	.003788	.0615	.01946
265	7 02 25	16.2788	51.4782	.003774	.0614	.01943
266	7 07 56	16.3095	51.5752	.003759	.0613	.01939
267	7 12 89	16.3401	51.6720	.003745	.0612	.01935
268	7 18 24	16.3707	51.7687	.003731	.0611	.01932
269	7 23 61	16.4012	51.8652	.003717	.0610	.01928
270	7 29 00	16.4317	51.9615	.003704	.0609	.01925
271	7 34 41	16.4621	52.0577	.003690	.0607	.01921
272	7 39 84	16.4924	52.1536	.003676	.0606	.01917
273	7 45 29	16.5227	52.2494	.003663	.0605	.01914
274	7 50 76	16.5529	52.3450	.003650	.0604	.01910
275	7 56 25	16.5831	52.4404	.003636	.0603	.01907
276	7 61 76	16.6132	52.5357	.003623	.0602	.01903
277	7 67 29	16.6433	52.6308	.003610	.0601	.01900
278	7 72 84	16.6733	52.7257	.003597	.0600	.01897
279	7 78 41	16.7033	52.8205	.003584	.0599	.01893
280	7 84 00	16.7332	52.9150	.003571	.0598	.01890
281	7 89 61	16.7631	53.0094	.003559	.0597	.01886
282	7 95 24	16.7929	53.1037	.003546	.0595	.01883
283	8 00 89	16.8226	53.1977	.003534	.0594	.01880
284	8 06 56	16.8523	53.2917	.003521	.0593	.01876
285	8 12 25	16.8819	53.3854	.003509	.0592	.01873
286	8 17 96	16.9115	53.4790	.003497	.0591	.01870
287	8 23 69	16.9411	53.5724	.003484	.0590	.01867
288	8 29 44	16.9706	53.6656	.003472	.0589	.01863
289	8 35 21	17.0000	53.7587	.003460	.0588	.01860
290	8 41 00	17.0294	53.8516	.003448	.0587	.01857
291	8 46 81	17.0587	53.9444	.003436	.0586	.01854
292	8 52 64	17.0880	54.0370	.003425	.0585	.01851
293	8 58 49	17.1172	54.1295	.003413	.0584	.01847
294	8 64 36	17.1464	54.2218	.003401	.0583	.01844
295	8 70 25	17.1756	54.3139	.003390	.0582	.01841
296	8 76 16	17.2047	54.4059	.003378	.0581	.01838
297	8 82 09	17.2337	54.4977	.003367	.0580	.01835
298	8 88 04	17.2627	54.5894	.003356	.0579	.01832
299	8 94 01	17.2916	54.6809	.003344	.0578	.01829
300	9 00 00	17.3205	54.7723	.003333	.0577	.01826

Table A-2 *Squares, square roots, and reciprocals (continued)*

N	N^2	$\sqrt{N}$	$\sqrt{10N}$	$1/N$	$1/\sqrt{N}$	$1/\sqrt{10N}$
301	9 06 01	17.3494	54.8635	.003322	.0576	.01823
302	9 12 04	17.3781	54.9545	.003311	.0575	.01820
303	9 18 09	17.4069	55.0454	.003300	.0574	.01817
304	9 24 16	17.4356	55.1362	.003289	.0574	.01814
305	9 30 25	17.4642	55.2268	.003279	.0573	.01811
306	9 36 36	17.4929	55.3173	.003268	.0572	.01808
307	9 42 49	17.5214	55.4076	.003257	.0571	.01805
308	9 48 64	17.5499	55.4977	.003247	.0570	.01802
309	9 54 81	17.5784	55.5878	.003236	.0569	.01799
310	9 61 00	17.6068	55.6776	.003226	.0568	.01796
311	9 67 21	17.6352	55.7674	.003215	.0567	.01793
312	9 73 44	17.6635	55.8570	.003205	.0566	.01790
313	9 79 69	17.6918	55.9464	.003195	.0565	.01787
314	9 85 96	17.7200	56.0357	.003185	.0564	.01785
315	9 92 25	17.7482	56.1249	.003175	.0563	.01782
316	9 98 56	17.7764	56.2139	.003165	.0563	.01779
317	10 04 89	17.8045	56.3028	.003155	.0562	.01776
318	10 11 24	17.8326	56.3915	.003145	.0561	.01773
319	10 17 61	17.8606	56.4801	.003135	.0560	.01771
320	10 24 00	17.8885	56.5685	.003125	.0559	.01768
321	10 30 41	17.9165	56.6569	.003115	.0558	.01765
322	10 36 84	17.9444	56.7450	.003106	.0557	.01762
323	10 43 29	17.9722	56.8331	.003096	.0556	.01760
324	10 49 76	18.0000	56.9210	.003086	.0556	.01757
325	10 56 25	18.0278	57.0088	.003077	.0555	.01754
326	10 62 76	18.0555	57.0964	.003067	.0554	.01751
327	10 69 29	18.0831	57.1839	.003058	.0553	.01749
328	10 75 84	18.1108	57.2713	.003049	.0552	.01746
329	10 82 41	18.1384	57.3585	.003040	.0551	.01743
330	10 89 00	18.1659	57.4456	.003030	.0550	.01741
331	10 95 61	18.1934	57.5326	.003021	.0550	.01738
332	11 02 24	18.2209	57.6194	.003012	.0549	.01736
333	11 08 89	18.2483	57.7062	.003003	.0548	.01733
334	11 15 56	18.2757	57.7927	.002994	.0547	.01730
335	11 22 25	18.3030	57.8792	.002985	.0546	.01728
336	11 28 96	18.3303	57.9655	.002976	.0546	.01725
337	11 35 69	18.3576	58.0517	.002967	.0545	.01723
338	11 42 44	18.3848	58.1378	.002959	.0544	.01720
339	11 49 21	18.4120	58.2237	.002950	.0543	.01718
340	11 56 00	18.4391	58.3095	.002941	.0542	.01715
341	11 62 81	18.4662	58.3952	.002933	.0542	.01712
342	11 69 64	18.4932	58.4808	.002924	.0541	.01710
343	11 76 49	18.5203	58.5662	.002915	.0540	.01707
344	11 83 36	18.5472	58.6515	.002907	.0539	.01705
345	11 90 25	18.5742	58.7367	.002899	.0538	.01703
346	11 97 16	18.6011	58.8218	.002890	.0538	.01700
347	12 04 09	18.6279	58.9067	.002882	.0537	.01698
348	12 11 04	18.6548	58.9915	.002874	.0536	.01695
349	12 18 01	18.6815	59.0762	.002865	.0535	.01693
350	12 25 00	18.7083	59.1608	.002857	.0535	.01690

Table A-2 *Squares, square roots, and reciprocals (continued)*

N	N²	$\sqrt{N}$	$\sqrt{10N}$	1/N	$1/\sqrt{N}$	$1/\sqrt{10N}$
351	12 32 01	18.7350	59.2453	.002849	.0534	.01688
352	12 39 04	18.7617	59.3296	.002841	.0533	.01685
353	12 46 09	18.7883	59.4138	.002833	.0532	.01683
354	12 53 16	18.8149	59.4979	.002825	.0531	.01681
355	12 60 25	18.8414	59.5819	.002817	.0531	.01678
356	12 67 36	18.8680	59.6657	.002809	.0530	.01676
357	12 74 49	18.8944	59.7495	.002801	.0529	.01674
358	12 81 64	18.9209	59.8331	.002793	.0529	.01671
359	12 88 81	18.9473	59.9166	.002786	.0528	.01669
360	12 96 00	18.9737	60.0000	.002778	.0527	.01667
361	13 03 21	19.0000	60.0833	.002770	.0526	.01664
362	13 10 44	19.0263	60.1664	.002762	.0526	.01662
363	13 17 69	19.0526	60.2495	.002755	.0525	.01660
364	13 24 96	19.0788	60.3324	.002747	.0524	.01657
365	13 32 25	19.1050	60.4152	.002740	.0523	.01655
366	13 39 56	19.1311	60.4979	.002732	.0523	.01653
367	13 46 89	19.1572	60.5805	.002725	.0522	.01651
368	13 54 24	19.1833	60.6630	.002717	.0521	.01648
369	13 61 61	19.2094	60.7454	.002710	.0521	.01646
370	13 69 00	19.2354	60.8276	.002703	.0520	.01644
371	13 76 41	19.2614	60.9098	.002695	.0519	.01642
372	13 83 84	19.2873	60.9918	.002688	.0518	.01640
373	13 91 29	19.3132	61.0737	.002681	.0518	.01637
374	13 98 76	19.3391	61.1555	.002674	.0517	.01635
375	14 06 25	19.3649	61.2372	.002667	.0516	.01633
376	14 13 76	19.3907	61.3188	.002660	.0516	.01631
377	14 21 29	19.4165	61.4003	.002653	.0515	.01629
378	14 28 84	19.4422	61.4817	.002646	.0514	.01627
379	14 36 41	19.4679	61.5630	.002639	.0514	.01624
380	14 44 00	19.4936	61.6441	.002632	.0513	.01622
381	14 51 61	19.5192	61.7252	.002625	.0512	.01620
382	14 59 24	19.5448	61.8061	.002618	.0512	.01618
383	14 66 89	19.5704	61.8870	.002611	.0511	.01616
384	14 74 56	19.5959	61.9677	.002604	.0510	.01614
385	14 82 25	19.6214	62.0484	.002597	.0510	.01612
386	14 89 96	19.6469	62.1289	.002591	.0509	.01610
387	14 97 69	19.6723	62.2093	.002584	.0508	.01607
388	15 05 44	19.6977	62.2896	.002577	.0508	.01605
389	15 13 21	19.7231	62.3699	.002571	.0507	.01603
390	15 21 00	19.7484	62.4500	.002564	.0506	.01601
391	15 28 81	19.7737	62.5300	.002558	.0506	.01599
392	15 36 64	19.7990	62.6099	.002551	.0505	.01597
393	15 44 49	19.8242	62.6897	.002545	.0504	.01595
394	15 52 36	19.8494	62.7694	.002538	.0504	.01593
395	15 60 25	19.8746	62.8490	.002532	.0503	.01591
396	15 68 16	19.8997	62.9285	.002525	.0503	.01589
397	15 76 09	19.9249	63.0079	.002519	.0502	.01587
398	15 84 04	19.9499	63.0872	.002513	.0501	.01585
399	15 92 01	19.9750	63.1664	.002506	.0501	.01583
400	16 00 00	20.0000	63.2456	.002500	.0500	.01581

Table A-2 *Squares, square roots, and reciprocals (continued)*

N	N^2	$\sqrt{N}$	$\sqrt{10N}$	$1/N$	$1/\sqrt{N}$	$1/\sqrt{10N}$
401	16 08 01	20.0250	63.3246	.002494	.0499	.01579
402	16 16 04	20.0499	63.4035	.002488	.0499	.01577
403	16 24 09	20.0749	63.4823	.002481	.0498	.01575
404	16 32 16	20.0998	63.5610	.002475	.0498	.01573
405	16 40 25	20.1246	63.6396	.002469	.0497	.01571
406	16 48 36	20.1494	63.7181	.002463	.0496	.01569
407	16 56 49	20.1742	63.7966	.002457	.0496	.01567
408	16 64 64	20.1990	63.8749	.002451	.0495	.01566
409	16 72 81	20.2237	63.9531	.002445	.0494	.01564
410	16 81 00	20.2485	64.0312	.002439	.0494	.01562
411	16 89 21	20.2731	64.1093	.002433	.0493	.01560
412	16 97 44	20.2978	64.1872	.002427	.0493	.01558
413	17 05 69	20.3224	64.2651	.002421	.0492	.01556
414	17 13 96	20.3470	64.3428	.002415	.0491	.01554
415	17 22 25	20.3715	64.4205	.002410	.0491	.01552
416	17 30 56	20.3961	64.4981	.002404	.0490	.01550
417	17 38 89	20.4206	64.5755	.002398	.0490	.01549
418	17 47 24	20.4450	64.6529	.002392	.0489	.01547
419	17 55 61	20.4695	64.7302	.002387	.0489	.01545
420	17 64 00	20.4939	64.8074	.002381	.0488	.01543
421	17 72 41	20.5183	64.8845	.002375	.0487	.01541
422	17 80 84	20.5426	64.9615	.002370	.0487	.01539
423	17 89 29	20.5670	65.0385	.002364	.0486	.01538
424	17 97 76	20.5913	65.1153	.002358	.0486	.01536
425	18 06 25	20.6155	65.1920	.002353	.0485	.01534
426	18 14 76	20.6398	65.2687	.002347	.0485	.01532
427	18 23 29	20.6640	65.3452	.002342	.0484	.01530
428	18 31 84	20.6882	65.4217	.002336	.0483	.01529
429	18 40 41	20.7123	65.4981	.002331	.0483	.01527
430	18 49 00	20.7364	65.5744	.002326	.0482	.01525
431	18 57 61	20.7605	65.6506	.002320	.0482	.01523
432	18 66 24	20.7846	65.7267	.002315	.0481	.01521
433	18 74 89	20.8087	65.8027	.002309	.0481	.01520
434	18 83 56	20.8327	65.8787	.002304	.0480	.01518
435	18 92 25	20.8567	65.9545	.002299	.0479	.01516
436	19 00 06	20.8806	66.0303	.002294	.0479	.01514
437	19 09 69	20.9045	66.1060	.002288	.0478	.01513
438	19 18 44	20.9284	66.1816	.002283	.0478	.01511
439	19 27 21	20.9523	66.2571	.002278	.0477	.01509
440	19 36 00	20.9762	66.3325	.002273	.0477	.01508
441	19 44 81	21.0000	66.4078	.002268	.0476	.01506
442	19 53 64	21.0238	66.4831	.002262	.0476	.01504
443	19 62 49	21.0476	66.5582	.002257	.0475	.01502
444	19 71 36	21.0713	66.6333	.002252	.0475	.01501
445	19 80 25	21.0950	66.7083	.002247	.0474	.01499
446	19 89 16	21.1187	66.7832	.002242	.0474	.01497
447	19 98 09	21.1424	66.8581	.002237	.0473	.01496
448	20 07 04	21.1660	66.9328	.002232	.0472	.01494
449	20 16 01	21.1896	67.0075	.002227	.0472	.01492
450	20 25 00	21.2132	67.0820	.002222	.0471	.01491

Table A-2 *Squares, square roots, and reciprocals (continued)*

N	N²	√N̄	√10N̄	1/N	1/√N̄	1/√10N̄
451	20 34 01	21.2368	67.1565	.002217	.0471	.01489
452	20 43 04	21.2603	67.2309	.002212	.0470	.01487
453	20 52 09	21.2838	67.3053	.002208	.0470	.01486
454	20 61 16	21.3073	67.3795	.002203	.0469	.01484
455	20 70 25	21.3307	67.4537	.002198	.0469	.01482
456	20 79 36	21.3542	67.5278	.002193	.0468	.01481
457	20 88 49	21.3776	67.6018	.002188	.0468	.01479
458	20 97 64	21.4009	67.6757	.002183	.0467	.01478
459	21 06 81	21.4243	67.7495	.002179	.0467	.01476
460	21 16 00	21.4476	67.8233	.002174	.0466	.01474
461	21 25 21	21.4709	67.8970	.002169	.0466	.01473
462	21 34 44	21.4942	67.9706	.002165	.0465	.01471
463	21 43 69	21.5174	68.0441	.002160	.0465	.01470
464	21 52 96	21.5407	68.1175	.002155	.0464	.01468
465	21 62 25	21.5639·	68.1909	.002151	.0464	.01466
466	21 71 56	21.5870	68.2642	.002146	.0463	.01465
467	21 80 89	21.6102	68.3374	.002141	.0463	.01463
468	21 90 24	21.6333	68.4105	.002137	.0462	.01462
469	21 99 61	21.6564	68.4836	.002132	.0462	.01460
470	22 09 00	21.6795	68.5565	.002128	.0461	.01459
471	22 18 41	21.7025	68.6294	.002123	.0461	.01457
472	22 27 84	21.7256	68.7023	.002119	.0460	.01456
473	22 37 29	21.7486	68.7750	.002114	.0460	.01454
474	22 46 76	21.7715	68.8477	.002110	.0459	.01452
475	22 56 25	21.7945	68.9202	.002105	.0459	.01451
476	22 65 76	21.8174	68.9928	.002101	.0458	.01449
477	22 75 29	21.8403	69.0652	.002096	.0458	.01448
478	22 84 84	21.8632	69.1375	.002092	.0457	.01446
479	22 94 41	21.8861	69.2098	.002088	.0457	.01445
480	23 04 00	21.9089	69.2820	.002083	.0456	.01443
481	23 13 61	21.9317	69.3542	.002079	.0456	.01442
482	23 23 24	21.9545	69.4262	.002075	.0455	.01440
483	23 32 89	21.9773	69.4982	.002070	.0455	.01439
484	23 42 56	22.0000	69.5701	.002066	.0455	.01437
485	23 52 25	22.0227	69.6419	.002062	.0454	.01436
486	23 61 96	22.0454	69.7137	.002058	.0454	.01434
487	23 71 69	22.0681	69.7854	.002053	.0453	.01433
488	23 81 44	22.0907	69.8570	.002049	.0453	.01431
489	23 91 21	22.1133	69.9285	.002045	.0452	.01430
490	24 01 00	22.1359	70.0000	.002041	.0452	.01429
491	24 10 81	22.1585	70.0714	.002037	.0451	.01427
492	24 20 64	22.1811	70.1427	.002033	.0451	.01426
493	24 30 49	22.2036	70.2140	.002028	.0450	.01424
494	24 40 36	22.2261	70.2851	.002024	.0450	.01423
495	24 50 25	22.2486	70.3562	.002020	.0449	.01421
496	24 60 16	22.2711	70.4273	.002016	.0449	.01420
497	24 70 09	22.2935	70.4982	.002012	.0449	.01418
498	24 80 04	22.3159	70.5691	.002008	.0448	.01417
499	24 90 01	22.3383	70.6399	.002004	.0448	.01416
500	25 00 00	22.3607	70.7107	.002000	.0447	.01414

Table A-2 *Squares, square roots, and reciprocals (continued)*

N	N²	$\sqrt{N}$	$\sqrt{10N}$	1/N	$1/\sqrt{N}$	$1/\sqrt{10N}$
501	25 10 01	22.3830	70.7814	.001996	.0447	.01413
502	25 20 04	22.4054	70.8520	.001992	.0446	.01411
503	25 30 09	22.4277	70.9225	.001988	.0446	.01410
504	25 40 16	22.4499	70.9930	.001984	.0445	.01409
505	25 50 25	22.4722	71.0634	.001980	.0445	.01407
506	25 60 36	22.4944	71.1337	.001976	.0445	.01406
507	25 70 49	22.5167	71.2039	.001972	.0444	.01404
508	25 80 64	22.5389	71.2741	.001969	.0444	.01403
509	25 90 81	22.5610	71.3442	.001965	.0443	.01402
510	26 01 00	22.5832	71.4143	.001961	.0443	.01400
511	26 11 21	22.6053	71.4843	.001957	.0442	.01399
512	26 21 44	22.6274	71.5542	.001953	.0442	.01398
513	26 31 69	22.6495	71.6240	.001949	.0442	.01396
514	26 41 96	22.6716	71.6938	.001946	.0441	.01395
515	26 52 25	22.6936	71.7635	.001942	.0441	.01393
516	26 62 56	22.7156	71.8331	.001938	.0440	.01392
517	26 72 89	22.7376	71.9027	.001934	.0440	.01391
518	26 83 24	22.7596	71.9722	.001931	.0439	.01389
519	26 93 61	22.7816	72.0417	.001927	.0439	.01388
520	27 04 00	22.8035	72.1110	.001923	.0439	.01387
521	27 14 41	22.8254	72.1803	.001919	.0438	.01385
522	27 24 84	22.8473	72.2496	.001916	.0438	.01384
523	27 35 29	22.8692	72.3187	.001912	.0437	.01383
524	27 45 76	22.8910	72.3878	.001908	.0437	.01381
525	27 56 25	22.9129	72.4569	.001905	.0436	.01380
526	27 66 76	22.9347	72.5259	.001901	.0436	.01379
527	27 77 29	22.9565	72.5948	.001898	.0436	.01378
528	27 87 84	22.9783	72.6636	.001894	.0435	.01376
529	27 98 41	23.0000	72.7324	.001890	.0435	.01375
530	28 09 00	23.0217	72.8011	.001887	.0434	.01374
531	28 19 61	23.0434	72.8697	.001883	.0434	.01372
532	28 30 24	23.0651	72.9383	.001880	.0434	.01371
533	28 40 89	23.0868	73.0068	.001876	.0433	.01370
534	28 51 56	23.1084	73.0753	.001873	.0433	.01368
535	28 62 25	23.1301	73.1437	.001869	.0432	.01367
536	28 72 96	23.1517	73.2120	.001866	.0432	.01366
537	28 83 69	23.1733	73.2803	.001862	.0432	.01365
538	28 94 44	23.1948	73.3485	.001859	.0431	.01363
539	29 05 21	23.2164	73.4166	.001855	.0431	.01362
540	29 16 00	23.2379	73.4847	.001852	.0430	.01361
541	29 26 81	23.2594	73.5527	.001848	.0430	.01360
542	29 37 64	23.2809	73.6206	.001845	.0430	.01358
543	29 48 49	23.3024	73.6885	.001842	.0429	.01357
544	29 59 36	23.3238	73.7564	.001838	.0429	.01356
545	29 70 25	23.3452	73.8241	.001835	.0428	.01355
546	29 81 16	23.3666	73.8918	.001832	.0428	.01353
547	29 92 09	23.3880	73.9594	.001828	.0428	.01352
548	30 03 04	23.4094	74.0270	.001825	.0427	.01351
549	30 14 01	23.4307	74.0945	.001821	.0427	.01350
550	30 25 00	23.4521	74.1620	.001818	.0426	.01348

Table A-2   *Squares, square roots, and reciprocals* (*continued*)

N	N²	$\sqrt{N}$	$\sqrt{10N}$	1/N	1/$\sqrt{N}$	1/$\sqrt{10N}$
551	30 36 01	23.4734	74.2294	.001815	.0426	.01347
552	30 47 04	23.4947	74.2967	.001812	.0426	.01346
553	30 58 09	23.5160	74.3640	.001808	.0425	.01345
554	30 69 16	23.5372	74.4312	.001805	.0425	.01344
555	30 80 25	23.5584	74.4983	.001802	.0424	.01342
556	30 91 36	23.5797	74.5654	.001799	.0424	.01341
557	31 02 49	23.6008	74.6324	.001795	.0424	.01340
558	31 13 64	23.6220	74.6994	.001792	.0423	.01339
559	31 24 81	23.6432	74.7663	.001789	.0423	.01338
560	31 36 00	23.6643	74.8331	.001786	.0423	.01336
561	31 47 21	23.6854	74.8999	.001783	.0422	.01335
562	31 58 44	23.7065	74.9667	.001779	.0422	.01334
563	31 69 69	23.7276	75.0333	.001776	.0421	.01333
564	31 80 96	23.7487	75.0999	.001773	.0421	.01332
565	31 92 25	23.7697	75.1665	.001770	.0421	.01330
566	32 03 56	23.7908	75.2330	.001767	.0420	.01329
567	32 14 89	23.8118	75.2994	.001764	.0420	.01328
568	32 26 24	23.8328	75.3658	.001761	.0420	.01327
569	32 37 61	23.8537	75.4321	.001757	.0419	.01326
570	32 49 00	23.8747	75.4983	.001754	.0419	.01325
571	32 60 41	23.8956	75.5645	.001751	.0418	.01323
572	32 71 84	23.9165	75.6307	.001748	.0418	.01322
573	32 83 29	23.9374	75.6968	.001745	.0418	.01321
574	32 94 76	23.9583	75.7628	.001742	.0417	.01320
575	30 06 25	23.9792	75.8288	.001739	.0417	.01319
576	33 17 76	24.0000	75.8947	.001736	.0417	.01318
577	33 29 29	24.0208	75.9605	.001733	.0416	.01316
578	33 40 84	24.0416	76.0263	.001730	.0416	.01315
579	33 52 41	24.0624	76.0920	.001727	.0416	.01314
580	33 64 00	24.0832	76.1577	.001724	.0415	.01313
581	33 75 61	24.1039	76.2234	.001721	.0415	.01312
582	33 87 24	24.1247	76.2889	.001718	.0415	.01311
583	33 98 89	24.1454	76.3544	.001715	.0414	.01310
584	34 10 56	24.1661	76.4199	.001712	.0414	.01309
585	34 22 25	24.1868	76.4853	.001709	.0413	.01307
586	34 33 96	24.2074	76.5506	.001706	.0413	.01306
587	34 45 69	24.2281	76.6159	.001704	.0413	.01305
588	34 57 44	24.2487	76.6812	.001701	.0412	.01304
589	34 69 21	24.2693	76.7463	.001698	.0412	.01303
590	34 81 00	24.2899	76.8115	.001695	.0412	.01302
591	34 92 81	24.3105	76.8765	.001692	.0411	.01301
592	35 04 64	24.3311	76.9415	.001689	.0411	.01300
593	35 16 49	24.3516	77.0065	.001686	.0411	.01299
594	35 28 36	24.3721	77.0714	.001684	.0410	.01297
595	35 40 25	24.3926	77.1362	.001681	.0410	.01296
596	35 52 16	24.4131	77.2010	.001678	.0410	.01295
597	35 64 09	24.4336	77.2658	.001675	.0409	.01294
598	35 76 04	24.4540	77.3305	.001672	.0409	.01293
599	35 88 01	24.4745	77.3951	.001669	.0409	.01292
600	36 00 00	24.4949	77.4597	.001667	.0408	.01291

Table A-2 *Squares, square roots, and reciprocals (continued)*

N	N^2	$\sqrt{N}$	$\sqrt{10N}$	$1/N$	$1/\sqrt{N}$	$1/\sqrt{10N}$
601	36 12 01	24.5153	77.5242	.001664	.0408	.01290
602	36 24 04	24.5357	77.5887	.001661	.0408	.01289
603	36 36 09	24.5561	77.6531	.001658	.0407	.01288
604	36 48 16	24.5764	77.7174	.001656	.0407	.01287
605	36 60 25	24.5967	77.7817	.001653	.0407	.01286
606	36 72 36	24.6171	77.8460	.001650	.0406	.01285
607	36 84 49	24.6374	77.9102	.001647	.0406	.01284
608	36 96 64	24.6577	77.9744	.001645	.0406	.01282
609	37 08 81	24.6779	78.0385	.001642	.0405	.01281
610	37 21 00	24.6982	78.1025	.001639	.0405	.01280
611	37 33 21	24.7184	78.1665	.001637	.0405	.01279
612	37 45 44	24.7386	78.2304	.001634	.0404	.01278
613	37 57 69	24.7588	78.2943	.001631	.0404	.01277
614	37 69 96	24.7790	78.3582	.001629	.0404	.01276
615	37 82 25	24.7992	78.4219	.001626	.0403	.01275
616	37 94 56	24.8193	78.4857	.001623	.0403	.01274
617	38 06 89	24.8395	78.5493	.001621	.0403	.01273
618	38 19 24	24.8596	78.6130	.001618	.0402	.01272
619	38 31 61	24.8797	78.6766	.001616	.0402	.01271
620	38 44 00	24.8998	78.7401	.001613	.0402	.01270
621	38 56 41	24.9199	78.8036	.001610	.0401	.01269
622	38 68 84	24.9399	78.8670	.001608	.0401	.01268
623	38 81 29	24.9600	78.9303	.001605	.0401	.01267
624	38 93 76	24.9800	78.9937	.001603	.0400	.01266
625	39 06 25	25.0000	79.0569	.001600	.0400	.01265
626	39 18 76	25.0200	79.1202	.001597	.0400	.01264
627	39 31 29	25.0400	79.1833	.001595	.0399	.01263
628	39 43 84	25.0599	79.2465	.001592	.0399	.01262
629	39 56 41	25.0799	79.3095	.001590	.0399	.01261
630	39 69 00	25.0998	79.3725	.001587	.0398	.01260
631	39 81 61	25.1197	79.4355	.001585	.0398	.01259
632	39 94 24	25.1396	79.4984	.001582	.0398	.01258
633	40 06 89	25.1595	79.5613	.001580	.0397	.01257
634	40 19 56	25.1794	79.6241	.001577	.0397	.01256
635	40 32 25	25.1992	79.6869	.001575	.0397	.01255
636	40 44 96	25.2190	79.7496	.001572	.0397	.01254
637	40 57 69	25.2389	79.8123	.001570	.0396	.01253
638	40 70 44	25.2587	79.8749	.001567	.0396	.01252
639	40 83 21	25.2784	79.9375	.001565	.0396	.01251
640	40 96 00	25.2982	80.0000	.001562	.0395	.01250
641	41 08 81	25.3180	80.0625	.001560	.0395	.01249
642	41 21 64	25.3377	80.1249	.001558	.0395	.01248
643	41 34 49	25.3574	80.1873	.001555	.0394	.01247
644	41 47 36	25.3772	80.2496	.001553	.0394	.01246
645	41 60 25	25.3969	80.3119	.001550	.0394	.01245
646	41 73 16	25.4165	80.3714	.001548	.0393	.01244
647	41 86 09	25.4362	80.4363	.001546	.0393	.01243
648	41 99 04	25.4558	80.4984	.001543	.0393	.01242
649	42 12 01	25.4775	80.5605	.001541	.0393	.01241
650	42 25 00	25.4951	80.6226	.001538	.0392	.01240

Table A-2 *Squares, square roots, and reciprocals (continued)*

N	N²	√N	√10N	1/N	1/√N	1/√10N
651	42 38 01	25.5147	80.6846	.001536	.0392	.01239
652	42 51 04	25.5343	80.7465	.001534	.0392	.01238
653	42 64 09	25.5539	80.8084	.001531	.0391	.01237
654	42 77 16	25.5734	80.8703	.001529	.0391	.01237
655	42 90 25	25.5930	80.9321	.001527	.0391	.01236
656	43 03 36	25.6125	80.9938	.001524	.0390	.01235
657	43 16 49	25.6320	81.0555	.001522	.0390	.01234
658	43 29 64	25.6515	81.1172	.001520	.0390	.01233
659	43 42 81	25.6710	81.1788	.001517	.0390	.01232
660	43 56 00	25.6905	81.2404	.001515	.0389	.01231
661	43 69 21	25.7099	81.3019	.001513	.0389	.01230
662	43 82 44	25.7294	81.3634	.001511	.0389	.01229
663	43 95 69	25.7488	81.4248	.001508	.0388	.01228
664	44 08 96	25.7682	81.4862	.001506	.0388	.01227
665	44 22 25	25.7876	81.5475	.001504	.0388	.01226
666	44 35 56	25.8070	81.6088	.001502	.0387	.01225
667	44 48 89	25.8263	81.6701	.001499	.0387	.01224
668	44 62 24	25.8457	81.7313	.001497	.0387	.01224
669	44 75 61	25.8650	81.7924	.001495	.0387	.01223
670	44 89 00	25.8844	81.8535	.001493	.0386	.01222
671	45 02 41	25.9037	81.9146	.001490	.0386	.01221
672	45 15 84	25.9230	81.9756	.001488	.0386	.01220
673	45 29 29	25.9422	82.0366	.001486	.0385	.01219
674	45 42 76	25.9615	82.0975	.001484	.0385	.01218
675	45 56 25	25.9808	82.1584	.001481	.0385	.01217
676	45 69 76	26.0000	82.2192	.001479	.0385	.01216
677	45 83 29	26.0192	82.2800	.001477	.0384	.01215
678	45 96 84	26.0384	82.3408	.001475	.0384	.01214
679	46 10 41	26.0576	82.4015	.001473	.0384	.01214
680	46 24 00	26.0768	82.4621	.001471	.0383	.01213
681	46 37 61	26.0960	82.5227	.001468	.C383	.01212
682	46 51 24	26.1151	82.5833	.001466	.0383	.01211
683	46 64 89	26.1343	82.6438	.001464	.0383	.01210
684	46 78 56	26.1534	82.7043	.001462	.0382	.01209
685	46 92 25	26.1725	82.7647	.001460	.0382	.01208
686	47 05 96	26.1916	82.8251	.001458	.0382	.01207
687	47 19 69	26.2107	82.8855	.001456	.0382	.01206
688	47 33 44	26.2298	82.9458	.001453	.0381	.01206
689	47 47 21	26.2488	83.0060	.001451	.0381	.01205
690	47 61 00	26.2679	83.0662	.001449	.0381	.01204
691	47 74 81	26.2869	83.1264	.001447	.0380	.01203
692	47 88 64	26.3059	83.1865	.001445	.0380	.01202
693	48 02 49	26.3249	83.2466	.001443	.0380	.01201
694	48 16 36	26.3439	83.3067	.001441	.0380	.01200
695	48 30 25	26.3629	83.3667	.001439	.0379	.01200
696	48 44 16	26.3818	83.4266	.001437	.0379	.01199
697	48 58 09	26.4008	83.4865	.001435	.0379	.01198
698	48 72 04	26.4197	83.5464	.001433	.0379	.01197
699	48 86 01	26.4386	83.6062	.001431	.0378	.01196
700	49 00 00	26.4575	83.6660	.001429	.0378	.01195

Table A-2 *Squares, square roots, and reciprocals (continued)*

N	N²	√N	√10N	1/N	1/√N	1/√10N
701	49 14 01	26.4764	83.7257	.001427	.0378	.01194
702	49 28 04	26.4953	83.7854	.001425	.0377	.01194
703	49 42 09	26.5141	83.8451	.001422	.0377	.01193
704	49 56 16	26.5330	83.9047	.001420	.0377	.01192
705	49 70 25	26.5518	83.9643	.001418	.0377	.01191
706	49 84 36	26.5707	84.0238	.001416	.0376	.01190
707	49 98 49	26.5895	84.0833	.001414	.0376	.01189
708	50 12 64	26.6083	84.1427	.001412	.0376	.01188
709	50 26 81	26.6271	84.2021	.001410	.0376	.01188
710	50 41 00	26.6458	84.2615	.001408	.0375	.01187
711	50 55 21	26.6646	84.3208	.001406	.0375	.01186
712	50 69 44	26.6833	84.3801	.001404	.0375	.01185
713	50 83 69	26.7021	84.4393	.001403	.0375	.01184
714	50 97 96	26.7208	84.4985	.001401	.0374	.01183
715	51 12 25	26.7395	84.5577	.001399	.0374	.01183
716	51 26 56	26.7582	84.6168	.001397	.0374	.01182
717	51 40 89	26.7769	84.6759	.001395	.0373	.01181
718	51 55 24	26.7955	84.7349	.001393	.0373	.01180
719	51 69 61	26.8142	84.7939	.001391	.0373	.01179
720	51 84 00	26.8328	84.8528	.001389	.0373	.01179
721	51 98 41	26.8514	84.9117	.001387	.0372	.01178
722	52 12 84	26.8701	84.9706	.001385	.0372	.01177
723	52 27 29	26.8887	85.0294	.001383	.0372	.01176
724	52 41 76	26.9072	85.0882	.001381	.0372	.01175
725	52 56 25	26.9258	85.1469	.001379	.0371	.01174
726	52 70 76	26.9444	85.2056	.001377	.0371	.01174
727	52 85 29	26.9629	85.2643	.001376	.0371	.01173
728	52 99 84	26.9815	85.3229	.001374	.0371	.01172
729	53 14 41	27.0000	85.3815	.001372	.0370	.01171
730	53 29 00	27.0185	85.4400	.001370	.0370	.01170
731	53 43 61	27.0370	85.4985	.001368	.0370	.01170
732	53 58 24	27.0555	85.5570	.001366	.0370	.01169
733	53 72 89	27.0740	85.6154	.001364	.0369	.01168
734	53 87 56	27.0924	85.6738	.001362	.0369	.01167
735	54 02 25	27.1109	85.7321	.001361	.0369	.01166
736	54 16 96	27.1293	85.7904	.001359	.0369	.01166
737	54 31 69	27.1477	85.8487	.001357	.0368	.01165
738	54 46 44	27.1662	85.9069	.001355	.0368	.01164
739	54 61 21	27.1846	85.9651	.001353	.0368	.01163
740	54 76 00	27.2029	86.0233	.001351	.0368	.01162
741	54 90 81	27.2213	86.0814	.001350	.0367	.01162
742	55 05 64	27.2397	86.1394	.001348	.0367	.01161
743	55 20 49	27.2580	86.1974	.001346	.0367	.01160
744	55 35 36	27.2764	86.2554	.001344	.0367	.01159
745	55 50 25	27.2947	86.3134	.001342	.0366	.01159
746	55 65 16	27.3130	86.3713	.001340	.0366	.01158
747	55 80 09	27.3313	86.4292	.001339	.0366	.01157
748	55 95 04	27.3496	86.4870	.001337	.0366	.01156
749	56 10 01	27.3679	86.5448	.001335	.0365	.01155
750	56 25 00	27.3861	86.6025	.001333	.0365	.01155

N	N^2	$\sqrt{N}$	$\sqrt{10N}$	$1/N$	$1/\sqrt{N}$	$1/\sqrt{10N}$
751	56 40 01	27.4044	86.6603	.001332	.0365	.01154
752	56 55 04	27.4226	86.7179	.001330	.0365	.01153
753	56 70 09	27.4408	86.7756	.001328	.0364	.01152
754	56 85 16	27.4591	86.8332	.001326	.0364	.01152
755	57 00 25	27.4773	86.8907	.001325	.0364	.01151
756	57 15 36	27.4955	86.9483	.001323	.0364	.01150
757	57 30 49	27.5136	87.0057	.001321	.0363	.01149
758	57 45 64	27.5318	87.0632	.001319	.0363	.01149
759	57 60 81	27.5500	87.1206	.001318	.0363	.01148
760	57 76 00	27.5681	87.1780	.001316	.0363	.01147
761	57 91 21	27.5862	87.2353	.001314	.0362	.01146
762	58 06 44	27.6043	87.2926	.001312	.0362	.01146
763	58 21 69	27.6225	87.3499	.001311	.0362	.01145
764	58 36 96	27.6405	87.4071	.001309	.0362	.01144
765	58 52 25	27.6586	87.4643	.001307	.0362	.01143
766	58 67 56	27.6767	87.5214	.001305	.0361	.01143
767	58 82 89	27.6948	87.5785	.001304	.0361	.01142
768	58 98 24	27.7128	87.6356	.001302	.0361	.01141
769	59 13 61	27.7308	87.6926	.001300	.0361	.01140
770	59 29 00	27.7489	87.7496	.001299	.0360	.01140
771	59 44 41	27.7669	87.8066	.001297	.0360	.01139
772	59 59 84	27.7849	87.8635	.001295	.0360	.01138
773	59 75 29	27.8029	87.9204	.001294	.0360	.01137
774	59 90 76	27.8209	87.9773	.001292	.0359	.01137
775	60 06 25	27.8388	88.0341	.001290	.0359	.01136
776	60 21 76	27.8568	88.0909	.001289	.0359	.01135
777	60 37 29	27.8747	88.1476	.001287	.0359	.01134
778	60 52 84	27.8927	88.2043	.001285	.0359	.01134
779	60 68 41	27.9106	88.2610	.001284	.0358	.01133
780	60 84 00	27.9285	88.3176	.001282	.0358	.01132
781	60 99 61	27.9464	88.3742	.001280	.0358	.01132
782	61 15 24	27.9643	88.4308	.001279	.0358	.01131
783	61 30 89	27.9821	88.4873	.001277	.0357	.01130
784	61 46 56	28.0000	88.5438	.001276	.0357	.01129
785	61 62 25	28.0179	88.6002	.001274	.0357	.01129
786	61 77 96	28.0357	88.6566	.001272	.0357	.01128
787	61 93 69	28.0535	88.7130	.001271	.0356	.01127
788	62 09 44	28.0713	88.7694	.001269	.0356	.01127
789	62 25 21	28.0891	88.8257	.001267	.0356	.01126
790	62 41 00	28.1069	88.8819	.001266	.0356	.01125
791	62 56 81	28.1247	88.9382	.001264	.0356	.01124
792	62 72 64	28.1425	88.9944	.001263	.0355	.01124
793	62 88 49	28.1603	89.0505	.001261	.0355	.01123
794	63 04 36	28.1780	89.1067	.001259	.0355	.01122
795	63 20 25	28.1957	89.1628	.001258	.0355	.01122
796	63 36 16	28.2135	89.2188	.001256	.0354	.01121
797	63 52 09	28.2312	89.2749	.001255	.0354	.01120
798	63 68 04	28.2489	89.3308	.001253	.0354	.01119
799	63 84 01	28.2666	89.3868	.001252	.0354	.01119
800	64 00 00	28.2843	89.4427	.001250	.0354	.01118

Table A-2 *Squares, square roots, and reciprocals (continued)*

N	N²	√N	√10N	1/N	1/√N	1/√10N
801	64 16 01	28.3019	89.4986	.001248	.0353	.01117
802	64 32 04	28.3196	89.5545	.001247	.0353	.01117
803	64 48 09	28.3373	89.6103	.001245	.0353	.01116
804	64 64 16	28.3549	89.6660	.001244	.0353	.01115
805	64 80 25	28.3725	89.7218	.001242	.0352	.01115
806	64 96 36	28.3901	89.7775	.001241	.0352	.01114
807	65 12 49	28.4077	89.8332	.001239	.0352	.01113
808	65 28 64	28.4253	89.8888	.001238	.0352	.01112
809	65 44 81	28.4429	89.9444	.001236	.0352	.01112
810	65 61 00	28.4605	90.0000	.001235	.0351	.01111
811	65 77 21	28.4781	90.0555	.001233	.0351	.01110
812	65 93 44	28.4956	90.1110	.001232	.0351	.01110
813	66 09 69	28.5132	90.1665	.001230	.0351	.01109
814	66 25 96	28.5307	90.2219	.001229	.0350	.01108
815	66 42 25	28.5482	90.2774	.001227	.0350	.01108
816	66 58 56	28.5657	90.3327	.001225	.0350	.01107
817	66 74 89	28.5832	90.3881	.001224	.0350	.01106
818	66 91 24	28.6007	90.4434	.001222	.0350	.01106
819	67 07 61	28.6182	90.4986	.001221	.0349	.01105
820	67 24 00	28.6356	90.5539	.001220	.0349	.01104
821	67 40 41	28.6531	90.6091	.001218	.0349	.01104
822	67 56 84	28.6705	90.6642	.001217	.0349	.01103
823	67 73 29	28.6880	90.7193	.001215	.0349	.01102
824	67 89 76	28.7054	90.7744	.001214	.0348	.01102
825	68 06 25	28.7228	90.8295	.001212	.0348	.01101
826	68 22 76	28.7402	90.8845	.001211	.0348	.01100
827	68 39 29	28.7576	90.9395	.001209	.0348	.01100
828	68 55 84	28.7750	90.9945	.001208	.0348	.01099
829	68 72 41	28.7924	91.0494	.001206	.0347	.01098
830	68 89 00	28.8097	91.1043	.001205	.0347	.01098
831	69 05 61	28.8271	91.1592	.001203	.0347	.01097
832	69 22 24	28.8444	91.2140	.001202	.0347	.01096
833	69 38 89	28.8617	91.2688	.001200	.0346	.01096
834	69 55 56	28.8791	91.3236	.001199	.0346	.01095
835	69 72 25	28.8964	91.3783	.001198	.0346	.01094
836	69 88 96	28.9137	91.4330	.001196	.0346	.01094
837	70 05 69	28.9310	91.4877	.001195	.0346	.01093
838	70 22 44	28.9482	91.5423	.001193	.0345	.01092
839	70 39 21	28.9655	91.5969	.001192	.0345	.01092
840	70 56 00	28.9828	91.6515	.001190	.0345	.01091
841	70 72 81	29.0000	91.7061	.001189	.0345	.01090
842	70 89 64	29.0172	91.7606	.001188	.0345	.01090
843	71 06 49	29.0345	91.8150	.001186	.0344	.01089
844	71 23 36	29.0517	91.8695	.001185	.0344	.01089
845	71 40 25	29.0689	91.9239	.001183	.0344	.01088
846	71 57 16	29.0861	91.9783	.001182	.0344	.01087
847	71 74 09	29.1033	92.0326	.001181	.0344	.01087
848	71 91 04	29.1204	92.0869	.001179	.0343	.01086
849	72 08 01	29.1376	92.1412	.001178	.0343	.01085
850	72 25 00	29.1548	92.1954	.001176	.0343	.01085

Table A-2 *Squares, square roots, and reciprocals (continued)*

N	N²	√N	√10N	1/N	1/√N	1/√10N
851	72 42 01	29.1719	92.2497	.001175	.0343	.01084
852	72 59 04	29.1890	92.3038	.001174	.0343	.01083
853	72 76 09	29.2062	92.3580	.001172	.0342	.01083
854	72 93 16	29.2233	92.4121	.001171	.0342	.01082
855	73 10 25	29.2404	92.4662	.001170	.0342	.01081
856	73 27 36	29.2575	92.5203	.001168	.0342	.01081
857	73 44 49	29.2746	92.5743	.001167	.0342	.01080
858	73 61 64	29.2916	92.6283	.001166	.0341	.01080
859	73 78 81	29.3087	92.6823	.001164	.0341	.01079
860	73 96 00	29.3258	92.7362	.001163	.0341	.01078
861	74 13 21	29.3428	92.7901	.001161	.0341	.01078
862	74 30 44	29.3598	92.8440	.001160	.0341	.01077
863	74 47 69	29.3769	92.8978	.001159	.0340	.01076
864	74 64 96	29.3939	92.9516	.001157	.0340	.01076
865	74 82 25	29.4109	93.0054	.001156	.0340	.01075
866	74 99 56	29.4279	93.0591	.001155	.0340	.01075
867	75 16 89	29.4449	93.1128	.001153	.0340	.01074
868	75 34 24	29.4618	93.1665	.001152	.0339	.01073
869	75 51 61	29.4788	93.2202	.001151	.0339	.01073
870	75 69 00	29.4958	93.2738	.001149	.0339	.01072
871	75 86 41	29.5127	93.3274	.001148	.0339	.01071
872	76 03 84	29.5296	93.3809	.001147	.0339	.01071
873	76 21 29	29.5466	93.4345	.001145	.0338	.01070
874	76 38 76	29.5635	93.4880	.001144	.0338	.01070
875	76 56 25	29.5804	93.5414	.001143	.0338	.01069
876	76 73 76	29.5973	93.5949	.001142	.0338	.01068
877	76 91 29	29.6142	93.6483	.001140	.0338	.01068
878	77 08 84	29.6311	93.7017	.001139	.0337	.01067
879	77 26 41	29.6479	93.7550	.001138	.0337	.01067
880	77 44 00	29.6648	93.8083	.001136	.0337	.01066
881	77 61 61	29.6816	93.8616	.001135	.0337	.01065
882	77 79 24	29.6985	93.9149	.001134	.0337	.01065
883	77 96 89	29.7153	93.9681	.001133	.0337	.01064
884	78 14 56	29.7321	94.0213	.001131	.0336	.01064
885	78 32 25	29.7489	94.0744	.001130	.0336	.01063
886	78 49 96	29.7658	94.1276	.001129	.0336	.01062
887	78 67 69	29.7825	94.1807	.001127	.0336	.01062
888	78 85 44	29.7993	94.2338	.001126	.0336	.01061
889	79 03 21	29.8161	94.2868	.001125	.0335	.01061
890	79 21 00	29.8329	94.3398	.001124	.0335	.01060
891	79 38 81	29.8496	94.3928	.001122	.0335	.01059
892	79 56 64	29.8664	94.4458	.001121	.0335	.01059
893	79 74 49	29.8831	94.4987	.001120	.0335	.01058
894	79 92 36	29.8998	94.5516	.001119	.0334	.01058
895	80 10 25	29.9166	94.6044	.001117	.0334	.01057
896	80 28 16	29.9333	94.6573	.001116	.0334	.01056
897	80 46 09	29.9500	94.7101	.001115	.0334	.01056
898	80 64 04	29.9666	94.7629	.001114	.0334	.01055
899	80 82 01	29.9833	94.8156	.001112	.0334	.01055
900	81 00 00	30.0000	94.8683	.001111	.0333	.01054

N	N^2	$\sqrt{N}$	$\sqrt{10N}$	$1/N$	$1/\sqrt{N}$	$1/\sqrt{10N}$
901	81 18 01	30.0167	94.9210	.001110	.0333	.01054
902	81 36 04	30.0333	94.9737	.001109	.0333	.01053
903	81 54 09	30.0500	95.0263	.001107	.0333	.01052
904	81 72 16	30.0666	95.0789	.001106	.0333	.01052
905	81 90 25	30.0832	95.1315	.001105	.0332	.01051
906	82 08 36	30.0998	95.1840	.001104	.0332	.01051
907	82 26 49	30.1164	95.2365	.001103	.0332	.01050
908	82 44 64	30.1330	95.2890	.001101	.0332	.01049
909	82 62 81	30.1496	95.3415	.001100	.0332	.01049
910	82 81 00	30.1662	95.3939	.001099	.0331	.01048
911	82 99 21	30.1828	95.4463	.001098	.0331	.01048
912	83 17 44	30.1993	95.4987	.001096	.0331	.01047
913	83 35 69	30.2159	95.5510	.001095	.0331	.01047
914	83 53 96	30.2324	95.6033	.001094	.0331	.01046
915	83 72 25	30.2490	95.6556	.001093	.0331	.01045
916	83 90 56	30.2655	95.7079	.001092	.0330	.01045
917	84 08 89	30.2820	95.7601	.001091	.0330	.01044
918	84 27 24	30.2985	95.8123	.001089	.0330	.01044
919	84 45 61	30.3150	95.8645	.001088	.0330	.01043
920	84 64 00	30.3315	95.9166	.001087	.0330	.01043
921	84 82 41	30.3480	95.9687	.001086	.0330	.01042
922	85 00 84	30.3645	96.0208	.001085	.0329	.01041
923	85 19 29	30.3809	96.0729	.001083	.0329	.01041
924	85 37 76	30.3974	96.1249	.001082	.0329	.01040
925	85 56 25	30.4138	96.1769	.001081	.0329	.01040
926	85 74 76	30.4302	96.2289	.001080	.0329	.01039
927	85 93 29	30.4467	96.2808	.001079	.0328	.01039
928	86 11 84	30.4631	96.3328	.001078	.0328	.01038
929	86 30 41	30.4795	96.3846	.001076	.0328	.01038
930	86 49 00	30.4959	96.4365	.001075	.0328	.01037
931	86 67 61	30.5123	96.4883	.001074	.0328	.01036
932	86 86 24	30.5287	96.5401	.001073	.0328	.01036
933	87 04 89	30.5450	96.5919	.001072	.0327	.01035
934	87 23 56	30.5614	96.6437	.001071	.0327	.01055
935	87 42 25	30.5778	96.6954	.001070	.0327	.01034
936	87 60 96	30.5941	96.7471	.001068	.0327	.01034
937	87 79 69	30.6105	96.7988	.001067	.0327	.01033
938	87 98 44	30.6268	96.8504	.001066	.0327	.01033
939	88 17 21	30.6431	96.9020	.001065	.0326	.01032
940	88 36 00	30.6594	96.9536	.001064	.0326	.01031
941	88 54 81	30.6757	97.0052	.001063	.0326	.01031
942	88 73 64	30.6920	97.0567	.001062	.0326	.01030
943	88 92 49	30.7083	97.1082	.001060	.0326	.01030
944	89 11 36	30.7246	97.1597	.001059	.0325	.01029
945	89 30 25	30.7409	97.2111	.001058	.0325	.01029
946	89 49 16	30.7571	97.2625	.001057	.0325	.01028
947	89 68 09	30.7734	97.3139	.001056	.0325	.01028
948	89 87 04	30.7896	97.3653	.001055	.0325	.01027
949	90 06 01	30.8058	97.4166	.001054	.0325	.01027
950	90 25 00	30.8221	97.4679	.001053	.0324	.01026

N	N²	$\sqrt{N}$	$\sqrt{10N}$	1/N	$1/\sqrt{N}$	$1/\sqrt{10N}$
951	90 44 01	30.8383	97.5192	.001052	.0324	.01025
952	90 63 04	30.8545	97.5705	.001050	.0324	.01025
953	90 82 09	30.8707	97.6217	.001049	.0324	.01024
954	91 01 16	30.8869	97.6729	.001048	.0324	.01024
955	91 20 25	30.9031	97.7241	.001047	.0324	.01023
956	91 39 36	30.9192	97.7753	.001046	.0323	.01023
957	91 58 49	30.9354	97.8264	.001045	.0323	.01022
958	91 77 64	30.9516	97.8775	.001044	.0323	.01022
959	91 96 81	30.9677	97.9285	.001043	.0323	.01021
960	92 16 00	30.9839	97.9796	.001042	.0323	.01021
961	92 35 21	31.0000	98.0306	.001041	.0323	.01020
962	92 54 44	31.0161	98.0816	.001040	.0322	.01020
963	92 73 69	31.0322	98.1326	.001038	.0322	.01019
964	92 92 96	31.0483	98.1835	.001037	.0322	.01019
965	93 12 25	31.0644	98.2344	.001036	.0322	.01018
966	93 31 56	31.0805	98.2853	.001035	.0322	.01017
967	93 50 89	31.0966	98.3362	.001034	.0322	.01017
968	93 70 24	31.1127	98.3870	.001033	.0321	.01016
969	93 89 61	31.1288	98.4378	.001032	.0321	.01016
970	94 09 00	31.1448	98.4886	.001031	.0321	.01015
971	94 28 41	31.1609	98.5393	.001030	.0321	.01015
972	94 47 84	31.1769	98.5901	.001029	.0321	.01014
973	94 67 29	31.1929	98.6408	.001028	.0321	.01014
974	94 86 76	31.2090	98.6914	.001027	.0320	.01013
975	95 06 25	31.2250	98.7421	.001026	.0320	.01013
976	95 25 76	31.2410	98.7927	.001025	.0320	.01012
977	95 45 29	31.2570	98.8433	.001024	.0320	.01012
978	95 64 84	31.2730	98.8939	.001022	.0320	.01011
979	95 84 41	31.2890	98.9444	.001021	.0320	.01011
980	96 04 00	31.3050	98.9949	.001020	.0319	.01010
981	96 23 61	31.3209	99.0454	.001019	.0319	.01010
982	96 43 24	31.3369	99.0959	.001018	.0319	.01009
983	96 62 89	31.3528	99.1464	.001017	.0319	.01009
984	96 82 56	31.3688	99.1968	.001016	.0319	.01008
985	97 02 25	31.3847	99.2472	.001015	.0319	.01008
986	97 21 96	31.4006	99.2975	.001014	.0318	.01007
987	97 41 69	31.4166	99.3479	.001013	.0318	.01007
988	97 61 44	31.4325	99.3982	.001012	.0318	.01006
989	97 81 21	31.4484	99.4485	.001011	.0318	.01006
990	98 01 00	31.4643	99.4987	.001010	.0318	.01005
991	98 20 81	31.4802	99.5490	.001009	.0318	.01005
992	98 40 64	31.4960	99.5992	.001008	.0318	.01004
993	98 60 49	31.5119	99.6494	.001007	.0317	.01004
994	98 80 36	31.5278	99.6995	.001006	.0317	.01003
995	99 00 25	31.5436	99.7497	.001005	.0317	.01003
996	99 20 16	31.5595	99.7998	.001004	.0317	.01002
997	99 40 09	31.5753	99.8499	.001003	.0317	.01002
998	99 60 04	31.5911	99.8999	.001002	.0317	.01001
999	99 80 01	31.6070	99.9500	.001001	.0316	.01001
1,000	1 00 00 00	31.6228	100.0000	.001000	.0316	.01000

Table A-3   *Random digits**

00000	10097 32533	74520 13586	34673 54876	80959 09117	39292 74945
00001	37542 04805	64894 74296	24805 24037	20636 10402	00822 91665
00002	08422 68953	19645 09303	23209 02560	15953 34764	35080 33606
00003	99019 02529	09376 70715	38311 31165	88676 74397	04436 27659
00004	12807 99970	80157 36147	64032 36653	98951 16877	12171 76833
00005	66065 74717	34072 76850	36697 36170	65813 39885	11199 29170
00006	31060 10805	45571 82406	35303 42614	86799 07439	23403 09732
00007	85269 77602	02051 65692	68665 74818	73053 85247	18623 88579
00008	63573 32135	05325 47048	90553 57548	28468 28709	83491 25624
00009	73796 45753	03529 64778	35808 34282	60935 20344	25273 88435
00010	98520 17767	14905 68607	22109 40558	60970 93433	50500 73998
00011	11805 05431	39808 27732	50725 68248	29405 24201	52775 67851
00012	83452 99634	06288 98083	13746 70078	18475 40610	28711 77817
00013	88685 40200	86507 58401	36766 67951	90364 76493	29609 11062
00014	99594 67348	87517 64969	91826 08928	93785 61368	23478 34113
00015	65481 17674	17468 50950	58047 76974	73039 57186	40218 16544
00016	80124 35635	17727 08015	45318 22374	21115 78253	14385 53763
00017	74350 99817	77402 77214	43236 00210	45521 64237	96286 02655
00018	69916 26803	66252 29148	36936 87203	76621 13990	94400 56418
00019	09893 20505	14225 68514	46427 56788	96297 78822	54382 14598
00020	91499 14523	68479 27686	46162 83554	94750 89923	37089 20048
00021	80336 94598	26940 36858	70297 34135	53140 33340	42050 82341
00022	44104 81949	85157 47954	32979 26575	57600 40881	22222 06413
00023	12550 73742	11100 02040	12860 74697	96644 89439	28707 25815
00024	63606 49329	16505 34484	40219 52563	43651 77082	07207 31790
00025	61196 90446	26457 47774	51924 33729	65394 59593	42582 60527
00026	15474 45266	95270 79953	59367 83848	82396 10118	33211 59466
00027	94557 28573	67897 54387	54622 44431	91190 42592	92927 45973
00028	42481 16213	97344 08721	16868 48767	03071 12059	25701 46670
00029	23523 78317	73208 89837	68935 91416	26252 29663	05522 82562
00030	04493 52494	75246 33824	45862 51025	61962 79335	65337 12472
00031	00549 97654	64051 88159	96119 63896	54692 82391	23287 29529
00032	35963 15307	26898 09354	33351 35462	77974 50024	90103 39333
00033	59808 08391	45427 26842	83609 49700	13021 24892	78565 20106
00034	46058 85236	01390 92286	77281 44077	93910 83647	70617 42941
00035	32179 00597	87379 25241	05567 07007	86743 17157	85394 11838
00036	69234 61406	20117 45204	15956 60000	18743 92423	97118 96338
00037	19565 41430	01758 75379	40419 21585	66674 36806	84962 85207
00038	45155 14938	19476 07246	43667 94543	59047 90033	20826 69541
00039	94864 31994	36168 10851	34888 81553	01540 35456	05014 51176
00040	98086 24826	45340 28404	44999 08896	39094 73407	35441 31880
00041	33185 16232	41941 50949	89435 48581	88695 41994	37548 73043
00042	80951 00406	96382 70774	20151 23387	25016 25298	94624 61171
00043	79752 49140	71961 28296	69861 02591	74852 20539	00387 59579
00044	18633 32537	98145 06571	31010 24674	05455 61427	77938 91936

* Source: Rand (1955), 1–4. Used by permission of the publisher.

Table A-3 *Random digits (continued)*

00045	74029 43902	77557 32270	97790 17119	52527 58021	80814 51748
00046	54178 45611	80993 37143	05335 12969	56127 19255	36040 90324
00047	11664 49883	52079 84827	59381 71539	09973 33440	88461 23356
00048	48324 77928	31249 64710	02295 36870	32307 57546	15020 09994
00049	69074 94138	87637 91976	35584 04401	10518 21615	01848 76938
00050	09188 20097	32825 39527	04220 86304	83389 87374	64278 58044
00051	90045 85497	51981 50654	94938 81997	91870 76150	68476 64659
00052	73189 50207	47677 26269	62290 64464	27124 67018	41361 82760
00053	75768 76490	20971 87749	90429 12272	95375 05871	93823 43178
00054	54016 44056	66281 31003	00682 27398	20714 53295	07706 17813
00055	08358 69910	78542 42785	13661 58873	04618 97553	31223 08420
00056	28306 03264	81333 10591	40510 07893	32604 60475	94119 01840
00057	53840 86233	81594 13628	51215 90290	28466 68795	77762 20791
00058	91757 53741	61613 62269	50263 90212	55781 76514	83483 47055
00059	89415 92694	00397 58391	12607 17646	48949 72306	94541 37408
00060	77513 03820	86864 29901	68414 82774	51908 13980	72893 55507
00061	19502 37174	69979 20288	55210 29773	74287 75251	65344 67415
00062	21818 59313	93278 81757	05686 73156	07082 85046	31853 38452
00063	51474 66499	68107 23621	94049 91345	42836 09191	08007 45449
00064	99559 68331	62535 24170	69777 12830	74819 78142	43860 72834
00065	33713 48007	93584 72869	51926 64721	58303 29822	93174 93972
00066	85274 86893	11303 22970	28834 34137	73515 90400	71148 43643
00067	84133 89640	44035 52166	73852 70091	61222 60561	62327 18423
00068	56732 16234	17395 96131	10123 91622	85496 57560	81604 18880
00069	65138 56806	87648 85261	34313 65861	45875 21069	85644 47277
00070	38001 02176	81719 11711	71602 92937	74219 64049	65584 49698
00071	37402 96397	01304 77586	56271 10086	47324 62605	40030 37438
00072	97125 40348	87083 31417	21815 39250	75237 62047	15501 29578
00073	21826 41134	47143 34072	64638 85902	49139 06441	03856 54552
00074	73135 42742	95719 09035	85794 74296	08789 88156	64691 19202
00075	07638 77929	03061 18072	96207 44156	23821 99538	04713 66994
00076	60528 83441	07954 19814	59175 20695	05533 52139	61212 06455
00077	83596 35655	06958 92983	05128 09719	77433 53783	92301 50498
00078	10850 62746	99599 10507	13499 06319	53075 71839	06410 19362
00079	39820 98952	43622 63147	64421 80814	43800 09351	31024 73167
00080	59580 06478	75569 78800	88835 54486	23768 06156	04111 08408
00081	38508 07341	23793 48763	90822 97022	17719 04207	95954 49953
00082	30692 70668	94688 16127	56196 80091	82067 63400	05462 69200
00083	65443 95659	18288 27437	49632 24041	08337 65676	96299 90836
00084	27267 50264	13192 72294	07477 44606	17985 48911	97341 30358
00085	91307 06991	19072 24210	36699 53728	28825 35793	28976 66252
00086	68434 94688	84473 13622	62126 98408	12843 82590	09815 93146
00087	48908 15877	54745 24591	35700 04754	83824 52692	54130 55160
00088	06913 45197	42672 78601	11883 09528	63011 98901	14974 40344
00089	10455 16019	14210 33712	91342 37821	88325 80851	43667 70883

00090	12883	97343	65027	61184	04285	01392	17974	15077	90712	26769
00091	21778	30976	38807	36961	31649	42096	63281	02023	08816	47449
00092	19523	59515	65122	59659	86283	68258	69572	13798	16435	91529
00093	67245	52670	35583	16563	79246	86686	76463	34222	26655	90802
00094	60584	47377	07500	37992	45134	26529	26760	83637	41326	44344
00095	53853	41377	36066	94850	58838	73859	49364	73331	96240	43642
00096	24637	38736	74384	89342	52623	07992	12369	18601	03742	83873
00097	83080	12451	38992	22815	07759	51777	97377	27585	51972	37867
00098	16444	24334	36151	99073	27493	70939	85130	32552	54846	54759
00099	60790	18157	57178	65762	11161	78576	45819	52979	65130	04860
00100	03991	10461	93716	16894	66083	24653	84609	58232	88618	19161
00101	38555	95554	32886	59780	08355	60860	29735	47762	71299	23853
00102	17546	73704	92052	46215	55121	29281	59076	07936	27954	58909
00103	32643	52861	95819	06831	00911	98936	76355	93779	80863	00514
00104	69572	68777	39510	35905	14060	40619	29549	69616	33564	60780
00105	24122	66591	27699	06494	14845	46672	61958	77100	90899	75754
00106	61196	30231	92962	61773	41839	55382	17267	70943	78038	70267
00107	30532	21704	10274	12202	39685	23309	11061	68829	55986	66485
00108	03788	97599	75867	20717	74416	53166	35208	33374	87539	08823
00109	48228	63379	85783	47619	53152	67433	35663	52972	16818	60311
00110	60365	94653	35075	33949	42614	29297	01918	28316	98953	73231
00111	83799	42402	56623	34442	34994	41374	70071	14736	09958	18065
00112	32960	07405	36409	83232	99385	41600	11133	07586	15917	06253
00113	19322	53845	57620	52606	66497	68646	78138	66559	19640	99413
00114	11220	94747	07399	37408	48509	23929	27482	45476	85244	35159
00115	31751	57260	68980	05339	15470	48355	88651	22596	03152	19121
00116	88492	99382	14454	04504	20094	98977	74843	93413	22109	78508
00117	30934	47744	07481	84828	73788	06533	28597	20405	94205	20380
00118	22888	48893	27499	98748	60530	45128	74022	84617	82037	10268
00119	78212	16993	35902	91386	44372	15486	65741	14014	87481	37220
00120	41849	84547	46850	52326	34677	58300	74910	64345	19325	81549
00121	46352	33049	69248	93460	45305	07521	61318	31855	14413	70951
00122	11087	96294	14013	31792	59747	67277	76503	34513	39663	77544
00123	52701	08337	56303	87315	16520	69676	11654	99893	02181	68161
00124	57275	36898	81304	48585	68652	27376	92852	55866	88448	03584
00125	20857	73156	70284	24326	79375	95220	01159	63267	10622	48391
00126	15633	84924	90415	93614	33521	26665	55823	47641	86225	31704
00127	92694	48297	39904	02115	59589	49067	66821	41575	49767	04037
00128	77613	19019	88152	00080	20554	91409	96277	48257	50816	97616
00129	38688	32486	45134	63545	59404	72059	43947	51680	43852	59693
00130	25163	01889	70014	15021	41290	67312	71857	15957	68971	11403
00131	65251	07629	37239	33295	05870	01119	92784	26340	18477	65622
00132	36815	43625	18637	37509	82444	99005	04921	73701	14707	93997
00133	64397	11692	05327	82162	20247	81759	45197	25332	83745	22567
00134	04515	25624	95096	67946	48460	85558	15191	18782	16930	33361

00135	83761 60873	43253 84145	60833 25983	01291 41349	20368 07126
00136	14387 06345	80854 09279	43529 06318	38384 74761	41196 37480
00137	51321 92246	80088 77074	88722 56736	66164 49431	66919 31678
00138	72472 00008	80890 18002	94813 31900	54155 83436	35352 54131
00139	05466 55306	93128 18464	74457 90561	72848 11834	79982 68416
00140	39528 72484	82474 25593	48545 35247	18619 13674	18611 19241
00141	81616 18711	53342 44276	75122 11724	74627 73707	58319 15997
00142	07586 16120	82641 22820	92904 13141	32392 19763	61199 67940
00143	90767 04345	13574 17200	69902 63742	78464 22501	18627 90872
00144	40188 28193	29593 88627	94972 11598	62095 36787	00441 58997
00145	34414 82157	86887 55087	19152 00023	12302 80783	32624 68691
00146	63439 75363	44989 16822	36024 00867	76378 41605	65961 63488
00147	67049 09070	93399 45547	94458 74284	05041 49807	20288 34060
00148	79495 04146	52162 90286	54158 34243	46978 35482	59362 95938
00149	91704 30552	04737 21031	75051 93029	47665 64382	99782 93478
00150	94015 46874	32444 48277	59820 96163	64654 25843	41145 42820
00151	74108 88222	88570 74015	25704 91035	01755 14750	48968 38603
00152	62880 87873	95160 59221	22304 90314	72877 17334	39283 04149
00153	11748 12102	80580 41867	17710 59621	06554 07850	73950 79552
00154	17944 05600	60478 03343	25852 58905	57216 39618	49856 99326
00155	66067 42792	95043 52680	46780 56487	09971 59481	37006 22186
00156	54244 91030	45547 70818	59849 96169	61459 21647	87417 17198
00157	30945 57589	31732 57260	47670 07654	46376 25366	94746 49580
00158	69170 37403	86995 90307	94304 71803	26825 05511	12459 91314
00159	08345 88975	35841 85771	08105 59987	87112 21476	14713 71181
00160	27767 43584	85301 88977	29490 69714	73035 41207	74699 09310
00161	13025 14338	54066 15243	47724 66733	47431 43905	31048 56699
00162	80217 36292	98525 24335	24432 24896	43277 58874	11466 16082
00163	10875 62004	90391 61105	57411 06368	53856 30743	08670 84741
00164	54127 57326	26629 19087	24472 88779	30540 27886	61732 75454
00165	60311 42824	37301 42678	45990 43242	17374 52003	70707 70214
00166	49739 71484	92003 98086	76668 73209	59202 11973	02902 33250
00167	78626 51594	16453 94614	39014 97066	83012 09832	25571 77628
00168	66692 13986	99837 00582	81232 44987	09504 96412	90193 79568
00169	44071 28091	07362 97703	76447 42537	98524 97831	65704 09514
00170	41468 85149	49554 17994	14924 39650	95294 00556	70481 06905
00171	94559 37559	49678 53119	70312 05682	66986 34099	74474 20740
00172	41615 70360	64114 58660	90850 64618	80620 51790	11436 38072
00173	50273 93113	41794 86861	24781 89683	55411 85667	77535 99892
00174	41396 80504	90670 08289	40902 05069	95083 06783	28102 57816
00175	25807 24260	71529 78920	72682 07385	90726 57166	98884 08583
00176	06170 97965	88302 98041	21443 41808	68984 83620	89747 98882
00177	60808 54444	74412 81105	01176 28838	36421 16489	18059 51061
00178	80940 44893	10408 36222	80582 71944	92638 40333	67054 16067
00179	19516 90120	46759 71643	13177 55292	21036 82808	77501 97427

Table A-4 *Areas and ordinates of the standard normal distribution, ordered by z-scores**

(1) z_i Standard score	(2) Area from mean to z_i	(3) Area in larger portion	(4) Area in smaller portion	(5) y Ordinate at z_i
0.00	.0000	.5000	.5000	.3989
0.01	.0040	.5040	.4960	.3989
0.02	.0080	.5080	.4920	.3989
0.03	.0120	.5120	.4880	.3988
0.04	.0160	.5160	.4840	.3986
0.05	.0199	.5199	.4801	.3984
0.06	.0239	.5239	.4761	.3982
0.07	.0279	.5279	.4721	.3980
0.08	.0319	.5319	.4681	.3977
0.09	.0359	.5359	.4641	.3973
0.10	.0398	.5398	.4602	.3970
0.11	.0438	.5438	.4562	.3965
0.12	.0478	.5478	.4522	.3961
0.13	.0517	.5517	.4483	.3956
0.14	.0557	.5557	.4443	.3951
0.15	.0596	.5596	.4404	.3945
0.16	.0636	.5636	.4364	.3939
0.17	.0675	.5675	.4325	.3932
0.18	.0714	.5714	.4286	.3925
0.19	.0753	.5753	.4247	.3918
0.20	.0793	.5793	.4207	.3910
0.21	.0832	.5832	.4168	.3902
0.22	.0871	.5871	.4129	.3894
0.23	.0910	.5910	.4090	.3885
0.24	.0948	.5948	.4052	.3876
0.25	.0987	.5987	.4013	.3867
0.26	.1026	.6026	.3974	.3857
0.27	.1064	.6064	.3936	.3847
0.28	.1103	.6103	.3897	.3836
0.29	.1141	.6141	.3859	.3825
0.30	.1179	.6179	.3821	.3814
0.31	.1217	.6217	.3783	.3802
0.32	.1255	.6255	.3745	.3790
0.33	.1293	.6293	.3707	.3778
0.34	.1331	.6331	.3669	.3765

* Source: From *Statistical Methods*, Second Edition, by Allen L. Edwards. Copyright 1954, ©
1967 by Allen L. Edwards. Reprinted by permission of Holt, Rinehart and Winston, Inc. Directions
for the use of this table are given in Section 7-3.

(1) z_i Standard score	(2) Area from mean to z_i	(3) Area in larger portion	(4) Area in smaller portion	(5) y Ordinate at z_i
0.35	.1368	.6368	.3632	.3752
0.36	.1406	.6406	.3594	.3739
0.37	.1443	.6443	.3557	.3725
0.38	.1480	.6480	.3520	.3712
0.39	.1517	.6517	.3483	.3697
0.40	.1554	.6554	.3446	.3683
0.41	.1591	.6591	.3409	.3668
0.42	.1628	.6628	.3372	.3653
0.43	.1664	.6664	.3336	.3637
0.44	.1700	.6700	.3300	.3621
0.45	.1736	.6736	.3264	.3605
0.46	.1772	.6772	.3228	.3589
0.47	.1808	.6808	.3192	.3572
0.48	.1844	.6844	.3156	.3555
0.49	.1879	.6879	.3121	.3538
0.50	.1915	.6915	.3085	.3521
0.51	.1950	.6950	.3050	.3503
0.52	.1985	.6985	.3015	.3485
0.53	.2019	.7019	.2981	.3467
0.54	.2054	.7054	.2946	.3448
0.55	.2088	.7088	.2912	.3429
0.56	.2123	.7123	.2877	.3410
0.57	.2157	.7157	.2843	.3391
0.58	.2190	.7190	.2810	.3372
0.59	.2224	.7224	.2776	.3352
0.60	.2257	.7257	.2743	.3332
0.61	.2291	.7291	.2709	.3312
0.62	.2324	.7324	.2676	.3292
0.63	.2357	.7357	.2643	.3271
0.64	.2389	.7389	.2611	.3251
0.65	.2422	.7422	.2578	.3230
0.66	.2454	.7454	.2546	.3209
0.67	.2486	.7486	.2514	.3187
0.68	.2517	.7517	.2483	.3166
0.69	.2549	.7549	.2451	.3144
0.70	.2580	.7580	.2420	.3123
0.71	.2611	.7611	.2389	.3101
0.72	.2642	.7642	.2358	.3079
0.73	.2673	.7673	.2327	.3056
0.74	.2704	.7704	.2296	.3034

(1) z_i Standard score	(2) Area from mean to z_i	(3) Area in larger portion	(4) Area in smaller portion	(5) y Ordinate at z_i
0.75	.2734	.7734	.2266	.3011
0.76	.2764	.7764	.2236	.2989
0.77	.2794	.7794	.2206	.2966
0.78	.2823	.7823	.2177	.2943
0.79	.2852	.7852	.2148	.2920
0.80	.2881	.7881	.2119	.2897
0.81	.2910	.7910	.2090	.2874
0.82	.2939	.7939	.2061	.2850
0.83	.2967	.7967	.2033	.2827
0.84	.2995	.7995	.2005	.2803
0.85	.3023	.8023	.1977	.2780
0.86	.3051	.8051	.1949	.2756
0.87	.3078	.8078	.1922	.2732
0.88	.3106	.8106	.1894	.2709
0.89	.3133	.8133	.1867	.2685
0.90	.3159	.8159	.1841	.2661
0.91	.3186	.8186	.1814	.2637
0.92	.3212	.8212	.1788	.2613
0.93	.3238	.8238	.1762	.2589
0.94	.3264	.8264	.1736	.2565
0.95	.3289	.8289	.1711	.2541
0.96	.3315	.8315	.1685	.2516
0.97	.3340	.8340	.1660	.2492
0.98	.3365	.8365	.1635	.2468
0.99	.3389	.8389	.1611	.2444
1.00	.3413	.8413	.1587	.2420
1.01	.3438	.8438	.1562	.2396
1.02	.3461	.8461	.1539	.2371
1.03	.3485	.8485	.1515	.2347
1.04	.3508	.8508	.1492	.2323
1.05	.3531	.8531	.1469	.2299
1.06	.3554	.8554	.1446	.2275
1.07	.3577	.8577	.1423	.2251
1.08	.3599	.8599	.1401	.2227
1.09	.3621	.8621	.1379	.2203
1.10	.3643	.8643	.1357	.2179
1.11	.3665	.8665	.1335	.2155
1.12	.3686	.8686	.1314	.2131
1.13	.3708	.8708	.1292	.2107
1.14	.3729	.8729	.1271	.2083

(1) z_i Standard score	(2) Area from mean to z_i	(3) Area in larger portion	(4) Area in smaller portion	(5) y Ordinate at z_i
1.15	.3749	.8749	.1251	.2059
1.16	.3770	.8770	.1230	.2036
1.17	.3790	.8790	.1210	.2012
1.18	.3810	.8810	.1190	.1989
1.19	.3830	.8830	.1170	.1965
1.20	.3849	.8849	.1151	.1942
1.21	.3869	.8869	.1131	.1919
1.22	.3888	.8888	.1112	.1895
1.23	.3907	.8907	.1093	.1872
1.24	.3925	.8925	.1075	.1849
1.25	.3944	.8944	.1056	.1826
1.26	.3962	.8962	.1038	.1804
1.27	.3980	.8980	.1020	.1781
1.28	.3997	.8997	.1003	.1758
1.29	.4015	.9015	.0985	.1736
1.30	.4032	.9032	.0968	.1714
1.31	.4049	.9049	.0951	.1691
1.32	.4066	.9066	.0934	.1669
1.33	.4082	.9082	.0918	.1647
1.34	.4099	.9099	.0901	.1626
1.35	.4115	.9115	.0885	.1604
1.36	.4131	.9131	.0869	.1582
1.37	.4147	.9147	.0853	.1561
1.38	.4162	.9162	.0838	.1539
1.39	.4177	.9177	.0823	.1518
1.40	.4192	.9192	.0808	.1497
1.41	.4207	.9207	.0793	.1476
1.42	.4222	.9222	.0778	.1456
1.43	.4236	.9236	.0764	.1435
1.44	.4251	.9251	.0749	.1415
1.45	.4265	.9265	.0735	.1394
1.46	.4279	.9279	.0721	.1374
1.47	.4292	.9292	.0708	.1354
1.48	.4306	.9306	.0694	.1334
1.49	.4319	.9319	.0681	.1315
1.50	.4332	.9332	.0668	.1295
1.51	.4345	.9345	.0655	.1276
1.52	.4357	.9357	.0643	.1257
1.53	.4370	.9370	.0630	.1238
1.54	.4382	.9382	.0618	.1219

Table A-4 *Areas and ordinates of the standard normal distribution, ordered by z-scores (continued)*

(1) z_i Standard score	(2) Area from mean to z_i	(3) Area in larger portion	(4) Area in smaller portion	(5) y Ordinate at z_i
1.55	.4394	.9394	.0606	.1200
1.56	.4406	.9406	.0594	.1182
1.57	.4418	.9418	.0582	.1163
1.58	.4429ˈ	.9429	.0571	.1145
1.59	.4441	.9441	.0559	.1127
1.60	.4452	.9452	.0548	.1109
1.61	.4463	.9463	.0537	.1092
1.62	.4474	.9474	.0526	.1074
1.63	.4484	.9484	.0516	.1057
1.64	.4495	.9495	.0505	.1040
1.65	.4505	.9505	.0495	.1023
1.66	.4515	.9515	.0485	.1006
1.67	.4525	.9525	.0475	.0989
1.68	.4535	.9535	.0465	.0973
1.69	.4545	.9545	.0455	.0957
1.70	.4554	.9554	.0446	.0940
1.71	.4564	.9564	.0436	.0925
1.72	.4573	.9573	.0427	.0909
1.73	.4582	.9582	.0418	.0893
1.74	.4591	.9591	.0409	.0878
1.75	.4599	.9599	.0401	.0863
1.76	.4608	.9608	.0392	.0848
1.77	.4616	.9616	.0384	.0833
1.78	.4625	.9625	.0375	.0818
1.79	.4633	.9633	.0367	.0804
1.80	.4641	.9641	.0359	.0790
1.81	.4649	.9649	.0351	.0775
1.82	.4656	.9656	.0344	.0761
1.83	.4664	.9664	.0336	.0748
1.84	.4671	.9671	.0329	.0734
1.85	.4678	.9678	.0322	.0721
1.86	.4686	.9686	.0314	.0707
1.87	.4693	.9693	.0307	.0694
1.88	.4699	.9699	.0301	.0681
1.89	.4706	.9706	.0294	.0669
1.90	.4713	.9713	.0287	.0656
1.91	.4719	.9719	.0281	.0644
1.92	.4726	.9726	.0274	.0632
1.93	.4732	.9732	.0268	.0620
1.94	.4738	.9738	.0262	.0608

(1) z_i Standard score	(2) Area from mean to z_i	(3) Area in larger portion	(4) Area in smaller portion	(5) y Ordinate at z_i
1.95	.4744	.9744	.0256	.0596
1.96	.4750	.9750	.0250	.0584
1.97	.4756	.9756	.0244	.0573
1.98	.4761	.9761	.0239	.0562
1.99	.4767	.9767	.0233	.0551
2.00	.4772	.9772	.0228	.0540
2.01	.4778	.9778	.0222	.0529
2.02	.4783	.9783	.0217	.0519
2.03	.4788	.9788	.0212	.0508
2.04	.4793	.9793	.0207	.0498
2.05	.4798	.9798	.0202	.0488
2.06	.4803	.9803	.0197	.0478
2.07	.4808	.9808	.0192	.0468
2.08	.4812	.9812	.0188	.0459
2.09	.4817	.9817	.0183	.0449
2.10	.4821	.9821	.0179	.0440
2.11	.4826	.9826	.0174	.0431
2.12	.4830	.9830	.0170	.0422
2.13	.4834	.9834	.0166	.0413
2.14	.4838	.9838	.0162	.0404
2.15	.4842	.9842	.0158	.0396
2.16	.4846	.9846	.0154	.0387
2.17	.4850	.9850	.0150	.0379
2.18	.4854	.9854	.0146	.0371
2.19	.4857	.9857	.0143	.0363
2.20	.4861	.9861	.0139	.0355
2.21	.4864	.9864	.0136	.0347
2.22	.4868	.9868	.0132	.0339
2.23	.4871	.9871	.0129	.0332
2.24	.4875	.9875	.0125	.0325
2.25	.4878	.9878	.0122	.0317
2.26	.4881	.9881	.0119	.0310
2.27	.4884	.9884	.0116	.0303
2.28	.4887	.9887	.0113	.0297
2.29	.4890	.9890	.0110	.0290
2.30	.4893	.9893	.0107	.0283
2.31	.4896	.9896	.0104	.0277
2.32	.4898	.9898	.0102	.0270
2.33	.4901	.9901	.0099	.0264
2.34	.4904	.9904	.0096	.0258

(1) z_i Standard score	(2) Area from mean to z_i	(3) Area in larger portion	(4) Area in smaller portion	(5) y Ordinate at z_i
2.35	.4906	.9906	.0094	.0252
2.36	.4909	.9909	.0091	.0246
2.37	.4911	.9911	.0089	.0241
2.38	.4913	.9913	.0087	.0235
2.39	.4916	.9916	.0084	.0229
2.40	.4918	.9918	.0082	.0224
2.41	.4920	.9920	.0080	.0219
2.42	.4922	.9922	.0078	.0213
2.43	.4925	.9925	.0075	.0208
2.44	.4927	.9927	.0073	.0203
2.45	.4929	.9929	.0071	.0198
2.46	.4931	.9931	.0069	.0194
2.47	.4932	.9932	.0068	.0189
2.48	.4934	.9934	.0066	.0184
2.49	.4936	.9936	.0064	.0180
2.50	.4938	.9938	.0062	.0175
2.51	.4940	.9940	.0060	.0171
2.52	.4941	.9941	.0059	.0167
2.53	.4943	.9943	.0057	.0163
2.54	.4945	.9945	.0055	.0158
2.55	.4946	.9946	.0054	.0154
2.56	.4948	.9948	.0052	.0151
2.57	.4949	.9949	.0051	.0147
2.58	.4951	.9951	.0049	.0143
2.59	.4952	.9952	.0048	.0139
2.60	.4953	.9953	.0047	.0136
2.61	.4955	.9955	.0045	.0132
2.62	.4956	.9956	.0044	.0129
2.63	.4957	.9957	.0043	.0126
2.64	.4959	.9959	.0041	.0122
2.65	.4960	.9960	.0040	.0119
2.66	.4961	.9961	.0039	.0116
2.67	.4962	.9962	.0038	.0113
2.68	.4963	.9963	.0037	.0110
2.69	.4964	.9964	.0036	.0107
2.70	.4965	.9965	.0035	.0104
2.71	.4966	.9966	.0034	.0101
2.72	.4967	.9967	.0033	.0099
2.73	.4968	.9968	.0032	.0096
2.74	.4969	.9969	.0031	.0093

Table A-4 *Areas and ordinates of the standard normal distribution, ordered by z-scores (continued)*

(1) z_i Standard score	(2) Area from mean to z_i	(3) Area in larger portion	(4) Area in smaller portion	(5) y Ordinate at z_i
2.75	.4970	.9970	.0030	.0091
2.76	.4971	.9971	.0029	.0088
2.77	.4972	.9972	.0028	.0086
2.78	.4973	.9973	.0027	.0084
2.79	.4974	.9974	.0026	.0081
2.80	.4974	.9974	.0026	.0079
2.81	.4975	.9975	.0025	.0077
2.82	.4976	.9976	.0024	.0075
2.83	.4977	.9977	.0023	.0073
2.84	.4977	.9977	.0023	.0071
2.85	.4978	.9978	.0022	.0069
2.86	.4979	.9979	.0021	.0067
2.87	.4979	.9979	.0021	.0065
2.88	.4980	.9980	.0020	.0063
2.89	.4981	.9981	.0019	.0061
2.90	.4981	.9981	.0019	.0060
2.91	.4982	.9982	.0018	.0058
2.92	.4982	.9982	.0018	.0056
2.93	.4983	.9983	.0017	.0055
2.94	.4984	.9984	.0016	.0053
2.95	.4984	.9984	.0016	.0051
2.96	.4985	.9985	.0015	.0050
2.97	.4985	.9985	.0015	.0048
2.98	.4986	.9986	.0014	.0047
2.99	.4986	.9986	.0014	.0046
3.00	.4987	.9987	.0013	.0044
3.01	.4987	.9987	.0013	.0043
3.02	.4987	.9987	.0013	.0042
3.03	.4988	.9988	.0012	.0040
3.04	.4988	.9988	.0012	.0039
3.05	.4989	.9989	.0011	.0038
3.06	.4989	.9989	.0011	.0037
3.07	.4989	.9989	.0011	.0036
3.08	.4990	.9990	.0010	.0035
3.09	.4990	.9990	.0010	.0034
3.10	.4990	.9990	.0010	.0033
3.11	.4991	.9991	.0009	.0032
3.12	.4991	.9991	.0009	.0031
3.13	.4991	.9991	.0009	.0030
3.14	.4992	.9992	.0008	.0029

Table A-4 *Areas and ordinates of the standard normal distribution, ordered by z-scores (continued)*

(1) z_i Standard score	(2) Area from mean to z_i	(3) Area in larger portion	(4) Area in smaller portion	(5) y Ordinate at z_i
3.15	.4992	.9992	.0008	.0028
3.16	.4992	.9992	.0008	.0027
3.17	.4992	.9992	.0008	.0026
3.18	.4993	.9993	.0007	.0025
3.19	.4993	.9993	.0007	.0025
3.20	.4993	.9993	.0007	.0024
3.21	.4993	.9993	.0007	.0023
3.22	.4994	.9994	.0006	.0022
3.23	.4994	.9994	.0006	.0022
3.24	.4994	.9994	.0006	.0021
3.30	.4995	.9995	.0005	.0017
3.40	.4997	.9997	.0003	.0012
3.50	.4998	.9998	.0002	.0009
3.60	.4998	.9998	.0002	.0006
3.70	.4999	.9999	.0001	.0004

Table A-5 *Standard scores (or deviates) and ordinates corresponding to divisions of the area under the normal curve into a larger proportion (B) and a smaller proportion (C); also the value $\sqrt{BC}$**

B The larger area	z Standard score	y Ordinate	$\sqrt{BC}$	C The smaller area
.500	.0000	.3989	.5000	.500
.505	.0125	.3989	.5000	.495
.510	.0251	.3988	.4999	.490
.515	.0376	.3987	.4998	.485
.520	.0502	.3984	.4996	.480
.525	.0627	.3982	.4994	.475
.530	.0753	.3978	.4991	.470
.535	.0878	.3974	.4988	.465
.540	.1004	.3969	.4984	.460
.545	.1130	.3964	.4980	.455
.550	.1257	.3958	.4975	.450
.555	.1383	.3951	.4970	.445
.560	.1510	.3944	.4964	.440
.565	.1637	.3936	.4958	.435
.570	.1764	.3928	.4951	.430
.575	.1891	.3919	.4943	.425
.580	.2019	.3909	.4936	.420
.585	.2147	.3899	.4927	.415
.590	.2275	.3887	.4918	.410
.595	.2404	.3876	.4909	.405
.600	.2533	.3863	.4899	.400
.605	.2663	.3850	.4889	.395
.610	.2793	.3837	.4877	.390
.615	.2924	.3822	.4867	.385
.620	.3055	.3808	.4854	.380
.625	.3186	.3792	.4841	.375
.630	.3319	.3776	.4828	.370
.635	.3451	.3759	.4814	.365
.640	.3585	.3741	.4800	.360
.645	.3719	.3723	.4785	.355
.650	.3853	.3704	.4770	.350
.655	.3989	.3684	.4754	.345
.660	.4125	.3664	.4737	.340
.665	.4261	.3643	.4720	.335
.670	.4399	.3621	.4702	.330
.675	.4538	.3599	.4684	.325
.680	.4677	.3576	.4665	.320
.685	.4817	.3552	.4645	.315
.690	.4959	.3528	.4625	.310
.695	.5101	.3503	.4604	.305

* Source: From *Fundamental Statistics in Psychology and Education*, by J. P. Guilford. Copyright © 1965 by McGraw-Hill, Inc. Used with permission of McGraw-Hill Book Company.

Table A-5 *Standard scores (or deviates) and ordinates corresponding to divisions of the area under normal curve into a larger proportion (B) and a smaller proportion (C); also the value $\sqrt{BC}$ (continued)*

B The larger area	z Standard score	y Ordinate	$\sqrt{BC}$	C The smaller area
.700	.5244	.3477	.4583	.300
.705	.5388	.3450	.4560	.295
.710	.5534	.3423	.4538	.290
.715	.5681	.3395	.4514	.285
.720	.5828	.3366	.4490	.280
.725	.5978	.3337	.4465	.275
.730	.6128	.3306	.4440	.270
.735	.6280	.3275	.4413	.265
.740	.6433	.3244	.4386	.260
.745	.6588	.3211	.4359	.255
.750	.6745	.3178	.4330	.250
.755	.6903	.3144	.4301	.245
.760	.7063	.3109	.4271	.240
.765	.7225	.3073	.4240	.235
.770	.7388	.3036	.4208	.230
.775	.7554	.2999	.4176	.225
.780	.7722	.2961	.4142	.220
.785	.7892	.2922	.4108	.215
.790	.8064	.2882	.4073	.210
.795	.8239	.2841	.4037	.205
.800	.8416	.2800	.4000	.200
.805	.8596	.2757	.3962	.195
.810	.8779	.2714	.3923	.190
.815	.8965	.2669	.3883	.185
.820	.9154	.2624	.3842	.180
.825	.9346	.2578	.3800	.175
.830	.9542	.2531	.3756	.170
.835	.9741	.2482	.3712	.165
.840	.9945	.2433	.3666	.160
.845	1.0152	.2383	.3619	.155
.850	1.0364	.2332	.3571	.150
.855	1.0581	.2279	.3521	.145
.860	1.0803	.2226	.3470	.140
.865	1.1031	.2171	.3417	.135
.870	1.1264	.2115	.3363	.130
.875	1.1503	.2059	.3307	.125
.880	1.1750	.2000	.3250	.120
.885	1.2004	.1941	.3190	.115
.890	1.2265	.1880	.3129	.110
.895	1.2536	.1818	.3066	.105

Table A-5 *Standard scores (or deviates) and ordinates corresponding to divisions of the area under the normal curve into a larger proportion (B) and a smaller proportion (C); also the value* $\sqrt{BC}$ *(continued)*

B The larger area	z Standard score	y Ordinate	$\sqrt{BC}$	C The smaller area
.900	1.2816	.1755	.3000	.100
.905	1.3106	.1690	.2932	.095
.910	1.3408	.1624	.2862	.090
.915	1.3722	.1556	.2789	.085
.920	1.4051	.1487	.2713	.080
.925	1.4395	.1416	.2634	.075
.930	1.4757	.1343	.2551	.070
.935	1.5141	.1268	.2465	.065
.940	1.5548	.1191	.2375	.060
.945	1.5982	.1112	.2280	.055
.950	1.6449	.1031	.2179	.050
.955	1.6954	.0948	.2073	.045
.960	1.7507	.0862	.1960	.040
.965	1.8119	.0773	.1838	.035
.970	1.8808	.0680	.1706	.030
.975	1.9600	.0584	.1561	.025
.980	2.0537	.0484	.1400	.020
.985	2.1701	.0379	.1226	.015
.990	2.3263	.0267	.0995	.010
.995	2.5758	.0145	.0705	.005
.996	2.6521	.0118	.0631	.004
.997	2.7478	.0091	.0547	.003
.998	2.8782	.0063	.0447	.002
.999	3.0902	.0034	.0316	.001
.9995	3.2905	.0018	.0224	.0005

Table A-6 *Critical values of the t and z distributions**

Statistic	Degrees of freedom	.05	.025	.01	.005	.0005	Alpha level for directional test†
		.10	.05	.02	.01	.001	Alpha level for nondirectional test
t	1	6.314	12.706	31.821	63.657	636.619	
	2	2.920	4.303	6.965	9.925	31.598	
	3	2.353	3.182	4.541	5.841	12.941	
	4	2.132	2.776	3.747	4.604	8.610	
	5	2.015	2.571	3.365	4.032	6.859	
	6	1.943	2.447	3.143	3.707	5.959	
	7	1.895	2.365	2.998	3.499	5.405	
	8	1.860	2.306	2.896	3.355	5.041	
	9	1.833	2.262	2.821	3.250	4.781	
	10	1.812	2.228	2.764	3.169	4.587	
	11	1.796	2.201	2.718	3.106	4.437	
	12	1.782	2.179	2.681	3.055	4.318	
	13	1.771	2.160	2.650	3.012	4.221	
	14	1.761	2.145	2.624	2.977	4.140	
	15	1.753	2.131	2.602	2.947	4.073	
	16	1.746	2.120	2.583	2.921	4.015	
	17	1.740	2.110	2.567	2.898	3.965	
	18	1.734	2.101	2.552	2.878	3.922	
	19	1.729	2.093	2.539	2.861	3.883	
	20	1.725	2.086	2.528	2.845	3.850	
	21	1.721	2.080	2.518	2.831	3.819	
	22	1.717	2.074	2.508	2.819	3.792	
	23	1.714	2.069	2.500	2.807	3.767	
	24	1.711	2.064	2.492	2.797	3.745	
	25	1.708	2.060	2.485	2.787	3.725	
	26	1.706	2.056	2.479	2.779	3.707	
	27	1.703	2.052	2.473	2.771	3.690	
	28	1.701	2.048	2.467	2.763	3.674	
	29	1.699	2.045	2.462	2.756	3.659	
	30	1.697	2.042	2.457	2.750	3.646	
	40	1.684	2.021	2.423	2.704	3.551	
	60	1.671	2.000	2.390	2.660	3.460	
	120	1.658	1.980	2.358	2.617	3.373	
z		1.645	1.960	2.326	2.576	3.291	

* Source: From *Biometrika Tables for Statisticians,* Vol. 1, E. S. Pearson and H. O. Hartley (eds.), 1966. Cambridge University Press. The original tables contain critical values for other significance levels as well. Use of this table is explained in Section 7-4.

† Note: for a directional test, tabulated values are to be preceded by either a + or a − sign. For a nondirectional test, precede tabulated values by a ± to indicate sample values at either end of the distribution can be used to reject the null hypothesis.

Table A-7 *Critical values of the chi-square distribution**

	Lower tail of distribution				Upper tail of distribution			
α, one-tail test	.005	.01	.025	.05	.05	.025	.01	.005
α, two-tail test	.01	.02	.05	.10	.10	.05	.02	.01
df	(Tabulated value is upper bound to alpha)				(Tabulated value is lower bound to alpha)			
1	$392704 \cdot 10^{-10}$	$157088 \cdot 10^{-9}$	$982069 \cdot 10^{-9}$	$393214 \cdot 10^{-8}$	3.84146	5.02389	6.63490	7.87944
2	0.0100251	0.0201007	0.0506356	0.102587	5.99146	7.37776	9.21034	10.5966
3	0.0717218	0.114832	0.215795	0.351846	7.81473	9.34840	11.3449	12.8382
4	0.206989	0.297109	0.484419	0.710723	9.48773	11.1433	13.2767	14.8603
5	0.411742	0.554298	0.831212	1.145476	11.0705	12.8325	15.0863	16.7496
6	0.675727	0.872090	1.23734	1.63538	12.5916	14.4494	16.8119	18.5476
7	0.989256	1.239043	1.68987	2.16735	14.0671	16.0128	18.4753	20.2777
8	1.34441	1.64650	2.17973	2.73264	15.5073	17.5345	20.0902	21.9550
9	1.73493	2.08790	2.70039	3.32511	16.9190	19.0228	21.6660	23.5894
10	2.15586	2.55821	3.24697	3.94030	18.3070	20.4832	23.2093	25.1882
11	2.60322	3.05348	3.81575	4.57481	19.6751	21.9200	24.7250	26.7568
12	3.07382	3.57057	4.40379	5.22603	21.0261	23.3367	26.2170	28.2995
13	3.56503	4.10692	5.00875	5.89186	22.3620	24.7356	27.6882	29.8195
14	4.07467	4.66043	5.62873	6.57063	23.6848	26.1189	29.1412	31.3194
15	4.60092	5.22935	6.26214	7.26094	24.9958	27.4884	30.5779	32.8013
16	5.14221	5.81221	6.90766	7.96165	26.2962	28.8454	31.9999	34.2672
17	5.69722	6.40776	7.56419	8.67176	27.5871	30.1910	33.4087	35.7185
18	6.26480	7.01491	8.23075	9.39046	28.8693	31.5264	34.8053	37.1565
19	6.84397	7.63273	8.90652	10.1170	30.1435	32.8523	36.1909	38.5823
20	7.43384	8.26040	9.59078	10.8508	31.4104	34.1696	37.5662	39.9968
21	8.03365	8.89720	10.28293	11.5913	32.6706	35.4789	38.9322	41.4011
22	8.64272	9.54249	10.9823	12.3380	33.9244	36.7807	40.2894	42.7957
23	9.26043	10.19567	11.6886	13.0905	35.1725	38.0756	41.6384	44.1813
24	9.88623	10.8564	12.4012	13.8484	36.4150	39.3641	42.9798	45.5585
25	10.5197	11.5240	13.1197	14.6114	37.6525	40.6465	44.3141	46.9279
26	11.1602	12.1981	13.8439	15.3792	38.8851	41.9232	45.6417	48.2899
27	11.8076	12.8785	14.5734	16.1514	40.1133	43.1945	46.9629	49.6449
28	12.4613	13.5647	15.3079	16.9279	41.3371	44.4608	48.2782	50.9934
29	13.1211	14.2565	16.0471	17.7084	42.5570	45.7223	49.5879	52.3356
30	13.7867	14.9535	16.7908	18.4927	43.7730	46.9792	50.8922	53.6720
40	20.7065	22.1643	24.4330	26.5093	55.7585	59.3417	63.6907	66.7660
50	27.9907	29.7067	32.3574	34.7643	67.5048	71.4202	76.1539	79.4900
60	35.5345	37.4849	40.4817	43.1880	79.0819	83.2977	88.3794	91.9517
70	43.2752	45.4417	48.7576	51.7393	90.5312	95.0232	100.425	104.215
80	51.1719	53.5401	57.1532	60.3915	101.879	106.629	112.329	116.321
90	59.1963	61.7541	65.6466	69.1260	113.145	118.136	124.116	128.299
100	67.3276	70.0649	74.2219	77.9295	124.342	129.561	135.807	140.169

* Source: Pearson and Hartley, 1966. Reprinted by permission of author and publisher. Use of this table is explained in Section 7.6

Degrees of freedom for denominator	Degrees of freedom for numerator																		
	1	2	3	4	5	6	7	8	9	10	12	15	20	24	30	40	60	120	∞
1	161.4	199.5	215.7	224.6	230.2	234.0	236.8	238.9	240.5	241.9	243.9	245.9	248.0	249.1	250.1	251.1	252.2	253.3	254.3
2	18.51	19.00	19.16	19.25	19.30	19.33	19.35	19.37	19.38	19.40	19.41	19.43	19.45	19.45	19.46	19.47	19.48	19.49	19.50
3	10.13	9.55	9.28	9.12	9.01	8.94	8.89	8.85	8.81	8.79	8.74	8.70	8.66	8.64	8.62	8.59	8.57	8.55	8.53
4	7.71	6.94	6.59	6.39	6.26	6.16	6.09	6.04	6.00	5.96	5.91	5.86	5.80	5.77	5.75	5.72	5.69	5.66	5.63
5	6.61	5.79	5.41	5.19	5.05	4.95	4.88	4.82	4.77	4.74	4.68	4.62	4.56	4.53	4.50	4.46	4.43	4.40	4.36
6	5.99	5.14	4.76	4.53	4.39	4.28	4.21	4.15	4.10	4.06	4.00	3.94	3.87	3.84	3.81	3.77	3.74	3.70	3.67
7	5.59	4.74	4.35	4.12	3.97	3.87	3.79	3.73	3.68	3.64	3.57	3.51	3.44	3.41	3.38	3.34	3.30	3.27	3.23
8	5.32	4.46	4.07	3.84	3.69	3.58	3.50	3.44	3.39	3.35	3.28	3.22	3.15	3.12	3.08	3.04	3.01	2.97	2.93
9	5.12	4.26	3.86	3.63	3.48	3.37	3.29	3.23	3.18	3.14	3.07	3.01	2.94	2.90	2.86	2.83	2.79	2.75	2.71
10	4.96	4.10	3.71	3.48	3.33	3.22	3.14	3.07	3.02	2.98	2.91	2.85	2.77	2.74	2.70	2.66	2.62	2.58	2.54
11	4.84	3.98	3.59	3.36	3.20	3.09	3.01	2.95	2.90	2.85	2.79	2.72	2.65	2.61	2.57	2.53	2.49	2.45	2.40
12	4.75	3.89	3.49	3.26	3.11	3.00	2.91	2.85	2.80	2.75	2.69	2.62	2.54	2.51	2.47	2.43	2.38	2.34	2.30
13	4.67	3.81	3.41	3.18	3.03	2.92	2.83	2.77	2.71	2.67	2.60	2.53	2.46	2.42	2.38	2.34	2.30	2.25	2.21
14	4.60	3.74	3.34	3.11	2.96	2.85	2.76	2.70	2.65	2.60	2.53	2.46	2.39	2.35	2.31	2.27	2.22	2.18	2.13
15	4.54	3.68	3.29	3.06	2.90	2.79	2.71	2.64	2.59	2.54	2.48	2.40	2.33	2.29	2.25	2.20	2.16	2.11	2.07
16	4.49	3.63	3.24	3.01	2.85	2.74	2.66	2.59	2.54	2.49	2.42	2.35	2.28	2.24	2.19	2.15	2.11	2.06	2.01
17	4.45	3.59	3.20	2.96	2.81	2.70	2.61	2.55	2.49	2.45	2.38	2.31	2.23	2.19	2.15	2.10	2.06	2.01	1.96
18	4.41	3.55	3.16	2.93	2.77	2.66	2.58	2.51	2.46	2.41	2.34	2.27	2.19	2.15	2.11	2.06	2.02	1.97	1.92
19	4.38	3.52	3.13	2.90	2.74	2.63	2.54	2.48	2.42	2.38	2.31	2.23	2.16	2.11	2.07	2.03	1.98	1.93	1.88
20	4.35	3.49	3.10	2.87	2.71	2.60	2.51	2.45	2.39	2.35	2.28	2.20	2.12	2.08	2.04	1.99	1.95	1.90	1.84
21	4.32	3.47	3.07	2.84	2.68	2.57	2.49	2.42	2.37	2.32	2.25	2.18	2.10	2.05	2.01	1.96	1.92	1.87	1.81
22	4.30	3.44	3.05	2.82	2.66	2.55	2.46	2.40	2.34	2.30	2.23	2.15	2.07	2.03	1.98	1.94	1.89	1.84	1.78
23	4.28	3.42	3.03	2.80	2.64	2.53	2.44	2.37	2.32	2.27	2.20	2.13	2.05	2.01	1.96	1.91	1.86	1.81	1.76
24	4.26	3.40	3.01	2.78	2.62	2.51	2.42	2.36	2.30	2.25	2.18	2.11	2.03	1.98	1.94	1.89	1.84	1.79	1.73
25	4.24	3.39	2.99	2.76	2.60	2.49	2.40	2.34	2.28	2.24	2.16	2.09	2.01	1.96	1.92	1.87	1.82	1.77	1.71
26	4.23	3.37	2.98	2.74	2.59	2.47	2.39	2.32	2.27	2.22	2.15	2.07	1.99	1.95	1.90	1.85	1.80	1.75	1.69
27	4.21	3.35	2.96	2.73	2.57	2.46	2.37	2.31	2.25	2.20	2.13	2.06	1.97	1.93	1.88	1.84	1.79	1.73	1.67
28	4.20	3.34	2.95	2.71	2.56	2.45	2.36	2.29	2.24	2.19	2.12	2.04	1.96	1.91	1.87	1.82	1.77	1.71	1.65
29	4.18	3.33	2.93	2.70	2.55	2.43	2.35	2.28	2.22	2.18	2.10	2.03	1.94	1.90	1.85	1.81	1.75	1.70	1.64
30	4.17	3.32	2.92	2.69	2.53	2.42	2.33	2.27	2.21	2.16	2.09	2.01	1.93	1.89	1.84	1.79	1.74	1.68	1.62
40	4.08	3.23	2.84	2.61	2.45	2.34	2.25	2.18	2.12	2.08	2.00	1.92	1.84	1.79	1.74	1.69	1.64	1.58	1.51
60	4.00	3.15	2.76	2.53	2.37	2.25	2.17	2.10	2.04	1.99	1.92	1.84	1.75	1.70	1.65	1.59	1.53	1.47	1.39
120	3.92	3.07	2.68	2.45	2.29	2.17	2.09	2.02	1.96	1.91	1.83	1.75	1.66	1.61	1.55	1.50	1.43	1.35	1.25
∞	3.84	3.00	2.60	2.37	2.21	2.10	2.01	1.94	1.88	1.83	1.75	1.67	1.57	1.52	1.46	1.39	1.32	1.22	1.00

* Source: Pearson and Hartley, 1966. Reprinted by permission of author and publisher. For directions on the use of this table, see Section 7-7.

Table A-8 Critical values of the F distribution (continued), $\alpha = .025$ for one-tail test, $\alpha = .05$ for two-tail test *

Degrees of freedom for denominator	Degrees of freedom for numerator																		
	1	2	3	4	5	6	7	8	9	10	12	15	20	24	30	40	60	120	∞
1	647.8	799.5	864.2	899.6	921.8	937.1	948.2	956.7	963.3	968.6	976.7	984.9	993.1	997.2	1001	1006	1010	1014	1018
2	38.51	39.00	39.17	39.25	39.30	39.33	39.36	39.37	39.39	39.40	39.41	39.43	39.45	39.46	39.46	39.47	39.48	39.49	39.50
3	17.44	16.04	15.44	15.10	14.88	14.73	14.62	14.54	14.47	14.42	14.34	14.25	14.17	14.12	14.08	14.04	13.99	13.95	13.90
4	12.22	10.65	9.98	9.60	9.36	9.20	9.07	8.98	8.90	8.84	8.75	8.66	8.56	8.51	8.46	8.41	8.36	8.31	8.26
5	10.01	8.43	7.76	7.39	7.15	6.98	6.85	6.76	6.68	6.62	6.52	6.43	6.33	6.28	6.23	6.18	6.12	6.07	6.02
6	8.81	7.26	6.60	6.23	5.99	5.82	5.70	5.60	5.52	5.46	5.37	5.27	5.17	5.12	5.07	5.01	4.96	4.90	4.85
7	8.07	6.54	5.89	5.52	5.29	5.12	4.99	4.90	4.82	4.76	4.67	4.57	4.47	4.42	4.36	4.31	4.25	4.20	4.14
8	7.57	6.06	5.42	5.05	4.82	4.65	4.53	4.43	4.36	4.30	4.20	4.10	4.00	3.95	3.89	3.84	3.78	3.73	3.67
9	7.21	5.71	5.08	4.72	4.48	4.32	4.20	4.10	4.03	3.96	3.87	3.77	3.67	3.61	3.56	3.51	3.45	3.39	3.33
10	6.94	5.46	4.83	4.47	4.24	4.07	3.95	3.85	3.78	3.72	3.62	3.52	3.42	3.37	3.31	3.26	3.20	3.14	3.08
11	6.72	5.26	4.63	4.28	4.04	3.88	3.76	3.66	3.59	3.53	3.43	3.33	3.23	3.17	3.12	3.06	3.00	2.94	2.88
12	6.55	5.10	4.47	4.12	3.89	3.73	3.61	3.51	3.44	3.37	3.28	3.18	3.07	3.02	2.96	2.91	2.85	2.79	2.72
13	6.41	4.97	4.35	4.00	3.77	3.60	3.48	3.39	3.31	3.25	3.15	3.05	2.95	2.89	2.84	2.78	2.72	2.66	2.60
14	6.30	4.86	4.24	3.89	3.66	3.50	3.38	3.29	3.21	3.15	3.05	2.95	2.84	2.79	2.73	2.67	2.61	2.55	2.49
15	6.20	4.77	4.15	3.80	3.58	3.41	3.29	3.20	3.12	3.06	2.96	2.86	2.76	2.70	2.64	2.59	2.52	2.46	2.40
16	6.12	4.69	4.08	3.73	3.50	3.34	3.22	3.12	3.05	2.99	2.89	2.79	2.68	2.63	2.57	2.51	2.45	2.38	2.32
17	6.04	4.62	4.01	3.66	3.44	3.28	3.16	3.06	2.98	2.92	2.82	2.72	2.62	2.56	2.50	2.44	2.38	2.32	2.25
18	5.98	4.56	3.95	3.61	3.38	3.22	3.10	3.01	2.93	2.87	2.77	2.67	2.56	2.50	2.44	2.38	2.32	2.26	2.19
19	5.92	4.51	3.90	3.56	3.33	3.17	3.05	2.96	2.88	2.82	2.72	2.62	2.51	2.45	2.39	2.33	2.27	2.20	2.13
20	5.87	4.46	3.86	3.51	3.29	3.13	3.01	2.91	2.84	2.77	2.68	2.57	2.46	2.41	2.35	2.29	2.22	2.16	2.09
21	5.83	4.42	3.82	3.48	3.25	3.09	2.97	2.87	2.80	2.73	2.64	2.53	2.42	2.37	2.31	2.25	2.18	2.11	2.04
22	5.79	4.38	3.78	3.44	3.22	3.05	2.93	2.84	2.76	2.70	2.60	2.50	2.39	2.33	2.27	2.21	2.14	2.08	2.00
23	5.75	4.35	3.75	3.41	3.18	3.02	2.90	2.81	2.73	2.67	2.57	2.47	2.36	2.30	2.24	2.18	2.11	2.04	1.97
24	5.72	4.32	3.72	3.38	3.15	2.99	2.87	2.78	2.70	2.64	2.54	2.44	2.33	2.27	2.21	2.15	2.08	2.01	1.94
25	5.69	4.29	3.69	3.35	3.13	2.97	2.85	2.75	2.68	2.61	2.51	2.41	2.30	2.24	2.18	2.12	2.05	1.98	1.91
26	5.66	4.27	3.67	3.33	3.10	2.94	2.82	2.73	2.65	2.59	2.49	2.39	2.28	2.22	2.16	2.09	2.03	1.95	1.88
27	5.63	4.24	3.65	3.31	3.08	2.92	2.80	2.71	2.63	2.57	2.47	2.36	2.25	2.19	2.13	2.07	2.00	1.93	1.85
28	5.61	4.22	3.63	3.29	3.06	2.90	2.78	2.69	2.61	2.55	2.45	2.34	2.23	2.17	2.11	2.05	1.98	1.91	1.83
29	5.59	4.20	3.61	3.27	3.04	2.88	2.76	2.67	2.59	2.53	2.43	2.32	2.21	2.15	2.09	2.03	1.96	1.89	1.81
30	5.57	4.18	3.59	3.25	3.03	2.87	2.75	2.65	2.57	2.51	2.41	2.31	2.20	2.14	2.07	2.01	1.94	1.87	1.79
40	5.42	4.05	3.46	3.13	2.90	2.74	2.62	2.53	2.45	2.39	2.29	2.18	2.07	2.01	1.94	1.88	1.80	1.72	1.64
60	5.29	3.93	3.34	3.01	2.79	2.63	2.51	2.41	2.33	2.27	2.17	2.06	1.94	1.88	1.82	1.74	1.67	1.58	1.48
120	5.15	3.80	3.23	2.89	2.67	2.52	2.39	2.30	2.22	2.16	2.05	1.94	1.82	1.76	1.69	1.61	1.53	1.43	1.31
∞	5.02	3.69	3.12	2.79	2.57	2.41	2.29	2.19	2.11	2.05	1.94	1.83	1.71	1.64	1.57	1.48	1.39	1.27	1.00

Degrees of freedom for numerator

Degrees of freedom for denominator	1	2	3	4	5	6	7	8	9	10	12	15	20	24	30	40	60	120	∞
1	4052	4999.5	5403	5625	5764	5859	5928	5981	6022	6056	6106	6157	6209	6235	6261	6287	6313	6339	6366
2	98.50	99.00	99.17	99.25	99.30	99.33	99.36	99.37	99.39	99.40	99.42	99.43	99.45	99.46	99.47	99.47	99.48	99.49	99.50
3	34.12	30.82	29.46	28.71	28.24	27.91	27.67	27.49	27.35	27.23	27.05	26.87	26.69	26.60	26.50	26.41	26.32	26.22	26.13
4	21.20	18.00	16.69	15.98	15.52	15.21	14.98	14.80	14.66	14.55	14.37	14.20	14.02	13.93	13.84	13.75	13.65	13.56	13.46
5	16.26	13.27	12.06	11.39	10.97	10.67	10.46	10.29	10.16	10.05	9.89	9.72	9.55	9.47	9.38	9.29	9.20	9.11	9.02
6	13.75	10.92	9.78	9.15	8.75	8.47	8.26	8.10	7.98	7.87	7.72	7.56	7.40	7.31	7.23	7.14	7.06	6.97	6.88
7	12.25	9.55	8.45	7.85	7.46	7.19	6.99	6.84	6.72	6.62	6.47	6.31	6.16	6.07	5.99	5.91	5.82	5.74	5.65
8	11.26	8.65	7.59	7.01	6.63	6.37	6.18	6.03	5.91	5.81	5.67	5.52	5.36	5.28	5.20	5.12	5.03	4.95	4.86
9	10.56	8.02	6.99	6.42	6.06	5.80	5.61	5.47	5.35	5.26	5.11	4.96	4.81	4.73	4.65	4.57	4.48	4.40	4.31
10	10.04	7.56	6.55	5.99	5.64	5.39	5.20	5.06	4.94	4.85	4.71	4.56	4.41	4.33	4.25	4.17	4.08	4.00	3.91
11	9.65	7.21	6.22	5.67	5.32	5.07	4.89	4.74	4.63	4.54	4.40	4.25	4.10	4.02	3.94	3.86	3.78	3.69	3.60
12	9.33	6.93	5.95	5.41	5.06	4.82	4.64	4.50	4.39	4.30	4.16	4.01	3.86	3.78	3.70	3.62	3.54	3.45	3.36
13	9.07	6.70	5.74	5.21	4.86	4.62	4.44	4.30	4.19	4.10	3.96	3.82	3.66	3.59	3.51	3.43	3.34	3.25	3.17
14	8.86	6.51	5.56	5.04	4.69	4.46	4.28	4.14	4.03	3.94	3.80	3.66	3.51	3.43	3.35	3.27	3.18	3.09	3.00
15	8.68	6.36	5.42	4.89	4.56	4.32	4.14	4.00	3.89	3.80	3.67	3.52	3.37	3.29	3.21	3.13	3.05	2.96	2.87
16	8.53	6.23	5.29	4.77	4.44	4.20	4.03	3.89	3.78	3.69	3.55	3.41	3.26	3.18	3.10	3.02	2.93	2.84	2.75
17	8.40	6.11	5.18	4.67	4.34	4.10	3.93	3.79	3.68	3.59	3.46	3.31	3.16	3.08	3.00	2.92	2.83	2.75	2.65
18	8.29	6.01	5.09	4.58	4.25	4.01	3.84	3.71	3.60	3.51	3.37	3.23	3.08	3.00	2.92	2.84	2.75	2.66	2.57
19	8.18	5.93	5.01	4.50	4.17	3.94	3.77	3.63	3.52	3.43	3.30	3.15	3.00	2.92	2.84	2.76	2.67	2.58	2.49
20	8.10	5.85	4.94	4.43	4.10	3.87	3.70	3.56	3.46	3.37	3.23	3.09	2.94	2.86	2.78	2.69	2.61	2.52	2.42
21	8.02	5.78	4.87	4.37	4.04	3.81	3.64	3.51	3.40	3.31	3.17	3.03	2.88	2.80	2.72	2.64	2.55	2.46	2.36
22	7.95	5.72	4.82	4.31	3.99	3.76	3.59	3.45	3.35	3.26	3.12	2.98	2.83	2.75	2.67	2.58	2.50	2.40	2.31
23	7.88	5.66	4.76	4.26	3.94	3.71	3.54	3.41	3.30	3.21	3.07	2.93	2.78	2.70	2.62	2.54	2.45	2.35	2.26
24	7.82	5.61	4.72	4.22	3.90	3.67	3.50	3.36	3.26	3.17	3.03	2.89	2.74	2.66	2.58	2.49	2.40	2.31	2.21
25	7.77	5.57	4.68	4.18	3.85	3.63	3.46	3.32	3.22	3.13	2.99	2.85	2.70	2.62	2.54	2.45	2.36	2.27	2.17
26	7.72	5.53	4.64	4.14	3.82	3.59	3.42	3.29	3.18	3.09	2.96	2.81	2.66	2.58	2.50	2.42	2.33	2.23	2.13
27	7.68	5.49	4.60	4.11	3.78	3.56	3.39	3.26	3.15	3.06	2.93	2.78	2.63	2.55	2.47	2.38	2.29	2.20	2.10
28	7.64	5.45	4.57	4.07	3.75	3.53	3.36	3.23	3.12	3.03	2.90	2.75	2.60	2.52	2.44	2.35	2.26	2.17	2.06
29	7.60	5.42	4.54	4.04	3.73	3.50	3.33	3.20	3.09	3.00	2.87	2.73	2.57	2.49	2.41	2.33	2.23	2.14	2.03
30	7.56	5.39	4.51	4.02	3.70	3.47	3.30	3.17	3.07	2.98	2.84	2.70	2.55	2.47	2.39	2.30	2.21	2.11	2.01
40	7.31	5.18	4.31	3.83	3.51	3.29	3.12	2.99	2.89	2.80	2.66	2.52	2.37	2.29	2.20	2.11	2.02	1.92	1.80
60	7.08	4.98	4.13	3.65	3.34	3.12	2.95	2.82	2.72	2.63	2.50	2.35	2.20	2.12	2.03	1.94	1.84	1.73	1.60
120	6.85	4.79	3.95	3.48	3.17	2.96	2.79	2.66	2.56	2.47	2.34	2.19	2.03	1.95	1.86	1.76	1.66	1.53	1.38
∞	6.63	4.61	3.78	3.32	3.02	2.80	2.64	2.51	2.41	2.32	2.18	2.04	1.88	1.79	1.70	1.59	1.47	1.32	1.00

* Source: Pearson and Hartley, 1966. Reprinted by permission of author and publisher. For directions on the use of this table, see Section 7-7.

Table A-8 Critical values of the F distribution (continued), $\alpha = .005$ for one-tail test, $\alpha = .01$ for two-tail test *

Degrees of freedom for denominator	Degrees of freedom for numerator																		
	1	2	3	4	5	6	7	8	9	10	12	15	20	24	30	40	60	120	∞
1	16211	20000	21615	22500	23056	23437	23715	23925	24091	24224	24426	24630	24836	24940	25044	25148	25253	25359	25465
2	198.5	199.0	199.2	199.2	199.3	199.3	199.4	199.4	199.4	199.4	199.4	199.4	199.4	199.4	199.5	199.5	199.5	199.5	199.5
3	55.55	49.80	47.47	46.19	45.39	44.84	44.43	44.13	43.88	43.69	43.39	43.08	42.78	42.62	42.47	42.31	42.15	41.99	41.83
4	31.33	26.28	24.26	23.15	22.46	21.97	21.62	21.35	21.14	20.97	20.70	20.44	20.17	20.03	19.89	19.75	19.61	19.47	19.32
5	22.78	18.31	16.53	15.56	14.94	14.51	14.20	13.96	13.77	13.62	13.38	13.15	12.90	12.78	12.66	12.53	12.40	12.27	12.14
6	18.63	14.54	12.92	12.03	11.46	11.07	10.79	10.57	10.39	10.25	10.03	9.81	9.59	9.47	9.36	9.24	9.12	9.00	8.88
7	16.24	12.40	10.88	10.05	9.52	9.16	8.89	8.68	8.51	8.38	8.18	7.97	7.75	7.65	7.53	7.42	7.31	7.19	7.08
8	14.69	11.04	9.60	8.81	8.30	7.95	7.69	7.50	7.34	7.21	7.01	6.81	6.61	6.50	6.40	6.29	6.18	6.06	5.95
9	13.61	10.11	8.72	7.96	7.47	7.13	6.88	6.69	6.54	6.42	6.23	6.03	5.83	5.73	5.62	5.52	5.41	5.30	5.19
10	12.83	9.43	8.08	7.34	6.87	6.54	6.30	6.12	5.97	5.85	5.66	5.47	5.27	5.17	5.07	4.97	4.86	4.75	4.64
11	12.23	8.91	7.60	6.88	6.42	6.10	5.86	5.68	5.54	5.42	5.24	5.05	4.86	4.76	4.65	4.55	4.44	4.34	4.23
12	11.75	8.51	7.23	6.52	6.07	5.76	5.52	5.35	5.20	5.09	4.91	4.72	4.53	4.43	4.33	4.23	4.12	4.01	3.90
13	11.37	8.19	6.93	6.23	5.79	5.48	5.25	5.08	4.94	4.82	4.64	4.46	4.27	4.17	4.07	3.97	3.87	3.76	3.65
14	11.06	7.92	6.68	6.00	5.56	5.26	5.03	4.86	4.72	4.60	4.43	4.25	4.06	3.96	3.86	3.76	3.66	3.55	3.44
15	10.80	7.70	6.48	5.80	5.37	5.07	4.85	4.67	4.54	4.42	4.25	4.07	3.88	3.79	3.69	3.58	3.48	3.37	3.26
16	10.58	7.51	6.30	5.64	5.21	4.91	4.69	4.52	4.38	4.27	4.10	3.92	3.73	3.64	3.54	3.44	3.33	3.22	3.11
17	10.38	7.35	6.16	5.50	5.07	4.78	4.56	4.39	4.25	4.14	3.97	3.79	3.61	3.51	3.41	3.31	3.21	3.10	2.98
18	10.22	7.21	6.03	5.37	4.96	4.66	4.44	4.28	4.14	4.03	3.86	3.68	3.50	3.40	3.30	3.20	3.10	2.99	2.87
19	10.07	7.09	5.92	5.27	4.85	4.56	4.34	4.18	4.04	3.93	3.76	3.59	3.40	3.31	3.21	3.11	3.00	2.89	2.78
20	9.94	6.99	5.82	5.17	4.76	4.47	4.26	4.09	3.96	3.85	3.68	3.50	3.32	3.22	3.12	3.02	2.92	2.81	2.69
21	9.83	6.89	5.73	5.09	4.68	4.39	4.18	4.01	3.88	3.77	3.60	3.43	3.24	3.15	3.05	2.95	2.84	2.73	2.61
22	9.73	6.81	5.65	5.02	4.61	4.32	4.11	3.94	3.81	3.70	3.54	3.36	3.18	3.08	2.98	2.88	2.77	2.66	2.55
23	9.63	6.73	5.58	4.95	4.54	4.26	4.05	3.88	3.75	3.64	3.47	3.30	3.12	3.02	2.92	2.82	2.71	2.60	2.48
24	9.55	6.66	5.52	4.89	4.49	4.20	3.99	3.83	3.69	3.59	3.42	3.25	3.06	2.97	2.87	2.77	2.66	2.55	2.43
25	9.48	6.60	5.46	4.84	4.43	4.15	3.94	3.78	3.64	3.54	3.37	3.20	3.01	2.92	2.82	2.72	2.61	2.50	2.38
26	9.41	6.54	5.41	4.79	4.38	4.10	3.89	3.73	3.60	3.49	3.33	3.15	2.97	2.87	2.77	2.67	2.56	2.45	2.33
27	9.34	6.49	5.36	4.74	4.34	4.06	3.85	3.69	3.56	3.45	3.28	3.11	2.93	2.83	2.73	2.63	2.52	2.41	2.29
28	9.28	6.44	5.32	4.70	4.30	4.02	3.81	3.65	3.52	3.41	3.25	3.07	2.89	2.79	2.69	2.59	2.48	2.37	2.25
29	9.23	6.40	5.28	4.66	4.26	3.98	3.77	3.61	3.48	3.38	3.21	3.04	2.86	2.76	2.66	2.56	2.45	2.33	2.21
30	9.18	6.35	5.24	4.62	4.23	3.95	3.74	3.58	3.45	3.34	3.18	3.01	2.82	2.73	2.63	2.52	2.42	2.30	2.18
40	8.83	6.07	4.98	4.37	3.99	3.71	3.51	3.35	3.22	3.12	2.95	2.78	2.60	2.50	2.40	2.30	2.18	2.06	1.93
60	8.49	5.79	4.73	4.14	3.76	3.49	3.29	3.13	3.01	2.90	2.74	2.57	2.39	2.29	2.19	2.08	1.96	1.83	1.69
120	8.18	5.54	4.50	3.92	3.55	3.28	3.09	2.93	2.81	2.71	2.54	2.37	2.19	2.09	1.98	1.87	1.75	1.61	1.43
∞	7.88	5.30	4.28	3.72	3.35	3.09	2.90	2.74	2.62	2.52	2.36	2.19	2.00	1.90	1.79	1.67	1.53	1.36	1.00

Table A-9 *Critical values of the Kruskal–Wallis H statistic**

N_1	N_2	N_3	$H_{.05}$	$H_{.01}$
2	1	1		
2	2	1		
2	2	2		
3	1	1		
3	2	1		
3	2	2	4.7143	
3	3	1	5.1429	
3	3	2	5.3611	
3	3	3	5.6000	7.2000
4	1	1		
4	2	1		
4	2	2	5.3333	
4	3	1	5.2083	
4	3	2	5.4444	6.4444
4	3	3	5.7273	6.7455
4	4	1	4.9667	6.6667
4	4	2	5.4545	7.0364
4	4	3	5.5985	7.1439
4	4	4	5.6923	7.6538
5	1	1		
5	2	1	5.0000	
5	2	2	5.1600	6.5333
5	3	1	4.9600	
5	3	2	5.2509	6.8218
5	3	3	5.6485	7.0788
5	4	1	4.9855	6.9545
5	4	2	5.2682	7.1182
5	4	3	5.6308	7.4449
5	4	4	5.6176	7.7604
5	5	1	5.1273	7.3091
5	5	2	5.3385	7.2692
5	5	3	5.7055	7.5429
5	5	4	5.6429	7.7914
5	5	5	5.7800	7.9800

* Adapted and abridged from Kruskal and Wallis (1952, 1953) with the kind permission of authors and publisher. An observed value of *H* is significant at the indicated level if it is greater than the tabulated value. Where blanks appear, significance cannot be achieved. Instructions for use of the table may be found in Section 14-6. More detailed tables can be found in the original sources and in Siegel (1956).

Table A-10 *Critical values of the Mann–Whitney U. $\alpha = .005$ for one-tail test, $\alpha = .01$ for two-tail test (Reject H_0 if observed U is less than or equal to the tabulated value. For calculations, see Section 14-3.)**

N of larger sample	N of smaller sample																			
	1	2	3	4	5	6	7	8	9	10	11	12	13	14	15	16	17	18	19	20
1	—																			
2	—	—																		
3	—	—	—																	
4	—	—	—	—																
5	—	—	—	—	0															
6	—	—	—	0	1	2														
7	—	—	—	0	1	3	4													
8	—	—	—	1	2	4	6	7												
9	—	—	0	1	3	5	7	9	11											
10	—	—	0	2	4	6	9	11	13	16										
11	—	—	0	2	5	7	10	13	16	18	21									
12	—	—	1	3	6	9	12	15	18	21	24	27								
13	—	—	1	3	7	10	13	17	20	24	27	31	34							
14	—	—	1	4	7	11	15	18	22	26	30	34	38	42						
15	—	—	2	5	8	12	16	20	24	29	33	37	42	46	51					
16	—	—	2	5	9	13	18	22	27	31	36	41	45	50	55	60				
17	—	—	2	6	10	15	19	24	29	34	39	44	49	54	60	65	70			
18	—	—	2	6	11	16	21	26	31	37	42	47	53	58	64	70	75	81		
19	—	0	3	7	12	17	22	28	33	39	45	51	57	63	69	74	81	87	93	
20	—	0	3	8	13	18	24	30	36	42	48	54	60	67	73	79	86	92	99	105
21	—	0	3	8	14	19	25	32	38	44	51	58	64	71	78	84	91	98	105	112
22	—	0	4	9	14	21	27	34	40	47	54	61	68	75	82	89	96	104	111	118
23	—	0	4	9	15	22	29	35	43	50	57	64	72	79	87	94	102	109	117	125
24	—	0	4	10	16	23	30	37	45	52	60	68	75	83	91	99	107	115	123	131
25	—	0	5	10	17	24	32	39	47	55	63	71	79	87	96	104	112	121	129	138
26	—	0	5	11	18	25	33	41	49	58	66	74	83	92	100	109	118	127	135	144
27	—	1	5	12	19	27	35	43	52	60	69	78	87	96	105	114	123	132	142	151
28	—	1	5	12	20	28	36	45	54	63	72	81	91	100	109	119	128	138	148	157
29	—	1	6	13	21	29	38	47	56	66	75	85	94	104	114	124	134	144	154	164
30	—	1	6	13	22	30	40	49	58	68	78	88	98	108	119	129	139	150	160	170
31	—	1	6	14	22	32	41	51	61	71	81	92	102	113	123	134	145	155	166	177
32	—	1	7	14	23	33	43	53	63	74	84	95	106	117	128	139	150	161	172	184
33	—	1	7	15	24	34	44	55	65	76	87	98	110	121	132	144	155	167	179	190
34	—	1	7	16	25	35	46	57	68	79	90	102	113	125	137	149	161	173	185	197
35	—	1	8	16	26	37	47	59	70	82	93	105	117	129	142	154	166	179	191	203
36	—	1	8	17	27	38	49	60	72	84	96	109	121	134	146	159	172	184	197	210
37	—	1	8	17	28	39	51	62	75	87	99	112	125	138	151	164	177	190	203	217
38	—	1	9	18	29	40	52	64	77	90	102	116	129	142	155	169	182	196	210	223
39	—	2	9	19	30	41	54	66	79	92	106	119	133	146	160	174	188	202	216	230
40	—	2	9	19	31	43	55	68	81	95	109	122	136	150	165	179	193	208	222	237

* Source: R. C. Milton, 1964. Used by permission of the author.

N of larger sample	N of smaller sample																			
	1	2	3	4	5	6	7	8	9	10	11	12	13	14	15	16	17	18	19	20
1	—																			
2	—	—																		
3	—	—	—																	
4	—	—	—	—																
5	—	—	—	—	0	1														
6	—	—	—	1	2	3														
7	—	—	0	1	3	4	6													
8	—	—	0	2	4	6	7	9												
9	—	—	1	3	5	7	9	11	14											
10	—	—	1	3	6	8	11	13	16	19										
11	—	—	1	4	7	9	12	15	18	22	25									
12	—	—	2	5	8	11	14	17	21	24	28	31								
13	—	0	2	5	9	12	16	20	23	27	31	35	39							
14	—	0	2	6	10	13	17	22	26	30	34	38	43	47						
15	—	0	3	7	11	15	19	24	28	33	37	42	47	51	56					
16	—	0	3	7	12	16	21	26	31	36	41	46	51	56	61	66				
17	—	0	4	8	13	18	23	28	33	38	44	49	55	60	66	71	77			
18	—	0	4	9	14	19	24	30	36	41	47	53	59	65	70	76	82	88		
19	—	1	4	9	15	20	26	32	38	44	50	56	63	69	75	82	88	94	101	
20	—	1	5	10	16	22	28	34	40	47	53	60	67	73	80	87	93	100	107	114
21	—	1	5	11	17	23	30	36	43	50	57	64	71	78	85	92	99	106	113	121
22	—	1	6	11	18	24	31	38	45	53	60	67	75	82	90	97	105	112	120	127
23	—	1	6	12	19	26	33	40	48	55	63	71	79	87	94	102	110	118	126	134
24	—	1	6	13	20	27	35	42	50	58	66	75	83	91	99	108	116	124	133	141
25	—	1	7	13	21	29	36	45	53	61	70	78	87	95	104	113	122	130	139	148
26	—	1	7	14	22	30	38	47	55	64	73	82	91	100	109	118	127	136	146	155
27	—	2	7	15	23	31	40	49	58	67	76	85	95	104	114	123	133	142	152	162
28	—	2	8	16	24	33	42	51	60	70	79	89	99	109	119	129	139	149	159	169
29	—	2	8	16	25	34	43	53	63	73	83	93	103	113	123	134	144	155	165	176
30	—	2	9	17	26	35	45	55	65	76	86	96	107	118	128	139	150	161	172	182
31	—	2	9	18	27	37	47	57	68	78	89	100	111	122	133	144	156	167	178	189
32	—	2	9	18	28	38	49	59	70	81	92	104	115	127	138	150	161	173	185	196
33	—	2	10	19	29	40	50	61	73	84	96	107	119	131	143	155	167	179	191	203
34	—	3	10	20	30	41	52	64	75	87	99	111	123	135	148	160	173	185	198	210
35	—	3	11	20	31	42	54	66	78	90	102	115	127	140	153	165	178	191	204	217
36	—	3	11	21	32	44	56	68	80	93	106	118	131	144	158	171	184	197	211	224
37	—	3	11	22	33	45	57	70	83	96	109	122	135	149	162	176	190	203	217	231
38	—	3	12	22	34	46	59	72	85	99	112	126	139	153	167	181	195	209	224	238
39	—	3	12	23	35	48	61	74	88	101	115	129	144	158	172	187	201	216	230	245
40	—	3	13	24	36	49	63	76	90	104	119	133	148	162	177	192	207	222	237	252

Table A-10 *Critical values of the Mann–Whitney U (continued).* α = .025 *for one-tail test,* α = .05 *for two-tail test*

N of larger sample	\multicolumn{20}{c}{N of smaller sample}
	1	2	3	4	5	6	7	8	9	10	11	12	13	14	15	16	17	18	19	20
1	—																			
2	—	—																		
3	—	—	—																	
4	—	—	—	0																
5	—	—	0	1	2															
6	—	—	1	2	3	5														
7	—	—	1	3	5	6	8													
8	—	0	2	4	6	8	10	13												
9	—	0	2	4	7	10	12	15	17											
10	—	0	3	5	8	11	14	17	20	23										
11	—	0	3	6	9	13	16	19	23	26	30									
12	—	1	4	7	11	14	18	22	26	29	33	37								
13	—	1	4	8	12	16	20	24	28	33	37	41	45							
14	—	1	5	9	13	17	22	26	31	36	40	45	50	55						
15	—	1	5	10	14	19	24	29	34	39	44	49	54	59	64					
16	—	1	6	11	15	21	26	31	37	42	47	53	59	64	70	75				
17	—	2	6	11	17	22	28	34	39	45	51	57	63	69	75	81	87			
18	—	2	7	12	18	24	30	36	42	48	55	61	67	74	80	86	93	99		
19	—	2	7	13	19	25	32	38	45	52	58	65	72	78	85	92	99	106	113	
20	—	2	8	14	20	27	34	41	48	55	62	69	76	83	90	98	105	112	119	127
21	—	3	8	15	22	29	36	43	50	58	65	73	80	88	96	103	111	119	126	134
22	—	3	9	16	23	30	38	45	53	61	69	77	85	93	101	109	117	125	133	141
23	—	3	9	17	24	32	40	48	56	64	73	81	89	98	106	115	123	132	140	149
24	—	3	10	17	25	33	42	50	59	67	76	85	94	102	111	120	129	138	147	156
25	—	3	10	18	27	35	44	53	62	71	80	89	98	107	117	126	135	145	154	163
26	—	4	11	19	28	37	46	55	64	74	83	93	102	112	122	132	141	151	161	171
27	—	4	11	20	29	38	48	57	67	77	87	97	107	117	127	137	147	158	168	178
28	—	4	12	21	30	40	50	60	70	80	90	101	111	122	132	143	154	164	175	186
29	—	4	13	22	32	42	52	62	73	83	94	105	116	127	138	149	160	171	182	193
30	—	5	13	23	33	43	54	65	76	87	98	109	120	131	143	154	166	177	189	200
31	—	5	14	24	34	45	56	67	78	90	101	113	125	136	148	160	172	184	196	208
32	—	5	14	24	35	46	58	69	81	93	105	117	129	141	153	166	178	190	203	215
33	—	5	15	25	37	48	60	72	84	96	108	121	133	146	159	171	184	197	210	222
34	—	5	15	26	38	50	62	74	87	99	112	125	138	151	164	177	190	203	217	230
35	—	6	16	27	39	51	64	77	89	103	116	129	142	156	169	183	196	210	224	237
36	—	6	16	28	40	53	66	79	92	106	119	133	147	161	174	188	202	216	231	245
37	—	6	17	29	41	55	68	81	95	109	123	137	151	165	180	194	209	223	238	252
38	—	6	17	30	43	56	70	84	98	112	127	141	156	170	185	200	215	230	245	259
39	0	7	18	31	44	58	72	86	101	115	130	145	160	175	190	206	221	236	252	267
40	0	7	18	31	45	59	74	89	103	119	134	149	165	180	196	211	227	243	258	274

Table A-10 *Critical values of the Mann–Whitney U (continued).* $\alpha = .05$ *for one-tail test,* $\alpha = .10$ *for two-tail test*

N of larger sample	N of smaller sample																			
	1	2	3	4	5	6	7	8	9	10	11	12	13	14	15	16	17	18	19	20
1	—																			
2	—	—																		
3	—	—	0																	
4	—	—	0	1																
5	—	0	1	2	4															
6	—	0	2	3	5	7														
7	—	0	2	4	6	8	11													
8	—	1	3	5	8	10	13	15												
9	—	1	4	6	9	12	15	18	21											
10	—	1	4	7	11	14	17	20	24	27										
11	—	1	5	8	12	16	19	23	27	31	34									
12	—	2	5	9	13	17	21	26	30	34	38	42								
13	—	2	6	10	15	19	24	28	33	37	42	47	51							
14	—	3	7	11	16	21	26	31	36	41	46	51	56	61						
15	—	3	7	12	18	23	28	33	39	44	50	55	61	66	72					
16	—	3	8	14	19	25	30	36	42	48	54	60	65	71	77	83				
17	—	3	9	15	20	26	33	39	45	51	57	64	70	77	83	89	96			
18	—	4	9	16	22	28	35	41	48	55	61	68	75	82	88	95	102	109		
19	0	4	10	17	23	30	37	44	51	58	65	72	80	87	94	101	109	116	123	
20	0	4	11	18	25	32	39	47	54	62	69	77	84	92	100	107	115	123	130	138
21	0	5	11	19	26	34	41	49	57	65	73	81	89	97	105	113	121	130	138	146
22	0	5	12	20	28	36	44	52	60	68	77	85	94	102	111	119	128	136	145	154
23	0	5	13	21	29	37	46	54	63	72	81	90	98	107	116	125	134	143	152	161
24	0	6	13	22	30	39	48	57	66	75	85	94	103	113	122	131	141	150	160	169
25	0	6	14	23	32	41	50	60	69	79	89	98	108	118	128	137	147	157	167	177
26	0	6	15	24	33	43	53	62	72	82	92	103	113	123	133	143	154	164	174	185
27	0	7	15	25	35	45	55	65	75	86	96	107	117	128	139	149	160	171	182	192
28	0	7	16	26	36	46	57	68	78	89	100	111	122	133	144	156	167	178	189	200
29	0	7	17	27	38	48	59	70	82	93	104	116	127	138	150	162	173	185	196	208
30	0	7	17	28	39	50	61	73	85	96	108	120	132	144	156	168	180	192	204	216
31	0	8	18	29	40	52	64	76	88	100	112	124	136	149	161	174	186	199	211	224
32	0	8	19	30	42	54	66	78	91	103	116	128	141	154	167	180	193	206	218	231
33	0	8	19	31	43	56	68	81	94	107	120	133	146	159	172	186	199	212	226	239
34	0	9	20	32	45	57	70	84	97	110	124	137	151	164	178	192	206	219	233	247
35	0	9	21	33	46	59	73	86	100	114	128	141	156	170	184	198	212	226	241	255
36	0	9	21	34	48	61	75	89	103	117	131	146	160	175	189	204	219	233	248	263
37	0	10	22	35	49	63	77	91	106	121	135	150	165	180	195	210	225	240	255	271
38	0	10	23	36	50	65	79	94	109	124	139	154	170	185	201	216	232	247	263	278
39	1	10	23	38	52	67	82	97	112	128	143	159	175	190	206	222	238	254	270	286
40	1	11	24	39	53	68	84	99	115	131	147	163	179	196	212	228	245	261	278	294

Table A-11 *Critical values of Wilcoxon T**

Number of pairs (N)	α for two-tail test				MAX
	.10	.05	.02	.01	
	α for one-tail test				
	.05	.025	.01	.005	
4					10
5	0				15
6	2	0			21
7	3	2	0		28
8	5	3	1	0	36
9	8	5	3	1	45
10	10	8	5	3	55
11	13	10	7	5	66
12	17	13	9	7	78
13	21	17	12	9	91
14	25	21	15	12	105
15	30	25	19	15	120
16	35	29	23	19	136
17	41	34	27	23	153
18	47	40	32	27	171
19	53	46	37	32	190
20	60	52	43	37	210
21	67	58	49	42	231
22	75	65	55	48	253
23	83	73	62	54	276
24	91	81	69	61	300
25	100	89	76	68	325
26	110	98	84	75	351
27	119	107	92	83	378
28	130	116	101	91	406
29	140	126	110	100	435
30	151	137	120	109	465
31	163	147	130	118	496
32	175	159	140	128	528
33	187	170	151	138	561
34	200	182	162	148	595
35	213	195	173	159	630
36	227	208	185	171	666
37	241	221	198	182	703
38	256	235	211	194	741
39	271	249	224	207	780
40	286	264	238	220	820
41	302	279	252	233	861
42	319	294	266	247	903
43	336	310	281	261	946
44	353	327	296	276	990
45	371	343	312	291	1035
46	389	361	328	307	1081
47	407	378	345	322	1128
48	426	396	362	339	1176
49	446	415	379	355	1225
50	466	434	397	373	1275
51	486	453	416	390	1326

* Reprinted from McCornack (1965) with the kind permission of author and publisher. Use of this table is discussed in Section 14-5.

Table A-11 *Critical values of Wilcoxon T (continued)*

Number of pairs (N)	α for two-tail test				
	.10	.05	.02	.01	
	α for one-tail test				
	.05	.025	.01	.005	MAX
52	507	473	434	408	1378
53	529	494	454	427	1431
54	550	514	473	445	1485
55	573	536	493	465	1540
56	595	557	514	484	1596
57	618	579	535	504	1653
58	642	602	556	525	1711
59	666	625	578	546	1770
60	690	648	600	567	1830
61	715	672	623	589	1891
62	741	697	646	611	1953
63	767	721	669	634	2016
64	793	747	693	657	2080
65	820	772	718	681	2145
66	847	798	742	705	2211
67	875	825	768	729	2278
68	903	852	793	754	2346
69	931	879	819	779	2415
70	960	907	846	805	2485
71	990	936	873	831	2556
72	1020	964	901	858	2628
73	1050	994	928	884	2701
74	1081	1023	957	912	2775
75	1112	1053	986	940	2850
76	1144	1084	1015	968	2926
77	1176	1115	1044	997	3003
78	1209	1147	1075	1026	3081
79	1242	1179	1105	1056	3160
80	1276	1211	1136	1086	3240
81	1310	1244	1168	1116	3321
82	1345	1277	1200	1147	3403
83	1380	1311	1232	1178	3486
84	1415	1345	1265	1210	3570
85	1451	1380	1298	1242	3655
86	1487	1415	1332	1275	3741
87	1524	1451	1366	1308	3828
88	1561	1487	1400	1342	3916
89	1599	1523	1435	1376	4005
90	1638	1560	1471	1410	4095
91	1676	1597	1507	1445	4186
92	1715	1635	1543	1480	4278
93	1755	1674	1580	1516	4371
94	1795	1712	1617	1552	4465
95	1836	1752	1655	1589	4560
96	1877	1791	1693	1626	4656
97	1918	1832	1731	1664	4753
98	1960	1872	1770	1702	4851
99	2003	1913	1810	1740	4950
100	2045	1955	1850	1779	5050

Table A-12 *Critical values of the Hartley F_{max} statistic**

N = Number of subjects per sample	$\alpha = .05$, J = number of samples										
	2	3	4	5	6	7	8	9	10	11	12
2	39.0	87.5	142	202	266	333	403	475	550	626	704
3	15.4	27.8	39.2	50.7	62.0	72.9	83.5	93.9	104	114	124
4	9.60	15.5	20.6	25.2	29.5	33.6	37.5	41.1	44.6	48.0	51.4
5	7.15	10.8	13.7	16.3	18.7	20.8	22.9	24.7	26.5	28.2	29.9
6	5.82	8.38	10.4	12.1	13.7	15.0	16.3	17.5	18.6	19.7	20.7
7	4.99	6.94	8.44	9.70	10.8	11.8	12.7	13.5	14.3	15.1	15.8
8	4.43	6.00	7.18	8.12	9.03	9.78	10.5	11.1	11.7	12.2	12.7
9	4.03	5.34	6.31	7.11	7.80	8.41	8.95	9.45	9.91	10.3	10.7
10	3.72	4.85	5.67	6.34	6.92	7.42	7.87	8.28	8.66	9.01	9.34
12	3.28	4.16	4.79	5.30	5.72	6.09	6.42	6.72	7.00	7.25	7.48
15	2.86	3.54	4.01	4.37	4.68	4.95	5.19	5.40	5.59	5.77	5.93
20	2.46	2.95	3.29	3.54	3.76	3.94	4.10	4.24	4.37	4.49	4.59
30	2.07	2.40	2.61	2.78	2.91	3.02	3.12	3.21	3.29	3.36	3.39
60	1.67	1.85	1.96	2.04	2.11	2.17	2.22	2.26	2.30	2.33	2.36
∞	1.00	1.00	1.00	1.00	1.00	1.00	1.00	1.00	1.00	1.00	1.00
	$\alpha = .01$, J = number of samples										
2	199	448	729	1036	1362	1705	2063	2432	2813	3204	3605
3	47.5	85	120	151	184	21(6)	24(9)	28(1)	31(0)	33(7)	36(1)
4	23.2	37	49	59	69	79	89	97	106	113	120
5	14.9	22	28	33	38	42	46	50	54	57	60
6	11.1	15.5	19.1	22	25	27	30	32	34	36	37
7	8.89	12.1	14.5	16.5	18.4	20	22	23	24	26	27
8	7.50	9.9	11.7	13.2	14.5	15.8	16.9	17.9	18.9	19.8	21
9	6.54	8.5	9.9	11.1	12.1	13.1	13.9	14.7	15.3	16.0	16.6
10	5.85	7.4	8.6	9.6	10.4	11.1	11.8	12.4	12.9	13.4	13.9
12	4.91	6.1	6.9	7.6	8.2	8.7	9.1	9.5	9.9	10.2	10.6
15	4.07	4.9	5.5	6.0	6.4	6.7	7.1	7.3	7.5	7.8	8.0
20	3.32	3.8	4.3	4.6	4.9	5.1	5.3	5.5	5.6	5.8	5.9
30	2.63	3.0	3.3	3.4	3.6	3.7	3.8	3.9	4.0	4.1	4.2
60	1.96	2.2	2.3	2.4	2.4	2.5	2.5	2.6	2.6	2.7	2.7
∞	1.00	1.0	1.0	1.0	1.0	1.0	1.0	1.0	1.0	1.0	1.0

* Reprinted from Pearson and Hartley (1966), 202. Use of this table is discussed in Section 11-4.

Table A-13 *Critical values of Friedman χ_R^2 statistic**

N	J = 3		J = 4	
	$\alpha = .05$	$\alpha = .01$	$\alpha = .05$	$\alpha = .01$
2	—	—	6.0	—
3	6.00	—	7.4	9.0
4	6.5	8.0	7.8	9.6
5	6.4	8.4		
6	7.00	9.00		
7	7.143	8.857		
8	6.25	9.00		
9	6.222	8.667		

* Adapted from Friedman, 1937. The Friedman test is discussed in Section 14-7. Reject the null hypothesis if $\chi_R^2 \geqq$ the tabulated value. Test significance by comparison with the standard chi-square distribution if $J = 3$ and $N > 9$, or $J = 4$ and $N > 4$, or $J > 4$.

Table A-14 *Critical values of S, the numerator of tau or gamma, when there are no ties**

N	.05	.025	.01	.005	α, one-tail test
	.10	.05	.02	.01	α, two-tail test
3	—	—	—	—	
4	6	—	—	—	
5	8	10	10	—	
6	11	13	13	15	
7	13	15	17	19	
8	16	18	20	22	
9	18	20	24	26	
10	21	23	27	29	

* Reproduced from *Rank Correlation Methods* (3rd Edition, 1962), M. G. Kendall, by permission of Publishers Charles Griffin & Company Ltd., London. Tau or gamma is significant at α if S(observed) $\geqq$ S(tabled). Tau and gamma are discussed in Section 17-2. When there are ties, use Table A-15.

Table A-15 *Critical values of S, the numerator of tau or gamma**

N	p_2	p_3	.05	.025	.01	.005	α, one-tail test
			.10	.05	.02	.01	α, two-tail test
3			—	—	—	—	
4			6	—	—	—	
5			8	10	10	—	
	1	9	9	9	—	—	
	2		8	—	—	—	
		1	7	—	—	—	
6			11	13	13	15	
	1		10	12	14	14	
	2		11	11	13	—	
	3		10	10	—	—	
		1	10	12	12	—	
	1	1	9	11	—	—	
		2	9	—	—	—	
7			13	15	17	19	
	1		12	14	16	18	
	2		13	15	17	17	
	3		12	14	16	18	
		1	12	14	16	18	
	1	1	13	15	15	17	
	2	1	12	14	16	16	
		2	11	13	15	—	

* Adapted from G. P. Silitto, "The Distribution of Kendall's τ Coefficient of Rank Correlation in Rankings Containing Ties," *Biometrika*, 1947, 34, pp. 36–40. Tau or Gamma is significant if S(observed) $\geqq$ S(tabled). Tau and gamma are discussed in Section 17-2. (Note: Values where there are no ties are the same as found in Table A-14.)

Table A-15 *Critical values of S, the numerator of tau or gamma (continued)*

			.05	.025	.01	.005	α, one-tail test
N	p_2	p_3	.10	.05	.02	.01	α, two-tail test
8			16	18	20	22	
	1		15	17	19	21	
	2		16	18	20	22	
	3		15	17	19	21	
	4		14	18	20	22	
		1	15	17	19	21	
	1	1	14	16	20	20	
	2	1	15	17	19	21	
		2	14	16	18	20	
	1	2	15	17	19	21	
9			18	20	24	26	
	1		17	21	23	25	
	2		18	20	24	26	
	3		17	21	23	25	
	4		18	20	22	26	
		1	17	21	23	25	
	1	1	18	20	22	24	
	2	1	17	19	23	25	
	3	1	18	20	22	24	
		2	18	20	22	24	
	1	2	17	19	23	25	
		3	17	19	23	23	
10			21	23	27	29	
	1		20	24	28	30	
	2		21	23	27	29	
	3		20	24	26	30	
	4		21	23	27	29	
	5		20	24	26	30	
		1	20	24	26	30	
	1	1	21	23	27	29	
	2	1	20	24	26	28	
	3	1	19	23	27	29	
		2	19	23	27	29	
	1	2	20	22	26	28	
	2	2	19	23	27	29	
		3	20	22	26	28	

Table A-16 *Table of critical values of S, in the Kendall coefficient of concordance W**

k	N					Additional values for $N = 3$	
	3†	4	5	6	7	k	s
	Values at the .05 level of significance						
3			64.4	103.9	157.3	9	54.0
4		49.5	88.4	143.3	217.0	12	71.9
5		62.6	112.3	182.4	276.2	14	83.8
6		75.7	136.1	221.4	335.2	16	95.8
8	48.1	101.7	183.7	299.0	453.1	18	107.7
10	60.0	127.8	231.2	376.7	571.0		
15	89.8	192.9	349.8	570.5	864.9		
20	119.7	258.0	468.5	764.4	1,158.7		
	Values at the .01 level of significance						
3			75.6	122.8	185.6	9	75.9
4		61.4	109.3	176.2	265.0	12	103.5
5		80.5	142.8	229.4	343.8	14	121.9
6		99.5	176.1	282.4	422.6	16	140.2
8	66.8	137.4	242.7	388.3	579.9	18	158.6
10	85.1	175.3	309.1	494.0	737.0		
15	131.0	269.8	475.2	758.2	1,129.5		
20	177.0	364.2	641.2	1,022.2	1,521.9		

* Adapated from Friedman, 1940.

† Notice that additional critical values of S for $N = 3$ are given in the right-hand column of this table.

Bibliography

Anderberg, M. R., *Cluster Analysis for Applications.* New York: Academic Press, Inc., 1973.

Andrews, F. M., L. Klem, T. N. Davidson, P. M. O'Malley, and W. L. Rodgers, *A Guide for Selecting Statistical Techniques for Analyzing Social Science Data.* Ann Arbor, Mich., Institute for Social Research, University of Michigan, 1974.

Baggaley, A. R., *Intermediate Correlation Methods.* New York: John Wiley & Sons, Inc., 1964.

————, *Mathematics for Introductory Statistics: A Programmed Review.* New York: John Wiley & Sons, Inc., 1969.

Bartlett, M. S., "Square Root Transformation in Analysis of Variance." *Suppl. J. Royal Statist. Soc., 3* (1936), 68–78.

————, "Properties of Sufficiency and Statistical Texts." *Proc. Royal Soc. London,* **A160** (1937a), 268–282.

————, "Some Examples of Statistical Methods of Research in Agriculture and Applied Biology." *Suppl. J. Royal Statist. Soc., 4* (1937b), 137–170.

————, "The Use of Transformations." *Biometrics, 3* (1947), 39–52.

Bashaw, W. L., *Mathematics for Statistics.* New York: John Wiley & Sons, Inc., 1969.

Campbell, D. P., "The Vocational Interests of American Psychological Association Presidents." *Amer. Psychologist, 20* (1965), 636.

Cohen, J., "Some Statistical Issues in Psychological Research," in B. Wolman (ed.), *Handbook of Clinical Psychology.* New York: McGraw-Hill Book Company, 1965.

————, "Multiple Regression as a General Data-Analytic System." *Psychol. Bull., 70* (1968), 426–443.

Cooley, W. W., and P. R. Lohnes, *Multivariate Procedures for the Behavioral Sciences.* New York: John Wiley & Sons, Inc., 1962.

Cramer, H., *Mathematical Methods of Statistics.* Princeton, N.J.: Princeton University Press, 1946.

Dixon, W. J., and F. J. Massey, Jr., *Introduction to Statistical Analysis,* 3rd ed. New York: McGraw-Hill Book Company, 1969.

Edwards, A. L., *Experimental Design in Psychological Research,* rev. ed. New York: Holt, Rinehart and Winston, Inc., 1960.

————, *Statistical Methods,* 3rd ed. New York: Holt, Rinehart and Winston, Inc., 1973.

Ezekiel, M., and K. A. Fox, *Methods of Correlation and Regression Analysis.* New York: John Wiley & Sons, Inc., 1966.

Freeman, L. C., *Elementary Applied Statistics for Students in Behavioral Science.* New York: John Wiley & Sons, Inc., 1965.

Freeman, M. F., and J. W. Tukey, "Transformations Related to the Angular and the Square Root." *Ann. Math. Statist., 21* (1950), 607–611.

Friedman, M., "The Use of Ranks to Avoid the Assumption of Normality Implied in the Analysis of Variance." *J. Amer. Statist. Assn., 32* (1937), 675–701.

———, "A Comparison of Alternative Tests of Significance for the Problem of *m* Rankings." *Ann. Math. Statist., 11* (1940), 86–92.

Games, P. A., and G. R. Klare, *Elementary Statistics: Data Analysis for the Behavioral Sciences.* New York: McGraw-Hill Book Company, 1967.

Glass, G. V., and A. R. Hakstian, "Measures of Association in Comparative Experiments: Their Development and Interpretation." *Amer. Educ. Research J., 6* (1969), 403–414.

Goodman, L. A., and W. H. Kruskal, "Measures of Association for Cross Classifications." *J. Amer. Statist. Assn., 49* (1954), 732ff.

———, "Measures of Association for Cross Classifications, III: Approximate Sampling Theory." *J. Amer. Statist. Assn., 58* (1963), 310ff.

Gorsuch, R. L., *Factor Analysis.* Philadelphia: W. B. Saunders Company, 1974.

Guilford, J. P., *Fundamental Statistics in Psychology and Education.* New York: McGraw-Hill Book Company, 1965.

Harshbarger, T. R., "Easy Numbers for Statistics Homework and Examination Problems." American Psychological Association, *Catalogue of Selected Documents in Psychology, 4* (1974) (MS No. 738).

Hart, W. L., *College Algebra,* 5th ed., Boston: D. C. Heath and Company, 1966.

Hartigan, J. A., *Clustering Algorithms.* New York: John Wiley & Sons, Inc., 1975.

Hartley, H. O., "The Maximum *F* Ratio as a Short-Cut Test for Homogeneity of Variance." *Biometrika, 37* (1950), 308–319.

Hays, W. L., *Statistics for Psychologists.* New York: Holt, Rinehart and Winston, Inc., 1963.

Hotelling, H., and L. R. Frankel, "The Transformation of Statistics to Simplify Their Distribution." *Ann. Math. Statist., 9* (1938), 87–96.

Huff, D., *How to Lie with Statistics.* New York: W. W. Norton & Company, Inc., 1954.

Hull, C. L., *Principles of Behavior: An Introduction to Behavior Theory.* New York: Appleton-Century-Crofts, Inc., 1943.

Hyman, H. H., and P. B. Sheatsley, "Attitudes Toward Desegregation." *Scientific American, 211* (1964), 18–19.

Institute of Life Insurance, *Life Insurance Fact Book.* New York, 1965.

Jaspen, N., "Serial Correlation." *Psychometrika, 11* (1946), 23–30.

Kelley, T. L., "An Unbiased Correlation Ratio Measure." *Proc. Nat. Acad. Sci.*, **21** (1935), 554–559.

Kendall, M. G., *The Advanced Theory of Statistics*. London: Charles Griffin & Company Ltd., 1947.

———, *Rank Correlation Methods*. London: Charles Griffin & Company Ltd., 1948.

Kerlinger, F. N., and E. J. Pedhazur, *Multiple Regression in Behavioral Research*. New York: Holt, Rinehart and Winston, Inc., 1974.

Kruskal, W. H., "A Nonparametric Test for the Several-Sample Problem." *Ann. Math. Statist.*, **23** (1952), 525–540.

———, and W. A. Wallis, "Use of Ranks in One-Criterion Variance Analysis." *J. Amer. Statist. Assn.*, **47** (1952), 583–621.

———, "Errata." *J. Amer. Statist. Assn.*, **48** (1953), 910.

Lazarsfeld, P. F., and M. Rosenberg, *The Language of Social Research*. New York: The Free Press of Glencoe, 1955.

Li, J. C. R., *Statistical Inference*, Vol. 1. Ann. Arbor, Mich.: J. W. Edwards, Publisher, Inc., 1964.

Linton, M., and P. S. Gallo, *The Practical Statistician: Simplified Handbook of Statistics*. Monterey, Calif., Brooks/Cole Publishing Company, 1975.

Lipset, S. M., "Opinion Formation in a Crisis Situation." *Publ. Opin. Q.*, **17** (1953), 20–46.

Mann, H. B., and D. R. Whitney, "On a Test of Whether One of Two Random Variables Is Stochastically Larger Than the Other." *Ann. Math. Statist.*, **18** (1947), 50–60.

Mattson, D. E., "A Generalization of the Median Test." *Educ. Psychol. Measurement*, **25** (1965), 1023–1028.

McCornak, R. L., "Extended Tables of the Wilcoxon Matched Pair Signed Rank Statistic." *J. Amer. Statist. Assn.*, **60** (1965), 864–870.

McNemar, Q., *Psychological Statistics*, 4th ed. New York: John Wiley & Sons, Inc., 1969.

Milton, R. C., "Extended Table of Critical Values for the Mann–Whitney (Wilcoxon) Two-Sample Statistic." *J. Amer. Statist. Assn.*, **59** (1964), 925–933.

Moran, P. A. P., "Partial and Multiple Rank Correlation." *Biometrika*, **38** (1951), 26–32.

Mueller, C. G., "Numerical Transformations in the Analysis of Experimental Data." *Psychol. Bull.*, **46** (1949), 198–223.

Mulaik, S. A., *The Foundations of Factor Analysis*. New York: McGraw-Hill Book Company, 1972.

National Education Association, "Sampling Study of the Teaching Faculty in Higher Education." *N. E. A. Research Bull.*, **44** (1966), 3–10.

Osgood, C. E., G. G. Suci, and P. H. Tannenbaum, *The Measurement of Meaning*. Urbana, Ill.: The University of Illinois Press, 1957.

Overall, J. E., and C. J. Klett, *Applied Multivariate Analysis*. New York: McGraw-Hill Book Company, 1972.

Pearson, E. S., and H. O. Hartley (eds.), *Biometrika Tables for Statisticians*, Vol. 1. Cambridge University Press, 1966.

Peatman, J. G., *Introduction to Applied Statistics*. New York: Harper & Row, Publishers, 1963.

Peters, C. C., and W. R. Van Voorhis, *Statistical Procedures and Their Mathematical Bases*. New York: McGraw-Hill Book Company, 1940.

Rand Corporation, *A Million Random Digits with 100,000 Normal Deviates*. New York: The Free Press of Glencoe, 1955.

Ray, W. S., *Statistics in Psychological Research*. New York: Macmillan Publishing Co., Inc., 1962.

Rosenzweig, M. R., "Environmental Complexity, Cerebral Change, and Behavior." *Amer. Psychol.*, **21** (1966), 321–332.

Siegel, S., *Nonparametric Statistics for the Behavioral Sciences*. New York: McGraw-Hill Book Company, 1956.

Silitto, G. P., "The Distribution of Kendall's τ Coefficient of Rank Correlation in Rankings Containing Ties." *Biometrika*, **34** (1947), 36–40.

Skinner, B. F., "'Superstition' in the Pigeon." *J. Exper. Psychol.*, **38** (1948), 168–172.

Snedecor, G. W., *Statistical Methods*. Ames, Iowa: The Iowa State University Press, 1956.

Tate, M. W., and L. A. Hyer, "Inaccuracy of the χ^2 Test of Goodness of Fit When Expected Frequencies Are Small." *J. Amer. Statist. Assn.*, **68** (1973), 836–841.

Tatsuoka, M. M., *Multivariate Analysis: Techniques for Educational and Psychological Research*. New York: John Wiley & Sons, Inc., 1971.

Walker, H. M., *Mathematics Essential for Elementary Statistics*, rev. ed. New York: Holt, Rinehart and Winston, Inc., 1951.

——, and J. Lev, *Statistical Inference*. New York: Holt, Rinehart and Winston, Inc., 1953.

Walsh, J. E., *A Handbook of Nonparametric Statistics*. New York: D. Van Nostrand Company, Inc., 1962.

Welch, B. L., "The Generalization of 'Student's' Problem When Several Different Population Variances Are Involved." *Biometrika*, **34** (1947), 28–35.

Winer, B. J., *Statistical Principles in Experimental Design*. New York: McGraw-Hill Book Company, 1962.

Wolmann, B., *Handbook of Clinical Psychology*. New York: McGraw-Hill Book Company, 1965.

Index

[1] Other correlation coefficients discussed in this text (including biserial, coefficient of concordance, contingency coefficient, eta, gamma, lambda, multiserial, Pearson, phi, phi-prime, point-biserial, rho, tau, and various multiple coefficients) are indexed under individual headings and may also be found in the maps for Chaps. 16, 17, and 18.

[2] For application of the concept of theoretical distributions, *see* Sections 7-4, 7-5, and 7-6. Applications to hypothesis testing are discussed with each individual test.